LUNAR
SETTLEMENTS

ADVANCES IN ENGINEERING

A SERIES OF REFERENCE BOOKS, MONOGRAPHS, AND TEXTBOOKS

Series Editor

Haym Benaroya

Department of Mechanical and Aerospace Engineering
Rutgers University

Published Titles:

Lunar Settlements, *Haym Benaroya*

Handbook of Space Engineering, Archaeology and Heritage, *Ann Darrin and Beth O'Leary*

Spatial Variation of Seismic Ground Motions: Modeling and Engineering Applications, *Aspasia Zerva*

Fundamentals of Rail Vehicle Dynamics: Guidance and Stability, *A. H. Wickens*

Advances in Nonlinear Dynamics in China: Theory and Applications, *Wenhu Huang*

Virtual Testing of Mechanical Systems: Theories and Techniques, *Ole Ivar Sivertsen*

Nonlinear Random Vibration: Analytical Techniques and Applications, *Cho W. S. To*

Handbook of Vehicle-Road Interaction, *David Cebon*

Nonlinear Dynamics of Compliant Offshore Structures, *Patrick Bar-Avi and Haym Benaroya*

LUNAR
SETTLEMENTS

EDITED BY

HAYM BENAROYA

CRC Press
Taylor & Francis Group
Boca Raton London New York

CRC Press is an imprint of the
Taylor & Francis Group, an **informa** business

CRC Press
Taylor & Francis Group
6000 Broken Sound Parkway NW, Suite 300
Boca Raton, FL 33487-2742

First issued in paperback 2017

ISBN 13: 978-1-138-11401-2 (pbk)
ISBN 13: 978-1-4200-8332-3 (hbk)

Library of Congress Cataloging-in-Publication Data

Lunar settlements / editor, Haym Benaroya.
    p. cm. -- (Advances in engineering)
  "A CRC title."
    Includes bibliographical references (p.   ).
  ISBN 978-1-4200-8332-3 (hardcover : alk. paper)
  1. Lunar bases. 2. Space industrialization. I. Benaroya, Haym, 1954- II. Title. III. Series.

TL795.7.L86 2010
629.45'4--dc22   2009038016

Visit the Taylor & Francis Web site at
http://www.taylorandfrancis.com

and the CRC Press Web site at
http://www.crcpress.com

Contents

Preface

It is a pleasure to present to you the culmination of many years of effort, a compilation of edited papers that evolved from the Rutgers Symposium on Lunar Settlements. Papers are written by symposium attendees as well as invited authors. While no volume can completely represent the views on our return to the Moon to stay, we can hope to present views that are significant to that return. This preface is written on 20 January 2009, on Inauguration Day for our new president, Barack Obama. The whole space community eagerly waits to learn his views on manned space and the Bush vision on lunar settlement.

Acknowledgments

There are many people whose efforts made the Symposium an excellent meeting. Certainly first on this list must be Mrs. Patricia Mazzucco, whose tireless efforts in working with our vendors and local organizational matters, as well as with many of the attendees, truly made the meeting come together in such a nice way. We are grateful to Vice President for Academic Affairs Philip Furmanski for his enthusiasm as well as financial support for the Symposium. Similarly we are grateful to Rutgers University, the institution, for being the exciting and dynamic place that it is. Elan Borenstein put forth significant effort to create our website and made sure we were properly set up for the Symposium. Tushar Saraf prepared the Abstract Book. Paul Bonness worked as a floor manager at the Symposium. Shefali Patel and Helene Press assisted and supported the Symposium preparations. Aiesha Jenkins was supportive during the Symposium. Kendra Cameron is thanked for her assistance in helping us gather promotional items for the attendees. Of course we are truly grateful to all the presenters who took the time and expense to come to Rutgers and offer us some of their expertise. Finally, a personal thanks to Ana Benaroya, my daughter, for her illustration of a future lunar settlement that became our Symposium logo.

Those who lent their time and expertise to the creation of the Proceedings Volume are all the authors, of course, Mrs. Patricia Mazzucco for working with the authors, Paul Bonness for pulling together all the manuscripts and formatting them in an efficient and beautiful way, Jonathan Plant, Senior Editor for Taylor & Francis, Amy Blalock, Project Coordinator, and Amy Rodriguez, Project Editor, for working on a difficult project, Professor Haim Baruh of Rutgers University and the NASA Space Grant Consortium for providing some of the financial resources needed to create this volume, and Ana Benaroya for the original cover.

A book such as this can only be dedicated to all those who make space

accessible to human and machine. To them we owe our future.

The Editor

Haym Benaroya is Professor of Mechanical and Aerospace Engineering at Rutgers University. He is also founder and director of the Center for Structures in Extreme Environments, a center focusing on the conceptualization and analysis of structures placed in challenging environments. We have considered offshore drilling structures, aircraft structures and luggage containers subjected to explosions, nanostructures and lunar surface structures for manned habitation. Professor Benaroya earned his BE degree from The Cooper Union in New York, and his MS and PhD (1981) from the University of Pennsylvania. Prior to joining Rutgers University in 1989, Professor Benaroya was a senior research engineer at Weidlinger Associates in New York for eight years, with responsibilities to a number of defense-related studies. While at Rutgers, Professor Benaroya has mentored nine students to their PhDs and a similar number to their MS degrees. He is the author of about 70 refereed journal publications, two textbooks—one on vibration and the other on probabilistic modeling—and two research monographs with two former PhD students on structural dynamics in the ocean. Professor Benaroya is a Fellow of the ASME and the British Interplanetary Society, an Associate Fellow of the AIAA, and was elected to the International Academy of Astronautics in 2007.

Section I

The Past and Future

1

Return to the Moon

Harrison H. Schmitt
University of Wisconsin-Madison

Harrison Schmitt, Apollo 17 Astronaut and former U.S. Senator, is Adjunct Professor of Engineering, University of Wisconsin-Madison, teaching "Resources from Space." Schmitt is a founder and serves as Chairman of Interlune Intermars Initiative, Inc., advancing the private sector's acquisition of lunar resources and its development of ^{3}He fusion power.

Schmitt, a native of Silver City, New Mexico, has the diverse experience of a geologist, pilot, astronaut, administrator, businessman, writer, and U.S. Senator. He married Teresa A. Fitzgibbon of Los Alamos, New Mexico in 1985. Schmitt received his B.S. from Caltech, studied as a Fulbright Scholar in Oslo, Norway, and attended graduate school at Harvard University, Cambridge, Massachusetts. Geological field studies in Norway formed the basis of his PhD in 1964. As a civilian, Schmitt received Air Force jet pilot wings in 1965 and Navy helicopter wings in 1967, logging more than 2100 hours of flying time.

Selected for the Scientist-Astronaut program in 1965, Schmitt organized the lunar science training for the Apollo astronauts, represented the crews during the development of hardware and procedures for lunar surface exploration, and oversaw the final preparation of the Apollo 11 Lunar Module descent stage. He served as mission scientist in support of the Apollo 11 mission. After training as backup Lunar Module pilot for Apollo 15, Schmitt flew in space as Lunar Module pilot for Apollo 17—the last Apollo mission to the moon. On December 11, 1972, he landed in the Valley of Taurus-Littrow as the only scientist and the last of 12 men to step on the Moon.

In 1975, after two years managing NASA's Energy Program office, Schmitt fulfilled a longstanding personal commitment by entering politics. Elected in 1976, he served a 6-year term in the U.S. Senate beginning in 1977. Senator Schmitt, the only "natural scientist" in the Senate since Thomas Jefferson, was Vice President of the United States and president of the Senate, worked as a member of the Senate Commerce, Banking, Appropriations, Intelligence, and Ethics Committees. In his last two years in the Senate, Schmitt held the position of chairman of the Commerce Subcommittee on Science, Technology, and Space and of the Appropriations Subcommittee on Labor, Health and

Human Services, and Education. He later served on the President's Foreign Intelligence Advisory Board, the President's Commission on Ethics Law Reform, the Army Science Board, as co-chairman of the International Observer Group for the 1992 Romanian elections, and as vice chairman of the U.S. delegation to the 1992 World Administrative Radio Conference in Spain. He is on the Maguire Energy Institute's board of advisors, and served as co-chair of NASA's Human Planetary Landing Systems Capabilities road-mapping effort in 2004–2005.

Harrison Schmitt became chairman of the NASA Advisory Council in November 2005, and served until October 2008. He led the council's deliberations on issues related to aeronautics, audit and finance, biomedicine, exploration (human flight systems development), human capital, science, and space operations. He also consults, speaks, and writes on policy issues of the future, the science of the moon and planets, history of space flight and geology, space exploration, space law, climate change, and the American Southwest. He presently is chair emeritus of The Annapolis Center (risk assessment) and is adjunct professor of engineering, University of Wisconsin-Madison, teaching Resources from Space. Schmitt became a consultant to the Fusion Technology Institute at the University of Wisconsin in 1986, advising on the economic geology of lunar resources and the engineering, operational, and financial aspects of returning to the moon. He is on the staff of the Institute for Human and Machine Cognition of Pensacola, Florida. Current board memberships include Orbital Sciences Corporation, Edenspace Systems Corporation, and PhDx Systems, Inc., and, as a retired director, he continues as an emeritus member of the corporation of the Draper Laboratory. He also has served as a member of the Energy Department's Laboratory Operations Board. In 1997, Schmitt co-founded and became chairman of Interlune-Intermars Initiative, Inc., advancing the private sector's acquisition of lunar resources and helium-3 fusion power and clinical use of medical isotopes produced by fusion-related processes. He is the author of *Return to the Moon* (Springer-Praxis, 2006), which describes a private enterprise approach to providing lunar helium-3 fusion energy resources for use on Earth.

Schmitt's honors include the 1973 Arthur S. Fleming Award; 1973 Distinguished Graduate of Caltech; 1973 Caltech Sherman Fairchild Scholar; 1973 Manned Spacecraft Center Superior Achievement Award; 1973 NASA Distinguished Service Award; 1973 First Extraterrestrial Field Geologist Award of the Geological Society of America; 1977 Fellow of the AIAA; Honorary Member of the American Association of Petroleum Geologists, Norwegian Geographical Society, New Mexico Geological Society, and Geological Association of Canada; 1981 Engineer of the Year Award from the National Society of Professional Engineers; 1981 National Security Award; 1982 Public Service Award of the American Association of Petroleum Geologists; 1989 Lovelace Award (space biomedicine); 1989 G.K. Gilbert Award (planetology); 2002 Aviation Week Legend Award; 2008 American Association of State Geologists Pick and Gavel Award; and Honorary Fellow of the Geological Society of America; American Institute of Mining, Metallurgical and Petroleum Engineers; and Geological Society of London. Dr. Schmitt has been inducted into the Astronaut Hall of Fame and the International Space Hall of Fame and has received honorary degrees from nine U.S. and Canadian Universities. In recognition of past service, the U.S. Department of State in July 2003 established the Harrison H. Schmitt Leadership Award for U.S. Fulbright Fellowship awardees. In 2007, Schmitt was awarded the first Eugene M. Shoemaker Memorial Award by Arizona State University, and is the first recipient of the National Space Society's Gerard K. O'Neill Memorial Space Settlement Award.

The financial, environmental, and national security carrot for a Return to the Moon consists of access to low cost lunar helium-3 fusion power. Helium-3 fusion represents an environmentally benign means of helping to meet an anticipated eight-fold or higher increase in energy demand by 2050. Not available in other than research quantities on Earth, this light isotope of ordinary helium-4 reaches the Moon as a component of the solar wind, along with hydrogen, helium-4, carbon and nitrogen. Embedded continuously in the lunar dust over almost four billion years of time, concentrations have reached levels that can legitimately be considered to be of economic interest. Two square kilometers of titanium-debris or regolith covering large areas of the lunar surface, to a depth of three meters, contains 100 kg (220 pounds) of helium-3, i.e., more than enough to power a 1000 megawatt (one gigawatt) fusion power plant for a year. Strong evidence exists that the concentration of helium-3 in the polar regions reaches three times that in the mare regolith. In 2008, helium-3's energy equivalent value relative to $5.00 per million BTU industrial coal equaled about $2800 million a metric tonne. One metric tonne (2200 pounds) of helium-3, fused with deuterium, a heavy isotope of hydrogen, has enough energy to supply a U.S. city of 10 million or one/sixth of the United Kingdom with a year's worth of electricity or over 10 gigawatts of power for that year.

The United States has two basic options for both assuring results from and continuation of a "sustained commitment" to deep space exploration and settlement. On the one hand, it could continue to restructure and revitalize NASA under the Vision for Space Exploration articulated by President George W. Bush and to provide that Vision with a guarantee of continued funding sufficient to do the job. A tough order in the current national political environment, but one the President and Congress have directed NASA and its Administrator to undertake. Alternatively, the country's entrepreneurial sector could persuade national and international investors to make sustaining commitments based on the economic potential of lunar resources. Not easy, but at least predictable in terms of what conditions investors require to be met relative to other competitive uses of their capital. The option of rebuilding NASA is highly unpredictable and its sustainability may depend

on the appearance of a set of national and world circumstances comparable to those facing the Congress and Presidents Eisenhower, Kennedy, and
Johnson in the late 1950s and throughout the 1960s.

Although it fundamentally has an investor-driven economy, America has a
tradition of parallel commercial and public technological endeavors, ranging
from transportation to agriculture to communication to medicine. Such
activities have often involved international partnerships and investors. Private
and public endeavors operating together clearly have been far more productive
then either would have been acting alone. In this vein, private space-related
initiatives can benefit from the research and technology development funded
by NASA and vice versa. The 20th century, particularly since World War II and
American stimulation of European and Asian post-war economic development,
has seen research and technology development in other nations become
positioned to participate in a privately led Return to the Moon initiative. That
initiative also can supplement, support, and, if necessary, pick up the baton of
space settlement if not carried forward by government.

The initial financial threshold for a private sector initiative to Return to the
Moon is low: about $15 million. This investment would initiate the first fusion-
based bridging business, that is, production of medical isotopes for point-of-
use support of diagnostic procedures using positron-emission tomography
(PET). In contrast, the funding threshold for the United States government
would be significantly higher: $800 million proposed for 2005 and build-
ing to an average annual addition of close to $1 billion. This latter estimate
assumes both a repetitively willing Congress and Administration as well as
a space agency capable of efficiently using this money. The Government, of
course, would not benefit directly from the retained earnings of the fusion-
based bridging businesses that are a natural consequence of the private sector approach although Americans would benefit from the new services and
tax base that the private sector could provide.

International law relative to outer space, specifically the Outer Space Treaty
of 1967, permits properly licensed and regulated commercial endeavors.
Under the treaty, lunar resources can be extracted and owned, but national
sovereignty cannot be asserted over the resource area. History clearly shows
that a system of internationally sanctioned private property, consistent with
the treaty, would encourage lunar settlement and development far more than
the establishment of a lunar "commons" as envisioned by the largely unrati-
fied 1979 Moon Agreement. Legal systems encompassing the recognition of
private property have provided far more benefit to the world than those that
attempt to manage common ownership.

The entrepreneurial private sector has an obligation to support a Return to
the Moon to stay, as articulated by President Bush. We also have an obligation to follow our own path to get there in order to be additive to the overall
goals of settling the solar system and improving lives for those who remain
on Earth. Traversing that path, with an ideally funded business plan, would
require about $15 billon and 15 years.

Whenever and however a Return to the Moon occurs, one thing is certain: That return will be historically comparable to the movement of our species out of Africa about 50,000 years ago. Further, if led by an entity representing the democracies of the Earth, a Return to the Moon to stay will be politically comparable to the first permanent settlement of North America by European immigrants.

Reference

Schmitt, H. H., 2006, *Return to the Moon*, Springer, New York, 335p.

2

Rutgers 2007 Symposium on Lunar Settlements

Paul D. Lowman, Jr.
Goddard Space Flight Center
Greenbelt, Maryland

Paul D. Lowman Jr. is a geologist in the Planetary Geodynamics Laboratory at the NASA Goddard Space Flight Center. The first geologist hired by NASA, in 1959, he has been involved in most major manned flight programs, including Mercury, Gemini, Apollo, Skylab, and the Shuttle. His main research area is comparative planetology as applied to the origin of the Earth's continental crust. He has worked in lunar sample analysis, orbital photography, and orbital radar. From 1973 to 2004 he shifted to global tectonics, producing a map of global tectonic and volcanic activity of the last one million years. He is author of several books, most recently *Exploring Space, Exploring Earth* (Cambridge University Press, 2002).

ABSTRACT   This paper summarizes the major themes and ideas presented at a 4 1/2 day symposium on lunar settlements held at Rutgers University in June, 2007, sponsored by the Department of Mechanical and Aerospace Engineering and organized by Haym Benaroya of that department. Presentations covered the political and economic aspects of lunar settlements, structures and transportation, the lunar environment, energy and power, medical aspects of lunar settlements, outpost site selection, and use of the Moon as a platform for astronomy. A keynote paper by H.H. Schmitt covered topics including legal aspects of lunar settlements and possible economic products such as helium-3, demonstrably valuable for production of short-lived medical isotopes and perhaps for thermonuclear energy. Potential medical problems, discussed by several speakers, including J. Logan and W. Rowe, included radiation exposure and the effect of long-term hypogravity. Lunar resources discussed by several speakers include lunar water from possible polar ice deposits, hydrogen, helium-3, and oxygen from the lunar regolith. Outpost site selection has been narrowed to the south polar regions, with Shackleton Crater and Malapert Mountain the most-favored candidates. Lunar settlement shelters were proposed by several speakers, a consensus being that these must be largely underground because of the radiation problem. The feasibility of agriculture in lunar settlements has been demonstrated by operating greenhouses in the Antarctic.

The overall result of the symposium was a summary of the problems, prospects, and practicality of such settlements, now benefiting from three decades to assimilate the lunar experience of the 6 Apollo landing missions and the many robotic missions, American and Soviet.

Introduction and Overview

A Symposium on Lunar Settlements was held at Rutgers University during the first week of June 2007. In the planning stages for about two years, the Symposium was organized by Professor Haym Benaroya of the Mechanical and Aerospace Engineering Department. While his research focused primarily on lunar structures, the Symposium was organized so that the focus was the lunar settlement, and as such all relevant disciplines were included. The Symposium was a single-track 4-day meeting, with a half-day on the 5th day used for a retrospective and summary group meeting with thoughts on both content and format. Details on this symposium as well as copies of presentations are available on the website http://www.lunarbase.rutgers.edu/. It is possible to also review the book of abstracts and to find contact information for the presenters.

Key Presentations

The Rutgers Symposium was 4 1/2 days in duration. It was a productive and stimulating meeting that will be summarized here. It was organized and run by the Center for Structures in eXtreme Environments. An abstracts volume is available at the symposium website http://www.lunarbase.rutgers.edu. In order of importance, here are summaries for a few of the main talks.

Harrison H. Schmitt, Apollo 17 Astronaut

Schmitt gave the keynote talk "Return to the Moon: Expanding the Earth's Economic Sphere." His first point was that a return to the Moon will help insure the long-term survival of the human species. Humanity has benefited from exploration for some 40,000 years, referring to migration of mankind from a point of origin into successively larger areas, for living space and resources. Exploration of America eventually permitted extension of free institutions to a new continent.

Economic factors affecting a permanent return to the Moon include initial capitalization, economic self-sufficiency, affordable access, and compatible space law. He cited government estimates of $150B for settlement of the Moon, but said we should stress annual costs, not cumulative ones.

Schmitt discussed He-3 at length, comparing it to "steam coal." However, as in his book *Return to the Moon*, he stressed the near-term potential of He-3 for fusion reactors to produce short-lived medical isotopes. U. of Wisconsin is apparently close to achieving D-He-3 fusion at a level high enough to produce these isotopes. There was much more, but most of what he said about this is in his book.

The question of mining He-3 on the Moon brought up the question of legality. Schmitt said the 1967 Outer Space Treaty prohibited territorial claims on the Moon, but did permit mining. He recommended ignoring the 1979 Moon Treaty, which like the Law of the Sea (LOS) would make lunar resources the common heritage of mankind. He said the LOS is still under study, and that it will be important to see if it restricts sea-bed mining and by implication lunar mining.

In the question period, Schmitt discussed the ISS, which he had barely mentioned. He said the ISS could be a valuable medical research facility for study of osteoporosis, and that he had been trying to persuade NASA to cooperate with NIH to explore this possibility. He raised the interesting point that space station astronauts can generally recover their lost bone mass in a few months, but that old people on earth with the condition do not. This has triggered considerable interest in the medical community.

Haym Benaroya, CSXE, Rutgers University

This was a very broad talk, "Lunar Structures," covering all aspects of the return to the Moon. Benaroya made the point that we must get off the planet to insure the long-term survival of humanity. It will also be a great educational stimulus, attracting young Americans to the hard subjects—engineering, science, and mathematics—where enrollment is declining. He covered the political aspects of the Vision for Space Exploration (VSE), referring to manifest destiny in relation to emerging space powers such as China.

The Moon also offers a site for development of "space legs," i.e., to learn how to live on other planets, which cannot be done only in LEO. Benaroya argued that a return to the Moon would be a great antidote to "societal pessimism," by providing a positive and uplifting vision for humanity.

There are many problems remaining before the Moon can be settled, but the basic knowledge of "lunar structures" is growing continually, and a small but "exciting" initial settlement is clearly feasible. Major problems remaining include radiation and long-term low gravity.

James S. Logan, Space Medicine Associates

This talk, on "biomedical showstoppers for long duration lunar habitation," was an extremely important one for the lunar outpost concept. Logan has a long career in aerospace medicine, and is now at JSC, but attended this symposium at his own expense and stressed that his views were strictly his own, not NASA's.

Logan gave a detailed summary of the medical problems on the Moon: lunar dust, radiation, hypogravity, and probable synergistic effects. The discussion of lunar dust was unusually interesting because terrestrial experience was cited. Within 5 years of the drilling of the Hawk's Nest Tunnel through Gauley Mountain in West Virginia in the 1930s, hundreds of miners died of silicosis even though they had been exposed for only a few months. The incident was arguably one of the biggest occupational health disasters in U.S. history. During the dry drilling process miners had been exposed to fine quartz dust kicked into the air. Medical experts are concerned activated fine lunar regolith may possess similar reactivity because of weathering by solar wind and repeated vapor deposition due to micrometeorite impacts. The issue will be to empirically determine the toxicity of activated lunar regolith compared to known terrestrial hazards such as $TiO2$ (minimally reactive) and quartz (highly reactive).

Logan's discussion of radiation was equally informative. He pointed out that permissible exposure limits have been steadily lowered in recent years, citing the recent NRC study of space radiation. He concluded that short sorties on the lunar surface are feasible, but longer-term occupation is much more problematic due to elevated radiation exposures over time. Prolonged lunar surface habitation will effectively be precluded without substantial shielding. Because the space suit provides virtually no radiation protection, multiple lunar EVAs (by the same crew members) will most likely be severely constrained.

He discussed "hypogravity," i.e., the low lunar gravity (1/6 Earth gravity). No permanent deleterious effects were observed in Apollo missions but exposure times were very short. (Lunar missions were approximately two weeks in total duration with no more than 3 days on the lunar surface.) Longer duration exposures (6 months) to microgravity have demonstrated significant effects, most of which appear to be reversible upon return to earth. However, recent data utilizing more sensitive measuring techniques suggest the rate of bone demineralization may be almost twice initial estimates. Worse, these changes may not be fully reversible in all crew members. He pointed out that lunar gravity may be very deleterious for the developing fetus, infant or toddler. If confirmed, microgravity effects could prohibit permanent "settlement" (i.e., men, women, children and multiple generations) of the Moon or even Mars.

Logan's warning against multiple EVAs has direct implications for possible human service missions to the JWST, which at a Lagrangian point will be far beyond the magnetosphere. Dan Lester had an imaginative artist's concept of a JWST service mission in a Nov. 2006 *Physics Today* article, but Logan's discussion indicates that such missions would be problematic if they involved significant cumulative exposures during multiple EVAs.

Logan did not go into detail on "synergistic effects," beyond warning that there would be such effects. All in all, this talk was perhaps the most important of the symposium in its implications for long-term settlement of the Moon or Mars, as outlined in the VSE.

In the question period, Larry Taylor said the space medicine people argued for either equatorial or polar base locations because they permitted return to Earth at any time. Lowman has pointed out in several papers that such locations permit abort from the surface to an orbiting spacecraft without any plane changes. (This is why the early Apollo missions were equatorial.)

William Rowe, Medical University of Ohio at Toledo

This was another important paper by a medical doctor with a background in space medicine: "Moon dust may simulate vascular hazards of urban pollution." Rowe discussed several space flight factors with possible cardiovascular effect, by producing changes in the endothelium (artery lining). Fine urban dust has been found to produce vascular constriction and high diastolic blood pressure. Rowe studied the Apollo 15 mission, which put great stress on Scott and Irwin, from which they took weeks to recover. Rowe suggested that the lunar dust might have been a contributing factor. He also suggested that a magnesium deficiency could have had an effect. In the question period, it was mentioned that JSC had prescribed massive doses of orange juice for the following missions in the belief that a potassium deficiency might have been responsible; he replied that both potassium and magnesium were probably involved.

Werner Grandl, Consulting Architect and Civil Engineer

Grandl presented "Lunar Base 2015: A Preliminary Design Study." This was actually a detailed and specific proposal for a modular lunar station made of six cylindrical modules, each weighing 10.2 tons. These would be double-walled, with regolith for shielding. This base could be established with 11 Arianne 5 launches. This paper was interesting in that it showed how a lunar base could be developed with an existing launch vehicle. An obvious implication is that given the funding and motivation, Europe could produce a lunar base on its own.

Charles Lundquist, University of Alabama at Huntsville

Lundquist discussed "Apollo Knowledge Transfer." This is a major and broad-based effort by the University of Alabama at Huntsville to bridge the 40-year gap since Apollo by archiving all sorts of data from the Apollo era. It includes all sorts of Apollo documents from retirees, oral histories, and many computer archives. The problem of changing these archives into formats readable with today's equipment is being worked on. Lundquist said that there is a Lunar Orbiter image recovery program going on at Ames, and that LO tapes have been sent from Goddard to Ames.

In the question period, Paul Lowman said that the National Space Science Data Center is working on the same problem, reformatting lunar data such as that from the ALSEPs into modern versions for re-analysis.

Larry Taylor, University of Tennessee at Knoxville

Taylor gave an excellent broad review of "In Situ Resource Utilization on the Moon." He stressed the interrelation of science and exploration, and warned against separating the two, as in recent trends. Apollo was a success partly because of close relations between the engineers and scientists.

He presented well-illustrated discussions of potential oxygen, hydrogen, helium 3, carbon, and nitrogen resources, all solar wind-implanted. He also stressed the value of high-Ti basalts, since oxygen can be produced from ilmenite by hydrogen reduction. He discussed the problem of lunar dust, suggesting that a magnetic filter could remove part of the fine fraction, which has metallic iron.

Gregory Konesky, SGK Nanostructures, Inc.

Konesky discussed "hierarchical roving," telerobotic lunar rovers of various sizes. His company has built working models, with a large rover about the size of a wheelbarrow carrying smaller ones to be deployed from it. He showed slides of Mars from various rovers, pointing out that there are arguments for large and small rovers depending on the nature of the terrain. Sojourner, for example, landed on a very rocky terrain, but was able to drive between the rocks better than a large rover might have. A large lunar rover could serve as an anchor point from which small tethered rovers could go down steep slopes.

There was discussion of letting the public back on Earth drive such rovers. Lowman has recommended this as a means of arousing public interest by letting students and others take direct part in exploration of the Moon.

Roger Launius, National Air and Space Museum, Smithsonian

This talk also emphasized the problem we face in returning to the Moon, chiefly public attitudes and budget limits. Launius showed a number of graphs illustrating public opinion about space exploration over the last several decades. Right after Sputnik, support was at an all-time high because of fear of the Soviets. However, it dropped off sharply. A 2007 Harris poll asking people how to reduce the federal deficit had 51% saying cut the space program, right at the top of the list. An earlier poll showed that 14–18-year-olds have little or no interest in a return to the Moon, and about 27% in this age range think the Apollo landings were faked.

Launius said that the primary driver for a new lunar program would have to be, like Apollo, national prestige and geopolitics. He quoted John Kennedy as saying that he was not really interested in space. The motivation for his proposal to go to the Moon was the Bay of Pigs fiasco followed shortly by Gagarin's flight.

Launius showed that we got 3.8% of the federal budget at the peak of Apollo. He warned that we would probably have to live with a NASA budget under 0.9%. In summary, we really have a job ahead of us to sell the public on our programs.

Brent Sherwood, Jet Propulsion Laboratory

"What Will We Actually Do on the Moon?" was Sherwood's title, and his reply was a broad summary of various ideas put forth at recent NASA meetings. Two of these were to establish simple observatories on the Moon, and stimulate public interest with "high-fidelity telepresence." The latter sounds like Lowman's recommendations for publicly accessible robotic telescopes and telerobotic rovers to be driven by the public. Like Launius, Sherwood said the public has little interest in the VSE.

Yuriy Gulak, CSXE, Rutgers University

In this talk, "Heat Pipes: How to Increase the Capillary Heat Transfer Limit," Gulak outlined work on increasing the effectiveness of heat pipes by using a wick whose porosity varies along the length of the pipe. This would be applicable to Earth, Moon, and zero g conditions, where it is necessary to control internal temperatures in spacecraft.

Gregory Konesky, SGK Nanostructures, Inc.

This was the second paper by Konesky. It was a detailed quantitative discussion of optical data links, on the Moon and from the Moon to Earth. Such data links point to point on the Moon are limited chiefly by the Moon's topography, in the absence of an atmosphere.

Konesky presented a detailed quantitative study of Moon-Earth optical links. The power requirements are in principle modest: a 1-watt laser on the Moon, sent from a 1-meter aperture, could be received on Earth with a 1-meter dish. However, the actual requirements would be much harsher, because of the atmosphere, clouds, etc. He discussed the problems of an orbital optical link.

Gene Giacomelli, University of Arizona, Tucson

In terms of sheer novelty, this was one of the most interesting papers in the symposium. The title, "Development of a Lunar Habitat Demonstrator," refers to work being done at the U. of Arizona in Tucson and at the South Pole Amundsen-Scott base. This group has established at the South Pole a greenhouse in which hydroponics farming under high intensity sodium lights produces continuous supply of fresh vegetables. The greenhouse is tended partly by volunteers, who enjoy the warm humid atmosphere. There is a small café/lounge attached where station personnel can meet for meals and conversation. It has tables with tablecloths, dishes, and cutlery. The greenhouse generates oxygen, not needed on Earth but of obvious importance for a lunar base. The U. of Arizona group is now building a similar structure on Tucson designed specifically for the Moon.

Tom Taylor, Lunar Transportation Systems, Inc.

In "Lunar Commercial Logistics Transportation," Taylor drew lessons from transportation to remote locations on Earth, such as Prudhoe Bay, Alaska and proposed a commercial lunar transportation architecture that is modular and flexible. He estimates that the Moon is 50 times more remote and 100–1,000 times more expensive than the Arctic, as well as more severe an environment. This architecture is based on refueling a fleet of fully reusable spacecraft at several locations in cislunar space, which creates a two-way highway between the Earth and the Moon. Coming from one who is a key investor and venture capitalist for space ventures, his views are valuable.

Paul Eckert, Integrated Defense Systems, The Boeing Co.

Eckert as well is a major player to identify business opportunities within the Return to the Moon, for both aerospace and other companies. The title "Attracting Private Investment for Lunar Commerce: Toward Economically Sustainable Development" summarizes the focus of his efforts over a number of years. He has led a number of meetings of business leaders who are working to encourage development of truly commercial, self-sustaining space activity involving non-government customers, rather than simply extending government contracting. It is clear that NASA or the Federal Government alone cannot develop space. We will be fortunate if they begin the development of the transportation system and the initial lunar infrastructure upon which businesses can build their plans.

Steve Durst, Space Age Publishing Company

Durst discussed the International Lunar Observatory Association, a group incorporated in Hawaii as a 501c to promote establishment as early as 2010 of a lunar observatory, the ILO. The ILOA is administered by Space Age Publishing Company/Lunar Enterprise Corporation, based in Waimea, HI. The ILO is planned initially as a robotic mission with optical and radio astronomy capability, on Malapert Mountain, which can eventually be visited by a human servicing mission. The ILOA involves the large astronomical community on the Big Island, where there are 13 observatories on Mauna Kea with control centers in Hilo and Waimea. Space agencies in India, China, and Japan have expressed interest in the ILO, which has been in the planning and organization stage since 2003.

Manny Pimenta, Lunar Explorer, LLC

This paper was titled "Malapert Base," which gives its essence briefly. Pimenta's group has designed a large modular base to be built underground at Malapert Mountain. It would be designed for maximum habitability, with large spaces reminding us of the underground shopping malls in Toronto. Pimenta has had no contact with the first author, so his recommendation of Malapert was an independent one, probably based on David Schrunk's presentation at the 2003 Waikoloa meeting.

Paul Lowman, NASA Goddard

This paper, "Malapert Mountain: A Recommended Site for a South Polar Outpost," was given on the first day, and there was time to hear reaction during the remaining days. It was apparently well received though there was only one question. However, several other speakers beside Pimenta mentioned Malapert. Durst's proposed ILO would be located on the top of Malapert.

An aspect not mentioned in this presentation is that an outpost in continual sunlight, and in continual view of the Earth, would be more benign psychologically than other recommended sites. Since several speakers at the Rutgers symposium brought up psychological stress problems, a few words about this aspect might have been added to this paper.

Summary and Conclusions

This was a very serious and somewhat sobering meeting, attended chiefly by very senior and high-level people. The main points from it appear to be the following.

1. The VSE has little public support. Most people have little interest in a return to the Moon, in particular the 14–18 age group, from which the scientists and engineers of the next generation must come. Part of the problem is that the public has no idea at all how little NASA spends. The first author suggests that this problem must be countered through two channels. First, the press must be persuaded to

   stop emphasizing the cost of NASA missions. Second, the educational community must be brought into the VSE, both teachers and
   students.

2. The problems of the lunar segment of the VSE are now well defined.
   They are in order of priority radiation, dust, and hypogravity. The
   combination of these will probably have synergistic effects. The dust
   problem (for manned missions) is well understood and probably
   controllable. However, radiation outside the magnetosphere remains
   a severe problem for long-duration lunar surface EVAs and surface
   structures. Hypogravity (1/6 g for the Moon) is no problem for short
   missions, but may well be a severe one for permanent settlements.
   It may in fact prohibit settlements involving large populations and
   families.

3. Lunar surface missions up to a few months are possible with
   present technology and knowledge, but "settlements" may not be
   for the foreseeable future. Manned Mars missions are probably
   decades down the road from a return to the Moon, the radiation
   and hypogravity (in space and on Mars) problems being major
   obstacles.

4. There is a real requirement for a robotic soft-landing lunar program
   before the return of humans to the Moon. No one brought this out
   specifically, except for Durst, but manned missions to regions very
   poorly known, specifically the South Pole, would be extremely risky
   without precursor robotic missions. The Apollo missions were pre-
   ceded by seven Surveyor ones, five of which were successful.

5. The commercial space sector is now seriously interested in a return
   to the Moon, and has much of the necessary technology in hand
   or nearly so. The Apollo Program was a government-funded and
   administered effort. However, the commercial opportunities of a
   return to the Moon are now much clearer. There appear to be no
   major legal barriers to private sector lunar programs, such as He-3
   extraction and export to Earth.

To summarize this summary, it isn't 1961 any more, when the lunar program
was carried along on a wave of public support. So we have a lot of work to do
to reignite this support.

3

Krafft Ehricke's Moon:
The Extraterrestrial Imperative

Marsha Freeman

Author, Technology Editor of Executive Intelligence Review

Marsha Freeman has been the technology editor of the weekly magazine *Executive Intelligence Review* since 1982. She was the Washington editor of *Fusion Magazine* from 1980-1987, and has been an associate editor of *21st Century Science & Technology Magazine* since 1988. She has written articles on all aspects of the U.S. space program, the history of the German space pioneers, the Soviet and Russian space programs, space medicine and biology, and the Japanese and European space programs, various fields of energy conversion, and nuclear and fusion energy technology.

She is the author of *How We Got to the Moon: The Story of the German Space Pioneers* (21st Century Science Associates, 1993), and the German translation, *Hin Zu Neuen Welten: Die Geschichte der Deutschen Raumfahrt-Pioniere* (Dr. Bottiger Verlags-GmbH, 1995); *Challenges of Human Space Exploration* (Springer/Praxis, 2000); and a biography of German space pioneer, Krafft Ehricke, to be published in early 2009, by Apogee Books.

Introduction

Space visionary Krafft Ehricke devoted more than three decades of his life to developing a comprehensive program for the industrial development of the Moon. He described Earth's companion as our planet's seventh continent. His devotion to the movement of human civilization into space began as a teenager in Germany in the late 1920s, and remained his life's work until his death in 1984. He became well known for his statement: "It has been said, 'If God wanted man to fly, He would have given man wings.' Today we can say, 'If God wanted man to become a spacefaring species, He would have given man a moon.'"

"Ours is a binary system," Krafft Ehricke explained. "There is no reason that only half of it should be inhabited, merely because life originated there... Instead of searching for and speculating about life elsewhere, we will put it there."

To Krafft Ehricke, the exploration and industrial development of the Moon was not simply a worthwhile undertaking, but an "extraterrestrial imperative." By this, he meant that man could not continue to grow and develop within the closed system of the Earth. He traced the roots of his concept of the extraterrestrial imperative to the European Renaissance, when the celebration of man's creativity led to the exploration of the New World, and to the breakthroughs in natural science that laid the basis for the modern era.

Krafft Ehricke believed that making clear the philosophical basis for mankind's coming exploration of space would provide the necessary foundation for all of his future extraterrestrial activities. In 1957, his *Anthropology of Astronautics* was published, outlining three laws that Ehricke believed should govern man's exploration of space:

First Law: "Nobody and nothing under the natural laws of this universe impose any limitations on man except man himself." This perspective became especially important in the late 1960s, when the limits to growth movement proposed that mankind must cut back its consumption and reduce its standard of living, because it was using up "limited" resources.

Second Law: "Not only the Earth, but the entire solar system, and as much of the universe as he can reach under the laws of nature, are man's rightful field of activity." Exploration and exploitation of the resources of the Moon and other heavenly bodies, Krafft Ehricke proposed, would eliminate the idea of limits to growth, by creating a "new open world" for mankind.

Third Law: "By expanding through the universe, man fulfills his destiny as an element of life, endowed with the power of reason and the
wisdom of moral law within himself." In this way, Krafft Ehricke saw
the spread of human civilization, culture, and science throughout
the Solar System, as the natural progression of humanity's billions-
years development, to its rightful place in the universe.

Today, space exploration programs are too often seen as simply another
line-item in an already overburdened federal budget. But Krafft Ehricke
made clear that the "imperative" arose from the understanding that there
would be grave consequences to trying to limit mankind's activity to his
home planet. In a diagram prepared in 1970, he contrasted a growth versus a no-growth future. A perspective for growth, which would necessitate
expansion beyond the Earth, would bring about international cooperation,
scientific developments, a global industrial revolution, and ultimately, the
preservation of civilization and of human growth potential.

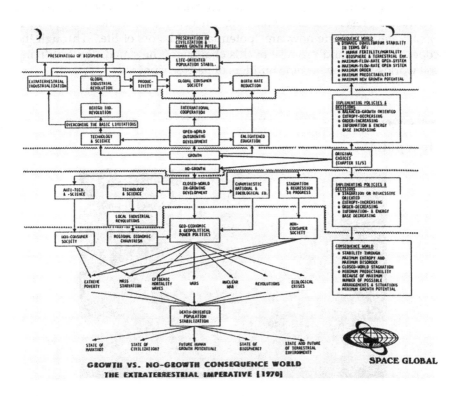

**GROWTH VS. NO-GROWTH CONSEQUENCE WORLD
THE EXTRATERRESTRIAL IMPERATIVE [1970]**

In contrast, a no-growth, closed-world pathway would lead to irrational
anti-science movements, geopolitical power politics, regional chauvinism,

and eventually wars over diminishing reserves of natural resources, many of which are problems we see today.

What space exploration and exploitation offered to mankind, in terms of practical results, Krafft Ehricke demonstrated, were advancements in the life and medical sciences, Earth and astronomical sciences, as well as advances in navigation, information, energy, and all aspects of advanced technology.

Krafft Ehricke concentrated great effort in studying and describing the development of the Moon, as the first step in what he called a "polyglobal civilization." However, he never limited his thinking to the nearby Moon. In 1948, while working for the United States Army, testing rockets and missiles, he wrote a fictional work on how Mars would be colonized, which he titled, "Expedition Ares." It was published for the first time in the Spring 2003 issue of *21st Century Science & Technology* magazine. He fully expected that the Moon would just be the first step in mankind's greatest adventure.

Krafft Ehricke described the Earth as a large spaceship, but not one that is alone, but one which "travels in a convoy." But since the other planets in the Solar System "are not unique in the same sense that Earth is," he wrote in 1969, "they are expendable and transformable." On Earth, we have the restraints of preserving and enlarging our ecosphere, while other planets, probably devoid of life now, are "potential incubators of life." Through the "scope, grandeur, and promise of this challenge," he stated, "by fertilizing these worlds, we may become builders of new ecospheres."

Selenopolis

The exploration and settlement of the Moon, Krafft Ehricke proposed, must be done on a grand scale. He envisioned the development of a city with thousands of citizens, which he named Selenopolis. The city would be built over time, and grow along with the development of lunar industries. Selenopolis would establish the industrialization and urbanization of the Moon, with the creation of a self-sufficient lunar biosphere. The mining and manufacturing industries would provide raw materials and minerals, and later semi- and finished goods, for use on the Moon, for export to the Earth, and for spacefarers heading to destinations beyond the Moon. The city of Selenopolis would be powered by fusion power plants, because only such an energy-dense energy source could energize a city that had to create its own environment, grow its own food, and provide educational, cultural, and entertainment activities for the people who would be developing the resources of the Moon. As Krafft Ehricke explained: "For Selenopolis, fusion energy is as indispensable and fundamental as the Sun's energy is for the terrestrial biosphere. Selenopolis cannot be built with yesterday's technology." The city on the Moon will not be an outpost, where a few people would brave the unknown, living in austere circumstances for short periods of time, but provide the basis for the movement of human civilization into space.

Although Selenopolis would have to be covered with lunar soil, in order to be well shielded from radiation, natural sunlight would be reflected into the city, through a series of mirrors atop the dome. There thus would be no indication that one were living virtually underground. Living on the Moon should not be all work, Krafft Ehricke stated. Inside the city of Selenopolis, in addition to living quarters, like any large city, there would be museums, schools, an elevated electric rail transport to travel from one venue to another, greenhouses, sports activities, and sections of the city with an array of different climates, such as are found on the Earth. For recreation, selenarians could spend time in the winter-climate sector, going ice skating, or they could relax on a beach, in a tropical environment. Life in Selenopolis would mirror, as much as possible, the very best aspects of life on Earth.

Krafft Ehricke likened the building of Selenopolis to the construction of the great cathedrals of the Middle Ages. And he advised that, like those cathedrals, it would be the work of generations.

Working on the Moon

In order to illustrate his concepts for living and working on the Moon, Krafft Ehricke created many of his own paintings, some of which are reproduced here. The development of the Moon would take place in stages, with increased capabilities in each stage, building upon the past accomplishments. The first stage would consist of detailed prospecting of the Moon, using automated, or robotic, probes. Although the Apollo program revolutionized our understanding of the history and composition of the Moon, many questions remain unanswered about its origin and evolution, and detailed resource mapping is still needed. This is indeed underway today, through probes being sent to the Moon by Europe, Japan, China, and the United States.

Stage two would continue detailed prospecting from circumlunar orbit, and capital equipment would begin to be launched from the Earth to the Moon, to prepare for human visits. By stage three, initial industrial bases would be established, including a Lunar Operations Center, which would consist of a laboratory and habitation module. An inverted shape would maximize the shielding for the modules, and would optimize temperature control, at the Lunar Operations Center in equatorial regions, which are warm during the day and cold during the lunar night. This shape also serves as an umbrella, to provide shade for astronauts working on the surface, in the vicinity of the Center.

In stage four, industrial feeder stations would be installed at various locations on the Moon, which would provide raw materials for industrial processing. A geolunar space market would develop, for products from the Moon to be exported to facilities in lunar orbit, and to geosynchronous orbit around the Earth. As the lunar economy and population grew, Selenopolis would take shape. And in stage five, the Earth and Moon have a roughly equal trade balance, with the Earth providing industrial capital goods and materials not available on the Moon, and lunar industries exporting finished goods and raw materials back to Earth.

Ehricke carried out very detailed studies of large-scale lunar mining operations, needed to provide the citizens of Selenopolis, and the people back on the Earth, with a source of raw materials. Before fusion plants were available, nuclear fission energy would be the primary source of power for the growing lunar economy, and industrial-scale mining operations. In an early 1980s scheme, Ehricke pictured collector trucks delivering the mined raw material, or "lunar crude," to the processing complex.

The lunar soil would be fed by a conveyor belt system into an atomic oven, which would be created through small nuclear detonations. In these underground caverns, the heat of the pulsed nuclear blasts would be contained in the desiccated, low-heat-conducting lunar rock. Atomic blasts inside the oven would vaporize the volatiles in the soil, such as oxygen, which would then be collected in a near-by buffer cavern. The gas would egress to the surface, where it would be collected and housed in storage containers, for use in Selenopolis. Many other precious resources, including man-made ores of enriched materials, would also be extracted from the atomic oven, as it is "mined" for other reduced materials. Then the oven would be reused.

New Transportation Systems

In order to make development of the Moon efficient and economical, Krafft Ehricke proposed that an entire family of new transportation vehicles be created, each optimized to operate in its own specific environment. Winged vehicles, taking advantage of the braking potential of the atmosphere of the Earth, he proposed, would be well suited for transport from Earth to Earth-orbiting space stations, and back again. Between the orbit of the Earth, and the orbit of the Moon, or through cislunar space, a class of vehicles without aerodynamic requirements, since they would not encounter a planetary atmosphere, would be developed.

The Earth-Moon transport system should make maximum use of the materials found and processed on the Moon, Krafft Ehricke proposed, since, due to the lower gravity, it is more efficient to launch material into cislunar space from the surface of the Moon, than from the surface of the Earth. He therefore designed a huge, nuclear-powered Cislunar Freighter, fueled by lunar oxygen and aluminum powder, to carry "lunar bounty" to terrestrial markets.

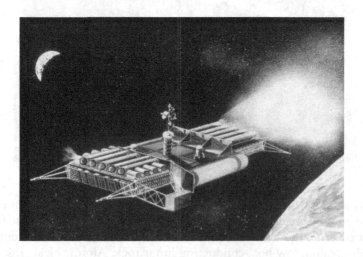

One of the primary cargoes carried by the Cislunar Freighter would be the isotope helium-3, which is very rare on the Earth, but has been deposited and has remained virtually undisturbed on the lunar surface, over eons by the solar wind. As has been recently emphasized by space scientists, particularly in Russia and China, this isotope will be the fuel for future terrestrial and lunar economies, used in advanced fusion power plants.

Landing on the Moon entails a different set of requirements than landing on the Earth, and specialized vehicles were designed by Krafft Ehricke for this purpose. For landing on the surface of the Moon, he invented an entirely new science—harenodynamics, from the Latin word for "sandy." This new science encompassed the determination of the dynamics of flow, boundary layer formation, pressure and temperature, and the effect of the release of oxygen, from the lunar sand. Harenodynamics is the substitution of the dynamic qualities of lunar sand, for the aerodynamic and hydrodynamic characteristics that enable man to fly, and land, in the atmosphere of the Earth. In this way, a vehicle could land like an airplane on the Moon, using the "drag" or friction of the sand as a substitute for an atmosphere, rather than use costly fuels for braking propulsion in rocket engines to slow down and drop gently to the surface.

Slide Landers would not require a paved runway, but it will be necessary to clear a path for the vehicles, by removing boulders and large rocks. Krafft Ehricke recommended that the lunar plains, the maria, provide suitable surface conditions. A nuclear-powered Mammoth Sweeper was designed by Ehricke, to prepare an 80-kilometer long lunar runway, or landing strip, with a smooth, although dusty, surface. The Slide Lander would approach the surface of the Moon from circumlunar orbit, to land on the runway prepared by the Sweeper. He made precise calculations, not only of the flow dynamics of this sandy medium, but also of the direction of the streams of sand created by the dynamic landing, so they would be propelled safely away from the vehicle.

Lunar mining operations would also require an extensive surface transportation system. Krafft Ehricke considered magnetic launchers and catapults for shorter distances, and systems similar to elevated light rail for distances beyond the horizon. Combinations of surface transport systems would be employed, depending upon the distances traversed over the surface, and the characteristics of what is being transported, such as people versus freight. During stage four, feeder stations, located in metal-rich provinces of the Moon, would be established. Using a ballistic delivery of this enriched "lunar crude" to central processing complexes, material would be hurled into a designated receiving crater, Ehricke proposed, from which it would be transported to the industrial processing facility. "The energy requirements for ballistic transmission are modest," he explained, "because of the low lunar gravity."

At the time Krafft Ehricke was designing his lunar city, the landings on the surface of the Moon, during the Apollo program, had been limited to equatorial regions. In order to extend the range of potential landing locations, and to increase the span of regions adaptable to the human exploration of the Moon, he developed a concept first proposed by German space pioneer Hermann Oberth, in the 1920s. Placed in the proper orbit around the Moon, a spacecraft could capture sunlight and reflect it to the surface. Such a Lunetta could illuminate the eternally-dark polar regions of the Moon, allowing the mining of what scientists hope are caches of lunar water ice at the poles. The orbiting mirror could also illuminate near-equatorial regions during the two-week lunar night, as well as provide light to the far side of the Moon during its night, which is not illuminated at all, even by earthshine. This global approach would be far superior to stringing electric power lines and placing street lights over the lunar surface, or carrying flashlights.

Space Should Be Therapeutic; and Fun!

On September 26, 1966, Krafft Ehricke appeared on national television with well-known television space reporter, Walter Cronkite. Ehricke used a set of models to explain to Mr. Cronkite, and the television audience, some of the ideas he was developing for exploring and living in space. While the Apollo lunar exploration program was in development, in preparation for the expected industrial development of the Moon, Ehricke suggested that space facilities in near-Earth orbit would provide a range of goods and services for the first space farers.

One such concept he described in the television presentation, was the design of an Earth-orbiting space hospital. It would be arranged as a series of separate but connected rotating cylinders, which would allow for the isolation of patients with different conditions. Access would be to each ward separately. Depending upon the distance from the center inside the rotating ward, the patient would experience varying levels of gravity. This would allow appropriate treatment for many maladies, such as heart conditions, or burns. Earlier the same year, Ehricke and B.D. Newsom, a research specialist in the life sciences at the Convair Division of General Dynamics, presented a paper on the "Utilization of Space Environment for Therapeutic Purposes," outlining the promise of Earth-orbital medicine.

In their paper, Ehricke and Newsom describe the creation of an ambulance launch vehicle, to transport seriously ill patients to the more benign microgravity environment of Earth orbit, and medical research and treatment facilities at the orbital hospital. Two categories of objectives were noted:

curative objectives, which would enable the patient to return to Earth in a better condition than he left; and alleviative objectives, which establish "the individual's behavioral freedom from the effects of a handicap by temporary or permanent transfer to a low-zero gravity environment." They note that while the microgravity environment had been shown to have deleterious effects when healthy human beings adapt to space, that environment can have curative effects, as well.

The reduction of stress on the heart and circulatory system are one noted potentially positive effect of microgravity, which can provide relief from hypertension. The absence of pressure on the musculo-skeletal system can release pinched nerves and compressed spinal disks. Space can be used as a means of physical therapy, through the controlled use of varying levels of gravity. Crippled, partially paralyzed, or otherwise handicapped people could benefit from the reduced environmental stress. Burn victims would likewise benefit from the lowered pressure to the skin in microgravity.

But space will not only be open for settlers, miners, engineers, and scientists, Krafft Ehricke proposed. It should also be fun! Decades before citizens who are not professional astronauts were lining up to experience space flight, Ehricke proposed in 1967, that tourism will be a natural extension of man in space. He said that people will go into space for the "sheer pleasure," just as they take trips on ocean liners and jet planes on vacation, on the Earth. New activities, only possible above the Earth, will provide relaxation, just as they do for astronauts today, when there is a window of opportunity, and window to peer out of, in their spaceship.

Obviously, Krafft Ehricke states, "space vacations offer attractions which literally are out of this world." Microgravity, itself, will provide a leisurely environment, removing the physical stresses of life on Earth. Global "sightseeing," as the astronauts do now, will be popular, and will include not only looking at the Earth, but also out at the universe, as "the sky is open" to the tourist "as it can never be on Earth."

Orbital hotels and tourist facilities will offer a range of entertainment, sports activities, and even orbital excursions (space walks, or extravehicular activities). One such facility for the vacationer is the Dynarium, which Krafft Ehricke describes as "the equivalent to the large swimming pool in modern resort hotels." The Dynarium is a large enclosure, where the "swimming pool" is three-dimensional, and guests can dart from wall to wall, gently float, tumble, roll, or fly under their own muscle power.

Orbital hotels will also include space zoos and botanic exhibitions, small laboratories, observation rooms, space walks, and space boats that can be rented for tours around the facilities.

Why will people want to spend their vacations in space? Krafft Ehricke poses the following questions, assuming that a thriving space tourism industry will eventually take explorers beyond the confines of Earth orbit: "Do we know what it is like to discern the manifold signatures of creation in space? Do we know what it is like to walk through the stillness of worlds

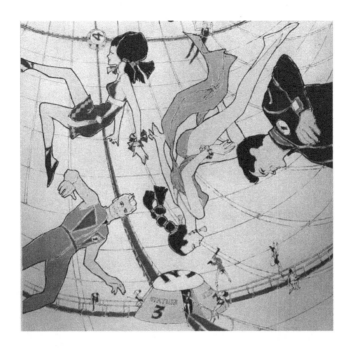

untrespassed by living beings? Do we know what it is like to touch the unchanged rocky texture of a column a billion or more years of age? Do we know what it is like to leap with seven-league boots on the Moon? Can we fathom the beauty of a Martian sunset? … to look at the mute traces of life somewhere that faded away eons ago?"

Recreation and enjoyment will not only be available for tourists who choose to spend their vacations in space. Inside the lunar city of Selenopolis, provision will be made, in the partial-Earth gravity of the Moon, for new experiences in sports activities, such as tennis. There will be opportunities for casual relaxation at "outdoor" cafes and restaurants, in areas of the city that are comparable to the "sunbelt" climates on Earth. Other parts of the city will mimic the colder climates or seasons on Earth, and Christmas in Selenopolis might include ice skating, and a visit to the Hall of Astronauts.

To Krafft Ehricke, the idea that there could be "limits to growth," which notion became popular during the late 1960s, was a repudiation of the human spirit, and the Renaissance view of man. He fought a constant battle to argue that only when the Earth is looked at as a closed system, could one consider that there are limits to resources. But, there is no limit to human creativity, and it is that resource, and the new knowledge it creates, that creates a new open world. This open world depends upon the exploration and exploitation by man of what is rightly his: the unlimited potential of space.

4

Looking Back at Apollo/Saturn: Planning Activities (1961–1965)

Excerpts of weekly notes of the director, future projects (NASA/MSFC) to the center director, Dr. Wernher von Braun

H.H. Koelle

H.H. Koelle was born 1925 in the former free state of Danzig. He was a pilot during World War II, founder of the postwar German Society of Space Research—GfW (1948), Dipl.-Ing. (MS) Mechanical Engineering, Technical University Stuttgart (1954); member of the Dr. W. von Braun team at Huntsville Alabama (1955–1965); Chief, Preliminary Design, U.S. Army Ballistic Missile Agency (ABMA), member of the launch crew of Explorer I, U.S. citizen (1961); Dr.-Ing. (PhD) Technical University, Berlin (1963); director, Future Projects Office NASA/MSFC (1961–1965), responsible for the preliminary design of the Saturn family of launch vehicles and planning of the MSFC share of the Apollo program; editor-in-chief of *Handbook of Astronautical Engineering* (McGraw-Hill, 1961); professor of space technology, Technical University Berlin (1965–1991); vice president of the International Astronautical Federation (1967–1969); dean, Department of Transportation, Technical University Berlin (1989–1991). He is a member of the International Academy of Astronautics (1966), chairman, IAA Subcommittee on Lunar Development (1985–1997), and has over 350 publications. He is recipient of the Medal of the Aeroclub of France, Hermann-Oberth Medal (DGRR), the Hermann Oberth Award (AIAA), the Eugen Saenger Medal (DGLR), the Patrick Moore Medal (BIS), and the Engineering Sciences Award of the International Academy of Astronautics. In 2003 he was elected Honorary Member of the International Academy of Astronautics; 2007 Space Pioneer Award of the National Space Society USA.

Introduction

It is interesting to note that the problems that were analyzed and discussed in the early sixties with respect to the future of the U.S. Space Program are back on the agenda again due to the new U.S. Space policy initiated by President

Bush in January 2004. Thus, a summary of respective items reported to Dr. von Braun by the director of the MSFC Future Projects Office indicates the problems occupying our minds some 40 years ago and are of relevance currently in planning the space exploration vision.

During the years 1960–1965 the author was director of the Future Projects Office at NASA/MSFC. These were the years where the Saturn/Apollo program was conceived and placed under development. I reported directly to Dr. Werner von Braun. People of his staff reporting to him directly had to submit by Friday noon each week one-page weekly notes, summarizing all that was happening and was expected for the following week and should get his attention. This was one of the management tools he used to keep things under control.

The "Weekly Notes" listed in this summary cover selected items that were of particular interest to Apollo/Saturn development to NASA management. This communication between Dr. von Braun and myself illustrates our aspirations as well as the spirit and thinking prevailing in 1961–1965, the years in which the Apollo program was taking shape and initiated.

Notes selected, illustrating the events, concepts and strategies, are limited to the highlights as seen from the vantage point of the year 2008. They are arranged by topics and dates originated.

System Analysis

10-2-61: Launch Vehicle Reliability

On September 20 a representation was given to the LLVPG (GOLOVIN committee) by H.H. Koelle, accompanied by J.W. Moody (M-Rel), on a method for reliability growth on large launch vehicles. The data presented was well received. Copy of presentation is available. (LLVPG Task No. 003.)

10-30-61: Reports of MSFC

More than 200 reports have been submitted to this office for evaluation and inclusion into a 10-volume collection of technical reports, demonstrating MSFC systems capability. It is estimated that about 100 to 120 reports will be selected. Abstracts of the selected reports have been typed and will be compiled into an Abstract Volume. This Abstract Volume and a cover letter signed by the Director, MSFC, will be sent to Mr. B. Holmes (newly appointed associate administrator of the Office of Manned Space Flight!) later this week, stating among other things that he can have the whole collection. These volumes should be ready for your inspection Tuesday afternoon in my office. Suggest you put it on your schedule.

12-10-62: Cost Control

I would like to offer a suggestion concerning cost control on future and present projects. We have observed that costs are greatly and quickly altered with change in specifications. Our present system permits each individual engineer and designer to incorporate specifications and changes which fall in the total spectrum from "nice-to-have" to "must!" However, the individual who introduces the specifications is rarely asked for a cost estimate tied to this specification or change, nor for an evaluation of the benefits (such as reliability), accrued from such a change. Normally, it suffices to say: I want it! While this is not objectionable when resources are plentiful, it might not be the best mode of operation in case of limited resources. I would suggest that, in the future, the Change Control Board and the Project Director insist on a cost and benefit evaluation before such changes are approved. There should be a certain cost limit above which must be obtained from the Deputy Director R&D or the Director MSFC. I hope we can keep costs down, at least for new projects if we make these trade-off studies, while we are writing the specifications during the program definition phase. I believe our present rather loose cost control is responsible for a fair share of our cost overruns. Dr. Lange and Mr. Maus might want to join in a little study on the practicability of this suggestion.

5-27-63: Launch Vehicle System Cost Model

Six proposals were received on our Launch Vehicle System Cost Model Study. Two outstanding proposals were submitted by General Dynamics/Fort Worth and Lockheed California Company. We plan to initiate contract negotiations with both companies and expect to have contracts underway by mid June, unless you have objections. These two studies will give us parallel effort in the important area of vehicle costing. We will be glad to give you a short briefing on the objectives of this study and the related work we are doing in the cost area, if you like.

[Comment by Dr. v.B: *No objections, let's combine this with next tech briefing.*]

9-30-63: Cost Studies

We are continuing, within our limited authority and capabilities, to study the reasons for cost overruns and to develop tools for better cost projections. We are coordinating this activity with those individuals from other elements of MSFC who have demonstrated an interest. Our activities can be summarized as follows:

Resources Model

This will give us upper and lower limits of expected funds to be available for the development, procurement, and operation of (Jack Waite and crew) with the assistance of FPO.

Mission Model

This mission model relates the mission capabilities of individual launch vehicles with mission objectives, state-of-the-art, and initial weight requirements in low altitude orbit, with the main variable being change of emphasis on space activities between global operations, orbital operations, lunar operations, and planetary operations. We are trying to interest Aero-Astrodynamics Laboratory (Mr. Thomae) in assisting us in the development of this model.

Cost Projection Model for Saturn-Type Expendable Launch Vehicles

This Model is in an advanced state of development. We will attempt—in close cooperation with Industrial Operations—to mechanize the process of cost projections. However, this model will not substitute for the detailed budget exercise we have to make several times per year to cover the immediate future. It will give long-range trends and permit an evaluation of the influence of the major system parameters. This model will have a great deal of sophistication and hopefully will help us to better understand some of the cost parameters and to develop correction factors for overruns.

Universal Cost Model for Launch Vehicles

This is a more advanced cost estimating procedure for reusable launch vehicles, which we are developing under contract with Lockheed and General Dynamics/Fort Worth, and with some assistance from Rand.

Cost Effectiveness of National Booster Program

We do have now an operational program, which allows us to mix firing rates and performance of all launch vehicles which are presently in the national booster program. This model helps us to analyze the influence of either individual launch vehicles on the total program, or the effect of individual vehicle improvements. This will be helpful for Saturn V and NOVA comparisons.

[Comment by Dr. v.B.: *I suggest that you also study (and possibly crank in): CADLE report (blue book); SCHRIEVER "lessons" report ("yellow book"). Maus knows both and can get them for you.*]

4-6-64: President Johnson's Request

In a letter to Mr. Webb, dated January 30, 1964, The President requested a NASA position on the following two subjects:

- A statement on possible future space objectives, and,
- estimates of time, funds, and technology required to implement major missions related to future objectives.

In a memorandum, dated March 25, 1964 Mr. Webb issued a directive to Dr. Seamans on how this request should be implemented.

A task group was established under Frances B. Smith (on loan from Langley) with two panels:

- An objective panel (headed by Dr. Richard Head)
- A mission panel under Mr. W. Flemming. This panel has a sub-panel for launch vehicles under Milton Rosen.

I have been listed as the MSFC member on this sub-panel and was alerted that this will result in a major workload, primarily for people in my office. Mr. Webb has directed that this should be considered a top priority assignment [Comment by Dr. v.B.: *Absolutely!*]. I am now standing by for further instructions and will keep you informed through these Notes.

6-29-64: Weighted Objectives of the National Space Program

We conducted an "opinion poll" among 60 MSFC key personnel on the relative importance of individual objectives of the National Space Program. This resulted in the following weighted objective list, which we intend to use to determine the "relative worth" of individual project combinations within the present long range planning exercise:

Ranking	Objective	Weight (Percent)
(1)	Achieve and preserve U.S. leadership	8.2
(2)	Utilization of space knowledge and technologies for the benefit of mankind	8.0
(3)	Gain knowledge about the nature of the universe and life itself	6.1
(4)	Develop an industrial base which can quickly respond to security needs in space	6.1
(5)	Incentive for improved education	5.9
(6)	Promote international cooperation	5.8
(7)	Stimulate the nation as a whole	5.7
(8)	Stimulate economy (investments and employment)	5.6
(9)	Demonstrate space systems applicable to security needs of the U.S.	5.6
(10)	Increased knowledge about terrestrial and space environment	5.2
(11)	Strengthen educational facilities	5.2
(12)	Improve industrial base continuity, including management practices to develop very complex technical systems	4.85
(13)	Development of manned space transportation systems as a new dimension to terrestrial transportation systems	4.85
(14)	Improve government capability to handle complex systems	4.3

Ranking	Objective	Weight (Percent)
(15)	Provide quick inspection capability to police arms control agreements	4.0
(16)	Exploit extraterrestrial resources	3.9
(17)	Improve U.S. competitive position in foreign trade by use of new techniques and procedures	3.2
(18)	Aeronautical transportation systems will be improved as a fallout of space technology	2.7
(19)	Development of efficient unmanned space vehicles for scientific research	2.5
(20)	Space transportation systems will also permit rapid global transportation	

Do you think that this is a reasonable basis for some initial comparison of project mixes?

[Comment by Dr. v.B.: *My personal opinion jibes pretty closely with this poll, except for the relative ranking indicated by the arrows, which indicate my preference. Item 16 should be at the end of the list, items 18 and 20 should be placed behind item 13.*]

8-31-64: Answer to President's Letter

A preliminary draft of the answer letter is now in circulation among staff offices in Washington and plenty of surgery is applied. It is a very broad description of the entire program, avoiding specifics. However, the following three projects are singled out as the most probable new starts this year: Voyager, Alls and Apollo X. The draft is supposed to be coordinated with DOD, PSAC and Dr. Welsh before it will be submitted in a final version in December.

10-5-64: Marketing of Space Flight

There is growing evidence that we will have difficulties in the years to come to sustain the public support we enjoyed in the last years. This suggests—and I know that you have preached along these lines quite often recently—that we at MSFC should make an organized effort to come up with a better utilization of our space flight capabilities and take great care to justify in a better way, the use of newly developed hardware and new projects; e.g. the "why" of spaceflight. Up to now we have concentrated almost exclusively on the "how" of space flight. I would suggest a special brainstorming session in which we analyze how our MSFC talents can be brought to bear more on the problem of space flight marketing. I am thinking of increasing our present effort (which is a few hundred man-hours per year) by at least two orders of magnitude. This might require an organizational change or shift of some manpower. I am thinking of a group of 5 to 10 professional people doing nothing but to compile potential applications and develop the uses of space flight in an aggressive manner within NASA policies and in conjunction with other organizations. Many avenues of approach offer themselves in accomplishing such an objective.

[Comment by Dr. v.B.: *Let's discuss this again (between you, Weidner, Frank W. and myself).*]

[I expressed my feelings during the 9/25 Executive Session.]

11-9-64: NASA Five-Year Plan

In case you are wondering about my reaction to the new long range plan: The new NASA five-year plan is better and more realistic than anything we have ever had! However, from where I am sitting, I see the following shortcomings:

a. It does not appear to be the type of a plan the President requested.

b. It does not appear to be a plan that will insure that we will be ahead of the Russians one of these days.

c. The plan does not offer a real challenge to NASA, and particularly MSFC, once the Apollo mission has been accomplished.

d. The plan does not have the elements which it takes (in my mind) to compete successfully for its share of the federal budget. I believe that this kind of a plan does not offer enough "return on investment" to sustain a 5.5. to 6.0 billion dollar budget a year for the future. [Comment by Dr. v.B.: *Let's talk about this. I don't see it that way at all.*]

e. This plan, if not accompanied by program definition studies for the time period beyond 1971, neglects the fact that lead times of complex spacecraft and launch vehicles are longer than 5 years (in fact 5 to 10) and we will end up in the same box we are today, with too few attractive missions and suitable payloads to fly in 1972 and beyond. To reinitiate such studies two years from now is too late to catch the budget decline around 1970, as presently projected.

f. In a few years from now, I can see difficulties for MSFC to obtain a project assignment which is in line with our talents, facilities and strength. [Comment by Dr. v.B.: *I've heard that before!*]

g. The taxpayer might view this plan (if at all) as one which offers "just more of the same"; I would expect some difficulties in keeping the public really interested in and behind the space program, a trend which eventually will reduce available resources.

h. I am disappointed in how plans of this importance are developed "over night" with somebody pushing the panic button. They always result in plenty of confusion, as well as a waste of money and manpower. NASA could do its job a lot better and save a lot of money. [Comment by Dr. v.B.: *You couldn't be more wrong! That plan has been under discussion for over 6 months in one version or the other!*]

For all those reasons, I prefer to identify the new five-year NASA plan as "a plan to depart from" and would like to see us constantly improving it and do this with vigor!

(When I joined your team almost ten years ago, you told me that you did not expect me to be a "yes man," I am still adhering to your advice, as you can see.)

12-7-64: New NASA Five-Year Plan

There must be somebody else within NASA who is not completely happy with the present state of affairs in the area of long-range planning. I quote the following sentence from Aviation Week of 11-16-64: "A working level complaint at NASA these days is that to get money in the new budget the concept has to look like Apollo, feel like Apollo and smell like Apollo."

[Comment by Dr. v.B.: *We need both: Long-range planning and Apollo ad-ons. The trouble at the moment is that for the momentum built up with the main-stream Apollo program. There has been too little of the latter, with resulting immediate dangers to the stature of the program.*]

12-14-64: Return-on-Investment

We have developed a calculation procedure which attempts to assess the "expected return on investment" of any space program, and can suggest the relative standing of alternative space program plans. It is based on the "weighted objectives list" which was derived by group judgment of senior MSFC personnel. Forty-four yield indicators are available to correlate the yield of a program with the given objectives. Each of the objectives is represented by a "worth estimating relationship." The number of terms in each equation is proportional to the weight of the objective. The individual terms are non-linear and produce a devaluating effect with increasing time and transportation volume.

The calculation procedure is simple enough to be explained to people not specializing in the "numbers racket." We are now testing this procedure to find out how useful and valuable it might be as a planning tool. The purpose of this limited effort is to gain more insight in "what makes the space program tick," not to derive a new space program by analytical procedures.

If we think we learned something, we will let you know.

[Comment by Dr. v.B.: *Although I'm sure you know as well as I do that the political feasibility of space programs depends not only on logic, but on such unfathomable elements as President Kennedy's election or NASA/AF relations. I am in complete agreement with you that it is absolutely necessary that we proceed in our long range plans with as much logic and care we can muster. Let's make sure that we discuss progress in this area at regular intervals, and don't get discouraged.*]

Launch Vehicle Concept Analysis

10.16.61: LLVPG Thinking

Emphasis is being placed on a 240" solid motor for C-1, C-4, and possibly NOVA application. The solid propellant enthusiasts are pushing for a solid vehicle back-up to the liquid (propellant) C-4 (configuration). This week the funding picture will be studied in detail, since the worst solution would be two under-funded programs.

[Comment by Dr. v.B.: *Couldn't agree more!*]

12-11-61: SATURN-D Follow-On Studies

A follow-on statement of work for General Dynamics and Lockheed has been prepared and coordinated with Mrazek, Col. Fellows, and Schramm. These studies will provide input into the RIFT (Reactor-In-Flight-Test) program and the C-4/C-5 program so that upon completion of RIFT an early operational capability can be achieved without major redesign. Because of the competitive nature of RIFT, certain precautions must be taken in the Saturn D.

12-18-61: Studies Planned

A memo was forwarded to you on December 14, 1961 regarding desired action on the following:

a. Electric propulsion mission study
b. Application of hydrogen/fluorine for both C-1 third stage and Apollo lunar launch vehicle.
c. Early manned planetary mission study.

A draft for the proposed NOVA study effort has been completed and I would like to discuss it as well as the initiation of the "NOVA definition effort" at your earliest convenience.

[Comment of Dr. v.B.: *Let's first get the C-5 issue out of the way. "One emergency at the time!" But why don't you draft something, meanwhile?*]

2-12-62: Liquid versus Solid Propellant Rocket Motors

In order to be better prepared for congressional hearing, do you think it would be a good idea to personally invite Dr. Ritchie (Thiokol) for a detailed technical discussion with system specialists. If you agree, I think I can handle it.

[Comment Dr. v.B: *I wouldn't invite him under the heading of "forthcoming congressional hearings" because he has made too many rather opinionated statements already. But if you, as head of MSFC's Future Projects Office, invite him to a* discussion *("as one rocket man to another")* to clear up certain questions you have in the area of analyzing complete solid-powered vehicles *(including thrust vector control, transportation, emergency procedures, the "donor" explosion criteria problem etc. etc.) that'll be just fine. Say we'd like to get to the hard core of the controversy - the question is not that we doubt you can build big solid motors, but those thousands of other unanswered questions. Please keep me posted.*]

Position Paper on Solids versus Liquids

The status report on "solids," which you requested as preparation for congressional hearings, is now undergoing the 3rd iteration process. Next Monday we will have clean draft ready for review by you and the Division Directors. At the present time it is about 20 triple-spaced typewritten pages.

[Comments by Dr. v.B.: *Suggest (if you have not done it) to crank into this paper the class 9/class 2 (explosives) argument my position (backed by Seamans). We want no part of C-5 or NOVA solids unless they are officially accepted as class 2 hazard in presence of donor. Suggest you check also LOD re-siting problems in new Cape area.*]

4-8-63:   NOVA Redirection

As a result of our discussion (April 4) and your directives, we are now reorienting our study efforts in the direction of unconventional reusable NOVA concepts. We will issue new guidelines to the contractors this week along the following lines:

A. Sixty percent or more of the total study effort will be applied in the operation analysis and conceptual design leading to and approaching the greatest practical extent of an "ideal NOVA" defined as follows:

(1) NOVA must have a multiple mission capability, preferably in all of the following areas:
   (a)   Earth to low orbit heavy cargo delivery
   (b)   Earth to orbit cargo delivery in connection with dog-legging into high orbit inclinations and/or inter-orbital transfer to high altitude orbits
   (c)   Global logistic transport for cargo and personnel
   (d)   Lunar logistic transport for mixed cargo and personnel
   (e)   Planetary logistic for cargo and personnel
   (f)   High velocity space probes. The "ideal" NOVA concept might have most of the following features (or equivalent):
   (g)   (a) Single stage (with air augmentation, and/or tank staging and JATO)

    (h) (b) Land and sea recovery (with payload in case of global transport). The "ideal" NOVA concept might have most of the following features (or equivalent):

(2) The "ideal" NOVA concept might have most of the following features (or equivalent):

    (a) Single stage (with air augmentation, and/or tank staging and JATO)

    (b) Land and sea recovery (with payload in case of global transport)

    (c) Design lifetime of 100 flights

    (d) Terminal guidance

    (e) Wide payload range capability (larger than Saturn V up to approximately 500 tons to low orbit)

    (f) Acceptable acceleration limits in case of personnel transport

    (g) Compatibility with nuclear upper stages.

B. The rest of the effort will be used to update conventional expendable or partially reusable NOVA vehicles, in the latter case with first stage recovery as minimum goal. This data will be used to evaluate the advantages offered (and price to be paid) by the various "ideal" NOVAs we hope to come up with.

Does this formula interpret your instructions close enough so that we can proceed?

[Comments by Dr. v.B: *Precisely. You may proceed on this basis. Capt. Freitag also agrees. Make sure to get Shea on board also (through Doug Lord).*]

10-14-63: Global Transport

I have discovered a quote of Dr. Lloyd Berkner (for many years president of the Space Science Board of the National Academy of Science), which I consider to be important and interesting:

[From Hearings before the Committee on Aeronautical and Space Sciences, United States Senate, June 10 and 11, 1963]:

> The future use of rockets for quick point-to-point transport now is as certain as we now know scheduled 600-m.p.h. air transport had become after the demonstration at Kitty Hawk. Without question, technological problems of regular rocket transport are enormous. But there is no known natural or scientific dictum that stands in the way of such transport on an efficient and regular basis. So, we can confidently predict that fast rocket transport will come, eventually, to give us 20-minute trans-Atlantic schedules.

1-13-64:   Post-Apollo Saturn IB Market

During Ed GRAY's visit last week, I had an opportunity to inform him that now is the time to plan for missions of the Saturn IB for the post-Apollo time period (mid 1968). In my opinion, there is a great probability that we will not have a large enough market to keep the Saturn IB production line going [Comment by Dr. v.B.: *I Agree*]. New payloads of a complex nature for 1968 and 1969 are almost out of question due to lack of resources. Further, we have the competition of Titan III to expect and we should not forget that the Saturn V also is around. With a firing rate of two per year (which could be all we might be able to sell for 2968/69/70). The cost of the IB might be more than the Saturn V. I am greatly concerned about the prospects in this area and am trying to get some joint activity started among OMSF (GRAY), MSC and MSFC to complement the contractor effort, which just resulted in their first report (containing a little too much window dressing and not enough analysis).

5-25-64:   Post-SATURN Review

Last week we had a two-day review of our post-Saturn study activities, including vehicle trade-offs, mission analysis, test facilities at MTO, and Launch facilities at MILA. We now have enough information to concentrate our conceptual design on a "base line" configuration which can be described as follows:

Concept

Fully recoverable and reusable, all LH2/LO2 propellants, two stages to orbit, three stages to escape (modified S-II Stage is a suitable third stage).

> First stage: 18 modified hinged M-1 engines around plug-nozzle, recoverable, separate tanks
>
> Second stage: 12 high pressure engines in the 315 K thrust range, possibly toroidal airspike integrated propulsion system for maximum performance. Plug used as re-entry body.
>
> Vehicle launch weight: Approximately 18,000,000 lb
>
> Diameter: Approximately 75 ft
>
> Payload LEO: Approximately: 1,000,000 lb
>
> Operational target date: 1980
>
> Cost effectiveness to LEO: 60 $/lb direct cost; 120 $/lb total cost
>
> Facilities: Two test stands for each stage at MTF, 2nd stage sectional R&D testing at MSFC in our large test stand. Three launch positions at the cape in hybrid fashion (enclosed tower serves as assembly building and launch site).

We will make additional trade-off studies and probably modifications to this base line concept as we go along. However, the planning exercise for the President, as well as our desire to make competitive study for Saturn V doing the same missions, is the reason for selecting a base line concept at this time. It might not be the very best, but it will be one of the best concepts available. Many considerations went into the selection of the new base line. We will be happy to give you a briefing on this subject, if you desire.

It might be of interest to you that all the guidelines we have received from headquarters do require a post-Saturn launch vehicle by about 1980 for manned planetary landing and capture missions!

11-16-64: Nuclear System Comparison

During Dr. Mueller's last visit here (October 22, 1964) he requested that we prepare a staff paper with preliminary answers to the following two questions:

1. What research and development tasks have to be undertaken for each of the applicant systems to demonstrate its technical feasibility?
2. How do these systems compare, performance and cost wise, if put on a common basis?

We have set the machinery in motion to prepare this staff paper and have formed an ad hoc working group with FPO, P&VE (Jordan), RP (Dr. Shelton), and Aero (Thomae) participating. A plan of attack has been agreed upon, which is as follows:

a. Nominal mission comparison is done for a 1984 Mars launch window with a 500-day trip time, including a minimum of 20 days stopover. The crew size will be a minimum of eight, and the landing vehicle will weigh 80,000 lb.
b. We will use 10 different modes to compare various combinations of types of propulsion systems. Some of the lower performance will use the Venus swingby mission profile, the higher performance will not. Earth reentry speed will in all cases be below 50,000 ft/sec. There will be no cases where the spacecraft returns into an Earth capture orbit instead of direct atmospheric entry.
c. Each of these 10 cases will be investigated assuming:
   (1) Saturn V only
   (2) Post-Saturn and Saturn V as launch vehicles

d. The following parameters will be derived for comparison purposes:
   (1) Total mass per ship leaving Earth orbit
   (2) Dry mass per ship leaving Earth orbit

(3) Total departure mass in Earth orbit divided by useful payload carried to Mars orbit

(4) Total departure mass in Earth orbit divided by spacecraft mass returning to Earth

(5) Total departure mass in Earth orbit divided by man-days available on Mars surface

(6) Total project cost divided by man-days available on Mars surface

(7) Direct operating cost (excluding R&D) divided by man-days available on Mars surface.

[Comment by Dr. v.B.: *How about "Total number of earth-orbit logistic supply flights," considering the losses incurred by orbital operations - effect of cryogenics versus dry Orion propulsion - effect of Post Saturn versus Saturn V?*]

Apollo Mission Modes

10-2-61

On September 27 a program review was held on the "Orbital Launch Operations Study" at Chance Vought, Dallas, Texas. This effort was redirected to include the effects of C-4 (launch vehicle concept) on the orbital development program.

2-12-62: Weight/Performance Control

With lunar orbital operations getting more momentum, I feel we have to make a special effort to control our C-5 weight and performance in order to stay competitive. There seems to be a trend that too many people try to put their own safety margins in. This results in one reserve on top of many others and the performance deteriorates rapidly. I am trying to get together with Mrazek and Dr. Geisler in our Performance Review Board. An occasional encouraging word from your side might help. [Comment Dr. v.B: *Agree, you have my full blessing!*]

2-26-62: Apollo Mission Performance Dispersion

A concentrated effort is being made to obtain a better feeling for the total mission performance picture, with emphasis on determining the upper and lower limited of the total tolerance. The total velocity requirements, specific impulses of individual engines, and weight estimates have been varied within the tolerances to be expected. We believe it is not correct to assume that all parameters will build up to the upper limit of the unfavorable side. By doing so in the past we have "proven" that the C-5 is now too small. We have already done this investigation for the connecting mode and find that we can expect to bring down to the moon anywhere between 53,000 to 84,000 lb with 65,000 t0 70,000 lb as the most likely weight. This makes, however, one assumption that we will be successful in telling our engineers how the interrelationships work and that is not permissible for everybody to put in his personal padding. W adding these "reserves" arithmetically (which is not the proper way) we sell ourselves out of business. We also will investigate the other modes of interest and then prepare a summary report. We hope that this study will result in certain target data for individual design parameters which will distribute evenly, throughout the system, the burden of doing a good job. [Comment Dr. v.B: *Agree. But organized weight and performance control for all elements is equally important*]

4-9-62:   C-5 Direct Capability

I will organize a small effort to result in a precise determination of the "margin of error" for this mode. I consider this as a fifth working group within Dr. Geisler's overall effort. I suggest that this be placed on the agenda of the next board meeting as a 5-minute discussion item.,

5-14-62:   Light Weight Apollo Design

Dr. Eggers will present his proposal for a light weight Apollo capsule to Dr. Shea this week. I believe he has a good story and I expect that the C-5 direct mode will pick up momentum as a result of Dr. Eggers presentation. We expect that our calculations will show an attractive performance margin for the entire mission profile.

Space Station

6-11-62:   Program Planning

We are presently compiling all available data useful for a preliminary development plan (PDP). Our proposal is based on the modification of a S-IC LOX tank as the basic structure. We are studying the use of the Titan II-Gemini, Saturn-IB (also with a solid booster) and the Titan III as a basic supply vehicle. This effort is in support of MSC. We should be ready for a presentation to you in about four weeks. [Comment by Dr. v.B.: *O.K.*]

7-16-62:   Space Station Preliminary Development Plan (PDP)

We have proceeded with the draft of our proposed report to a point where we would like to make an informal presentation to you and/or the Board.

We have heard that this subject will be on the agenda of the next Council Meeting. There are two problems to be discussed:

a. If the simple non-rotating station is favored as the cheaper and simpler solution, we seem to have an option to make a bid for the basic structure of the station. This can be built at Huntsville with S-1C tooling. It is essentially a glorified first stage Lox tank.

b. If this project comes early, we are facing a situation where our present C-1B launch facilities would be saturated by the requirements of this project alone (12 firings per year). This will be on top of possible planetary missions.

8-27-62:   PDP

It is anticipated that MSFC will forward their official preliminary project development plan concerning a manned space station to headquarters this week and at the latest probably next week.

Ed Olling of MSC visited us last week and indicated that considerable support is being generated in Houston and at headquarters level. They now have about 12 people in the space station office and a 40- to 50-man team working full time on the space station. Olling indicated that he is sending an official request to Marshall for a space station coordinator at this center concerning Saturn launch vehicles and the total program. They are also in the process of establishing technical advisory committees as they have in the past. All centers will be visited to establish coordination and planning.

A special committee will also be formed to determine the types of research and development that must be conducted on the first space station. Olling was informed of our preference that the next approach be on a small scale of approximately 6 to 12 men and that the program should involve minimum funding at the outset. Olling is presently interested in getting Phase I approval only. This covers a $ 2 million study phase on the space station, Apollo modification for 6 men, and a reusable logistics spacecraft. After that they will worry about total program cost and space station configurations and more details of the total program, as the politics and engineering factors evolve.

We will stay in touch with him and have a low level effort going within MSFC to study the problems of a small station as you suggested, based on Saturn 1B capabilities following the C-1B orbital rendezvous program. If this satisfies you, I think we can skip the meeting you offered on this subject and reschedule it a few months from now.

[Comment by Dr. v.B: *I'm convinced that in view of NASA's overall funding situation this space station thing will not go into high gear in the next few years. Minimum C-1B approach is the only thing we can afford at this time.*]

9-9-63:   Space Station

Last week we had a preliminary design review with P&VE Laboratory on their in-house study on the space station project. Their project engineer is Mr. Cash. Our project engineers are Jim Carter and Lou Ball. This effort was initiated upon your request approximately four months ago. The reasons for this study are as follows:

- To study compatibility of space station concepts with Saturn launch vehicles.

- To consolidate briefing material for you in preparation of a pending decision on this project in the Management Council Meeting in the next few months.

- To study the possible workload on MSFC, resulting from the initiation of a space station project, assuming various degrees of participation.

- To see if there are any concepts which look attractive to us, and which have not be studied at other places.

The three-hour review brought out that a spinning space station (3-man, 4-month) is marginal even if two Saturn IBs are used. There is a preference for a space station transported by a Saturn V into orbit, which can house 6 people, has a growth to 12, can spin and de-spin and is supplied by a six-man Apollo spacecraft. This solution is non-marginal as far as the weight is concerned, and permits quite some leeway in solving problems by adding weight.

P&VE has done an outstanding job on this study and we are planning to give you a somewhat more polished presentation within the next 8 weeks or whenever the subject should become a hot issue.

We also discussed this subject and recoverable boosters for a full day with Dr. Yarymovych from OMSF, who is now in charge of these studies within Dr. Shea's office. We believe we have established good and effective relations with this office, which is planned to be staffed by about 25 people.

[Comment by Dr.v.B: *Keep this hot! There are growing indications that NASA hq (Top Trinity) is far more excited about a manned orbital laboratory (Space Station) as next step after Apollo, than about manned planetary expeditions. And for good reasons. They think we have to know more about long stay times in space – and support is easier to get!*]

3-2-64:   Engineering Experiments for Orbital Research Laboratory (ORL)

The Office of Advanced Studies (Orbital Systems), NASA HQ has asked us to submit a series of engineering experiments, suitable for use on an orbital research laboratory. We have agreed to help. Research Projects Laboratory will do the overall coordination. This office will be the point of contact between MSFC and Headquarters until other arrangements are deemed necessary.

This request is somewhat significant since our performance, even though on a small scale, should permit MSFC to supply major experimental suggestions necessary for us to develop technology in support of orbital launch operations, lunar base operations, and cryogenic technology in space for orbital launch vehicles. We hope excellent laboratory participation will result. We consider this a practical approach to enter the field of orbital operations as an active participant.

[Comments Dr.v.B.: *Suggest we emphasize zero-g effects, ullage control, gravity versus capillary forces, etc. This is vital for deep space propulsion development. Suggest you talk to BLUMRICH. I am also referring to a recent BOEING presentation along these lines.*]

Interplanetary Missions

10-9-61:   Contracts

A contract was signed with Lockheed on the Earth-Planetary Transportation System Study. The purpose of this study is to provide information necessary to assist NASA in determining future launch vehicle requirements. The main emphasis will be a survey of interplanetary trajectories of particular interest. The contractor orientation (meeting) is scheduled for October 18, 1961.

4-29-63:   Interplanetary Missions

One of our system studies sponsored by OART concerns the development of a mathematical model of planetary transportation systems with emphasis on evaluation of propulsion concepts for a large number of missions. We have received nine proposals. The evaluation committee headed by Mr. Gradecak (FPO) selected GD/A and Martin, Denver to make one study each for $ 75,000. NAA was a close third. Do you go along with our selection?

[Comments by Dr.v.B: *YES. During our 4-30-63 Management Council Meeting the following question came up (and was discussed extensively): "Can we honestly say that we'll never be able to launch a manned planetary expedition, unless we have a NOVA of some kind? Have we ever analyzed, in some detail, the problems involved in a multiple Saturn V earth-to-orbit logistics operation?" We should expect orbital docking to become a routine operation by 1970. Request comments and suggestions.*]

5-6-63: SATURN V Planetary Capabilities

On the margin of last week's Notes you asked the question what is being done in the area of Saturn V manned planetary capabilities. Here is a short summary:

a. Lockheed is performing a study on minimum energy profiles for Venus and Mars fly-by missions, which can be performed with a few Saturn V flights and orbital operations. Dr. Shea originally did not like this too much. I had to make an extra trip to Washington (in February) to convince him that this study is needed to fill a gap in our knowledge. P&C is now negotiating the next phase of this study contract.

b. We have a study with Chance-Vought ($350,000) entitled "Advanced Orbital Operations" in which we will identify and define individual development problems and packages, where the Saturn V refueling and docking operation will play a primary role.

c. We are presently trying to initiate a small in-house study (3 to 5 man-years) which will attempt to describe such a manned Saturn V planetary mission in enough detail that we can evaluate its problems, timing and cost. We will base this study primarily on Joe deFries' previous orbital mode study where very good material was developed. The planetary mission, however, will take place quite a few years later than the lunar mission studied by deFries.

We believe that these studies will give us a pretty good feel for the problems involved and the attractiveness of such a mission. On the other hand, we will not make an all-out effort in this area at this time. If you agree, we will increase our effort on this mission study early next fiscal year when new money becomes available.

[Comment by Dr. v.B: *These studies should include NERVA and electric propulsion for deep-space portion of trip, of course.*]

Future Projects Office Activities—Contracts and In-House

10-30-61: Study Contracts

The following contracts are presently actively pursued (with termination dates):

a. Launch Vehicle Size and Cost Analysis Study – (Nov. 30, 1961)
b. Study of Large Launch Vehicle Utilizing Solid Propellants (Dec. 29, 1961)

  c. Earth-Planetary Transportation System Study – (April 1962)

  d. Earth-Lunar Transportation System Study – (Nov. 30, 1962)

  e. Orbital Launch Operations Study – (Jan. 18, 1962)

  f. Analysis of Medium Class (Thor, Atlas) Launch Vehicle Systems – (Jan. 14, 1962)

  g. Saturn C-3 Launch Facility Study – (Nov. 30, 1961)

  h. Saturn D (Nuclear Third Stage) Design Study – (Dec. 15, 1961)

We are presently negotiating an extension of the Martin contract (item "d") for a very detailed study on the problem of storability of propellants on the lunar surface.

1-15-62: Study Contracts

Approximately 30 study contracts are to be let during the remainder of this fiscal year, based on the funds released in the 3rd Quarterly Review. Some 20 work statements are in the final draft stage. The first requests for contracts are now with Procurements and Contracts. About 40 percent of these 30 contracts will be negotiated as follow-on or sole-source contracts; the rest will be awarded on the basis of open bid competition. To what degree do you want to participate in this effort?

  a. Do you want to see and approve each work statement?

  b. Do you want to see only those work statements which I feel you might be interested in?

  c. Do you want to participate in the selection of contractors?

  d. Do you want to limit yourself to attending the final contractor presentations?

  e. Do you want to restrict yourself to special cases which I cannot solve and hear only about the highlights through these notes?

[Comment Dr. v.B.: *Suggest we discuss the whole complex thoroughly one-1 hr. Then I won't interfere in detail implementation. Let's do that soon.*]

1-6-64: 1963 Meetings

You might be interested in how much time you have spent with me (and others participating) in discussing future projects during the past year. Here are my statistics:

You have participated in a total of 24 discussions and presentations, lasting 32 hours all together. You requested 28 of these meetings, I requested four of them. In the first 6 months we had 14 meetings, during the last 6 months there were 10 meetings. I believe this was adequate to keep you properly informed and I would be happy if we could continue at the same rate this year, which is probably optimistic. [Comment by Dr. v.B.: *WHY?*]

1-27-64: FPO Statistics

It might be of interest to those who feel that there is too much effort spent in future projects to know that the time charges to our studies have dropped 50 percent during the first 6 months [Comment Dr. v.B.: *I don't have that feeling at all!*]. It was an equivalent of 170 men in July 1963 and only 110 in December. I hope we can arrest this trend now, if not, I don't think we will be able to meet our obligations for the future.

It might be also interesting to note that we have a labor overhead, in terms of man-hours, of almost 199 percent and a cost overhead of 170%. This brings our in-house engineering man-year cost to $29,650, which is about the average we have to pay for contracted studies. The output per man-year in industry appears to be larger, since we control this effort quite closely and efficiently.

7-13-64: Study Contracts

The following tables summarize the fiscal year 1964 contracts for advanced studies, conducted or sponsored by our office, most of which have just been initiated:

Total: 28 study projects resulting in 40 contracts with 16 companies or 21 corporate divisions.

Contract expenditures by study area:

Operations Analysis and Supporting Studies	1,025,000
Orbital Systems	700,000
Lunar Systems	2,445,000
Planetary Systems	860,000
Launch Vehicles	4,900,000
Total	$9,930,000

12-28-64: Statistics

This year, according to my statistics, you called or attended 22 meetings at MSFC dealing with "Future Projects," which totaled 36 hours. You initiated 16 of the meetings taking 26 hours; I was responsible for the rest. I thought early this year that you would not be able to spend that much time in this area being, through sustainable economic growth.

5

Apollo Knowledge Transfer: Preserving and Transferring the Apollo Legacy to a New Generation

Charles A. Lundquist
Research Institute, University of Alabama
Huntsville, Alabama

Dennis Ray Wingo
Skycorp
Huntsville, Alabama

Charles A. Lundquist received a BS in engineering physics in 1949 from South Dakota State University and a PhD in physics in 1954 from the University of Kansas. During the 1953–1954 academic year, he was an assistant professor of engineering research at the Pennsylvania State University. He served in the Army for the following two years, assigned to the Technical Feasibility Study Office, Redstone Arsenal, Alabama. From 1956 to 1960, he was chief, Physics and Astrophysics Branch, Research Projects Division, Army Ballistic Missile Agency, Alabama. In this position, he participated in planning, launch, and analysis of Explorer 1 and other early U.S. satellites. Following the creation of the NASA Marshall Space Flight Center in July 1960, he was chief, Physics and Astrophysics Branch, Research Projects Office of MSFC. From 1962 to 1973, he was assistant director for science at the Astrophysical Observatory of the Smithsonian Institution, Cambridge, Massachusetts and research associate, Harvard College Observatory. While at the Astrophysical Observatory, he was a member of the NASA Group for Lunar Exploration Planning (GLEP) throughout the Apollo program. He was co-editor, with G. Veis, of the 1966 *Smithsonian Standard Earth*, the first comprehensive solution for the gravity field of the earth and the corresponding global coordinate system based on satellite tracking data. Subsequently, in1973, Dr. Lundquist returned to the Marshall Center as director of the Space Sciences Laboratory. In 1981 he joined the University of Alabama in Huntsville, first as director of research and later as associate vice president for research until 1996. Concurrently from 1985, he was director, Consortium for Materials Development in Space, a NASA and industry-sponsored commercial space center. Dr. Lundquist formally retired in 1999, but continues his efforts at UAH on a part-time basis as director, Interactive Projects Office in the Research Institute, University of Alabama, Huntsville.

ABSTRACT   A perplexing issue is how to convey knowledge and experience from the Apollo Program in a way that is efficiently helpful to present-day teams planning the return missions to the Moon. Such a transfer of knowhow is hard even when there is a continuity of workforce, but the transfer is exceedingly difficult when an interval of some 40 years must be bridged. Present-day lunar team members and students who will become team members have grown up in the era of computer databases. They are skilled at accessing such information. This suggests that one obvious option is to provide Apollo knowledge and experience in a computer searchable format. The Archives and Special Collections Department in the library at the University of Alabama in Huntsville (UAH) is among many organizations that have recognized and implemented this option. At UAH, the document holdings relevant to lunar exploration include a Saturn V collection, a Lunar Roving Vehicle Collection, an Apollo Missions Collection, and documents from the Apollo Group for Lunar Exploration Planning (GLEP). An online catalog has been produced for these materials and each entry has at least an abstract and sometimes a full online text. Further, many video history interviews have been conducted with original participants in the Apollo Program. These are also accessible online. Other information includes reformatted electronic records from the Apollo era and surveys or reviews of Apollo documentation. Given the scope of the past, present and future lunar operations and recognizing the large number of organizations involved, the Apollo information preservation and transfer task is indeed challenging.

The Knowledge Transfer Problem

In response to a space policy announced by President George W. Bush, the United States is again preparing to send people to the Moon. Current plans envision extended stays at lunar bases equipped appropriately for lengthy occupation. This objective requires the design and production of a new family of large rocket vehicles comparable to the Saturn rockets of the Apollo era. It also requires design and procurement of structures, facilities and equipment for use on the lunar surface. The return to the Moon is scheduled for the second decade of the century.

A perplexing issue is how to convey knowledge and experience from the Apollo Program in a way that is efficiently helpful to present-day teams planning the return missions to the Moon. Such a transfer of know-how is hard even when there is a continuity of work-force, but the transfer is exceedingly difficult when an interval of some forty years must be bridged.

Surely historical documents from the Apollo Program exist in many places and forms that are accessible with enough effort. Realistically, the present government and contractor team members are so pressed with current issues that they may feel they can devote but little time to primary literature searches. They can be aided by the broader space community.

Online Information Access

Present-day lunar team members have grown up in the era of computer data bases. They are skilled at accessing online data. One obvious way to aid them is to provide access to Apollo knowledge and experience in computer searchable data bases. Unfortunately, when the Apollo Program ended, little of its technical documentation and human experience was preserved in a computerized format.

For the knowledge from the Apollo Program to be accessible online, old documents must be cataloged and digitized, data in obsolete computer formats must be translated into current formats and oral or video material must be presented in digital form.

Clearly, the effective transfer of Apollo knowhow will not be accomplished by any single entity. Cooperation between national agencies, private companies, universities, libraries and other entities will be required.

The UAH Role

An acknowledged role of any university is to provide a repository of knowledge and to convey that knowledge. Recognizing its long-standing relationship with the NASA Marshall Space Flight Center and the Huntsville space contractor community [1], the University of Alabama in Huntsville (UAH) has embraced a role as a repository for space information deserving preservation in a publicly accessible archive. In the UAH Salmon Library, the Archives and Special Collections Department has this repository role [2].

The space collections in the Archives got a significant start in 1969 when the book collection of Willy Ley was placed in the UAH Library. Willy Ley had been a space enthusiast and an associate of Wernher von Braun in Germany before World War II. However, Willy immigrated to the United States before the war. When he died, his large library was sold by his widow.

Donations from Huntsville residents provided the money to buy this collection for UAH.

As the Saturn V Program neared completion, UAH accepted a contract from NASA to assemble program documents as a reference source for a history of the rockets used by Apollo. The history was written by R.E. Bilstein and published by NASA [3]. Subsequently, this very substantial reference collection was deposited in the UAH Archives and cataloged. This was the second large space collection in the Archives.

A more recent function of the Archives has been acceptance of personal materials donated by retirees from the Marshall Center and other individuals. This function is in part a result of an understanding between UAH and the Marshall Retirees Association. Many of the retirees who have donated their holdings are individuals who initiated the U.S. space program and executed the Apollo Program. The materials so donated have been very substantial and inclusive. Large collections have been organized and abstracts of each item prepared for online finding guides. Some of the most important documents have been scanned and made available in full.

The initial motivation for the UAH Archives was to provide a resource for historians and other research scholars. This objective remains intact. However, UAH personnel now recognize that the collections can serve further as a reference to specific space knowledge and experience that can be useful in the new space programs.

The document collections that are most pertinent as sources of information for the return to the Moon are listed in Table 5.1.

As already mentioned, the Saturn V Collection was assembled at the end of the Saturn Program. The collection was cataloged then and a finding guide prepared. Recently, many of the significant Saturn V documents have been scanned and their full text is available online. An Apollo Mission Collection was prepared in 2007 from materials donated by NASA retirees. Currently, this initial collection is being augmented by further donations. The Lunar Roving Vehicle Collection was completed in 2006. It contains extensive documentation preserved by participants in the vehicle design, fabrication and utilization. During the Apollo Program, NASA established the Group for Lunar Exploration Planning (GLEP) to provide analysis and advice concerning exploration and scientific operations planned for the successive missions to the Moon [4]. This collection contains many documents and working papers generated and used by the GLEP.

TABLE 5.1

UAH Archive Collections Most Pertinent for the Return to the Moon

Saturn V Collection
Apollo Missions Collection
Lunar Roving Vehicle Collection
Group for Lunar Exploration Planning Collection

Interviews and Talks

In the 1980s, Professor Donald Tartar of UAH recorded video interviews with a number of early members of the rocket team directed by Wernher von Braun [5]. These interviews included some ten members of the original von Braun group who came to the United States to work for the Army immediately after World War II. In conducting these interviews, Dr. Tartar was joined by Conrad Dannenberg, himself a leading member of the original German team. Recently, when the UAH Archives began to expand its online collections, these early interviews were re-transcribed in a digital format appropriate for access by personal computers.

Beginning in 2006, the UAH staff undertook to augment these interviews with comparable online interviews with other early members of the rocketry activities in Huntsville. By the end of 2007, some 80 people had been interviewed. Many of those interviewed were scientists and engineers born in the United States who were recruited, starting in about 1950, to join the von Braun team. These individuals had significant roles in the Apollo Program. To guide current efforts to revisit the Moon, it is instructive to extract from these interviews what factors were cited as contributing to the outstanding success of the Apollo Program. A list of the factors identified is shown in Table 5.2.

In addition to video interviews, the UAH Archives had tapes of vocal interviews with many individuals. One large collection of these was taped as resource information for the Bilstein book *Stages to Saturn*. This tape collection was a counterpart to the Saturn V document collection. Again, the format of the original tapes was not suitable for online computer access, and the tapes had to be digitized and reformatted. Also some old interviews were recorded on 33 rpm plastic discs. They of course had to be reformatted.

Still further, the Archives had access to video recordings of various talks, space symposia, forums and memorial events. These were translated from their original forms into a digital form for online access. A sampling of the many recorded talks by Wernher von Braun was put online by UAH. As a guide to

TABLE 5.2

Factors for Apollo Success Distilled from Oral Histories

Unequivocal political and popular support
Adequate budget
Recognized skillful leadership
Experienced team
Feasible plan
Schedule discipline
Commitment to testing
Open personal communications

finding remarks by particular individuals, an alphabetical list of speakers was compiled with information citing in which events each speaker participated.

Electronic Records

There are very few surviving electronic records from the Apollo era, although many magnetic tapes were produced. Those that survive are in obsolete digital or analog formats. Also the machines that originally read them do not exist or have not been functional for decades. To recover these data sets, dedicated efforts must be undertaken by skilled engineers and by retirees originally involved in the pertinent operations.

The Lunar Orbiter analog data tapes are an informative example. The highest resolution Lunar Orbiter images of the lunar surface were transmitted to Earth and recorded on 2 inch analog tapes. Initially, only a small, selected fraction of these tapes were processed to obtain the maximum resolution of about one meter for sites near the equator. Computer capabilities at the time made this process time consuming and expensive. The lunar surface areas processed for maximum resolution were primarily the potential and actual Apollo landing sites.

Fortunately, the full collection of Lunar Orbiter magnetic tapes was preserved by NASA under controlled environmental conditions. However, the tape drives to generate maximum resolution are 40 years old and have not operated in over 20 years. A set of surplused drives has been located and is available for refurbishment. An effort to accomplish the refurbishment and to demonstrate tape processing for highest resolution is being pursued.

The data potentially available from reprocessing the Lunar Orbiter tapes is a further informative example of how Apollo era information can support the renewed lunar program. The Lunar Reconnaissance Orbiter planned for an early mission to the Moon will have image resolution comparable to the best Lunar Orbiter image resolution. There is great value in having comparable images of the lunar surface separated by some forty years. A comparison will allow identification of craters with a diameter greater than a few meters that have been caused by meteoroid impacts during the years between Lunar Orbiter and Lunar Reconnaissance Orbiter.

The Apollo experience found that the nature of the lunar regolith is dominated by the continuing cultivation and mixing caused by meteoroid impacts. Hence the phenomena associated with meteoroid impacts are keys to understanding the regolith. Craters and ejecta from cratering are easily observable lunar surface features. Recent craters are prime sites for exploration missions because they expose material from beneath the lunar surface.

Reviews

Besides online access to original literature, another aid to present-day lunar team members is recourse to comprehensive surveys and reviews of earlier experience and understanding. A number of useful reviews have been prepared and can be accessible online.

A typical example of a recent review is "Lunar Outpost Development and the Role of Mechanical Systems for Payload Handling [6]." This document consolidates the findings of a number of previous studies. It is accessible online.

Conclusions

The Apollo era data sets, documents, and experience can provide the USA with valuable insights as well as guidance on how to maximize results from the return to the Moon. Providing this information online via professional archival methods brings an orderly and timely resource to the nation today and for future generations.

The authors fully recognize that providing online and in-library access to information is only one option in preserving and conveying the Apollo experience. Given the scope of the past, present and future lunar exploration programs, and recognizing the large number of organizations involved, the information preservation and transfer task is indeed a challenging problem.

References

[1] Charles A. Lundquist, "The Wernher von Braun Research Hall at the University of Alabama in Huntsville," pp. 114–115, 50 Years of Rockets and Spacecraft in the Rocket City, Huntsville, Alabama, Turner Publishing Company, Paducah KY, 2002.

[2] Anne M. Coleman, Charles A. Lundquist and David L. Christensen, "Organizational History of the Space Collections at the University of Alabama in Huntsville," IAC-04-IAA.6.15.2.07, International Astronautical Congress, 2004.

[3] R. E. Bilstein, Stages to Saturn: A Technological History of the Apollo/Saturn Launch Vehicles, National Aeronautics and Space Administration, Washington, D. C., 1980.

[4] The initial members of the Group for Lunar Exploration Planning, 1967, were W.N. Hess, chair, E. King, P. Gast, J. Arnold, E. Shoemaker, R. Jahns, F. Press,

C. Lundquist, M. Calvin, F. Johnson, D. Williams, N. Roman, P. Culbertson, R. Allenby, M. Faget, W. Stoney, H. Gartrell, H. Schmitt.

[5] Anne Coleman, Robert L. Middleton, Charles A. Lundquist and David L. Christensen, "The Oral History Tradition at the University of Alabama in Huntsville," International Astronautical Congress, 2006.

[6] Dennis Ray Wingo, Gordon Woodcock and Mark Maxwell, "Lunar Outpost Development and the Role of Mechanical Systems for Payload Handling," 112 pages, report by Skycorp Inc., 10 Feb. 2007.

6

Working in Space

Terry Hart

Professor of Practice, Lehigh University and Former NASA Astronaut

Terry Hart graduated in 1964 from Mt. Lebanon High School in Pittsburgh, Pennsylvania. He received a Bachelor of Science degree in mechanical engineering from Lehigh University in 1968, a Master of Science degree in mechanical engineering from the Massachusetts Institute of Technology in 1969, a Master of Science degree in electrical engineering from Rutgers University in 1978, and an honorary Doctorate of Engineering from Lehigh University in 1988. Dr. Hart began his career with Bell Telephone Laboratories in 1968 and was initially responsible for the mechanical design of magnetic tape transport systems. In 1969 he began a 4-year leave of absence with the Air Force, where he flew over 3200 hours in high-performance fighters before retiring from the Air National Guard as a Lieutenant Colonel in 1990. After returning from active duty in the Air Force in 1973, he worked in the Electronic Power Systems Laboratory where he was responsible for the mechanical and electrical design of power converters. He received two patents, one for a mechanical safety device and another for a noise suppression circuit. In January 1978, Hart was selected by NASA as one of 35 astronaut candidates, and after one year of training and evaluation, he was assigned to the support crew of the first three Space Shuttle missions. On April 6, 1984, he lifted off from the Kennedy Space Center in the *Challenger* on the eleventh flight of the Space Shuttle program, during which he was responsible for the rendezvous navigation and targeting, the remote manipulator system operation, and the IMAX camera operation. The mission objectives of deploying the long duration exposure facility and repairing the solar maximum mission satellite were successfully accomplished and demonstrated new capabilities in manned space flight. Upon his return from NASA in 1984, Hart held a variety of management positions in Bell Labs and led several projects in systems and software engineering, including an assignment as director of the Government Data Systems Division of AT&T in Stockholm, Sweden. He returned from Europe in 1991 as director of Satellite Engineering and Operations for AT&T and in 1997 became president of Loral Skynet when AT&T sold its satellite division to Loral Space and Communications. Hart retired from Skynet in 2004 and is currently teaching aerospace engineering at his alma mater, Lehigh University. Dr. Hart is a member of the American Institute of Aeronautics and Astronautics, the Institute of Electrical and Electronics Engineers, the Tau Beta Pi engineering society, the Sigma Xi scientific society, and the Aviation Hall of Fame in New Jersey. Hart has been awarded the New Jersey Distinguished Service Medal, the Air Force Commendation Medal, the NASA Space Flight Medal, the Delta Upsilon Medallion, the Rutgers University Medal, the Pride of Pennsylvania Medal, and was inducted into the Rutgers Hall of Distinguished Alumni. Lehigh University and Former NASA Astronaut.

FIGURE 6.1
Ed White proved people can work in space.

Since the first cosmonauts and astronauts walked in space, we have been climbing a continuous learning curve of how people can work productively in space. The effects of weightlessness and the physical limitations of pressure suits and spacecraft designs continue to challenge crews as ever-more sophisticated tasks are being accomplished.

And while we have come a long way in our ability to work in space, much needs to be done if we are to return to the moon with a permanent presence and venture on to Mars. Such long-duration missions will put new challenges on engineers and crews to adjust to the physical and psychological demands of these missions. With international cooperation, these challenges will be met and crews will learn to work effectively as we establish a permanent presence in space.

In 1960 at a medical conference on the anticipated effects of weightlessness on astronauts, several papers were presented by researchers who were concerned about the ability of humans to survive in space. One paper expressed doubt that the first astronauts could breathe effectively; another stated that they certainly could not eat solid food and may not be able to swallow water. A third dismally projected that sleep would be difficult, if not impossible, in weightlessness, and the lack of it would possibly lead to delusionary behavior and even psychosis.

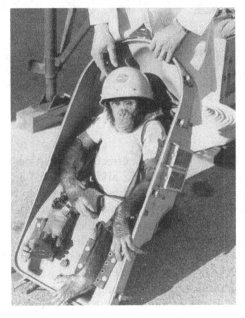

FIGURE 6.2
Ham, the first American in space.

As a result of this overwhelming pessimism, the first American in space was … Ham, the chimpanzee. Fortunately, Ham did just fine and not only survived, but was able to perform all of the tasks he was trained to do with no obvious ill-effects from the space environment. You would think that the medical community would have backed off after this good news, but their overly conservative attitudes about the ability of people to work effectively in space continued until the astronauts conclusively demonstrated that the human body can adapt and even prosper in this new environment.

The "Original 7" astronauts (in front of their F-106 training aircraft) flew six missions in the Mercury capsules and showed beyond a doubt that

FIGURE 6.3
The original seven.

people could survive and even work effectively in space. But much was yet to be learned about working in space, and some hard lessons lay ahead. It was during the two-man Gemini flights that we began to understand that weightlessness would pose a host of new problems. Ed White became the first American to "walk" in space, and while this first attempt was a success, a hint of future problems came as we learned just how easily cables and tethers become tangled. Early thoughts of using hand-held gas guns to easily propel ourselves through space (Figure 6.5) were quickly dispelled as we learned the difficulties of maneuvering in weightlessness. Spacewalkers became exhausted and visors fogged up as they struggled to overcome the awkwardness of working without the help of gravity.

But we learned quickly how to work effectively in space with help of the design of a variety of tools and support fixtures to assist the astronauts in each task they encountered. By the time we began the Apollo program, we had honed these skills and were able to successfully operate between the Earth and the Moon. And with the lunar landings came a new environment: one-sixth gravity on the Moon. It was difficult to simulate this environment on Earth, but the use of a variety of cables and pulleys gave the astronauts some idea of what they would need to do to balance themselves and move about the surface of the Moon. When we began to operate on the Moon, the astronauts learned just how quickly we could adapt to this environment. The famous photo of John Young saluting the flag as he jumps three feet off the surface of the Moon with several hundred pounds of equipment demonstrated our ability to adjust to new environments once again.

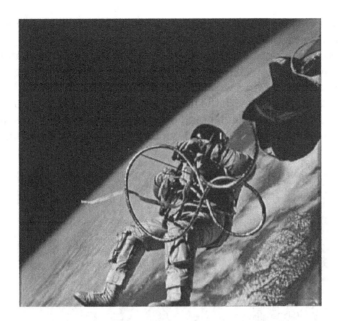

FIGURE 6.4
Tethers provide safety, but become a problem.

FIGURE 6.5
Self propelling devices proved unuseable.

And while we undoubtedly underestimated man's ability to adjust to some aspects of spaceflight, we did not anticipate a host of other problems that had to be solved in real time. The ingenious design of the lunar rover allowed the astronauts to explore much more of the Moon's surface, but no one had anticipated how the rover's tires would fling the lunar soil into the "air" and cover the crew and the vehicle with the highly abrasive dust. The crew improvised by using a checklist as a fender, and once again man's ability to solve problems in real-time was demonstrated with the help of a little duct tape.

FIGURE 6.6
John Young jumps three feet into the "air."

Apollo was followed with three long-term missions to Skylab, but before these missions could be flown, Skylab needed some repair. A solar panel had torn off during launch leaving the laboratory overheated and without sufficient power. The first flight became a rescue mission and the crew successfully repaired the damage and created a sunshield that stabilized the temperatures inside the laboratory. The role of astronauts in overcoming unanticipated problems in space with improvised solutions was now well-accepted, quite a change from the early concerns about man's ability to even survive in space.

As we moved into the Space Shuttle program, ever more sophisticated tools and training techniques were developed to support the astronauts as

FIGURE 6.7
Lunar dust kicks up from the rover's tires.

FIGURE 6.8
An improvised sunscreen was installed on Skylab to lower its interior temperature and save the mission.

they undertook new tasks in space. The most effective training environment on Earth has been the Weightless Environment Training Facility (WETF), a full underwater mockup of the Space Shuttle's payload bay. This facility has enabled engineers and crews to design the handholds, footholds, and tools they have needed to accomplish evermore complex tasks in space. Long duration missions in the Russian Mir space station similarly have greatly added to our knowledge and our ability to work in space.

FIGURE 6.9
The WETF provides the best weightless training on Earth.

FIGURE 6.10
The International Space Station.

Now the exploration of space has become a truly international endeavor with astronauts and cosmonauts from many nations routinely living and working on the International Space Station. We will return to the Moon someday soon, and go on to Mars as well. The long-term effects of weightlessness and the radiation of space will pose new problems that will need creative solutions, and there will be many new lessons to learn before we can maintain a permanent presence beyond the Earth orbit – but there is no doubt now that people can adjust and even prosper in space. And when historians look back on our times, they may well conclude that the most significant benefit of the space program has been the unifying effort of the nations of the world to live and work in space for the benefit of all of us on the good Earth.

FIGURE 6.11
Crews from many nations live and work effectively together in space.

Section II

Lunar Development

7

Attracting Private Investment for Lunar Commerce: Toward Economically Sustainable Development

Paul Eckert

International and Commercial Strategist
The Boeing Company "IDS" Space Exploration

Paul Eckert, PhD holds the position of international and commercial strategist within the Space Exploration division of The Boeing Company. In this role, Dr. Eckert develops strategies to strengthen global business relationships and explore new commercial markets. He serves as coordinator of the international Space Investment Summit Coalition, which presents events linking investors and entrepreneurs in order to encourage investment in entrepreneurial innovation. Within the U.S. Chamber of Commerce, Dr. Eckert chairs the Emerging Markets Working Group of the Space Enterprise Council. He also chairs the Entrepreneurship and Investment Technical Committee of the International Astronautical Federation and acts as commercial coordinator for the Lunar Exploration Analysis Group, which is chartered by the NASA Advisory Council. Having joined The Boeing Company in 2003, Dr. Eckert's prior roles have involved space exploration planning, infrastructure design, Earth observation, space science, government relations, and communications. Previously, within the U.S. Department of Commerce, he helped promote the growth of the commercial space industry, as part of the Office of Space Commercialization. Earlier, in the NASA Office of Legislative Affairs, Dr. Eckert coordinated liaison with the U.S. Congress involving space and aeronautics research, information technology, systems engineering, and technology transfer to industry. Prior to this, he served as science and technology advisor to U.S. Senator John Breaux, a key member of the Senate Commerce Committee, with jurisdiction over NASA. Dr. Eckert holds a bachelor's degree with high honors from Harvard University and a doctoral degree from Michigan State University.

Introduction: From Lunar Commerce to Space Commerce

A series of roundtable discussions involving industry, government, and academia has helped shed light on issues relevant to lunar commerce.[1,2,3] The first Lunar Commerce Roundtable, organized by a mixture of established and startup companies, from within as well as outside the space sector, took place in Dallas, Texas in June of 2005. Like the roundtables to follow, it involved between 75 and 100 participants, emphasizing quality of dialogue and with only limited availability for press coverage. The June 2005 roundtable surveyed a broad range of lunar-related business opportunities. There followed in October of the same year a second roundtable, this time in Houston, focusing specifically on lunar-related opportunities in four areas: solar power, propellant production, media, and robotics as an enabling technology.

Las Vegas welcomed the third roundtable, in July of 2006, this time with a new name "space commerce roundtable" instead of the former lunar focus. The dialogue shifted to an entire Earth-Moon economic system concept, recognizing that lunar commerce could not develop except as an extension of commerce closer to home. There was considerable emphasis on interconnections and synergies among multiple market areas, involving the Moon or other locales. Also of interest were strategically targeted public and private sector options for making good use of identified synergies, by using a single initiative to impact multiple sectors simultaneously. Five areas were addressed, including: space facilities, surface facilities, services (e.g., transportation and fuels, communication/navigation), space solar power, and experience-oriented activities (e.g., advertising, entertainment, education, tourism).

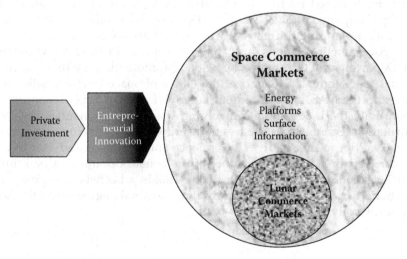

FIGURE 7.1
Private investment funding entrepreneurial innovation to fuel market growth.

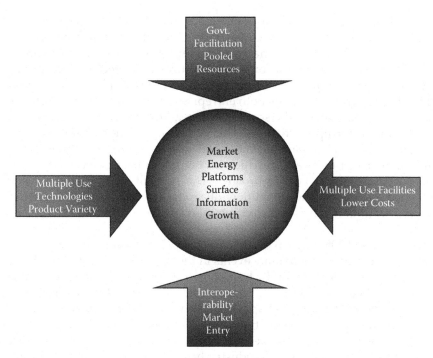

FIGURE 7.2
Examples of factors facilitating space-related market growth.

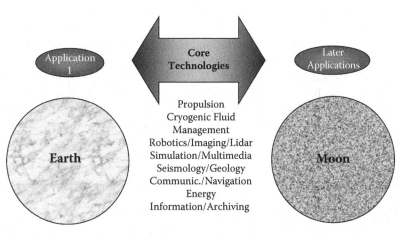

FIGURE 7.3
Multiple-use core technologies, with terrestrial followed by lunar use.

In early 2007, yet another name change occurred, with "space commerce roundtables" becoming "space investment summits." The relatively small-scale, dialogue oriented format was preserved but with a shift toward focusing on entrepreneurial business ventures and seed and early stage investors. Each summit was envisaged by an industry planning group as involving an effort to educate both entrepreneurs and investors concerning experience from outside the space sector, regarding successful investment and business startups. Although initially the goal of the summits was primarily educational, regarding increased understanding by space-related entrepreneurs of the ways in which investors evaluate potential opportunities, the ultimate goal was broader. The summit series was ultimately intended to increase the actual flow of capital into space-related startups, thereby stimulating space-related commercial innovation and fueling new market growth.

Since June of 2005, a global group of sponsors has chosen to provide financial sponsorship for one or more events in the evolving series just described, including some of the world's best-known aerospace companies. Among these have been notable firms from outside the United States, including EADS Astrium, Mitsubishi Corporation, and Thales Alenia Space (formerly Alcatel Alenia Space, in conjunction with Alenia Spazio North America). Also involved have been smaller non-U.S. firms, such as MDA, Mansat/SpaceIsle. com, and Odyssey Moon. U.S. industry has been well represented, including The Boeing Company, Lockheed Martin Corp., Northrop Grumman Corp., Honeywell International, United Space Alliance, Space Systems Loral, Wyle, and AGI, as well as entrepreneurial firms such as Ecliptic Enterprises, Lunar Transportation Systems, and Transformational Space Corporation. Non-profit groups and organizations contributing in-kind support initially included the Space Commerce Roundtable, the Space Frontier Foundation, and the National Space Society (in conjunction with leadership provided by Innovarium Ventures, under the direction of Dr. Burton Lee). Later, additional organizations became involved, including Space Foundation, California Space Authority, X Prize Foundation, Eisenhower Center for Space and Defense Studies, VC Private Equity Roundtable, Alliance for Commercial Enterprises and Education in Space (ACES), Eighth Continent, and the Space Tourism Society. By mid 2007, what had begun as a small roundtable effort had grown into a much larger group of company sponsors and supporting non-profit groups, known collectively as the Space Investment Summit Coalition.

Insights from Roundtable and Summit Dialogue

The series of roundtable and summit events has yielded insights regarding how to encourage space commerce and thereby help fuel economic growth in general, with all the benefits to society that business expansion and job growth can bring. First, it is important to involve in dialogue all key stakeholders who stand to contribute to, as well as benefit from, economic growth. This international stakeholder community includes established and startup space companies, non-space companies, investors, and additional players such as government, academia, and professional/trade associations. Markets meriting their attention include energy, platforms, surface activities, services (e.g., transportation, communications, navigation) and information/experience (e.g., advertising, branding, sponsorship). As noted earlier, it was the July 2006 roundtable that most effectively drove home the significance of this interconnected mixture of potential commercial activities. In such a context, a key principle that became increasingly evident was the importance—for investors and innovators alike—of an *incremental* approach to reducing risk, so as to justify gradually increasing levels of investment.[4]

Beyond dialogue, additional steps in an incremental process of entrepreneurial initiative include research, to subject general ideas generated by dialogue to empirical validation, demonstration projects, to test out alternatives before selecting the best technical approach, and—only after all these steps have been applied effectively—commitment of major investment to full-scale business projects. This is not to assert that only by following such steps can business success be achieved. The key assertion is that innovative initiatives appear to have a better chance of technical and financial success if such an incremental approach is followed. Beyond this, multiple-use technologies—which can be commercialized in a variety of products and services—may reduce risk more than single "all-or-nothing" commercial applications of a particular technical capability. Throughout the foregoing discussion, the concept of entrepreneurial innovation is paramount—not developing new technology but rather applying existing technology to meet customer needs profitably in the marketplace.

The Need for Increased Private Investment

A clear conclusion from the roundtables in late 2006 was, as noted previously, that private investment must be significantly increased in order to fuel entrepreneurial innovation with sufficient capital to help create and serve new markets. In a word, problems facing space commerce efforts are as much, if not more, financial than technical. The benefits of commerce—such as lower cost and increased choice of products and services—depend on attracting investment from investors willing to consider early-stage, high-risk, but also high-opportunity initiatives. "Angel" investors are often the most likely to take an interest in such opportunities. However, such independent, wealthy individuals are quite difficult to attract, because they listen more to other investors, whom they already know, than to enthusiastic strangers from outside the investment community. Add to this that angels often develop highly specialized interests and have many opportunities from which to choose, in addition to whatever space-related ventures may cross their paths.

In any case, contacts with these and other types of investors have begun to clarify some of their most important considerations in making decisions about where to direct the capital they control.

Key risks they would like to limit include: insufficient market size, accessibility and flexibility of customer demand, insufficient salvageable value of assets if a project fails, dominance of a market by competitors, unavailability of sufficient financial investment, inadequate management skill, ineffective technical approaches, and constraining legal and regulatory factors.

The Search for Entrepreneurial Innovation

In an effort to minimize such risks while maximizing size and proximity of returns, investors want evidence that entrepreneurial innovation is actually occurring, in development of products and services that effectively meet customer demand. In this quest for profitable innovation, they recognize that entrepreneurial innovation and value creation associated with it involve the "what" of innovation (i.e., product or service) and the "how" of innovation (i.e., market assessment and creation of an industrial entity to serve the market), as well as the "who and why" of innovation (i.e., the personalities that enable creative individuals to pursue entrepreneurship constructively).[5,6] The "what" can involve factors such as cost, performance, safety and reliability, with an improvement in an existing product or service potentially having as much market impact as the entry of a new product or service offering. The "how" is a multidimensional factor, including such factors as: identifying an opportunity, accumulating the necessary resources to address it, producing attractive products and service, marketing these to customers, building an organization, and responding to the requirements and reactions of government and society. The "who and why" factor is exceptionally rich in content, and in-depth studies have addressed complex psychological phenomena such as intrinsic vs. extrinsic motivation, as well as the fragile equilibrium balancing risk-taking with a pragmatic sense of business reality.

The Central Role of Market Demand

Above all, investors look for actual or potential market demand, without which even the best concepts will never be profitable. Space certainly offers a plethora of potential markets, although many exhibit major challenges when subjected to rigorous evaluation on investor criteria. The market categories already noted are only suggestive of the full breadth of potential commercial applications that might eventually attract investor interest.

Some specific examples may prove helpful in illustrating the breadth and scope of potential opportunities. Solar power might involve launch from Earth and assembly in space of orbiting satellites to beam power to the surface. Also envisaged is the use of lunar surface resources to produce solar power generation facilities to supply power for lunar surface use, beam power back to Earth, or contribute components that might lower the cost of constructing Earth-orbiting satellites. Supplying propellant for rocket-based chemical propulsion systems might involve sending fuel from Earth to orbiting propellant depots. An additional possibility is the production of propellant from lunar surface materials, for use not only to fuel vehicles operating in the vicinity of the Moon but also as a source of supply for the Earth-orbiting depots—cheaper than transporting fuel from Earth because of the Moon's weaker gravity. Communications and navigation applications might involve not only increasingly capable Earth-orbiting satellites but also a cislunar network of satellites, extending basic services deeper into space.

The domain of observation, similar to communication/navigation, offers opportunities not only for imaging Earth, from orbiting satellites, but also observing the planet from lunar and other space-based locations. Beyond this, commercial imaging and analysis services might also contribute to assessment of availability of valuable chemical and metallic resources on lunar, planetary, and asteroid surfaces.

Facilities might be operated in space or on the lunar surface, offering human habitats as well as bases for a variety of space activities. In addition, robotics, as exemplified by the Orbital Express satellite servicing demonstration project and the Mars rovers, might offer a number of applications potentially amenable to commercial exploitation. And media-related applications, involving robotic cameras and broadband signal transmission to produce content for advertising, entertainment, and the like, could be among the most attractive initial commercial activities. There is little or no need in such activities for on-site human presence or large-scale infrastructure—provided that at least minimal sensor/camera and signal relay resources can be put into place.

How far we are today from the "tipping point" of various markets—the point at which various preliminary economic steps finally culminate in an upward spike customer demand and investor interest—is impossible to assess.[7] But in any case, the central importance of encouraging, and responding to, market demand is one of the foundational principles of promoting lunar commercial development.

Some Key Market Enablers

Transportation

For virtually all existing or potential space commerce markets, some form of space transportation is an essential enabler. In the next decade, commercial activities on board the International Space Station (ISS) utilization could stimulate demand for transportation services, just as improved availability of transportation could promote ISS utilization.[8] The same might be true of other space platforms in development, which could benefit from ISS as an initial pathfinder for technical and economic approaches. The NASA Commercial Orbital Transportation Services (COTS) program may hold promise in providing lower cost and higher reliability transportation services within the next five years.[9] In fact, some entrepreneurs view Earth-to-orbit transportation as a stepping stone to commercial efforts involving the Moon and even beyond. The development of transportation as a space-related business enabler also has international implications, since robust utilization of ISS and other space platforms could create significant demand for non-U.S. as well as U.S. transportation services, thereby expanding market size and capitalization. To the extent that transportation involves a genuinely "open architecture," with modular and interchangeable components and subsystems, it will be more possible for a variety of international and commercial participants to enter the emerging marketplace.[10]

Interoperability

Evident in the domain of transportation but also relevant in many other areas, a market growth enabler of particular importance is "interoperability" of components, subsystems, and systems, so that more individuals and companies can participate in new markets as they grow.[11] Interoperability in this context involves standardized interfaces as well as interchangeable items. Standardized interfaces might involve a wide range of commonly-used items, such as:

- Pressure vessels (any functional volume pressurized for human habitation: landers, habitats, rovers, logistics modules, etc.)
- Electrical outlets and plugs

- Liquid connectors for each type (e.g., water)
- Gas connectors for each type (e.g., oxygen)
- Data (e.g., Ethernet cables, RF/WiFi)
- Airlock control panels
- Spacesuit connections with habitat life support and environment control systems (servicing, recharging, umbilical operations)
- Robotic manipulating mechanisms (robot to system interfaces)

Interchangeable items might involve both assemblies and components. Assembly examples include: life support, power (management, generation, & storage), waste management, thermal management, crew interface panel (e.g., command & control station), airlocks, and command and data handling systems. Interchangeable components could be items such as microprocessors, valves, tanks, filters, keyboards, software, fans, computers, pumps, batteries, solar arrays, circuit boards, and antennas.

By enabling development of a larger supplier and customer base due to the opportunity of multiple entities to use common standards to enable buying and selling, interoperability can lead to both increased supply and increased demand for space-related products and services. Several factors can help free up resources for new uses by reducing the cost of basic items. Such factors include: shared infrastructure and consumables, increased competition among providers, decreased logistics overhead (i.e., fewer unique parts), and enhanced reuse and reconfigurability (e.g., lunar lander converted to surface-based propellant depot). In addition, by using common standards for life support and other vital services, interoperability can enhance safety by facilitating rescue and repair, once again reducing risks that might repel investors.

Government Facilitation

Government—at the national, regional, and in some cases even the local level—can encourage early market growth, within limits.[12] Discussion above of private investment's importance should not obscure the significant role to be played by government investment. In fact, because of the technical and financial hurdles that must be overcome for many space-related entrepreneurial ventures to succeed, a combination of public and private investment may often be necessary. It may be that a transitional model of public private partnerships is applicable. Here government investment dominates in the earlier and higher-risk stages of a commercial activity and then private investment takes on an increasingly important role as risk is incrementally retired.

Public sector promotion of space-related commerce can take a variety of forms. For example, government purchase of commercial products and services, especially when done in advance to secure supply for an extended period, can serve as a potent stimulus to encourage private investment. The government commitment in effect assures private investors that at least an initial market will in fact exist. Of course, government has also traditionally been recognized as funding research and development activities, on which industry can draw in developing commercial applications. As noted earlier with regard to transportation, NASA's COTS program exemplifies government assistance with development and demonstration of initial commercially-relevant capabilities, with the intent of helping create what will eventually be self-sustaining commercial transportation services, with multiple customers in addition to government.

Beyond purchase and R&D, government can use financial incentives, such as investment tax credits, which can encourage private investors because, unlike the case of an income tax credit, the investment credit has economic value even for startup ventures that are not yet profitable. Loan guarantees, subsidized consulting services, and "business incubator" venues represent additional tools that government may choose to employ. Finally, a uniquely governmental function with significant impact on entrepreneurial innovation involves legal/regulatory framework development. Key factors facilitating or constraining business innovation involve such matters as limitation of liability and protection of property rights.

Multiple-Use Technologies

Multiple-use technologies can bring market profitability closer in time.[13] The concept of a multiple-use technology is straightforward, involving using the same core technical capability to produce products and services that can yield near-term profits in already established Earth markets, while at the same time gradually moving toward applications in more speculative space-related markets. For example, robotics applications already profitably in use to help automate Earth-based mining and undersea operations could later be applied to commercial lunar surface robotic missions. Companies that might find it untenable to invest corporate resources directly in somewhat speculative space initiatives at the outset might better tolerate initial exposure in well-known terrestrial markets. Profits gained through technology applications in established markets could then serve as a financial base from which to undertake ventures involving space-related activity.

Multiple-Use Facilities

Multiple-use facilities, whether in space or on the surface of the Moon and elsewhere, could facilitate commercial success by attracting more customers into markets.[14] As in shopping centers and office complexes on Earth, with a few large tenants initially moving in to provide a financial base, there may follow the entry of additional tenants that would have been too financially weak to "anchor" a facility alone. By having multiple users of a facility shoulder together the expense of shared infrastructure (e.g., transportation access, communications and navigation connectivity, power generation, life support functions), significant cost reduction could result, stretching the impact of limited investment dollars.

Conclusion

Over the last three years, industry dialogue about promotion of space-related economic growth has suggested a number of significant insights. First, lunar commerce cannot be addressed in isolation but only as an integrated part of a larger space commerce system, involving activities on Earth, in Earth orbit, in cislunar space, and eventually extending outward into the Solar System. An incremental process of attracting increased private investment, coupled with a milestone-based approach to business plans for entrepreneurial ventures, could facilitate development of lunar commerce, as well as other kinds of space-related business activity. It is essential to attract private investment into commercial space efforts, because government funding alone will never be adequate to stimulate development of an economically self-sustaining marketplace.

To gain the interest, and ultimately the capital, of seed and early-stage investors, leaders of entrepreneurial ventures must recognize that the effective management of risk is perhaps the single most important consideration of those with substantial money to spend. Clearly, the ultimate goal of most investors in managing risk is to succeed in gaining an acceptable financial return by profitably serving customers in a marketplace. Investors' focus on markets should be a strong signal to entrepreneurs to remain similarly focused. In fact, the very definition of entrepreneurial innovation involves effective application of technology to serve customers at a profit.

Key enablers for effectively serving customers, and hence for promoting market growth, constitute a multidimensional portfolio of public and private sector activity. For example, government could facilitate expansion of commercial activity through use of a policy toolbox including legal/regulatory clarification, financial incentives, targeted research and development, public private partnerships, and new business incubation services. Beyond this, an open-architecture approach to transportation services, made possible by industry and public sector collaboration, could encourage market entry by multiple providers. Interoperability in general, based on standardized interfaces and interchangeable components and subsystems, could enable a variety of companies from across the globe to participate in an international interplay of supply and demand. Such factors, approached in an integrated manner, could help hasten the day when a thriving Earth-Moon economic sphere increasingly contributes to societal well-being, through sustainable economic growth.

References

1. Space Investment Summit website, www.spaceinvestmentsummit.com [cited 13 April 2008].
2. Eckert, P. "Financing Entrepreneurship: Outreach to Non-Space Investors." *AIAA Space 2007 Conference*, 18–20 September 2007. URL: http://www.aiaa.org [cited 13 April 2008].
3. Eckert, P. "Innovation, Entrepreneurship, and Investment: Funding the Future." *International Astronautical Congress*, Hyderabad, India, 24–28 September 2007. Paper archive URL: http://www.aiaa.org/iacpapers/ [cited 13 April 2008].
4. Eckert, P. and Mankins, J. (eds.), *Bridging the Gap: From Earth Markets to New Space Markets*, Report on the Third Lunar Commerce Executive Roundtable. URL: http://www.lunarcommerceroundtable.com/lcr3_report.html [cited 20 August 2006].
5. Eckert, P., "Linking Entrepreneurial Innovators through Dialogue," 25th International Space Development Conference, National Space Society, Los Angeles, CA, 4–7 May 2006. URL: http://isdc.xisp.net/~kmiller/isdc_archive/isdc.php?link=submissionSelectArchive&sort= [cited 20 August 2006].
6. Fayolle, A., *Introduction à l'Entrepreneuriat*, Dunod, Paris, 2005, Chap. 1.
7. Gladwell, M., *The Tipping Point: How Little Things Can Make A Big Difference*. Little, Brown, New York, 2000.
8. Eckert, P. "Expanded ISS Utilization: Catalyst for Commercial Cislunar Transportation." DGLR International Symposium, *To Moon and Beyond*, Bremen, Germany, 14–16 March 2007.
9. NASA Commercial Orbital Transportation Services (COTS) acquisition information website. URL: http://procurement.jsc.nasa.gov/cots/ [cited 5 March 2006].
10. Eckert, P. "Open Architecture for Sustainable Exploration: Infrastructure Commonality and Flexibility." *Moonbase: Challenge for Humanity*, Moscow, Russia, 16 November 2006. URL: http://www.moonbase-russia.org/ [cited 10 February 2007].
11. Eckert, P. "Interoperability, Exploration, and Commerce: Expanding Industry's Scope." Center for Strategic and International Studies, *Interoperability and Space Exploration*, Arlington, Virginia, 6 September 2006. URL: http://www.csis.org/component/option,com_csis_events/task,view/id,1042/ [cited 4 February 2007].
12. Eckert, P. and Lavitola, M. "Promoting Space Commerce through Public-Private Risk Sharing," *AIAA Space Operations Conference*, Rome, Italy, 19–23 June, 2006. URL: http://www.aiaa.org/content.cfm?pageid=2 [cited 4 February 2007].
13. Eckert, P. and Barboza, M., "Attracting Nonspace Industry into Space: A Catalyst for Lunar Commercialization," *International Lunar Conference 2005*, Toronto, Canada, 18–23 Sept 2005.
14. Lenard, R. "ISRU Sub-Team Summary," in Eckert, P., and Mankins, J. (eds.), *Bridging the Gap: From Earth Markets to New Space Markets*, Report on the Third Lunar Commerce Executive Roundtable. URL: http://www.lunarcommerceroundtable.com/lcr3_report.html [cited 20 August 2006].

8

The Future Role of Human Resource Management in Non-Terrestrial Settlements: Some Preliminary Thoughts[1]

Ida Kutschera
Bellarmine University
Department of Business Administration
Louisville, Kentucky
(502) 452-8444
ikutschera@bellarmine.edu

Mike H. Ryan
Bellarmine University
Department of Business Administration
Louisville, Kentucky

Ida Kutschera, PhD (University of Oregon) is assistant professor of management at Bellarmine University's W. Fielding Rubel School of Business in Louisville, Kentucky. She is the co-author of several articles and a member of the Academy of Management and the Society for Human Resource Management.

Mike H. Ryan, PhD (University of Texas–Dallas) is professor of management at Bellarmine University's W. Fielding Rubel School of Business in Louisville, Kentucky. He is the author or editor of six books and author or co-author of numerous articles dealing with space-based business issues, senior editor for an ongoing research book series, International Research in the Business Disciplines, a member of the editorial board of the *Comparative Technology Transfer and Society Journal* and a Fellow of the British Interplanetary Society.

---

[1] Segments of this paper are based on areas initially discussed in Mike H. Ryan and Ida Kutschera, "Lunar-Based Enterprise Infrastructure—Hidden Keys for Long-term Business Success," *Space Policy*, Volume 23, Issue 1, February 2007, pp. 45–52.

Human Resource Management—Moving into the Future

Human Resource Management (HRM) professions have expanded significantly in direct response to the needs faced by modern organizations. It is not unreasonable to assume that as new needs or requirements move into the workplace, the role of the HRM office will expand in response. Moreover, it is not only likely but probable that the training and expertise of tomorrow's HRM professionals would become broader to deal with the range of new requirements and expectations. Nowhere is this move more likely than when private organizations begin to operate outside the terrestrial environment. Working on the Moon or in orbital space will provide some unique challenges.

Jump ahead 20 to 30 years. What will have happened to the role of the HRM professional? What additional responsibilities and concerns will represent the focus of the HRM office? And, just to make things interesting let us also assume that the pattern of space development currently underway has continued. For example, with sufficient funds it is now possible to become a space tourist and visit the International Space Station. A variety of competitive prizes are now being offered to generate interest in commercial space activity with the intent of providing a foundation for future developments. Recently, Bigelow Aerospace launched a test prototype of their orbiting space facility known to be a scaled down version of a proposed space hotel. These are not fantastic stories created to engender discussion. These are real attempts by the private sector to jump start a variety of space-based opportunities. Extrapolation of current space efforts does suggest that it is not unreasonable to predict that there will be active commercial operations on the Moon with a significant population of civilian employees. What might that future look like in terms of HRM?

A Taste of Things to Come

There are several ways of thinking about operations in a remote environment such as the Moon. One obvious analogy might be similar to that of a very large cruise ship. Centralized services, one-point shopping, multiple venues staffed by employees, with the passengers representing the actual business work force, have some appeal as examples of a self-sufficient work environment. The ship is just like a small floating city, so there is employment to match almost all skills or competencies. The threat of disaster is real and the need for advance preparation is clear. The limited duration of most cruises and the relative ease of access to assistance and resupply make using cruise ships less satisfactory as a model from one perspective. Modern cruise ships do, however, have many of the requirements for self-sufficiency in that they must provide their own power, water, sewage, air quality and internal food distribution and preparation. The larger vessels often carry well-trained medical staff and have state-of-the-art medical bays and medical communication equipment. There are even ships that operate as condominium communities where people own their own ship-borne homes that travel from port to port. These communities are, in fact, mobile environments with all the issues of a small city.

With the population of a small city, health emergencies are fairly common. Given the age range of the passengers, medical problems range from acute cardiac care to the typical injuries sustained by crew and people out to have a good time. Cruise physicians and nursing personnel are expected to be well versed in trauma care as well as the necessary life support protocols needed to keep patients stable prior to transport to an on-shore facility if necessary. Telemedicine, high speed communications and facilities that anticipate probable medical needs are the norm. So in one sense, we have the basis for medical care in remote locations and a direct example for future reference.

Cruise ships also have two additional interesting features in the person of a cruise director and ship's purser. These individuals have significant responsibility for many of the non-operational activities that take place in the vessel. Cruise director responsibilities range from the entertainment provided on board and can extend to the direct welfare of the passengers and even the crew. On modern-day passenger ships, the purser has evolved into a multiperson office that handles fees and charges, currency exchange, and any other money-related needs of the passengers and crew. Both of these positions require a significant amount of discretionary authority because of the impact they have on the overall welfare of the passengers and consequently the overall success of the cruise itself. Although it may not be appropriate to directly link these positions to our current HRM specialists, they do suggest the relative importance of indirect business issues to the success of a specific venture. But given that the activities in question occur in relative isolation in an environment that is largely self-sufficient, these two positions provide an interesting clue for future HRM activities in remote locations. The resident HRM personnel in a remote facility might well be expected to manage and operate an extremely wide range of services and activities. Remote locations, even with good communications, do not always lend themselves to being managed from a distance.

Another possible model to consider for operations in a remote environment might be that of large off-shore platforms used for drilling and oil production. These facilities are much like small cities on stilts, frequently located far from shore and in relatively remote locations. To encourage employees to endure the relatively isolated locations and difficult working environment, significant effort goes into making the living conditions palatable. Working aboard an oil rig is hard work; however, the facilities are often excellent. Many of the accommodation wings (called Floatels) meet hotel standards, despite being located in the middle of the ocean. Recent designs include individual cabins with private bathrooms and sufficient sound proofing to ensure not only privacy but essential peace and quiet critical to good sleeping. A well rested crew is less prone to making mistakes, better at maintaining good judgment and more able to keep a clear focus on the job at hand.

All food, board, and laundry expenses are usually met by the employer—and the food available is generally of exceptional quality. In such locations firms have come to view food preparation, choice, and quality as critical components to maintaining morale. With the 24/7 work conditions, many large platforms have chefs and other support personnel whose sole responsibility is that of making it easier or less burdensome for the oil workers to do their difficult jobs. This is in addition to the excellent pay and benefits which are often assumed by employers to be a necessary condition to secure competent employees for such facilities. Even transfer and traveling expenses to and from offshore oil rig drilling jobs are covered by the employer. Another considerable benefit is that most personnel work in a 14/21 day rotation. This means you will work for 14 days, and will get 21 off. This translates to holidays for 3/5 of the year.[2] The amount of down time is a direct result of the harsh working conditions and the need to ensure fresh, motivated, and non-accident prone employees in an environment that is regarded as one of the most dangerous places to work. Relative ease in transporting employees to and from off shore facilities makes such accommodations possible.

In any prototype lunar facility, even assuming dramatic improvements in transportation, large blocks of time away from work might not fare as well given the relative cost of getting employees to the facility. Long-duration stays are likely to become the norm there rather than the exception. It is the probability that lunar-based personnel will reside at a facility for prolonged periods of time that necessitates the shift in our future HRM specialist job requirements. Effectively, we are talking about a small corporate or business-based community that is remote in terms of distance, dangerous in terms of environment, and requires relative self sufficiency in order to operate effectively. As many of the essential requirements for safe and efficient operations require very specialized personnel and the living environment itself will have unique attributes, it may be critical to centralize in a single place or office the oversight needed to make things work. Much in the manner that present day HRM operations have expanded to medical care, financial planning and even environmental monitoring, our future HRM specialists on the Moon may find themselves responsible to ensuring the quality of health care and its delivery, establishing and maintaining many quality of life considerations, deciding how best to organize facilities for work and human interaction, and, of course, assisting in the selection, training and deployment of

---

[2] http://www.oil-rig-jobs.com

lunar based personnel. The HMR office door could become the point of entry for some of the most important issues facing firms that operate in a lunar environment. In time it might very well become the single most influential department as its responsibilities expand to cover an ever widening range of personnel and basic human needs considerations.

Extrapolating the HRM Association with Health Care

Distance and danger alone suggest that health considerations will have some serious implications for lunar-based activity. There are many things requisite to a comfortable living and working environment. As our facility could quickly grow beyond its industrial outpost origins, it must be equipped with safety and medical equipment compliant with whatever the highest standard exists at the time. Realistically, the distance and transportation difficulties alone will necessitate a medical facility equivalent to what we might now describe as a level one trauma center. Operating within a 1/6 gravity environment may be medically challenging but it would be far less troublesome than attempting to do so in the zero gee of freefall typical of a transit to Earth. That also assumes the patient could sustain the associated problems of reentry. After all, this is an industrial facility in a hostile environment with significant limits on getting a patient to a better equipped facility. *Self-sufficiency is critical.*[3] For example, dentists can typically treat between 40–50 patients daily depending on emergencies and the overall number of difficult procedures. Therefore a single dentist could conceivably provide dental care for upwards of 1000 people. This does not consider new treatment options that might either reduce the need for dentists or might increase their operational capacity. The operation of a dental clinic would have to be orchestrated by someone and the HRM office might become the obvious place. The overlap of medical and other skills and abilities will by necessity be greater than that of a terrestrial facility with easy access to other facilities. While cross training and the selection of uniquely qualified physicians and surgeons will help avoid overstaffing, from a practical perspective one must assume the probability of both emergencies and isolation. Although one might expect the medical decisions to be made and overseen by medical personnel, some agency needs to ensure that the overlap of necessary skills unique to a remote 24/7 facility is covered. It might make a great deal of sense to centralize the organization and selection of critical personnel in our lunar HRM office.

---

[3] This does not even consider pathology or autopsies as one might hope they would be needed infrequently. Again practical experience with construction projects and other high risk environments suggests that accidents alone might account for several fatalities per year as consistent with industrial facilities.

Repatriation of ill or injured individuals is unlikely to produce satisfactory results in severe situations even if circumstances permit transportation of those needing advanced medical treatment. In contrast to some health care providers that try to balance cost and efficacy, the lunar HMO would by necessity need to balance accessibility and timeliness with the cost of failing to provide an efficacious treatment option. Holding managers accountable for failing to anticipate potentially obvious situations could become even more serious on the Moon than in Earth-based operations. Even the most uninformed individuals could anticipate many of the difficulties inherent in operations on the Moon. Alternatively, some employees might actually consider employment in lunar businesses because of the generous and cutting edge health care that would by necessity be available to those working on the Moon. In time it may become a common factor for the HRM office to balance the medical and non-medical tradeoffs inherent in operating in a distant remote location. It is even conceivable that our future HRM director might be required to have a medical degree or training much as many current HRM directors now have legal degrees or training on top of the other HRM related credentials.

HRM Moves into Home Design

Realizing that space will be at a premium at a lunar facility, it is essential for the comfort and well-being of the expatriates to have their own personal space, however small it may be. Allocation of space will need to be even handed and fair if resentments are to be minimized. Where better to do this than through your on-site HRM office.

The initial dorm-like environment that might be expected will almost certainly give way to other more private and extensive accommodations. Employees need some place like a studio apartment that they can call their own, simply a place where they can "get away" from everything and everyone if they feel the need to retreat. *Such design considerations are common even if the available space is very limited.* What goes for individuals is doubly critical for couples. Couples need privacy and the ability to create a home-like environment even on the Moon. Without getting into the politics of marriage, cohabitation, and same-sex partnerships etc., it should be fairly clear to most people that a measure of privacy is critical to sustained intimacy.[4] *And*, intimacy is usually necessary for a successful relationship that in turn makes for sustained good morale and happy, productive, safe employees. To

---

[4] See: Mike H. Ryan, "Entrepreneurship on the Moon," appears in Space Studies Institute (eds.), *Return to the Moon*, Vol. II. New York: (Space Front Press) 2000, pp. 202–207.

make any personal space a "home," employees need to be allowed to bring a few personal items (mementos, pictures, etc.) from Earth (in compliance with established weight limits that would be adjusted for the duration of the proposed stay).[5]

Pragmatic experience suggests that a wider range of personal items only be transported after an individual has passed some period of adjustment. Some individuals might not be able to adjust even having passed extensive testing and Earth-based familiarization. On the other hand, some personal items such as musical instruments could be given special consideration due to the positive affect they might have on the overall lunar community. Hobbies such as painting, wood working, and ceramics among others might also represent good community investments in the overall esthetics of a developing lunar culture. While it might be tempting to make all such decisions from the home office, employee satisfaction might be better served if as many of the decisions as possible could be handled locally.

Again, where better to locate these important but not necessarily business or operation critical activities but in our on-site HRM office, managed by our cadre of HRM specialists. In fact, some of the possible skills that would creep into the mix might include ergonomic considerations for living much as current HRM offices often take responsibility for ergonomic features of work stations, chairs, computers etc. In some sense, as one specialist in the HRM office of the future needs an additional degree in medicine, so might another need a degree in interior design and the related fields of human habitation. The ability to alter the work and living spaces to accommodate changing needs is again one of those tasks that might be better managed on site rather than from the home office.

HRM Moves into the Restaurant and Food Business

Given the unique importance that food and particularly food aroma can play in maintaining morale, some thought must be given to specific food requests. It would be impossible to fulfill every request. Yet, some mechanism ought to be available to fulfill some requests that can be altered over time. Comfort food can be debated but it clearly has a role in maintaining an employee's attitude in difficult environments. Centralizing specific food requests given the distance and supply considerations likely to operate in a very remote location such as a lunar facility makes sense particularly if one envisions a supply chain operative over time as well as distance. Related to this are the

---

[5] Mike H. Ryan and Ida Kutschera, "Lunar-Based Enterprise Infrastructure—Hidden Keys for Long-term Business Success," *Space Policy*, Volume 23, Issue 1, February 2007, p. 47.

questions of who prepares food and where. The dangers involved in uncontrolled heat sources are well known. Rules and regulations are common for the prevention of fires in business facilities, college dorms, and even apartment complexes. Regardless, the issue of food preparation by individuals other than food preparation personnel is going to evolve into an interesting issue if people are assigned to a lunar facility for a prolonged period. Not everyone is a chef to be certain but almost everyone has food that they like to prepare, their way. It is not unreasonable that that type of request be considered particularly as individuals are prone to seek unapproved solutions in the absence of specific permission. Our HMR office might be the place where people schedule their time to prepare food for themselves or their friends and colleagues for a special meal apart from the company kitchens.

Food preparation, while simple in concept, can be expected to become more complicated over time. Increased requests for food variations and supplements can be anticipated as both the number of people and cultures represented on site increases. In some cultures food and beverages are considered something other than just fuel for sustaining the human body. The Japanese tea ceremony is but one example where the role of a beverage is more than just the preparation of a drink. Certification of a kitchen for kosher food preparation is another example. To the extent that food can serve to increase morale it can also become divisive if the aroma is not viewed as pleasing to the majority of those present. In a closed environment, not all smells are equally desired by everyone regardless of their origin or the intent of the preparer. Ultimately who makes these decisions and how might one organize an appeal or special consideration etc? Community guidelines on food preparation could become one of the more interesting documents produced by our lunar HRM office working with dietary and food preparation personnel. Creation of food options will be one of the more politically difficult tasks since everyone has their favorite foods. Sustaining an operation that provides varied cuisine is an enterprise fraught with problems.

A single facility might not be capable of supporting the entire lunar population and operationally food services probably should not move in that direction. It might be better for the perceived overall quality of the facility to have several options for food services. People enjoy having a choice. Giving the inhabitants options to eat out, so-to-speak, clearly reflects the norms that they might expect if living in any other community on Earth. Presumably, our medical personnel would also operate as our health department at least in the beginning, but then who would organize the inspections, follow up on issues and ensure compliance? The obvious choice could again be our local HRM office that probably selected the kitchen personnel, chose the dietitians, and even picked the initial restaurant themes. Whether it is necessary for our HRM specialist to attend the equivalent of the Le Cordon Bleu for advanced cooking instruction might be debated. It should be clear that something more than a general interest in eating or food preparation would be in order. Running multiple food services 24/7 is not an inherent skill. It must be learned. One more interesting facet for our multi talented lunar HRM office staff that indicates the HRM office of the future will be very different than those that exist now.

HRM as a Social Planner and Community Leader

Obviously, community development has to be fostered within the lunar operation as well. Given that the expatriates will come from very different cultural backgrounds, there has to be a way to enable everyone to preserve their own culture (food, customs, religion, and holidays) within the newly established lunar culture common to all employees. Dealing with any kind of issues related to diversity is already one of the main tasks for today's HR Managers, and it can be expected to be taken to the extreme for HR Managers in lunar operations where employees will most certainly come from a wide variety of countries, cultures and backgrounds. In addition to the set of values, beliefs and behavior patterns that are common to one particular culture and might as such differ strongly between cultures and individuals, different personalities and their respective impact on the social system of the community have to be considered.

Clearly, elaborate selection processes will aim to pick those individuals that are most likely to succeed in a hostile work environment such as the Moon. Just as is the case with selecting people working in extreme and isolated environments like a station at the Antarctica or on a submarine, extensive personality tests and behavioral interviews will ensure a cultural "fit" with the environment (well beyond the technical skill requirements). However, even if the hiring process is highly selective and only chooses a particular kind of person, issues resulting from individual diversity are unavoidable. With many people working in a hostile and sealed environment, there will undoubtedly be the need for an independent negotiator and arbitrator to settle disagreements. Again, our future HR Managers might have to fill that position of lunar ombudsman.

To come back to our earlier analogy, the HR Manager will also have to be the "cruise director" and "chief animateur" for the operation. Since lunar employees won't have the luxury to literally "go home" for their vacation or time off, every aspect of social life has to be provided on location. Workout facilities and entertainment opportunities like movie theaters, play houses and clubs (unless true virtual reality environments are available at that point in time) have to be offered and managed. One could even imagine special events like concerts and shows taking place much like the ones that are used currently to boost morale of our troops engaged far away from their home.

Not only do social interactions with fellow lunar colleagues have to be facilitated, but contact with family and friends back at Earth also needs to be made easy. Equally important is to accommodate the need to be a "lunar loner." Some people just need to be able to be left alone to do their work. Some extreme personality types might even be of great interest to a lunar operation, as what might be considered overly obsessive on Earth might become a desirable quality for certain jobs in places such as the Moon. Studies involving Antarctica expeditioners have shown that the separation from family and friends is one of the biggest social stress factors.[6] Again, in practical terms, providing communication access from an employee's quarters or residence at no cost as part of the employment package probably makes the most sense and would be a reflection of current practice. Again, current operational practice by off shore facilities is moving in that direction by providing real-time communications including Internet access. Anyone who has ever ventured far from home and friends knows the importance of direct communication. Physical reminders of friends and family, whether in the form of cherished food items, pictures, or other memorabilia, has lasting benefits that should not be underestimated. Thus, the HR office might also fulfill the function of a communications officer that enables everyone to receive information from back home as well as communicate with everyone on a regular basis. This responsibility would also include checking on those without regular communications to or from home. In an environment where individual lapses of attention or absent mindedness due to concerns at home could be fatal, the supervisory nature of the activity, while discrete, would be of great importance.

The HRM Office as Lunar Concierge

Added to the obvious concerns of HRM personnel in a lunar environment are the less obvious considerations represented by barbers, grocery stores, bistros, kids and even pets. As a final example of factors that contribute to the overall quality of life, one might contemplate some fairly simple activities, such as getting your hair cut. Unless bald becomes more fashionable for both men and women it is likely that haircuts will be needed on the Moon just as in any other location. Equally mundane, but probably of some importance, would be the act of shopping or selecting fresh produce or simply wandering through the local version of the general store to see what "new" products might be available. It may be debatable as to whether shopping as it is now done would work on the Moon. Certainly, things might be more reminiscent of a convenience store initially with few products and only a few choices. Eventually, some form of commercial store would become an

---

[6] Harrison, A. A. *Spacefaring: The Human Dimension*. Los Angeles, CA: University of California Press, 2001, p. 121.

important element to encouraging people to accept long duration assignments. Some people that might be important to a growing facility or commercial undertaking would not want to feel that by accepting a lunar assignment that they were depriving themselves of common services etc. While it can be argued that such activities would not be necessary given the vast array of communications to each and every inhabitant of the lunar facility, commercial exchange is often viewed as another opportunity for social interaction. Seeing something in print or on a screen is not necessarily a substitute for seeing it in person. People like to shop to some degree rather than just take what they are given.

The same can be said of entertainment. Not everyone enjoys the same things. Most individuals agree however that a live performance is different and often more engaging than one that has been recorded. Even mistakes provide a uniqueness that is frequently lacking in prerecorded media. A venue or even part-time use of another space for live music or performances would represent an important step in normalizing life. Apart from the live music played by the musically inclined residents, it may also be of great value to sponsor artists to play on the Moon for the lunar community. Experimental entertainment has long been the subject of science fiction with stories that imagine how ballet, ball room dancing and other physical forms of entertainment might evolve within the Moon's 1/6 gravity. Tossing your partner into the air might never be the same. The first concert broadcast from the Moon may not engender the long-term attention of historians, but it would certainly signal that people have come to the Moon to stay.

Much as people moved to the suburbs in the 1950s and 1960s to find a better quality of life, improved schools, etc., so might some families come to view the Moon in the same context. A truly international project would also have the advantage of concentrating large numbers of interesting people in one place with significant overlapping interests and goals. That could very well make for an attractive living environment even with some limitations. Families with special skills, whose children were above some minimum age, might become ideal candidates for lunar habitation. As a lunar facility matures, the requisite safety and habitation factors are made more or less automatic, it may be possible to lower the age at which children are allowed. One might imagine that the HR Manager is responsible for the initial certification of safety that would permit children. Equally likely, it might be the HR Manager who also certifies the educational system whether or not it originates from Earth.

In some respects, it may actually be easier to allow pets than children. In fact, some pets may be highly desirable both as companions and helpers. Animal sensitivity to internal environmental characteristics might, in some circumstances, be better than instrumentation for warning citizens of dangerous changes. While one might laugh at the notion of pressure suits for dogs or cats, humans generally have always found a way to get their canine or feline companions just about everywhere. It is doubtful that the Moon will be any different in that regard. Besides, few things are as intrinsically rewarding to human beings in difficult situations as a cold nose, a wagging tail, or a small furry creature curled up in one's lap. Arranging for the importation of animals, whether for research or work considerations, will quickly place the HR Manager in an "interesting" position. It will remain to be seen whether or not HR Managers add dogcatcher to their repertoire of skills and responsibilities.

HRM and Lunar Employee Selection

Since stakes are high for this project, there will be a strict selection process where identification with the purpose of the project is one of the main selection criteria. Potential employees will have to undergo substantial psychological testing to ensure their fit with the projects operating on the Moon and with their fellow employees. Maladjusted individuals regardless of their talent have no place in an environment where mistakes may very well kill one or more people. Any expatriate has to be prepared for the new role he or she is facing in the new environment. Although there will be orientation sessions before people leave Earth to prepare them for the upcoming role change in their job on the Moon, there should also be an extensive and formalized orientation program once they actually are on the Moon.[7] In all likelihood the HR Manager would be responsible for developing and conducting such a program. To accommodate the role change and the socialization process for each individual, there has to be a well-defined organizational structure. Considering the probability that most of the employees will likely be highly individualistic, this structure should be designed in an organic rather than a mechanistic fashion. From an organizational standpoint, the big challenge will be to keep a balance between the necessary stable, institutionalized environment and a myriad of different, constantly changing individual needs. Monitoring this adaptation will be one of the more salient duties of the HR Office.

Every single employee should, over the course of his/her stay on the Moon, have access to an Employee Assistance Program (EAP) where—at no cost and confidentially—they receive help with adjustment to their new roles and dealing with stressors as they come up. Also, the implementation of a formal mentoring program might achieve this as it is another way to convey a

---

[7] In a "working abroad" context, equally important are issues of repatriation—how to prepare people to go back to a life in their home country (or planet in this case) without experiencing culture shock. In some cases, people may not want to go home and their repatriation may require special handling.

feeling of stability and reduce stressors. The goal of programs like the orientation program, the EAP, and the mentoring program is to build social cohesion beyond professional connectedness. For the most part, employees will be highly motivated coming into the project. In the beginning, a lot of the motivation comes from the exhilaration of being part of something new and utterly exciting. All of these programs would require monitoring and the occasional tweak. Again, it might make sense to locate these important activities within an office already established to monitor other characteristics and features of lunar life—the HRM office. So in one sense we have come full circle by including the lunar versions of common HRM procedures to gather information in order to improve employee success.

Future HRM Issues—A Final Thought

Our lunar HRM office will be not only a broad based set of inter-related operations but central to the long term success of human habitation on the Moon. Knowing that you have both the people and the capability to deal with major problems is an attribute that generally makes residents feel more confident in their community's leadership. Having a central point of contact for the pesky common problems that creep into work and life could certainly make things easier for employees. More importantly, our future HRM office may be the source of a critical peace of mind while operating in a difficult and distant place. Employees might very well respond positively to the knowledge that should something happen someone will be looking out for their welfare.

9

Lunar Commercial Logistics Transportation

Walter P. Kistler

President, Lunar Transportation Systems, Inc.
Bellevue, Washington

Bob Citron

CEO, Lunar Transportation Systems, Inc.
Bellevue, Washington

Thomas C. Taylor

Vice-President, Lunar Transportation Systems, Inc.
Las Cruces, New Mexico

Thomas C. Taylor is an entrepreneur, inventor and a professional civil engineer in the commercial aerospace industry. His goal is building commercial space projects including an unmanned transportation cargo service to and from the moon's surface with Lunar Transportation Systems, Inc. (www.lunartransportationsystems.com).

Since 1979, Tom has helped to form 22 different entrepreneurial aerospace startup companies with four successful commercial space startup companies raising a total of $1.2 billion in private equity financing. The successful startup ventures raised private equity money totaling $1.2B in private equity. Tom enjoyed working in the trenches 4 to 12 years with each of these commercial space successes with Walter Kistler and Bob Citron, usually the founders of most of the successful startups. Examples include: SPACEHAB, Inc., a pressurized payload module in the space shuttle offering manned tended mid-deck lockers to NASA and the public at a reduction of the cost by a factor of ten compared to a previous Spacelab Module operated by NASA and providing the same mid-deck locker microgravity service. Bob Citron and Tom co-patented a shuttle pressurized module based on U.S. Patent Number 4,867,395 and obtainable free in pdf format by typing in the patent number above without commas at www.pat2pdf.org. Walter Kistler invested the early money. SPACEHAB hardware has flown ~23 times on the space shuttle with ~$900m+ revenue. Kistler Aerospace Corporation offered a two-stage reusable launch vehicle (RLV) to low Earth orbit at a potential cost saving, roughly $516m over a 12-launch sequence compared to a Delta II. The company raised

$860m in private equity investment and was founded by Bob Citron and Walter Kistler, with Tom being their first employee again. After 12 years of development, the Kistler Aerospace Team industrial partners grew to include Lockheed Martin, Northrop Grumman, Aerojet, Irwin Aerospace, Draper Labs and others. Lunar Transportation Systems, Inc., an unmanned logistics service anticipating commercial cargo to the moon's surface at commercial rates with scalable hardware. Started in 2005 by the founders Walter Kistler, Bob Citron, plus first employee Tom Taylor, LTS proposes privately financed services for commercial lunar development. The goal is a sustainable commercial transportation system for the moon to permit NASA to explore Mars.

ABSTRACT   This paper offers a commercial perspective to new lunar transportation and proposes a logistics architecture that is designed to have sustainable growth over 50 years, financed by private sector partners and capable of cargo transportation in both directions in support of lunar resource recovery. The paper's perspective is from an author's experience at remote resource recovery sites on Earth and some of the problems experienced in logistics that didn't always work. The planning and control of the flow of goods and materials to and from the Moon's surface may be the most complicated logistics challenge yet to be attempted. The price paid, if a single system does not work well, is significant. On the Alaskan North Slope, we had four different logistics transportation systems and none worked successfully all the time. The lessons learned will be discussed and solutions proposed. The industrial sector has, in the past, invested large sums of risk money, $20 billion for example, in resource recovery ventures like the North Slope of Alaska, when the incentive to do so was sufficient to provide a return on the risk investment. Stimulating an even larger private investment is needed for the Moon's resource development. The development of the Moon can build on mankind's successes in remote logistics bases on Earth and can learn from the $20 billion in private sector funds used to recover oil assets above the Arctic Circle. The Moon is estimated to be 50 times more remote than Prudhoe Bay, Alaska, and the early transportation to the Moon is 100 to 1,000 times more expensive than to the Arctic. The lunar environment is more severe than the Arctic, but some of the logistics lessons learned in the Arctic can potentially work again on the Moon. The proposed commercial lunar transportation architecture uses new innovations for modularity and flexibility leading to reduced development and logistics costs, faster development schedule, and better evolvability. This new lunar trade route of mankind utilizes existing Expendable Launch Vehicles (ELVs) and a commercially financed small fleet of new trans-lunar and lunar lander vehicles. This architecture is based on refueling a fleet of fully reusable spacecraft at several locations in cislunar space, which creates a two-way highway between the Earth and the Moon. This architecture offers NASA and other exploring nations more than one way to meet their near term strategic objectives with commercial space transportation, including sending small payloads to the lunar surface in a few short years, sending larger payloads to the lunar surface in succeeding years, and sending crews to the Moon and back to the Earth by the middle of the next decade. Commercially, this new lunar logistics route permits capability and technology growth as the market grows, offers affordable transportation for the commercial sector and the later recovery of lunar resources. After NASA moves on to other destinations in our solar system, commercial markets and this "in place" commercial logistics system can service, stimulate and sustain a lunar commercial market environment.

Mankind's Driver—Trade Route Expansion for Commercial Resources

Trade Route Transportation

Most of mankind's expansion on this planet over 40,000 years has produced benefits in the form of new lands, trade route commerce, new resource development and knowledge.[1] Trade routes are driven by commerce and exploration, which produce benefits for mankind in the form of new lands, trade route commerce expansion, increased resources and knowledge. Genuine innovation is needed to achieve the goals of affordability and sustainability called for by the President.

Lunar Transportation Systems, Inc. (LTS) proposes a commercial transportation system for the new lunar trade route with different innovation and architecture hardware that would be built with private capital and government cooperation. The goal is a basic and modest early commercial lunar transportation system with hardware that could be abandoned on the surface with 800 kg of useful cargo[2,3] and later evolve to 10 to 20 tons landed by a reusable vehicle with full transportation node support services.[4] This architecture is characterized by modularity, expandability, commercial sustainability and extreme flexibility leading to reduced development cost and better evolvability.[5,6]

Specific innovation includes the transfer of cryogenic propellant tanks instead of cryogenic propellant from one tank to another tank in microgravity. Further innovation includes payload transfer, nodes of transportation services, and eventually the return of resources to Earth. A hard look at this architecture will show that it enables NASA to meet its strategic objectives, including sending small payloads to the lunar surface in a few short years, sending larger payloads to the lunar surface in succeeding years, unmanned cargo transport in both directions and eventually with increased reliability, sending crews to the Moon and back to the Earth by the middle of the next decade.[7] The hardware shown in Figure 9.1 takes advantage of the

FIGURE 9.1
Lunar Transportation Systems concept for commercial logistics. The LTS Concept is still evolving, but the early conceptual architecture[2,3] proposed can be seen in video on the Website: http://www.lunartransportationsystems.com. Go to the YouTube.com website for an LTS 7 Minute Video on the Lunar Transportation Systems Concept:[2] http://www.youtube.com/watch?v=26Y5w0vqtIU

commercial first leg vehicles available from commercial sources. LTS uses commercially available launch vehicles for the Earth to LEO leg with payloads transported on a different vehicle. The strategy is based on refueling a fleet of fully reusable spacecraft at several locations in cislunar space, which create the equivalent of a two-way highway similar to the transcontinental railroad between the Earth and the Moon.

Logistics Is More Than Transportation

Lunar logistics is movement management or the planning, accomplishment and control of the flow of goods and materials to and from the Moon.[8] Figure 9.2 depicts a large payload Lunar Transportation Systems would like to transport. Lunar logistics flow over a fifty-year period evolves into reusable systems and includes more than just transportation vehicles and hardware. This later logistics system becomes a system of supply that includes transportation both on the surface of two celestial bodies and in the space between them.

In a remote resource recovery base on Earth it means the flow of mass away from a base in the form of useful resources must at some future time exceed the flow of mass into the base or there is little economic reason to set up the remote base commercially. At Prudhoe Bay Alaska, that mass break even point happened approximately 90 days after oil flow down the pipeline. In our lunar planning and development we seem to be focused on hardware that people want to build rather than our reasons for going to the Moon, which probably should be focused on the commerce, which the taxpayers can actually understand and support long term. This means commerce and the recovery of resources that are significant to build the support within our taxpayer stakeholders, our industrial partners and our partners overseas.

University of Wisconsin-Madison Miner

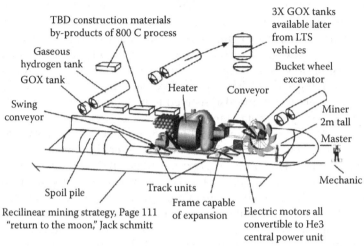

FIGURE 9.2
The University of Wisconsin-Madison Lunar Miner Mark III by Matthew Gadja.[9]

Commercial logistics means privately financed firms are providing the transportation services, including transportation vehicles and depots. In Alaska we tried many transportation methods and trade route concepts, but only four really worked most of the time with some short winter periods where none worked. If the base is remote (Prudhoe Bay was ~6,000 miles from the lower 48 states and most supplies) and the logistics tail is long, then transportation becomes more important and a larger, more expensive part of the logistics operation. The Moon is ~20 to 40 times further than any logistics support of a remote base on Earth. The lunar temperature extremes are significantly different than those on Earth and much of what we take for granted on Earth must be brought to the Moon by logistics, until we learn to "live off the land." A commercial logistics operation deals with the procurement, distribution, maintenance, and replacement of materiel and personnel with transportation understood to be a small part of the whole process.

Arctic exploration on Earth may provide some insight and lessons learned for future lunar development. The company towns created, private investment required, benefits generated through the resources recovered and other parameters of the Arctic Exploration fit reasonably well into a lunar scenario. Before we dream about the tourist phase, we need to understand that the early commercial humans on the lunar surface may be the people that build the facilities for the resource workers or miners. Before them come the resource locators in the form of geologists, resource extraction specialists, equipment designers and the dreamers capable of imagining the operations before they are created.

Figure 9.2 is a University of Wisconsin-Madison Miner by Matthew Gadja with some suggested innovation by LTS.[8] The miner processes the lunar regolith in place and extracts a multitude of near term and far term resources needed by lunar company towns, production facilities and mankind on Earth. The miner can be explored further at the University of Wisconsin-Madison.[10] Lunar Transportation Systems is proud to have contributed some funding for the miner evolving at the University of Wisconsin-Madison.[11] The bucket wheel excavator recovery technique also seemed to be validated at the recent NASA Excavator Challenge in Santa Maria, CA, where a bucket wheel excavator operates well in a NASA JSC1A stimulant box test. The bucket wheel was damaged in transport to the event and repaired, but failed after 11 minutes of a 30-minute run. It was excavating at a rate approximately one half of the rate required to meet the minimum required 150 kg. The University of Wisconsin-Madison miner excavates the lunar surface regolith, conveys it into the vehicle, heats the materials, drives off the volatiles for recovery and disposes the regolith in waste form and as a variety of building materials yet to be determined. The miner also starts the process that is sometimes called "living off the land."

"Living off the land" means developing an excavator that recovers gaseous oxygen and hydrogen that later is used to produce the cryogenic propellant for the LTS vehicle operation. This gaseous oxygen and hydrogen recovery is used for cryogenic propellant production in a plant requiring power and storage. This leads to a reduction in the cost of propellant for company use compared to propellant transported from Earth and propellant for sale in new commerce to others. These LTS cryogenic tanks are reused as gaseous oxygen and hydrogen tanks using a special valve capable of both liquid and gaseous oxygen and hydrogen use. On the Moon, the regolith can be microwaved into road surfaces, used as a concrete like material and can be heated to recover resources. The temperature required to accomplish different goals ranges from 800 degrees C for some materials of commercial value to ~2500 degrees C if the process gets serious about more complete oxygen recovery in gaseous form.

Remote Base Lessons Learned Can Help

The lessons learned in Alaska can possibly help lunar explorers create a logistics system capable of sustainable logistics support and growth with the lunar market.[12] The important lesson is how to get private industry to invest the kind of money they invested in Alaskan oil. Figure 9.3 depicts a partnership that may help.

The biggest lesson is probably money, specifically, private money by big oil to find, drill, develop and bring oil to the surface for $6.95/bbl, commonly called the wellhead price. The $20B in private risk capital was stimulated by a large return anticipated by the private oil company investors, but it took both majors and smaller companies pooling their risk capital to buy into the deal, which the author viewed as a 50-year deal starting with the state owned oil leases. The value of the pool of oil at Prudhoe Bay was calculated

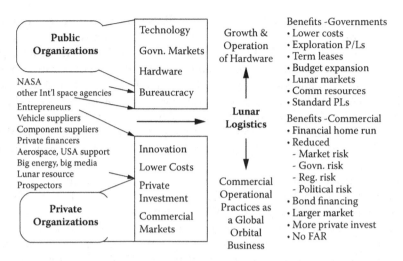

FIGURE 9.3
The public–private partnership is a method of cooperation between government and industry allowing each to bring skin in the game and derive benefits.

by the author and given the ~price of oil appeared to exceed one trillion.[13] Oil flowed down the pipeline in 1977. The majority of this investment was made when these same companies could buy oil from the Arabs at $0.50/bbl.

The company town created in Alaska provided living quarters called ATCO Buildings, reflecting the name of the manufacturing company and sheltered up to 12,000 construction workers at one time, and probably continues in use today, two decades after the initial work started in 1975. All vehicles burned benzene cracked from the oil field and Mother Nature forced much of the work planned for the summer months into the dark winter months.

Applying the same "Living off the Land" thinking could mean using an isotope of Helium, or He_3 as a power source for power grids, propellant recovery/processing and eventually vehicles on the Moon. Such an effort could change the He_3 effort on Earth from DOE, which has little incentive to turn away from their focus on other fusion research. It provides NASA or some other interested government agency with an incentive and the budget for the development of this potential power grid solution using the Moon's vacuum. Eventually the He_3 effort could evolve into solutions for the Earth power grid.

Other industries include Lunar Tourists who spend their own money to go the Moon and create added commercial markets. "Living off the Land" also includes reuse of equipment and packing crates and more.[14]

The Stimulation of Commerce Is One of NASA's Mandates

Our company hopes to use existing Expendable Launch Vehicles (ELVs), EELVs, RLVs or government vehicles. Our commercial lunar architecture utilizes current commercial ELVs and/or EELVs to bring a new fleet of reusable spacecraft, lunar payloads, propellants, and eventually crews from the Earth to Low Earth Orbit (LEO). Expendable launch vehicles and other types of launch vehicles are already commercial for this first leg of our lunar journey. The LTS reusable spacecraft could do the rest of the job and take payloads from LEO to the lunar surface and later bring payloads back to Earth from the Moon. This commercial strategic roadmap permits a "pay as you go" and a "technology development pathway" that allows NASA to achieve a series of its strategic objectives as funding and technology developments permit.

Our approach reduces recurring mission costs by advancing in-space transportation technology, and later, resource utilization, because this is less costly for us than investing in new ETO transportation.

NASA cooperation in the development of commerce can be an important goal for our government and may result in lower costs and increased sustainability of lunar development, permitting NASA to depart for Mars and beyond earlier.

The possibility of an innovative Public Private Partnership with NASA and other international governments could add a commercial market to the NASA vehicle use and provide potentially a 10 m or 33' payload diameter, which would be very attractive to commercial organizations. Public Private Partnerships (PPP) can be productive and are in use throughout the world to bring governments and private organizations together for their mutual benefit. Figure 9.3 provides a basic framework of cooperation and combined with a list of agreed milestones permits private sector financing and potentially lower costs for NASA.[15] Government can bring a government market and the start of a commercial market without narrowing the field to several winners. Private organizations can bring innovation and potentially lower costs. Each can bring much more if the PPP is truly creative and broad in its scope and cooperation.

One result may be the conversion of a government program that needs tax dollars to one that is commercial and pays taxes. New Zealand is one of many examples of where the PPP process and the dynamic effects it can produce for government can be shown.[16]

The Lunar Transportation startup team has in the past created commercial companies that have achieved an order of magnitude in cost reduction. A commercial microgravity service is available in the form of a Mid-Deck Locker service for $2m/locker at SPACEHAB and a similar potential magnitude at Kistler Aerospace in launch costs.[17,18] The same startup team has created Lunar Transportation Systems, Inc.[19]

Public Private Partnership Model

The Public Private Partnership has been used within NASA with success.[20] The Public Private Partnership has been explored[21] and now is ready for discussions with interested space agencies.[21]

Enabling Technologies

This In-Space transportation architecture, as described briefly in this paper, does not depend on the development of any new launch vehicles. It does depend on the development of five emerging technologies: 1) an autonomous rendezvous and docking system, 2) a new autonomous payload transfer system, 3) a new spacecraft to spacecraft cryogenic propellant tank transfer system, 4) an autonomous propellant tank tapping system, and 5) an autonomous lunar payload offload system. Developing these technologies is less risky and less costly than investments in ETO transportation or cryogenic propellant transfer technologies. These emerging technologies, except AR&D, are developable by ground tests and our program plan includes flight demonstration on early robotic missions to the Moon.

Lunar Payload Capabilities

The initial fleets of reusable spacecraft[4] are designed to fit the payload capabilities of Delta II Heavy class launch vehicles, commercial RLVs or other vehicles, but the Earth to Orbit vehicle payload bay defines the size of the LTS vehicle system. Basically, the larger the diameter of the initial payload, the more capable and efficient the LTS highway scales up. Lunar Lander spacecraft can deliver payloads of up to 8 metric tons from LEO to the lunar surface, depending on where and how frequently they are refueled on their way to the Moon. This architecture is capable of delivering 800 kg to the lunar surface directly from LEO without the need to refuel in space. It is capable of delivering payloads of 3.2 metric tons to the lunar surface with refueling at L1 only. Comparable payloads can be returned from the lunar surface to the Earth with refueling at one or more of those locations.

Existing Trade Routes created by Prudhoe Bay Oil are depicted graphically in Figure 9.4 showing the different methods used for the commercial logistics route to and from the Arctic. Figure 9.4 details the four logistics routes. The first trade route shown in blue in the original figure and the greatest tonnage at pennies per pound was transported on leased 300' long barges towed by ocean tugs from ports on the west coast of the United States to a point off shore of Point Barrow, AK. These barge fleets sometimes as large as 40 barges would circle a month or more waiting for Arctic Ocean ice to recede from the shoreline so these barges could race 300 miles to Prudhoe Bay to unload at the North Slope Oil Fields at Prudhoe Bay. In 1975–79, ice receded two out of four summers and unloading was accomplished, but for two summers the ice failed to provide the unloading opportunity and large Soviet icebreakers and other means were used. The second logistics route shown in green in the original figure was by commercial air for humans and critical cargo at approximately \$5 per pound. The third logistics route shown in violet in the original figure was overland truck with oversized fuel tanks, because

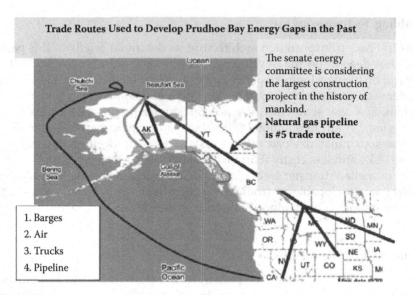

FIGURE 9.4
The logistics routes used on the development and recovery of resources at Prudhoe Bay, AK in the 1970-1980 timeframe. Trade route emerges, because of a logistics market driven by resource recovery by Mankind. Route 5 is emerging now after 40 years, because it is a market logistics driven link sensitive to the economics between the resources and the users.

Prudhoe had no diesel fuel and the truck logistics price varied greatly around a dollar per pound. The fourth system shown in dark blue was the pipeline itself, which took 8 days to transport oil 800 miles and cost pennies per pound to operate even when including the initial $8 billion of private money used to construct it.

The oil field development is also important as an indication of the amount of private money capable of being raised, if the return is significant. While this LTS initial logistics system is not meant to transport crews to and from the Moon, it is meant as a technology development testbed to prove reliability through repeated non-critical cargo mission of a later crewed Earth–Moon transportation system capable of sustaining the commercial development of the Moon and allowing NASA to move to destinations beyond Earth.

Scalability

This new Lunar Transportation System is scalable. A follow-on fleet of larger spacecraft, designed to fit the payload capabilities of Delta IV Heavy class launch vehicles can transport payloads of up to 30 metric tons from LEO to the lunar surface, depending on where and how frequently they are refueled on their way to the Moon.[5] These larger and later LTS spacecraft are capable of transporting crews to the lunar surface and returning them to the Earth. They also have the capability to provide heavy cargo transportation to support a permanent lunar base.

Cost Reduction

The non-recurring costs to develop this Earth–Moon transportation system are much lower than the cost of developing systems that use more traditional architectures because there are fewer unique developments and it relies on existing launch systems.

A significant reduction in lunar mission costs comes from the reusability of the major elements of this system.[5,6] The largest cost in operating this system is the delivery of the spacecraft, the propellants, and the lunar payloads from the Earth to LEO, which could be a government cost and an increase in NASA budgets. Our commercial company plans to bring the Earth–Moon transportation infrastructure from the Earth to LEO on existing expendable launch vehicles, but it could easily provide NASA with a second commercial customer for their new larger diameter (10 m) payload Ares vehicles. Big vehicles, like the space shuttle and the new exploration vehicles, require many launches per year to spread the overhead. Perhaps as much as 70% of the cost of each lunar mission will be to transport the LTS infrastructure from Earth to LEO. While these NASA vehicles are expensive to operate, the development cost of a significant new launch capability represents at least 100 launches of existing EELVs and many years of lunar transportation operations.

Our commercial company is prepared to start now using existing hardware. When propellants can be manufactured on the Moon, Earth–Moon mission costs may be reduced by 60% or more. If and when reusable Earth to LEO launch vehicles become available, lunar mission costs may be reduced by a further 60% or more.

Schedule

Because this system relies on existing technologies and existing ELVs and only requires the maturation of several enabling technologies, it can deliver payloads to the lunar surface relatively quickly and well within NASA's schedule for robotic and human lunar exploration, if a start can be immediate.[15]

The Bottom Line

This lunar architecture is based on concepts that reduce lunar mission life cycle costs and technical risks, improve reliability and crew safety, accelerate lunar mission schedules, and allow for the routine delivery of lunar payloads on a two-way highway between the Earth and the Moon.

Innovation on the Moon Reusing LTS Hardware

Lessons Learned Summary

Prepositioned Logistics

Before major resource recovery work is started on the Moon, these mining companies need incentives to spend the risk money to recover resources from the Moon. The Alaskan incentive was mining leases. The incentives used in the Transcontinental Railroad were mineral rights to land under the track as the track was laid. The 80′ soft coal seams mined 600 feet under these original Transcontinental Railroad mineral rights still have value as incentives.

Lunar incentive is or can be mining leases on the Moon. The North Slope of Alaska drew $20 B in private funds from the largest oil companies in the world by using drilling leases in similar manner. These same companies and/or other companies of similar risk takers asked various industries for studies on how to get the oil off the North Slope after it was found and developed. They also set up the design and materials to build the facilities and pipelines into the field and how to get the oil out. They asked submarine, pipeline, railroad, aerospace industries and others for suggestions, proposals and studies.

Each industry oddly enough suggested solutions using their industry. Submarine organizations suggested giant submarines, railroads suggested train tracks on the tundra, etc., only to be ruled out by a very shallow Arctic Ocean, difficult tundra foundation problems for rail lines, and other technical and cost barriers. The tundra, for example, thaws the top 5′ every year making gravel roads, runways and railroads very difficult or impossible.

Some industries suggested aircraft as an alternative, but the movement of mass cargo should be as economical as practically possible and it is not realistic to move mass cargo on the same hardware as people, given the economics involved. Most people can understand the choice for the oil flow out as the pipeline, but little was said about the four transportation systems used in construction for the flow of crew and materials into the oil field development and out. Each of these four logistics systems had a different cost, speed, safety, surge storage, weather sensitivity and mass transport capability.

Depending on any one transportation of these logistics systems for all cargo and human passenger transport would have caused many cost problems, plus no single system could operate all the time. All of the logistics routes were commercial and most were available with competitive pricing.

Logistics Transportation Systems and Storage

Figure 9.4 provides a brief depiction of the logistics transportation and storage methods used including: **First,** aircraft, 737 passenger and cargo jets, 3 flights per day, hauled up to 12,000 workers and high value cargo pallets of materials forgotten or critical to operations like welding rods, sockets, tools, etc., at about $5 per pound. Surge storage of critical cargo and materials plus staff located in Fairbanks, Anchorage and the lower 48. Winter travel was delayed for weeks at a time due to weather and the gravel strip didn't always remain serviceable in the summer thaw. **Second,** heavy highway trucks with 1500 gallon tanks north of Fairbanks required for the round trip, because Prudhoe didn't have gas or diesel, but burned benzene in all vehicles, which was cracked from the oil in a small refinery. Winter travel was too severe at times for trucks. Trucks cost about $1.00 per pound of cargo. **Third,** ocean going 300' long barges towed by ocean tugs with entire buildings on board, probably cost pennies per pound to transport, but only work well when Mother Nature blew the ice from 300 miles of Arctic shoreline for ~30 days in August. Figure 9.4 shows the Beaufort Sea 300-mile shoreline east of Point Barrow, which failed to open fully 2 out of 4 summers. Barges were used in only one direction and storage was in fabrication yards in Seattle and San Francisco. Half the time the ice and shallow Arctic Ocean were major problems forcing summer work into construction winter work. Finally the **fourth** logistics route, the pipeline, only works in one direction, takes eight days, 2.1 million barrels a day, but construction costs were increased by the permafrost foundation problems and difficult terrain. Some pump stations used natural gas, some oil for power. Large storage at the south end and transportation estimated cost pennies per barrel transported. A fifth trade route may emerge as the very expensive natural gas pipeline discussed in Congress.

Logistics Lessons Learned[12]

1. **Arctic engineering and construction are different than regular construction:** Recognize that lunar engineering and construction techniques will be ~10 times more severe and more different than the difference between normal engineering design and construction in the lower 48 compared to Arctic engineering and construction, impacting everything from packaging requirements to mass of large structures and equipment. Recommend Lunar Engineering be a course at interested universities. The lunar environment is more difficult.

2. **No single transportation or logistics system worked all the time:** Weather and ocean ice precluded each logistic transportation route at different times, so more than one was required and probably would be required for significant surface operations on the Moon. Recommend at least four different lunar logistics routes using different vehicles with different hardware to be considered, so the launch failure of one vehicle will not impact the safety of the lunar crew and

that pre-positioning of logistics be considered as part of any logistics plan. Suggest more than one system.

3. **No single transportation or logistics system offered the affordability for all materials:** Recommend multiple government agencies attempt to encourage multiple logistics transportation routes to and from the Moon. Smaller affordable systems can be commercial and more frequent in delivery.

4. **Sometimes you need something so quickly that transportation cost is overshadowed by need:** Recommend a last minute emergency cargo system be considered similar to throwing a part on a jet flight at the last minute and paying the increased cost. It happened every day on the slope with hundreds of companies filling about half of all passenger flights with cargo pallets and paying the cost. More frequent helps.

5. **Once a logistics system is created, commerce flourishes:** Recommend commercial alternatives be considered early. Non-workers or tourists would pay $1,000 to fly to Prudhoe Bay in the summer just to look at the tundra flowers and look at what $12 billion could build. The tundra flowers were smaller than a fingernail and underwhelming. Hotels were nonexistent and food was difficult. Plan for lunar tourists early.

6. **Some packing crates were used for a second purpose, because they were available[14]:** Recommend the use of Henry Ford's method of specifying crate materials and using them later as Henry did as oak running boards for the Model A.

7. **Even equipment used in transportation needs to have a second use at the destination:** Recommend reuse of tankage, and everything be considered for reuse in some form.

8. **Tooling and fasteners can be standardized and provide second use opportunities:** Recommend agreeing on fewer, but standard fasteners and the ability to take items apart and repair or reuse.

9. **The labor intensive work is done in the lower labor cost area and transported assembled to the high labor cost area when possible:** Recommend equipment, structures and other items be assembled by the future lunar worker at least once on Earth and "repairable thinking" be a part of all lunar thinking.

10. In the use of piece parts to be assembled on the slope, we suffered from single parts in structural steel, for example, not making it to the assembly site due to **logistics handling, mis-marking and storage** requiring a new fabrication on the slope for the part or adaptation of a mis-fabricated part. Recommend parts, boxes and all items be marked with bar codes plus "Plug-and-Play" information accessible from a space suit.

LTS Earth—Moon Transportation System Reuse

The lunar Transportation System proposed is for non-critical cargo initially and may become reliable enough for crew transportation to and from the lunar surface at some later date, when reliable unmanned flights are repeatable with sustained success, safety and comfort. The nearer term market includes a variety of missions to test and refine our in-space flight architecture and architecture support bases (propellant depots in LEO and Low Lunar Orbit (LLO)), LEO missions as an unmanned tug, missions in and beyond LEO to the L1 point in support of facilities in that part of space and missions to lunar orbit.

The size of the LTS vehicles may offer more trips to and from the lunar surface than the larger vehicles and provide some benefit in the frequency of flights. This proposed system builds a two-way transportation highway between Low Earth Orbit and the lunar surface, either from LEO directly to the lunar surface for smaller payloads, or from LEO by refueling in cislunar space for heavier payloads and for payloads returning from the Moon. The system uses a small fleet of reusable spacecraft, supported by a small fleet of expendable spacecraft, to transfer payloads in LEO, to transfer propellant tanks at specific locations in cislunar space, and to transport payloads to and from the lunar surface. The system uses existing ELVs to transport its entire infrastructure from the Earth to Low Earth Orbit and one could consider using government vehicles in some cost sharing manner suggested later.

Entrepreneurial Opportunity

The same team that brought researchers the mid-deck locker in microgravity offers a special small surface delivery package for your creativity, inventions and special equipment for the purpose of "prospecting" the lunar surface. Figure 9.5 is one idea.

On Earth we use FedEx, with package pickup at your door and no return delivery. To the Moon you can use the LTS vehicles and on the Moon's surface you are on your own, after we gently set you down. The mid-deck locker became an industry standard for researchers in the microgravity environment created by the SPACEHAB Module in low Earth orbit. The mid-deck locker was and still is a 2 cubic foot locker that costs $2 m and used an average of 125 watts of power and averaged 42 lbs flying on the space shuttle.

SPACEHAB still offers this smaller module version of an existing NASA Module called Spacelab started by an entrepreneurial startup company and manufactured in Italy by an early portion of the current Alenia Spazio Aerospace conglomerate. The space shuttle has launched the module and/ or their evolving components 22 times with the launch scheduled in the fall of 2007. Launched in this special commercial SPACEHAB Module are 60 or 80 individual lockers prepared and controlled by independent researchers with astronauts providing on orbit hands on services as required. This

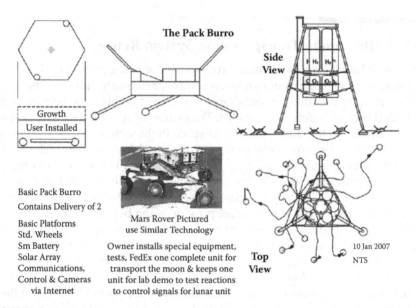

FIGURE 9.5
The Prospector's Pack Burro on the Moon's surface. Controlled remotely via the Internet, free to roam around and perform experiments.

cooperation between NASA and a small entrepreneurial company allows NASA to utilize the back half of the space shuttle payload compartment for other payloads. An estimated 2,000 such experiments have launched since the company started in 1983 and many of those 2 cubic foot experimental setups are ready to launch again. But the space shuttle has reached the end of its useful life and these experiments must transfer to Space Station and the Express Rack pricing and with a different launch vehicle and orbital accommodations.

If such an exploration package could be delivered to the lunar surface and landed effectively for use via the Internet, then what actions are required for its care and accommodation in this new environment besides transportation? First, it needs a name but it is not a mid-deck locker but has the same approximate size. It is not as cheap as an order of magnitude or two less than the traditional lander packages in cost. It does not just sit there on the surface, but goes out and prospects at the command and control of its "Prospector" on Earth. It does not sense the lunar environment with sensors, but travels out into the lunar environment and actually accomplishes things like prospecting and many other things at the direction of the owner. This platform needs locomotion, but rover technology should be able to provide a small platform with wheels and a solar array with some additional expense. In fact a whole list of equipment, sensors, solar cells and camera can be carried on the traveler rover for the prospecting of the lunar surface. Does positioning one's rover over a part of the Moon become a claim, because it is occupied? Well, that may not be entirely clear at this early stage.

Well, let's get back to a name. Maybe this little package is a bit like a miner's "burro." It carries equipment for the miner; it is packed by the miner with the needed tools and becomes an active assistant to the miner's activities. Could the "burro" come back to the LTS vehicle for food and water and reconfiguration; well maybe. Is there a limit to how far such a wireless "burro" could go; probably. If the burro came back to the "barn" could it be repaired, fed with new software and repacked? If the standard chassis were combined with standard bolt on components, it is realistic to think that such a feeding could take place at the "barn." Could universities all over the world have duplicates in their labs to test the software before it is fed to the burro; well maybe. Could students learn by doing and be inspired by the exercise; very likely. Could the "burro" ever come home to Earth; probably not for a while. It must work until worn out and wait to be picked up as a museum article for the university's museum.

What would such a "burro" look like? Not all burros would look alike. They have four feet and consume basic food and water, but each has a personality of its own. The Lunar "Burro" is likely to evolve to fit the lunar environment, much the way the Mars rovers evolved and became very adept at traveling across the Mars surface. A "burro" is the miner's friend and the miner has a mission, a life long mission of finding something of value that the miner can recover his or her investment in time, energy, creativity and money. The miner of the American West was a dreamer and explorer driven to find resources of value and to recover those resources for reward for the many years of effort.

Lunar Transportation Systems, Inc. proposes the "burros" for sale program. Figure 9.5 depicts the basic "Pack Burro" concept and outlines the basic platform and the anticipated hardware to be supplied for a price, which includes the one-way transportation to the lunar surface. The "Pack Burro" package includes a platform frame, some wheels and basic wireless communications, solar power and small batteries and a lifting eye. LTS expects to transport the "burros" to the lunar surface and place them on the regolith. The "Standby" ticket price on our vehicle is expected to reflect the mix of customers.

We don't know much about "burros," but other entrepreneurs do and we will help you sell your equipment in this out of this world miner's community. A platform with some wheels with communication gear with cameras might be a good start on a "burro." It has been suggested that small exploring and innovative unmanned devices operated by inventive creative people could establish a beachhead for resource miners on the Moon and the same technique can be used to explore the universe. Maybe LTS as an entrepreneurial company on the frontier of a new celestial body can be the "General Store" by helping to sell "gold pan like support equipment" and transporting "burros" to the region. Maybe LTS could get interested in setting up a corral, restoring the "burros" feet like a burro might get horse-shoes and some "hay" and new batteries, when they are hungry. This is not just our idea. We are not smart enough to think of all this by ourselves, but like the resource miner of the American West and countries all over the world, we see opportunity for great wealth in finding lunar equivalent to "gold" and the "burro" can help.

Reusability

A key feature of this Earth–Moon transportation system is that the two principal spacecraft, the lunar Lander and the Propellant Transporter are fully reusable. The lunar Lander transports payloads from LEO to the lunar surface and back. The Propellant Transporter transports cryogenic propellant tanks from LEO to any place in cislunar space where the lunar Landers need to be refueled. When approximately $100k per pound is spent transporting a vehicle to the lunar surface, then its value on the surface should be explored in all ways.

Shared Vehicles

Figure 9.6 is the LTS vehicle being serviced by a lunar service or utility vehicle. This surface utility vehicle is transported to the Moon on an LTS vehicle, assembled with a crew, uses various methods to make use of local materials and provides services and materials to other organizations as well as provides LTS with services. The Basic Frame of the surface vehicle comes without the mass required to provide the stability and non-tipping capabilities in the 1/6 gravity of the Moon. Rather, regolith and/or melted regolith are added to compensate for the massive counterweights, and other normal one gravity stability problems. The 1/6 gravity is different than one gravity in that it appears to increase the tendency to tip as a vehicle goes around a curve, because the mass holding it down on the road is greatly reduced. Crane counter weights need to be local materials, hopefully reduced in volume by melting. This counter weight mass can be local materials, but a revolution is required for the designers of the equipment. Local mass can be used to fill volumes of voids within the vehicle.

The utility and function of the "Live off the Land" vehicle uses Power Take Off Units (PTO) for each end, which save power from the basic vehicle, but add versatility and increased functions for the vehicle. The first vehicle will have to dig, plow, grade, microwave in place the regolith, carry large items like payloads, tanks, vehicles and building materials including lifting into place, plus some unknown functions yet to be determined.

The components of larger mining machines going to the Moon's surface must be small enough for LTS P/L bay, delivered in pieces and assembled on the lunar surface. The only real size constraint is the diameter of the originating vehicle delivering from Earth to an LTS vehicle in orbit. LTS expects

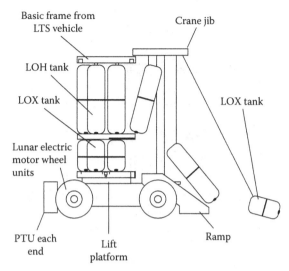

Basic frame from LTS vehicle

Crane jib

LOH tank

LOX tank

LOX tank

Lunar electric motor wheel units

PTU each end

Lift platform

Ramp

FIGURE 9.6
LTS uses utility vehicle for surface tasks.

this diameter to quickly expand to 10 meters or more with the advent of commercial resource recovery operations beyond Earth. Manned operation of the lunar surface utility vehicle is one of several methods of operation and would include a pressurized cabin with provisions for removal of exterior space suits and recharging of suit consumables plus a safe haven using the mass of the vehicle counterweight and frame as a radiation shield.

The remote control operation from Earth or anywhere in between is thought to be a part of surface vehicle hardware and capable of a graduated preplanned upgrade path of many segments. We expect a 30 to 50 year design life given the transportation cost.

The full reuse of hardware can be carried one step further to the component level. The Electric Wheel Units, for example, can be interchangeable and capable of repair on the Moon. Limiting the piece count and component count may seem to be limiting commerce and discouraging differences between commercial products from different companies, but reuse must, in the early development stages, take center stage for cost reasons.

A master mechanic, the usual innovator on remote construction projects shown in Figure 9.2, must be given a reasonable chance at keeping hardware operational and the more remote the base is the more difficult and expensive the "problem," even on Earth. Fasteners, wiring, tires, EVA suit items, batteries, solar arrays, containers, beds, reusable water bottles and a myriad of other well labeled logistics items fit into a general reuse policy of cost reduction using "Smart Logistics" techniques.

That is why lunar surface vehicles, like Earth vehicles, have hook locations, outriggers, tires of different types, weight shifting to increase traction and a

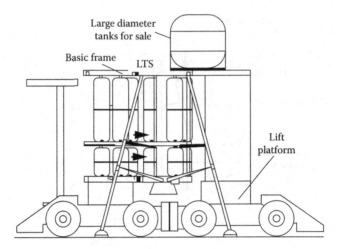

FIGURE 9.7
Surface utility vehicle loads payloads and propellant.

multitude of new challenges imposed by the lack of air, increased temperature extremes, long night operations, both manned and remote operations
and many more with all to be in the public glare. Special functions for the
possible LTS operations include a Platform Lift capable of reloading the LTS
vehicles with propellant and payloads as shown in Figure 9.7. Also needed
is the ability to reuse the LTS tanks and to collect and recover the gaseous
oxygen from the lunar miner.

Increased Payload Capabilities

The size of the payloads delivered to and from the Moon depends on where
and how many times lunar landers are refueled on their way to and from the
lunar surface. This system is capable of delivering 800 kg to the lunar surface
directly from LEO without the need to refuel in space. It is capable of delivering
3.2 metric tons to the lunar surface with refueling at L1 only. And it is capable
of delivering up to 10 metric tons to the lunar surface with refueling at MEO,
at L1, and in lunar orbit. Comparable payloads can be returned from the lunar
surface to LEO or to the Earth with refueling at one or more of those locations.

Cooperative Plants

Figure 9.8 is an early processing plant and again early operations are crude
and minimized until a market actually buys the product and services available. Then the commercial financing flows quickly to spend the money
required to meet the existing market, but upgradeable to more capability and
production, so as to lead or stimulate the early market and follow the market
with easier to raise private capital as the market grows or disappears.

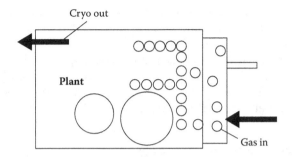

FIGURE 9.8
Surface propellant plant converts gaseous recovered resources to cryogenic propellants.

This means the NASA government funding of all up front budget expenditures are changed in the commercial world to the way most of the commercial world builds in a high cost remote environment. The plant basically converts gaseous oxygen stored and transported in reusable tanks into LOX in as efficient way as possible given the cost and schedule. The processing is thought to be easier if ice is present in significant quantities on the Moon and the plant can be planned and even conceptualized including transportation planning prior to the first trip to the Moon. This advanced thinking allows commercial space entrepreneurs the ability to raise the private capital and organize the transportation in parallel to early NASA efforts rather than in series and after traditional government space development.

The plant will require a lot of power, needs radiation protection and staff plus significant automation. This may be an opportunity for commercial sources to fall in love with a future market, develop the hardware, test the processes, determine the power source and provide the service plus related surface facilities and transportation with NASA encouragement in a Public Private Partnership.

The plan could include the developmental pull of He_3 into the venture rather than fighting for DOE budgets in an Earth environment where DOE has no real incentive to consider He_3 research. On the Moon the "Living off the Land" using He_3 becomes easier to sell and incentivize research to move in a different way. Incentivize in this meaning or case means to provide a motivation for moving in a new direction in a manner similar to the oil companies on the North Slope, who agreed to burn benzene in most equipment and power plants instead of a more expensive alternative gasoline and/or diesel. It was accepted practice in the Prudhoe Bay Oil Field that a small refinery would crack the oil on site and use benzene as a power source for almost everything. The cost was modifying the carburetors on the equipment and the saving was a small "cracking plant" refinery instead of a full refinery producing several blends of fuel. The decision for these commercial "big oil" companies was an easy one to make and had probably been made on every new remote oil field for the last few decades. Benzene is a colorless and flammable liquid with a sweet smell and a relatively high melting point. Benzene is also carcinogenic and when spilled tended to melt the threads of the Arctic Parkas before damaging the cloth itself.

He_3 is positive for health reasons and may be helped by the hard vacuum of the Moon. Research is a major economic driver to eventually use He_3 to power all lunar activities for the same reasons as benzene was used on the North Slope, availability and "living off the land."

Surface Transport Enhanced Use Leasing

Figure 9.9 depicts a Lunar Transportation Systems utility vehicle as a transportation service on the surface of the Moon using the surface utility vehicles and transporting empty, gaseous tanks and the full tanks (same tank, different mass and contents) of cryogenic propellant for a launch using resources recovered and processed into propellants on the Moon. Commerce continues into other commercial services financed by private money, because NASA and other space agencies see the value of saving their budget dollars and applying the techniques of Public Private Partnership agreements.

Other commercial services could include propellants for LTS use, "live off the land" services for company town, like oxygen, water, local transport services, camp construction and management including logistics, propellant for sale to others, incoming payload handling, return payload handling and other commerce.

Container Reuse

Figure 9.10 depicts containerized cargo. It took mankind 40,000 years of trade route development on Earth before containers evolved; why start at zero again? Traditional aerospace design includes an integration process between the payload and the vehicle, but no integration tie to vehicle can save money

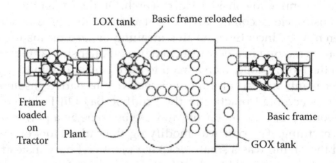

FIGURE 9.9
Surface utility vehicle transports recovered resources and production propellants to and from the plant.

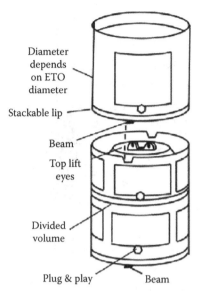

FIGURE 9.10
LTS container freight with reuse on several levels on the lunar surface.

and permit the most effective and affordable vehicles to be used on each different leg. LTS does not anticipate a complex integration process between the cargo container and the LTS vehicles except structural.

Our structural tie allows the quick transfer payload from one vehicle system to another, much like the containers transferred at ports, where a transfer occurs from ocean to land-based transportation. This changing of vehicle requirements also occurs at LEO where the ascent up from the Earth and through the gravity well requires a different vehicle. The no integration and standard container size is OK for 90% of the unmanned cargo and oversize equipment can be accommodated on a relatively affordable non-manned vehicle. In Alaska, for example, the human passenger and emergency cargo for them probably made up less than 1% of the total mass transported to the remote base and to pay the expense to fly everything would have been wasteful.

The container design could include some Special Features to assist crews on the Moon, including solar cells with batteries to provide a small amount of power for a plug and play content of the container output interface. Containers are customer sensitive and configured in many forms including liquid, dry, pressure, with hatches, cryogenic with mini coolers, and many others.

Special Mining Customers

Special customers are accommodated in a variety of ways including Figure 9.11, where the LTS basic vehicle is modified and is reused, because the LTS basic

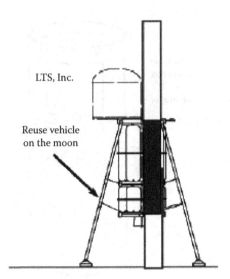

FIGURE 9.11
Special drilling accommodation on the LTS vehicle.

stack is used to land special drilling equipment. This service encourages innovation, permits a design to accommodate Drilling Operations and telescopes to cut the cost and increase functionality. Figure 9.12 depicts the drill operation using the LTS hardware as the initial structure and a mining system that uses a dragline to move regolith to a central processing plant. While this may be more expensive than processing regolith in place, it does provide a shaft operation, which is another option used by the mining industry on Earth. Figure 9.12 shows the Shaft Sinking/Drilling mining operation possible.

All LTS hardware needs commercial launch vehicles to launch each of the four types of LTS hardware from Earth to LEO. This first vehicle sets the diameter of the LTS hardware and can be on government vehicles, EELVs, ELVs, commercial vehicles and on vehicles yet to be created. The remaining passive minimum payload dispenser and the final stage are available for discard or reuse as mass for a propellant platform of other uses, because each pound of mass has $5k to $10k of valuable invested transportation energy in it in LEO.

To use this discarded item to replace something that must be launched at $10k per pound leads to cost reduction on a potentially significant scale. This discard mass with potential value could be viewed as the modern day equivalent of the land and mineral rights given to the Transcontinental Railroad builders as they built track. The track laying companies got every section of land they touched and in turn sold it for operating capital. Their customers were railroad customers, towns, cattle men and mining companies.

In the past the government has stimulated private investment with incentives. The Transcontinental Railroad, for example, gave two organizations land deeds and mineral rights for each mile of track laid. The issue was how

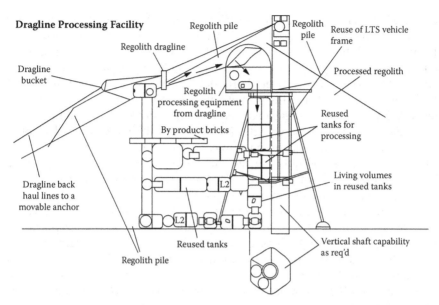

FIGURE 9.12
Reuse of the LTS vehicle structure telescopes into a shaft sinking/drilling mining operation.

to finance the railroad. The incentive worked so well that opposing track-laying crews went 200 miles past the other in Utah without connecting. In the 1840s western America was a remote wasteland between the Mississippi River and the West Coast. A gold rush in California helped make the railroad a reality. The railroad sold the land and the mineral rights to get the private investment to lay the track, buy the hardware and create future customers.

Conclusions

Commercial operations on the Moon's surface can happen and will happen sooner or later.[22] Lunar commerce can emerge in parallel with the President's Space Exploration vision, if government is prepared to join with commercial organizations to explore and implement realistic cooperative ventures benefiting both sides.

The Public Private Partnership may be the method of cooperation between government and commercial organizations, because both sides can benefit. Before private money flows into exploration activities, investors need to see an open ended "Financial Home Run" much in the way big oil developed Prudhoe Bay Oil field. Figure 9.2 depicts one method of mining and recovering lunar resources leading to the recovery of resources from the Moon. Figure 9.2 depicts a University of Wisconsin-Madison research concept[11] for recovering the lunar regolith by heating it to drive off the volatiles and recovering those volatiles for further processing and use in propellant sales and for use on the surface of the Moon. The recovered products can be near term consumables for the lunar outpost and later the propellants NASA requires to go to Mars and beyond. The regolith is excavated, continuously conveyed into the machine and heated to drive off the volatiles for recovery with the processed regolith returned to the surface and/or made into products for use on the Moon.

Figure 9.3 suggests a Public Private Partnership to stimulate private money flow into the recovery of resources. It is too early to predict the cost savings resulting from the use of lunar propellant, but for our company such production, once in place, could provide a 60% reduction in our operations cost, because propellant is a large part of the operations cost. Lunar Transportation Systems, Inc. is prepared to work with resource development organizations, willing to be the first customer for lunar produced propellant and abandon our vehicles on the surface to stimulate their reuse in surface facility cost reductions. Sharing the transportation from Earth to LEO with government may be the first step in cooperating on a large scale within a Public Private Partnership providing the underpinning of sustainable commerce supported by private investment.

Remote resource recovery bases on Earth have been on the leading edge of mankind's quest for resources on our planet and as we move beyond our planet, resource development is still a major driving force providing the economic and political sustainability for continued exploration.

The development of the Moon is like the North Pole. Early explorers want to plant the flag and gain prestige. When commercial developers came to the Arctic some 50 years later, they were interested in profits through resource development and recovery. They also brought $20 B in private risk capital in the 1970s and after finding 18 other oil fields they have invested several hundred additional billions coming out of profits.

References

1. Schmitt, Harrison, "Return to the Moon," Copernicus Books, New York, 2006, personal conversations, speeches and page 111.
2. Lunar Transportation Systems, Inc. Website, www.lunartransportationsystems. com With alternative 7 minutes of color early animation video at YouTube website http://www.youtube.com/watch?v=26Y5w0vqtIU.
3. Kistler, Walter P., "Conceptual Design of an Earth – Moon Spacecraft Fleet," Lunar Transportation Systems, Inc., Bellevue, Washington. January 15, 2004.
4. Kistler, Walter P., "The Design of Lunar Transportation Spacecraft," Lunar Transportation Systems, Inc., Bellevue, Washington, February 2nd, 2004.

5. Kistler, Walter P., "Some Design Details of the Propellant Dispenser, the Propellant Transporter, and the lunar Lander Spacecraft for an Earth–Moon Transportation System". Lunar Transportation Systems, Inc., Bellevue, Washington. Disclosure Document to the U.S. Patent Office, Washington, DC, February 10, 2004.
6. Taylor, Tom, "Lunar Transportation Spacecraft System Drawings for Patent Application". Lunar Transportation Systems, Inc., Bellevue, Washington, February 15, 2004.
7. Kistler, Walter P., Bob Citron, and Tom Taylor, "A New Earth – Moon Transportation System Concept," AIAA/NASA First Space Exploration Conference, Orlando, Florida, January 30th to February 1, 2005, in preparation.
8. Kistler, Walter P, Bob Citron, "Highway to the Moon," NASA Strategic Roadmap paper, Lunar Transportation Systems, Inc., Bellevue, Washington, December 2, 2004.
9. Walter P. Kistler, Bob Citron, and Tom Taylor, "To the Moon: Commercially," SOLE—The International Society of Logistics, 41st Annual International Conference and Exhibition, Dallas, 15–17 Aug 2006, also in condensed versions in Logistics Spectrum, Volume 40, Issue 4, Dec 06 ISSN 0024-5852, Lunar Transportation Systems, Inc. Bellevue, Washington 98004.
10. University of Wisconsin-Madison, College of Engineering, Advisor Gerald L. Kulcinski, Associate Dean for Research, Madison, WI 53706-1691, LTS support of Grad Student: Matthew Gadja, a second year graduate student in engineering physics at the University of Wisconsin-Madison. College of Engineering, Advisor Gerald L. Kulcinski, Associate Dean for Research, Madison, WI 53706-1691.
11. "A Lunar Volatiles Miner," www.nasa-academy.org/soffen, Matthew Gadja, a second year graduate student in engineering physics at the University of Wisconsin-Madison. Presented a talk on the design of a lunar miner at the International Conference on Engineering, Construction, and Operations in Challenging Environments in Houston, Texas. University of Wisconsin-Madison, Fusion Energy Institute.
12. Taylor, Thomas C., Four years on the North Slope of Alaska working four separate projects in the construction of the Prudhoe Bay Oil Field Facilities, 1975–1979.
13. Personal experience on the North Slope as a Civil Engineer and Supt. of crews in the assembly and construction of the Flow Station 1 and Flow Station 2 built in the lower 48 and assembled into a $1 billion facility to separate the seawater and natural gas from the oil.
14. Used six large packing crates for other structures such as tool sheds, crew rooms and lunch.
15. Public Private Partnerships, Public Private Partnerships, http://ncppp.org/, "Public Private Partnerships," "Reminders," The National Council for Public-Private Partnerships, 1660 L Street, NW, Suite 510, Washington, DC 20036, R. Norment, Exec. Director.
16. "Market Reform: Lessons from New Zealand", Rupert Darwall, http://www.policyreview.org/apr03/darwall.html, ONLINPolicy Review, 21 Dupont Circle, NW, Suite 310, Washington, DC 20036.
17. SPACEHAB, Inc., an entrepreneurial startup company in commercial space offering affordable access to microgravity via 22 space shuttle missions, www.spacehab.com, Start-up Team of Walter P. Kistler, Bob Citron and Tom Taylor, 1983, raised ~$300m in private funds, U.S. Patent 4,867,395, www.pat2pdf.org free copy.

18. Kistler Aerospace Corporation, an entrepreneurial startup company in commercial space offering affordable access to space via a 2 stage Reusable Launch Vehicle (RLV), www.rocketplanekistler.com, anticipated launch 2009-10 as winner of NASA COTS award of $207m, Start-up Team of Walter P. Kistler, Bob Citron and Tom Taylor, 1993, raised ~ $860m in private equity, U.S. Patent 5,927,653, 6,945,498.
19. Lunar Transportation Systems, Inc., an entrepreneurial startup company in commercial space offering potentially affordable access to the lunar surface via commercial logistics architecture, Start-up Team of Walter P. Kistler, Bob Citron and Tom Taylor, 2004, raised private equity, U.S. Patent 7,114,682, 7,118,077, 7,156,348.
20. Ball, J., "Public Private Partnership Opportunities at NASA," PPPs For Federal Properties Workshop, Washington, D.C. December 6, 2006, www.ncppp.org.
21. Taylor, T.C., "Maximizing the Commercial Space Transportation with a Public-Private Partnership," AIAA-2006-5883, AIAA Space Operations: Mission Management, Technologies, and Current Applications Book: Chapter 2 Space Operations, Revision of SpaceOps 06 Paper in Rome, same title.
22. Lunar Roundtable Conference, Las Vegas, NV, 17–19 July 2006, "Bridging the Gap From Earth Markets to New Space Markets" Report, Page 27 of 75, available at www.lunarcommerceroundtable.com click on "Event Archive" and "Roundtable."

10

Rocks to Robots: Concepts for Initial Robotic Lunar Resource Development*

Lee Morin
Astronaut Office, NASA Johnson Space Center
Houston, Texas

Sandra Magnus
Astronaut Office, NASA Johnson Space Center
Houston, Texas

Stanley Love
Astronaut Office, NASA Johnson Space Center
Houston, Texas

Donald Pettit
Astronaut Office, NASA Johnson Space Center
Houston, Texas

Mary Lynne Dittmar
Dittmar Associates, Inc.
Houston, Texas

Lee Morin, M.D., PhD, received a BS degree in mathematical/electrical science from the University of New Hampshire in 1974; an MS degree in biochemistry from New York University in 1978; doctorate of medicine and microbiology degrees from New York University in 1981 and 1982, respectively; a master of public health degree from the University of Alabama at Birmingham in 1988; and an MS degree in physics from the University of Houston–Clear Lake in 2009. His space flight experience is STS-110 Atlantis (April 8–19, 2002), the 13th Shuttle mission to visit the International Space Station, during which the crew delivered and installed the S-Zero Truss.

Sandra Magnus, PhD, NASA Astronaut received a BS in physics and an MS in electrical engineering from the University of Missouri-Rolla in 1986

---

*The opinions expressed are the personal views of the authors, and do not reflect current official NASA or U.S. Government policies or programs.

and 1990, respectively, and a doctorate from the School of Material Science and Engineering at the Georgia Institute of Technology in 1996. Selected by NASA in April 1996, Dr. Magnus reported to the Johnson Space Center in August 1996. Her space flight experience includes STS-112 Atlantis (October 7–18, 2002), an International Space Station assembly mission during which the crew delivered and installed the S-One Truss. On her second mission, she served as a flight engineer and NASA Space Station science officer for International Space Station Expedition 18 (November 2008–March 2009).

Stanley Love, PhD, NASA Astronaut earned a BS in physics from Harvey Mudd College, Claremont, California, in 1987. He attended the University of Washington in Seattle, receiving an MS in 1989 and a doctor of philosophy degree in 1993. Selected by NASA in June 1998, he reported for training in August 1998. His space flight experience is STS-122 Atlantis (February 7–20, 2008), the 24th Shuttle mission to visit the International Space Station, during which the crew delivered and installed the European Columbus Orbiting Facility.

Donald Pettit, PhD, NASA Astronaut received a Bachelor of Science degree in chemical engineering from Oregon State University in 1978 and a doctorate in chemical engineering from the University of Arizona in 1983. Selected by NASA in April 1996, Dr. Pettit reported to the Johnson Space Center in August 1996. His space flight experience is as follows. He completed his first space flight as Expedition 6 NASA Space Station Science Officer aboard the International Space Station (November 2002–May 2003). His second mission was aboard STS-126 Endeavour (November 14–30, 2008), the 27th Shuttle/ Station assembly mission.

Mary Lynne Dittmar, PhD, is president and CEO of Dittmar Associates, Inc. She received a BA in psychology and an MS in human factors psychology in 1980 and 1985, and a joint doctorate in experimental psychology and human factors engineering, 1989, all from the University of Cincinnati. Dr. Dittmar worked as NASA consultant in Human Factors between 1989 and 1995, while a faculty member at the University of Alabama in Huntsville. She managed Boeing Mission Operations and Astronaut/ Cosmonaut Training Integration for the International Space Station Program between 1995 and 2001, and carried out long range strategic planning for Boeing Space Systems between 2002 and 2004. She founded Dittmar Associates, Inc. in 2004, specializing in strategic planning, systems engineering, communications planning, training, and evaluation of emerging technologies. She has published more than 50 articles in science, aerospace, engineering, business, and the humanities, and is the author of *The Market Study for Space Exploration* (paperback; 2004, Dittmar Associates, Inc., Houston, Texas).

Introduction

Despite epochal exploration achievements a generation ago, lunar development has not yet materialized, and the vision of sustained colonization of other worlds has remained out of reach. In this paper, we propose a step-wise approach to the planning, design, development, building, and expansion of human/robotic presence on the Moon. The key concept underlying this approach is the use of bootstrapping—creating building blocks that in turn lead to more capability to make more building blocks—using local lunar resources to the greatest possible extent. We believe that this approach will yield lower-cost missions with high return—a productive materials processing capability that simultaneously generates increasingly more materials *and* more processing ability, leading to the rapid accumulation of construction materials and industrial manufacturing capacity. One of the main obstacles of reaching and re-supplying an off-world presence is the immense transportation cost, which is largely driven by the need to use impulsive chemical rockets to leave the Earth's gravity well. The problem is neatly described by the rocket equation, which determines the proportion of payload achievable given a desired velocity change. Contained in the equation is an exponential term which behaves in the same manner as compounded interest. At the high velocities required to establish and maintain an Earth orbit, the rocket equation dictates that vehicles consist of almost nine-tenths propellant, leaving only a small amount of up-mass possible for payloads. This fact, together with the need for precision machinery constructed of high grade metals and other exotic materials have contributed to the extreme costs of spaceflight.

The historical success of attempts to establish remote outposts has largely depended on the willingness and ability of explorers to utilize local resources and adapt to local conditions. The degree of dependence on distant motherlands inversely determined the long term viability of the outpost. Despite abundant or even lavish initial outfitting, expeditions not adapting to local conditions and optimizing the use of available resources have failed.

Today lunar exploration has become of prime interest to the increasing number of space-faring nations of the world.[1,2] The establishment of lunar outposts is just one of several goals planned or proposed by the United States,[3] as well as by other government and privately-held organizations.[4,5,6] Such remote outposts will provide many challenges, not only in technology, but also in sustainability and development. One hundred percent reliance on Earth resources will not only be prohibitively expensive, but will also ignore the historical outcomes outlined above. For a complex enterprise in such a remote location to be successful, lunar resources must be utilized.

With the abundant raw materials and sunlight available on the surface of the Moon,[7,8] exponential growth rates in materials and processing capacity can be attained. Done efficiently, development of the Moon will largely become an enterprise of transporting information rather than mass. By mainly using matter already on the Moon and developing infrastructure exponentially, we can overcome the limitations imposed by the unfavorable exponential of the rocket equation and eventually realize sustainable off-Earth presence.

We propose a series of low cost, unmanned, lunar robotic missions using standard expendable launch vehicles. We must carefully select material processing payloads and strategies that are scaled to be within the capabilities of the modest initial missions. Selected processes will be on a kilowatt-kilogram scale at most. They must utilize immediately-available materials and energy sources, and must immediately produce apparatus and knowledge that directly furthers the enterprise. The objective is to bootstrap our way to considerable lunar infrastructure and industrial capability without requiring new heavy Earth launch capability or waiting until manned outposts are established.

Some of the advantages to our proposal are:

- It results in permanent infrastructure on a stable platform (the Moon) that accumulates with "compound interest"—unlike facilities in Earth orbit, which are lost when orbits decay.

- It can be started quickly, and significant results can be obtained within the span of a few years.

- The technical risk is low. All the technical ingredients are well understood; it does not require a breakthrough invention.

- It can utilize existing expendable launch vehicles. No new launch vehicles are required.

- It offers tremendous near-term outreach, inspirational, and educational benefits.[9]

- It will stimulate graduate scientific and engineering education by providing ready access to the space environment for solving problems and developing technology for immediate space application.

- As the project develops it promises to make space very tangible to the public at large, increasing public appreciation of the value and relevance of our national investment in space.[10]

- Eventually many people will have a chance to experience lunar telepresence, which "puts you there."[11,12]

- It is affordable and is "pay as you go." Each mission is independent, and the cost of each is mainly that of an expendable launch vehicle.

- It is open ended. The more you do, the easier it is to do more.

- It fosters entrepreneurial and commercial space activity. The robots are on the Moon, but the business activity, knowledge and jobs are here on Earth.

- It opens a vast frontier the size of the continent of Africa. (The Moon has been called the "Eighth Continent.") People, especially children, will see a role for themselves there.

Establishing Robotic Telepresence

The first step in our approach is to establish robotic telepresence.[13] Dexterous robots controlled from Earth will be landed on the lunar surface and used to develop material science processes applicable to abundant lunar materials, particularly the lunar regolith.[14] Essentially we will be establishing a "lunar glovebox"—with the box on the Earth and the gloves on Moon.

Space-based robotics have been widely used.[15,16,17] Not only have we sent robots to distant worlds to explore and transmit data, but we also utilize dexterous, human-in-the-loop robots[18] as an integral part of the manned space program. By exploiting the experience gained over the many years of NASA robotic exploration and development and capitalizing on ever-advancing robotic technology, we plan on operating robots on the Moon by controlling them from Earth. Furthermore, the robots need not be as complex as top-of-the-line space or surgical robots, or previous NASA models. A simpler class of robots similar to industrial assembly robots, radioactive or explosive material remote handlers, or even a backhoe would be extremely useful on the Moon if operated by telepresence.

The operator, who sits in a control room on Earth, can view images down-linked in near real time, with a delay of only ~3 seconds required to transmit a signal round-trip from the Moon. Using virtual reality and feedback techniques the operator will feel immersed in the lunar environment; in essence be a "telepresence" on the lunar surface. This approach permits direct application of the full adaptive capability of the human mind to the remote task. Such a system allows for a much more adaptable and capable robot compared to one that is preprogrammed to execute tasks autonomously.

When complex robotic processes are required, telepresence robotic technology is currently highly advanced and capable of extreme dexterity.[19] Surgical robots are an excellent example of the ability to perform delicate tasks.[20] Medical systems are being developed that do not require the human controller, the surgeon, to be in the same room as the patient and robot.[21,22]

Of particular concern in such systems is the problem of latency. Studies are underway to understand the effects. In one case a mock vascular surgery has been performed remotely with latencies of nearly one second, with great interest in exploring even higher time delays.[23] Dexterity and tolerance of latency are relevant to our application. We seek to provide functional dexterity comparable to a human hand in a space-suit glove and to learn to work efficiently despite the ~3 second lunar latency. Techniques such as using graphical predictors of robotic motions displayed as images, together with tools and interfaces designed with latency in mind can optimize robotic control.

Robotic telepresence is central to the proposed plan. The use of telepresence robots will allow set-up, assembly, configuration change, and maintenance to be done remotely from Earth. With each subsequent mission additional telepresence robots will be commissioned. As the robotic presence on the moon increases, the cadre of experienced operators will also increase. Each phase will build on the previous phase as the inventory of tools, parts, stock, and know-how grows.

Applying Robotic Telepresence

The robotic telepresence will be applied to begin utilizing lunar resources to build a self-sustaining system. There are several key objectives that must be met in order to realize the concept of "exponential growth" in materials and materials processing. These include studying and understanding the bulk material properties of regolith in the lunar environment, developing appropriate processing techniques, and establishing a source of energy. Each goal builds on the previous one allowing development to proceed in an increasingly efficient and effective manner.

First, we must understand and learn to exploit the mechanical properties of bulk regolith in the lunar environment. We must be able to manipulate regolith mechanically. *Key capabilities include digging, grading, piling, entrenching, and tunneling.* We must devise techniques for digging into the hard packed material that is found below the first few centimeters of the lunar surface. In addition, we must be able to sort the regolith into components based on size and other properties.

We must also learn to exploit the electrical and magnetic properties of regolith so processes using these properties can be designed. For example, due to the presence of nanophase iron, the magnetic properties of regolith will be very useful. Regolith also exhibits similarities to abrasive materials used on Earth for grinding and polishing ceramics, glass, and other surfaces, and should find similar applications on the Moon. Finally, and of particular importance for ongoing operations, we must also create strategies to control the finest component of regolith, the lunar dust.

Although we have gaps in our knowledge, a great deal of information on the physical properties of regolith is already available as a result of Apollo and other programs. Using this information, various relevant material processes have already been developed. For example, if regolith is heated it melts and will fuse to form an opaque obsidian-like glass.[24,25,26,27] A key process that weakens glass is interaction with water molecules. However, because the Moon is so dry, lunar glass is expected to be much stronger than glass formed on Earth.[28,29]

Examples such as the one above demonstrate that the properties of regolith, in combination with the unique environmental characteristics of the Moon, offer tremendous potential for developing and testing technologies, materials and industrial processes leading to permanent habitation on the lunar surface. We have developed a series of key objectives—which may be thought of as milestones—to be accomplished early in the development of the program.

Key Objective 1: Melting the Regolith Using a Solar Furnace and Learning to Manipulate the Melt to Form Bricks, Bars, and Other Structural Elements

Because of the excellent vacuum on the Moon a very simple solar furnace can be built. It could consist of a light, metallized Mylar-type parabolic reflector focused on a crucible of regolith, or perhaps even a parabolic pit on the ground with a similar focusing method.

Key Objective 2: Sintering Regolith with Microwaves to Form Objects, and Characterizing Their Properties

Regolith contains nanophase iron which couples well with microwave energy, therefore microwaves can also be used to heat regolith efficiently.[30,31] If heated enough, regolith can be sintered with microwaves to form a solid.[32,33,34] This presents many possible applications including forming shaped objects, paving surfaces, perhaps even creating monuments of commercial value. It could also be the basis for a stereolithography-type process to form detailed objects.

Key Objective 3: Demonstration of Vapor Deposition Capability

The superb lunar vacuum allows vapor deposition to be a straightforward process. On the Moon, deposition can be performed without the overhead of vacuum pumps and chambers required on Earth. The reagents are heated and vaporized, and the plume is directly deposited on the substrate. Small quantities of reagents brought from Earth, such as aluminum, can be vaporized and deposited as thin films on surfaces to form coatings. Mirrors made in this manner can be used to concentrate solar energy or build additional solar furnaces. Likewise, metal deposition onto a spherical bowl-shaped excavation on the lunar surface can create an antenna dish, useful for enhancing communication with Earth. (The dish can be aimed without moving it, by positioning a feed horn like a miniature Arecibo.[35]) Vapor deposition of metals and thin films is an important industrial process for lunar development.

Key Objective 4: Demonstrate In-Situ Solar Cell Production Using Robotic Telepresence

Energy is a critical requirement. By applying semiconductor reagents (brought from Earth) as thin films on ceramic substrates, solar cells can be produced *in-situ*. Technical processes to produce solar cells in-situ from lunar materials have already been demonstrated in the laboratory on Earth.[36,37,38,39,40,41] Concentrated sunlight is used to melt and fuse the regolith in place, forming in place an eggshell-thin crust of glass on the lunar surface. This glass layer serves as the substrate for thin-film deposition of the semiconductor materials, forming solar cells. Thin film layers of only a few microns are adequate. Cells made in this manner yield efficiencies in the range of 6-10%. Even though these efficiencies are at the low end of what is technically feasible for photocells,[42] we have the vast surface of the Moon at our disposal, and in principle the solar cell area necessary to attain any desired power level can be achieved. The portion of such solar cells that must be brought from the Earth is small.

Key Objective 5: Demonstrate an Iron Producing Capability and Begin Scaling This Capability to Production Levels

One of the uses for an abundant source of energy is metal production. Iron is the metal most readily extractable from regolith. It is also the most useful single metal to have as a raw material due to its mechanical stiffness, magnetic properties, and the number of versatile alloys which can be made with it. The main drawback of iron, its tendency to corrode, will not be of concern in the dry, oxygen-free environment of the Moon.

Some iron (a fraction of one percent) is already present in regolith as reduced metal, namely as nanophase iron. Additional iron can be extracted from ferrous oxide (FeO) in the regolith. In the laboratory, FeO has been reduced to oxygen and iron by electrolysis at high temperatures, using a mixture of aluminum, calcium, and silicon oxides as a solvent—essentially the main components of regolith itself.[43,44,45,46,47,48] One drawback of this method is the corrosion of the anode caused by the hot oxygen created by the process. Since we are much more interested in producing iron than oxygen at this stage, we propose to redirect the hot oxygen to oxidize more FeO into Fe_3O_4. This disproportionation reaction, where FeO is simultaneously oxidized to Fe_3O_4 and reduced to iron, is well known.[49,50] The Fe_3O_4 produced is a potential feed stock for other uses including abrasive, magnetic and ferrite technology. In addition, bulk Fe_3O_4 provides for a very stable and compact form of oxygen storage. The oxygen can be recovered as water by heating with hydrogen in a reaction very similar to the well-known ilmenite process.

Once iron can be extracted at production levels, the stage is set for a broad array of follow-on technology including alloys, forming and machining techniques, powder metallurgy, electromagnetic machinery, and microwave and "vacuum-tube" electronic devices to name a few. As extraction activities scale up, additional regolith trace elements will become increasingly available, providing expanding opportunities for material applications.

Another process requiring energy that will eventually be developed is oxygen extraction from the regolith. However, handling oxygen as a compressed gas or cryogenic liquid is very difficult and requires complex precision machinery as well as strong, gas-tight vessels, seals, and plumbing to collect, compress, handle, and store the oxygen. While oxygen extraction is well beyond the initial capability of the proposed robotic development plan, a rudimentary demonstration may be possible during an early mission.

Key Objective 6: Set the Stage so Additional Telepresence Capability Can Be Produced *In Situ*

This is a longer term objective, but is the key to achieving the "compound interest" at the heart of this proposal. The energy and materials produced by the methods above are used first to supplement, and then gradually replace Earth content in subsequent telepresence facilities. By learning to turn regolith into telepresence workstations, or "Rocks to Robots", we close the circle to exponentially grow the most important commodity of our budding lunar information industry—lunar telepresence itself.

Early Construction

As the key objectives described above are progressively met, a source of building materials manufactured from local resources (regolith) will have been obtained. The development of basic construction techniques can begin. A series of key objectives concerns construction, including the creation of structures such as work surfaces, foundations, pathways, and fundamental architectural structures such as arches. These form the next "plateau" in our bootstrapping approach; once the mastery of rudimentary construction techniques is achieved, more complex projects can be undertaken.

One example of an early project might be the construction of a prototype lunar shelter, perhaps similar in shape to a miniature hanger based on a trench with paved walls. Once built, such a structure can be characterized for radiation shielding, thermal stability and other properties important for eventual human habitation. Additional regolith can be used on the outside of the structure to enhance desired characteristics such as radiation shielding and micrometeorite protection.

Another goal will be to learn how to build pressure vessels. To use pressure vessels we must also be able to produce and manipulate gases. Pressure vessels will be useful as prototypes for structures such as tanks, green-houses, and, of course, human habitats. Ideally before the first human lunar landing, we could have a decade of experience with lunar construction.

The First Mission[51]

The first mission will be a demonstration of fundamental technical capability. There are two main objectives. The first will establish a dexterous telepresence that can dig and manipulate regolith, assemble and operate apparatus, and process various samples. The second is creation of a regolith material processing lab that includes a solar furnace to bake and fuse regolith. This telepresence lab will also be able to characterize and analyze samples, microwave regolith, perform vapor and thin film deposition, make experimental solar cells, and perform experimental iron oxide reduction. Subsequent missions will build on these successes and extend these initial telepresence capabilities.

Summary

This proposal outlines a plan for the development toward a self-sustaining outpost on the Moon, and provides several key objectives that, if met, would see us well on our way to that goal. Starting with the creation of a telerobotic presence on the Moon and utilizing that capability to develop material processing techniques, the expansion of a lunar outpost will occur exponentially with minimal costs associated with transportation of materials from Earth. In addition, the missions can be scaled to fit within the available launch systems, thereby minimizing start up costs normally associated with new space projects.

Much of the technology proposed for use on the Moon is well understood and has been in use on Earth for decades; the challenge will lie in establishing and controlling our "lunar glove box" to apply these already mature techniques. With each success comes knowledge, experience, and materials useful in facilitating and accelerating the development and success of the next phase. We believe that such a modular approach, building on the success of previous missions, and utilizing the lunar resources is the most efficient and cost effective manner to establish and develop a permanent lunar presence. With Earth-to-Moon transportation costing many tens of thousands of dollars per pound, the value-add of lunar manufacturing is immense.

References

1. Lunar Enterprise Daily: http://www.spaceagepub.com/Daily.html
2. On India's First Lunar Probe: http://www.isro.org/chandrayaan/htmls/home.htm
3. On the President's Vision for Space Exploration: http://www.whitehouse.gov/space/renewed_spirit.html; http://www.whitehouse.gov/news/releases/2004/06/ 20040616-6.html
4. Eckart, P., *The Lunar Base Handbook—An Introduction to Lunar Base Design, Development, and Operations*, McGraw-Hill, 1999.
5. On the President's Vision for Space Exploration: http://www.whitehouse.gov/space/renewed_spirit.html; http://www.whitehouse.gov/news/releases/2004/06/20040616-6.html
6. Wingo, D. *Moonrush: Improving Life on Earth with the Moon's Resources*, Apogee Books Space Series, 2004.
7. Jolliff, B.L., Wieczorek, M.A., Shearer, C.K., Clive R. Neal, C.R. New Views of The Moon Mineralogical Society of America, *Reviews in Mineralogy and Geochemistry*, 2006.
8. Heiken, G., Vaniman, D., French, B.M., Schmitt, J. *Lunar Sourcebook: A User's Guide to the Moon*, Cambridge University Press, April 26, 1991.
9. Dittmar, M.L. Engaging the 18-25 Generation: Educational Outreach, Interactive Technologies, and Space. In *Proceedings of AIAA Space 2006*, September 19–21, San Jose, CA. Paper #AIAA 2006-7303. Washington, DC: AIAA, 2006.
10. Dittmar, M.L. Sustaining Exploration: Communications, Relevance, and Value. Parts 1 & 2. *The Space Review*, Nov. 12, 2007 and Nov. 19, 2007. Accessed at http://www.thespacereview.com/article/1000/1 and http://www.thespacereview.com/article/1005/1.
11. Craig, M.K. NASA's Value to the Nation: 50 Years of Lessons on Sustainability. In *Proceedings of AIAA Space 2007*, September 18–20, Long Beach, CA. Paper #2007-9931. Washington, DC: AIAA, 2007.
12. Dittmar, M.L. Gen Y and Space Exploration: A Desire for Interaction, Participation, and Empowerment. Paper presented at the AIAA 3rd Space Exploration Conference, Denver, CO, February 26–28, 2008. Accessed at www.nasa.gov/pdf/214675main_Dittmar.pdf.
13. Minsky, M., Telepresence, *Omni*, pp. 45–52, June 1980.
14. Regolith (Greek: "blanket rock") is a layer of loose, heterogeneous material covering solid rock. It includes dust, soil, broken rock, and other related materials. http://en.wikipedia.org/wiki/Regolith.
15. Heer, E. *Remotely Manned Systems: Exploration and Operation In Space*, California Institute of Technology, 1973.
16. Launius, R.D., McCurdy, H.E. *Robots in Space*, The Johns Hopkins University Press, 2008.
17. Space Automation, http://en.wikisource.org/wiki/Advanced_Automation_for_Space_Missions/Appendix_4D.
18. Robonaut references: http://robonaut.jsc.nasa.gov/; http://en.wikipedia.org/wiki/Robonaut.

19. Costello, A.J., Haxhimolla, H., Crowe, H., Peters, J.S. Installation of Telerobotic Surgery and Initial Experience with Telerobotic Radical Prostatectomy, *BJU Int.* Jul 2005; 96(1):34–38.
20. Rosen, J. and Hanaford, B. Doc at a Distance, *IEEE Spectrum NA* 39, October 2006.
21. Anvari, M. Remote Telepresence Surgery: The Canadian Experience, *Surgical Endoscopy* Apr 2007; 21(4):537–541, Epub 2007 Feb 6.
22. Anvari, M., McKinley, C., Stein, H. Establishment of the World's First Telerobotic Remote Surgical Service: for Provision of Advanced Laparoscopic Surgery in a Rural Community, *Ann Surg.* Mar 2005; 241(3):460–464.
23. These studies were performed on the NASA NEEMO 9 and NEEMO 12 missions, http://www.nasa.gov/mission_pages/NEEMO/index.html.
24. Kokh, P. Aboriginal Lunar Production of Glass, *Moon Miners' Manifesto* #96 June 1996, Section 6.9.3.2.096.of the Artemis Data Book.
25. Magoffin, M., Garvey, J. Lunar Glass Production Using Concentrated Solar Energy, AIAA-1990-3752 Space Programs and Technologies Conference, Huntsville, AL, Sept 25–27, 1990.
26. Allen, C.C. Bricks and Ceramics, *LPI Technical Report* 98-01, 1–2.
27. MacKenzie, J.D. and Claridge, R.C. Glass and Ceramics from Lunar Materials, Space Manufacturing III, *Proceedings of the 4th Princeton/AIAA Conference*, J. Grey and C. Krop, eds., pp. 14–17, Paper No. 79–1381 (May 1979).
28. Mendell, W. W. ed., Mechanical Properties of Lunar Materials under Anhydrous, Hard Vacuum Conditions: Applications of Lunar Glass Structural Components, in *Lunar Bases and Space Activities of the 21st Century*, Lunar and Planetary Inst., Houston TX, 487–495, 1986.
29. Carsley, J.E., Blacic, J.D. and B.J. Pletka, Vacuum Melting and Mechanical Testing of Simulated Lunar Glasses, in *Engineering, Construction and Operations in Space III*, edited by W.Z. Sadeh, S. Sture, and R.J. Miller, Amer. Soc. Civil Engin., 1219–1231, 1992.
30. Taylor, L.A., Hill, E., and Liu, Y. Unique Lunar Soil Properties for ISRU Microwave Processing, *Space Resources Roundtable VII*, Colo. Sch. Mines, ext. abstr. 2005.
31. Taylor, L.A. and Meek, T.T. Microwave Processing of Lunar Soil, *Proc. Int'l Lunar Conf.* 2003/Int'l Lunar Explor. Work. Grp. 5, Amer. Astronautical Soc. 108, 2004, pp.109–123.
32. Taylor, L.A. and Meek T.T. Microwave Sintering of Lunar Soil: Properties, Theory and Practice, *J. Aerospace Engr.*, July 2005.
33. Taylor, L.A., Schmitt, H.H., Carrier, W.D., and Nakagawa, M. The Lunar Dust Problem: From Liability to Asset, First Space Explor. Conf. Orlando 2005 AIAA 2005–2510.
34. Taylor, L.A., Hot-Pressed Iron from Lunar Soil, *Space Resources Roundtable II*, Colo. Sch. Mines, ext. abstr. 2000a.
35. Illingworth, V. and Clark, J.O.E. *Facts of File Dictionary of Astronomy* 4th Ed. Market House Books/Checkmark Books New York, p. 24, 2000.
36. Freundlich, A., Ignatiev, A., Horton, C., Duke, M., Curreri, P., Sibille, L. Manufacture of Solar Cells on the Moon, *Conference Record of the Thirty-First Photovoltaic Specialists Conference, 2005.* IEEE Volume, Issue, 3-7 pp 794–797, Jan 2005.
37. Ignatiev, A., Freundlich, A., Solar Cells for Lunar Applications by Vacuum Evaporation of Lunar Regolith Materials, NASA Report: IAF PAPER 92-0158, Accession Number: 92A55620; Document ID: 19920072996 (1992).

38. Criswell, D.R., Ignatiev, A. Production of Solar Photovoltaic Cells on the Moon, NASA Report: Accession Number: 91N26044; Document ID: 19910016730 (1991).
39. Ignatiev, A., Chu, C.W. Epitaxial Thin Film Growth in Outer Space, NASA Report: Accession Number: 89A51849; Document ID: 19890064478 (1988).
40. Ignatiev, A. Solar Cells for Lunar Applications by Vacuum Evaporation of Lunar Regolith Materials, NASA Report: Accession Number: 91N26047; Document ID: 19910016733 (1991).
41. Freundich, A., Ignatiev, A., Horton, C., Duke, M., Curren, P., and Sibille, L. Manufacture of Solar Cells on the Moon, NASA Report: Document ID: 20050110155 (2005).
42. Green, M.A. *Third Generation Photovoltaics-Advanced Solar Energy Conversion,* Springer-Verlag: Berlin, Heidelberg, pp. 3–4, 2003.
43. Curreri, P.A., Ethridge, E.C., Hudson, S.B., Miller, T.Y., Grugel, R.N., Sen, S., and Sadoway, D.R. Process Demonstration For Lunar In Situ Resource Utilization— Molten Oxide Electrolysis, NASA Report: NASA/TM—2006–214600.
44. Sadoway, D.R., Khetpal, D. From Oxygen Generation to Metals Production: In Situ Resource Utilization by Molten Oxide Electrolysis, In: *Materials Research at MIT Electrochemical Processing of Materials Research Reports 2003.* Sponsorship: NASA.
45. Sadoway, D.R. Towards Lunar Simulants Possessing Properties Critical to Research & Development of Extractive Processes, Lunar Regolith Simulant Materials Workshop, Huntsville, AL January 25, 2005, http://est.msfc.nasa.gov/workshops/lrsm2005_program.html.
46. Khetpal, D., Ducret, A.C., and Sadoway, D.R. From Oxygen Generation to Metals Production: In Situ Resource Utilization by Molten Oxide Electrolysis, Massachusetts Institute of Technology, http://amelia.db.erau.edu/nasacds/200307Disc1/research/20030060573_2003069287.pdf
47. Khetpal, D., Ducret, A. and Sadoway, D.R. Towards Carbon-Free Metals Production by Molten Oxide Electrolysis, TMS Meeting, Charlotte, NC March 17, 2004, http://web.mit.edu/dsadoway/www/MOE.pdf
48. Sadoway, D.R. Electrochemical Processing in Molten Salts: From "Green" Metals Extraction to Lunar Colonization, Department of Materials Science and Engineering, Massachusetts Institute of Technology, http://iusti.polytech.univ-mrs.fr/MOLTEN_SALTS/IUSTI_EDITION/70/edito70.pdf
49. Cornell, R.M. and Schwertmann, U. *The Iron Oxides* 2nd Ed. Wiley-VCH Weinheim, 2003.
50. Schwertmann, U. and Cornell, R.M. *Iron Oxides in the Laboratory* 2nd Ed. Wiley-VCH Weinheim, p. 13, 2000.
51. Morin, L. Rocks to Robots: A Biological Growth Approach to Rapid Lunar Industrialization, presented at NASA Ames Research Center Lunar Conference, June 2006.

11

Solar Cell Fabrication on the Moon from Lunar Resources

Alex Ignatiev and Alexandre Freundlich
Center for Advanced Materials, University of Houston
Houston, Texas

Klaus Heiss and Christopher Vizas
High Frontier
Alexandria, Virginia

Alex Ignatiev, PhD, is distinguished university professor of physics, chemistry, and electrical and computer engineering. Dr. Ignatiev is a graduate of the University of Wisconsin and Cornell University where he received his PhD in materials science in 1972. He is a former Fulbright Fellow, associate editor for *Vacuum,* and is the director of the Texas Center for Advanced Materials. He has been elected to the International Academy of Astronautics for his work in advanced materials development in space, and has been the recipient of the NSM Alumni Achievement Award, the Texas State Senate Recognition Award, and the City of Houston Science Recognition Award. He has developed the Wake Shield Facility space science payload that has flown three times on the Space Shuttle for the study of thin film growth in the vacuum of space. His research interests are focused on advanced thin film materials and device development and surface chemical interactions that form the basis for thin film growth. Recent efforts have been in the research of optical micro-detectors for artificial retina, thin film solar cells, thin film solid oxide fuel cells, thin film oxide resistive random access memory, and the fabrication of thin film solar cells on the Moon from lunar resources. He is the author of over 300 published research papers, holds 15 patents, and has been instrumental in the spinoff of five companies taking UH advanced materials technologies into the private sector.

ABSTRACT   The use of the indigenous resources of the Moon can result in the development of a power system on the Moon based on the fabrication of solar cells by thin film growth technology in the vacuum environment of the Moon. This can be accomplished by the deployment of a moderately-sized (~200 kg) crawler/rover on the surface of the Moon with the capabilities of preparation of the lunar regolith for use as a substrate, evaporation of the

143

appropriate semiconductor material for the solar cell structure, and deposition of metallic contacts and interconnects. This unique process will allow for the emplacement of a lunar electric power system that can reach a 1 MW capacity level in several years of crawler operation. This approach for the emplacement of an electric power system on the Moon would require the transportation of a much smaller mass of equipment to the Moon than would otherwise be required to install a complete electric power system brought to the Moon and emplaced there. It would also result in an electric power system that was repairable/replaceable through the simple fabrication of more solar cells.

Introduction

Energy is fundamental to nearly everything that humans would like to do in space, whether it is science, commercial development, or human exploration. If indigenous energy sources can be developed, a wide range of possibilities emerges for subsequent development. Some of these will lower the cost of future exploration, others will expand the range of activities that can be carried out, and some will reduce the risks of further exploration and development. This picture is particularly true for the Moon where significant electrical energy will be required for a number of lunar development scenarios, including science stations, lunar resource processing, and tourism. We present an approach to generate electrical energy on the Moon through the in-situ fabrication of thin film solar cells on the surface of the Moon. In supplying this electrical energy by in-situ fabricated solar cells, the costly transport and installation of an immense number of solar cells to support the energy need will not be required. The fabrication of solar cells on the surface of the Moon can be accomplished by the deployment on the Moon of a mobile solar cell fabricator, which will utilize the resources of the Moon to fabricate solar cells on location.

Lunar Solar Cells

The generation of electrical energy on the Moon can be undertaken through the fabrication of thin film solar cell directly on the surface of the Moon. The Moon has an ultra-high vacuum environment at it surface (~10E-10 Torr) and hence there is no requirement for vacuum chambers to undertake vacuum deposition of thin film materials and devices. The Moon's surface also contains all of the natural resources from which to fabricate thin film silicon solar cells: silicon, iron, magnesium, calcium, rutile, aluminum, etc., can be made available for thin film silicon solar cell production. The ultra-high vacuum environment at the surface of the Moon allows for the vacuum deposition of thin film silicon solar cells directly on the surface of the Moon without the need for vacuum chambers. As a result, thin film solar cells can be directly fabricated on the surface of the Moon through the integration of both a regolith processing step that is robotically undertaken to extract the needed raw materials for solar cell growth, and a solar cell vacuum deposition process undertaken by an autonomous robotic rover that lays down continuous ribbons of solar cells directly on the lunar regolith surface.

Regolith processing on the Moon to extract metallic and semiconducting elements needed for solar cell fabrications can be accomplished by a number of processes.[1-4] For silicon extraction, carbothermal reduction has been proposed for several abundant silicates, including anorthite ($CaAl_2Si_2O_8$) and pyroxene ($(Ca,Mg,Fe)SiO_3$). Anorthite is abundant in both maria and highlands rocks, pyroxene is most abundant in the maria. Most of these processes required a closed cycle process on the Moon to reduce the resupply of reagents from Earth. Electrolytic processing is also a possible approach to extraction of elements from regolith. Preliminary studies in the development of silicon solar cells from silicon extracted from simulated lunar regolith by electrolysis have been undertaken,[5] and show that such silicon can be used to fabricate thin film silicon solar cells through vacuum deposition.[6] It is well to note that although the electrolysis-processed regolith silicon was of moderate quality, i.e., not semiconductor grade, the vacuum deposition step for the thin film growth pre-purified the silicon through vacuum purification to yield moderate quality solar cells.[5]

Regolith reduction and silicon production however, are very energy intensive processes, and there is a number of competing reduction processes that can be applied to lunar environment. The major constraints for processes on the Moon are the need to encapsulate processes that use volatiles due to the Moon's vacuum environment, and minimization or elimination of materials/make-up materials to be brought from Earth (especially so for carbon containing materials that are not found on the Moon). With this in mind, magma electrolysis comes to the forefront as a viable method for application on the Moon. It requires no volatiles, and can utilize directly the regolith. Of principal concern for magma electrolysis is the need for stable electrodes and higher level of electrical energy. The former requires some R&D to optimize electrode performance, and the latter can be mitigated by the fabrication of lunar solar cells prior to deployment of the regolith processor. In such a "bootstrapping" approach the initial amount of material to be used for solar cell production on the Moon will need to be brought from Earth. This would then allow for the fabrication of an initial amount of thin film solar cells on the surface of the Moon (nominally ~50 kW capacity) the energy from which could then be utilized by follow-on missions which will focus on processing the regolith for materials extraction to continue to "feed" the thin film solar cell fabricators.

In a case where all of the source material may be brought from Earth, then a variety of semiconductor materials systems are available from which to

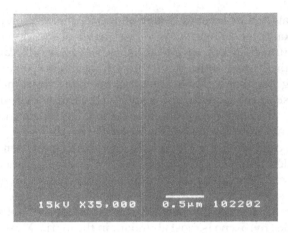

FIGURE 11.1
SEM micrograph of a thermally melted JSC-1 regolith simulant.

fabricate thin film solar cells. In such a case, it may be best to choose a system where the thin film solar cell fabrication power budget is as low as possible. However, for the purpose of fabricating thin film solar cell from in-situ materials focus needs to be maintained on the use of silicon, which is prevalent on the Moon, and thus the silicon solar cell structure becomes the structure of choice.

On Earth, silicon solar cells are not typically vacuum deposited on glass substrates but are principally fabricated from single crystal wafers not unlike those used for semiconductor device fabrication. However, as noted, the Moon possesses an ultra-high vacuum surface environment; hence vacuum deposition of silicon can be well used on the Moon. Terrestrially, thin film silicon solar cells, when vacuum deposited, are typically deposited in crystalline form on single crystal substrates. Vacuum deposition on glass is problematic due to atomic disorder in the grown films, and when cells are achieved, they typically have low efficiencies (~3-5%). This, however, may be acceptable for the lunar environment since large areas of low efficiency solar cells can be fabricated on the Moon to give the required TOTAL power needed for lunar use, i.e., quality can be traded off for quantity.

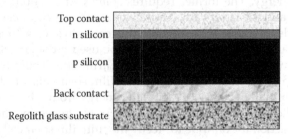

FIGURE 11.2
Schematic cross-section of lunar solar cell.

FIGURE 11.3
Artist's drawing of Cell Paver fabricating thin film solar cells on the surface of the Moon.

For a thin film solar cell, the substrate presents the most massive part of the structure. A common terrestrial substrate for thin film cells is glass. Nominal SiO_2 glass is not readily available on the surface of the Moon; however the lunar regolith can be melted to form a glass that is quite suitable as a solar cell substrate. The melted regolith simulant (JSC-1) exhibits a resistance of greater than $10^{11}\Omega$, and shows a smooth surface morphology consistent with good substrate material (Figure 11.1).

A typical lunar solar cell structure is shown in Figure 11.2, where melted lunar regolith is the substrate, a metallic back contact layer is evaporated onto the regolith substrate, p- and n- doped silicon is then evaporated onto the contact layer, a top contact is deposited onto the silicon through a contact mask and an antireflection coating can then be deposited on top of the whole cell. It has been shown that in addition to being a very favorable substrate layer for silicon solar cells, evaporated regolith has excellent optical transmission properties and can be used as an antireflection coating for enhancement of fabricated solar cell efficiency. In this manner, all the components of a thin film solar cell are available for fabrication of thin film silicon solar cells on the surface of the Moon.

These lunar solar cells would be fabricated by a facility deployed on the surface of the Moon. A movable "crawler"—a Cell Paver, of ~150 kg mass would traverse the lunar surface depositing solar cells[7] as part of the traverse (Figure 11.3). The Cell Paver wheeled vehicle could clear larger rocks and boulders from the terrain directly in front of it, thus preparing a bed for the fabrication of the lunar glass substrate onto which the solar cells would be directly fabricated by vacuum deposition.

The thermal energy required for each set of the evaporations in the above process would be obtained from direct solar energy collected by an array of small parabolic concentrators used to focus the solar energy and couple it into fiber optic bundles routed over the Cell Paver to whatever location it is needed. A large array of concentrators can be made so that the source material evaporation and regolith melting can be effectively undertaken. The fiber optic bundles would first focus energy to regolith melting to form the regolith glass substrate. They would then be moved to a metal evaporator for bottom electrode deposition. This would be followed by silicon (p- and n-doped) evaporation, top electrode evaporation, antireflection coating evaporation, and finally metallic interconnect evaporation to form the solar cell depicted in Figure 11.2.

The in-situ solar cells would be fabricated while the Cell Paver is moving. Individual cells would be connected in alternating series/parallel fashion by the deposition of thin film metallic strips (wires), and thus the interconnected cells would form arrays. In this manner, the crawler could continue to migrate over the lunar surface (maneuvering around large obstacles) and continuously lay down solar cells on an undulating landscape.

It is projected that in the initial version of the Cell Paver, silicon solar cells of ~1 m width and ~5% efficiency would be deposited on the lunar surface at a rate of approximately 1 m2 per hour giving the Cell Paver a motive speed of about 1m/hr. The cells would be integrated into a power system with both built in by-pass transistors also grown by thin film methods as part of the solar cell growth, and periodic array-grouping junctions which would be connected to a power management and distribution system to be brought from the Earth.

As noted, the initial set of lunar solar cells could be fabricated from raw materials brought from the Earth. Approximately 20 kg of raw materials

FIGURE 11.4
Artist's drawing of the Regolith Processor which would extract raw materials from the regolith by magma electrolysis.

would be required for the fabrication of ~50 kW of thin film solar cell electric power capacity. Beyond this amount, additional raw materials would need to be extracted to feed the Cell Paver. A Regolith Processor (Figure 11.4) would be required for this purpose. Noting that the initial run of the Cell Paver resulted in the fabrication of ~50 kW of solar cells, this power capacity could be used for regolith processing as well as follow-on development of the lunar site. The Regolith Processor would then be a second robotic vehicle[8] deployed on the Moon at the lunar solar cell production site. The Regolith Processor of ~150 kg mass would use power from the fabricated lunar solar arrays to process the lunar regolith to yield up to 200 kg of raw materials per year including silicon, iron-silicide, and aluminum among others. These raw materials would then be supplied to the solar cell crawler to continuously fabricate large numbers of silicon-based solar cells. These two vehicles would comprise the initial facility for the development of a lunar solar power system. The projected yield for this two-facility system is ~1MW capacity of solar cells fabricated over a 5-year period. It is clear that a series of such facilities deployed on the Moon could result in the generation of a significant amount of electrical energy on the surface of the Moon.

It is well to note here that this energy scenario for the Moon recognizes the economics of energy in space in that taking only the tools to the Moon to fabricate solar cells can be much more cost effective in the long run than just bringing the solar cells themselves and erecting them on the Moon. An initial cost analysis of the two scenarios indicates that at the current price of ~$1,000/W for space solar cells, and current projected lunar launch costs of ~$150,000/kg, the economic break-even point in bringing or fabricating solar cells on the Moon is approximately 125–150 kW. It is well to know, however, that this assumes only one-half year of operation for the Cell Paver/Regolith Processor. The projected lifetime for the Cell Paver and the Regolith Processor (based on past performance of deployable planetary robotic vehicles) is ~5 years. Hence, the deployment of the Cell Paver system for a near-term need for ~125 kW of electrical power capacity will result in a total of ~1 MW capacity after the five year operating period—a major benefit for timed expansion of a lunar base.

The longer-term development of lunar solar cell power systems would be driven by several factors including expanding needs for lunar electrical energy, energy needs for sis-lunar space, and electrical energy needs on Earth. The latter electrical energy need—energy for Earth—is one that will require a significant amount of lunar solar cell fabrication. To support this, a second generation of Cell Pavers and Regolith Processors would need to be developed that would benefit both from improvement of processes gained from first generation facilities operation on the Moon, and from advancement of the science and technology of thin film silicon solar cells.

Preliminary studies have indicated that thin film silicon cell efficiencies could reach up to 10% through the optimization and control of grain size in the thin film silicon to consistently yield grain sizes larger than 2 μm.[9]

The integration of higher efficiency thin film silicon solar cells (> 8%) with the ability to fabricate them at an increased rate (> 6 m^2/hr) would yield ~2 MW/yr growth capacity per Cell Paver II. One hundred of such second generation Cell Pavers deployed on the Moon could therefore result in 200 MW capacity of lunar solar cells fabricated in one year, and 1 GW of lunar solar cells fabricated in a 5-year period. The existence of such a large solar energy capacity on the Moon can now significantly impact the electrical energy environment on the Earth. Transporting 1 GW of electrical energy to the Earth from the Moon can be accomplished by energy beaming technologies. Both microwave and laser energy beaming have been discussed and proposed previously, especially in light of recent comments on space solar power generation and power beaming from artificial satellites containing massive solar cell arrays.[10] Such artificial solar cell satellite concepts are more complex and costly as compared to using the Moon and its resources as a solar power satellite, not only because it is an exceptionally stable platform on which to collect and transmit energy, but also because it is a platform on which the solar cell arrays can be built using in-situ resources. Even though a good deal has been said about microwave and laser power beaming, much is still needed in the technology development to realize high efficiency power beaming over large distances and at high power levels. The communications industry has done much to move the power beaming technology forward. However, their interests are to principally transmit a microwave signal to a widely dispersed area on the surface of the earth. For power beaming this concept needs to be inverted to address the transmission of a significant amount of power to a localized spot on the surface of the Earth. Progress in this arena is expected, with the promise of near-term realization of efficient and effective power beaming over large distances. This coupled with the ability to fabricate immense solar cell capacity on the Moon will enable a new lunar electrical energy source for Earth consumption.

Conclusion

The ability to "live off the land" brings a new paradigm to space exploration and utilization for the space programs of the world. Electrical energy will be required everywhere man or robots go in space. The development of space-fabricated solar power as described here can significantly lower the costs of operating in space, and will provide an energy-rich environment where rapid expansion of space activities can be undertaken. Furthermore, expansion of the scale of lunar solar cell array development to realize GWs of cell capacity can enable energy beaming to the Earth to help alleviate a portion of Earth's energy problems in the near future.

Acknowledgments

This work was supported in part by the NASA Cooperative Agreements NCC8-239, NAG9-1287, CAM through the State of Texas, and the R.A. Welch Foundation. The contributions of Drs. Michael Duke, Laurent Sibille, Peter Curreri, Sanders Rosenberg, and Charlie Horton are gratefully acknowledged.

References

1. S. D. Rosenberg, G. A. Guter, and F.R. Miller, "The On-Site Manufacture of Propellant Oxygen Utilizing Lunar Resources," Chem. Engr. Prog., 62, 228, (1964).
2. D. Bhogeswara Rao, "Extraction Processes for the Production of Aluminum, Titanium, Iron, Magnesium and Oxygen from Non-terrestrial Sources," Space Resources and Space Settlements, NASA SP-428, (1979).
3. C. Knudsen and M. Gibson, "Development of the Carbotek Process for Lunar Oxygen Production," Engineering, Construction and Operations in Space II, American Society of Civil Engineers, 357, (1990).
4. S. D. Rosenberg, P. Hermes, and E. E. Rice, "Carbothermal Reduction of Lunar Materials for Oxygen Production on the Moon," Final Report, In Space Propulsion, Ltd., Contract NAS 9-19080, (1996).
5. A. Ignatiev, T. Kubricht, and A. Freundlich, "Solar Cell Development on the Surface of the Moon," Proc. 49th International Astronautical Congress, IAA-98-IAA 13.2.03, (1998).
6. A. Ignatiev and A. Freundlich, "Lunar Regolith Thin Films: Vacuum Evaporation and Properties," AIP Proceedings 420, 660, (1998).
7. A. Ignatiev, A. Freundlich, C. Horton, M. Duke, S. Rosenberg, S. Carranza, and D. Makel, "CETDP Final Report," NASA (March, 2003).
8. A. Ignatiev, A. Freundlich, M. Duke, S. Rosenberg and D. Makel, "The Fabrication of Silicon Solar Cells on the Moon using In-Situ Resources," Proc. Intl. Astro. Cong. IAA-00-IAA 13.2.8, (2000).
9. Final Report NASA SBIR #NNM04AA65C, "Lunar In-Situ Fabrication: The Manufacturing of Thin Film Solar Clle on the Surface of the Moon," (July, 2006).
10. See as Example: Special Issue of Ad Astra, "Space-Based Solar Power," (Spring 2008).

Section III

Outer Space Habitat Design

12

Multidisciplinary Approach for User Reliability

Irene Lia Schlacht

Cand. PhD Human Machine System
Dipl. Industrial Designer
Technische Universität Berlin

Irene Lia Schlacht is an Italian designer born in Milan. After various experiences and several publications on the outer space field, she is now studying for a doctorate at the Technische Universität Berlin, under Prof. M. Rötting. The doctorate's title is "Visual Design for Outer Space Habitability." She earned a design master's degree in 2006 from the Politecnico di Milano; her thesis was titled "Colour Requirement in Outer Space Habitats" (Prof. D. Riccò). She was also an outer space human factor researcher at Università di Torino (Prof. M. Masali). After completing a report on outer space color design during her stage at Thales Alenia Space Italy, she directed an experiment on color perception in microgravity, taking parabolic flights with the European Space Agency (ESA) in September 2006. The goal of her research is to improve the working conditions of astronauts using design and ergonomics. Today, together with experts in several fields from several countries, including the Japanese space artist Ayako Ono, the Italian space anthropologist Prof. Melchiorre Masali, and the American Psychologist Scott Bates, she has created an outerspace research group called www.Extreme-Design.eu. Cand. PhD Human Machine System, Dipl. Industrial Designer, Technische Universität Berlin, Department of Human Machine System.

Introduction

A house is a machine for living,...one uses stone, wood, cement, and turns them into houses or palaces; that's constructions, call it for a skill. But, suddenly, you touch my heart; you make me feel good. I'm happy. I say it's beautiful. This is architecture. It is Art.

(Le Corbusier cit. Guilton, J., 1981, p. 17, 18)

ABSTRACT   The space habitat can be improved, to improve the living conditions and the safety of the astronauts, feeding-up human sensitivity, well-being and happiness with ergonomics, design, art and psychology.

Space habitats are artificial ecosystems designed for human space missions. With the current technology a mission to Mars will take about 3 years. Throughout that time the astronaut crew will be confined in their artificial habitat. In such long duration confinement the habitat design becomes a priority in order to guarantee the mental and physical well-being among space-travelers.

A multidisciplinary approach to Human Centered Design will be the correct methodology to build an artificial ecosystem able to give healthier support to the human life and its cultural expression.

Human factors, design, art and psychology should be considered from the preparatory stage of the habitat design in order to obtain a product fully adapted to the humans needs.

In this group of essays, experts from different fields suggest their viewpoints, solutions and designs to increase the habitability of outer space.

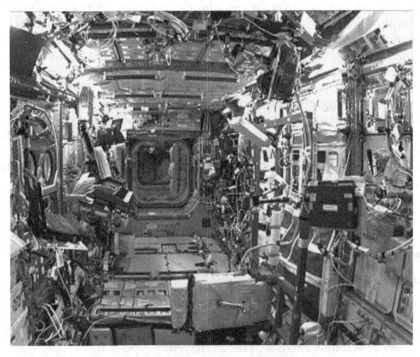

International Space Station, © NASA.

The Space Habitat

Looking to create habitats for long duration space mission, we must consider the thinking of Wright: "The nature of the design therefore should be something hand methods could do better than machinery. It was impossible to say how far we could go in any direction with machines, probably not very far." (Wright, F.L., 1955, p. 302)

The Space habitat is an artificial ecosystem with the maximum of self-sufficiency. Its aim is to sustain human life in the space hostile environment. In the ISS (International Space Station), for example, water is recycled with the earth ecosystem concept: from the wet wipes to the urine of the astronauts all the water of the station is filtered and lastly reprocessed also as drinkable water.

Now these habitats are far from being recognized as being comfortable, pleasant and fully efficient. People may mistakenly believe that the ISS is close to the *Star Trek* spacecraft, but that's not the reality today. The concept of "Space Habitability" and "Space Habitat Design" now are only applied to minimum human survival. A reflection on the goals of long time duration missions (like journeys to Mars or the lunar settlement), or even the increasing interest in the new activity of space tourism, leads to the obvious conclusion that space accessibility has to be improved.

"Outer space habitat design" is one that is focused on human factors and to increase the happiness of the user. The space habitat is an HT (high technology) complex habitation that at present is nearer to a machine and computer hardware than to a human habitation. User-machine interaction in a space habitat has to be considered from both sides: user and machine. The functionality of the human habitat has not to be misunderstood with one of the machine: to increase accessibility to the outer space extreme environment. We must also plan for the safety and the psychological well-being of the user. Living confined in an artificial environment without natural stimuli can change our needs and sense perception. Reality can be perceived differently and most importantly, we can have problems due to sensory deprivation. We need to stimulate the senses with a variety of stimuli, to maintain our consciousness, against the constancy of an artificial environment.

To understand the needs of outer space habitability, we can think of the space habitat as an artificial ecosystem like an aquarium, where all the elements placed inside are the product of human choice and forethought with the aim of obtaining a completely self-adequate environment. Each element is planned to sustain the equilibrium of the artificial ecosystem. Thus the question: Is the happiness of the fish part of the equilibrium? How can we fly into the Universe without Art, Nature and Design?

This paper considers not merely the economic factors (as an increase of the user's productivity), but also to remember that in terms of the interaction between humans and computers there is still a human being who needs to feed his spirits and needs to be happy after taking his eyes off the computer's monitor. That is the basis for the ability of doing good work.

Outer Space Habitat Design has as its goal not only to protect life but also to increase the comfort and the well-being of the crew, designing a habitat considering human factors such as feelings "to the advantage of greater productivity and higher level of the quality of life on board" (Dominoni, A., 2002, p. 33) and also for human cultural expression.

To present a different solution for the astronauts' psychological well-being, the following questions will be examined in this paper:

- What happens to the human being in the weightless Outer Space habitat?
- Which are the needs and the problems that have to be considered in a confined environment?
- Why the design of visual and perceptive factors has primary importance?
- How natural stimuli (plants, solar cycle, seasons, etc.) can improve human well-being?
- What is the role of Art in the enhancement of astronaut's well-being?

Methodology

Psychological, social, technical and mission-related aspects which are closely connected to the habitat architecture are further amplified in isolated and confined environments and can endanger the whole mission. With an incorporation of these issues into the design, stressors can be decreased or eliminated.

(Aguzzi et al., 2006, p. 4)

Multidisciplinarity

Outer Space Habitat Design is proposed here as a subset of Extreme Design that incorporates and connects multidisciplinary topics from the early phase of the design. The multidisciplinary approach is a methodology with the aim of creating the basis for mental and physical well-being in the space environment. The main disciplines considered are Anthropology, Architecture, Art, Communication, Design, Ergonomics, Philosophy, Physics, Physiology, Proxemics, Psychology and Semiotics, all of which are applied to the space environment. In this group of papers, different specialists from different fields and different countries present their designs, experiments and hypotheses in a way to give specific solutions on how to increase habitability in space.

Summary of This Group of Papers:

1. Melchiorre Masali, Marinella Ferrino: "Outer Space Anthropology"; human bio-cultural adaptation in space environment, human factors for the space habitability.
2. Irene Schlacht, Henrik Birke: "Outer Space Visual Design"; light color and biological stimuli to increase space habitability.
3. Ayako Ono: "Outer Space Art"; sound, Zen and art as natural elements for the astronaut's psychological support.
4. Scott Bates, Joshua Marquit: "Outer Space Psychology"; plant as natural stimuli to increase space habitability.

As a conclusion is presented "Open Space Technology", a methodology for the application of the multidisciplinary approach to space habitability design.

Approach

> Nature alone is truthful; it can inspire man-made works. But do not treat Nature as landscape painters do and show only his outward appearance. Search for the cause, the form, the animating spirit of things and synthesize this in the ornaments that you design.
>
> **(Le Corbusier cit. Guilton, 1981, p. 114)**

Natural Design has not its purpose to copy the habitat system present in our mother planet, but to synthesize the stimuli that can guarantee the human to be a terrestrial being also in space (not a space hybrid or human machine hybrid).

Outer Space Habitat Design is approached in a multidisciplinary way based on a Natural Design philosophy. This approach is not everlastingly linked to the re-creation of the natural mother planet system but by the use of bio-stimuli to guarantee the life, and the logic of the natural habitat context.

Natural Design is meant to have:

- Eco-mimesis[1] stimuli: re-creation of terrestrial psycho-physiological stimuli that humans as terrestrial animals need in order to have a satisfactory quality of life.
- Eco-mimicry: the habitat has to be developed not with Earth logic but with the logic of the natural context.

---

[1] Eco-mimesis: Eco from Greek "environment," "milieu" and Mimesis "simulation." Is a term to define milieu inspired Design, as a process of innovation for the habitability of extreme-environment. It involves mimicking the environmental stimuli to produce an ecosystem able to guarantee life needs in extreme environment. In common with Eco-mimicry it takes into account the local extreme environment characteristics. It can be applied to the outer space human habitability, it considers the Earth eco-stimuli (in which human species evolved) as a need for the survival in non terrestrial and/or artificial conditions. (Copyright: Melchiorre Masali, Irene Lia Schlacht, 28-11-2007, www.Extreme-Design.eu)

The basics of terrestrial stimuli that we need are: Variation and Variability, to obtain a state of consciousness, opposite of the inanimate artificial stimuli that create a sense of deprivation and a feeling of drowsiness.

Quoting Birren (2006), an artificial environment should stimulate the sense perception as it happens in Nature through the variation of color, light, sound, temperature, etc. The static condition is unnatural and perishable (Schlacht et al., March 2005).

As Ono, Schlacht, Masali stated at the 2007 International Astronautical Conference: "Natural Structures could be created to guarantee human well-being in an extreme context, such as a long duration mission on the Moon" (Ono et al., 2007).

In order to enhance the efficiency and well-being of the crew, in terms of the design of colors and interior décor, it is therefore necessary to recall the normal physical and psychical conditions whose characteristics are variety and variability in time (Bretania 2003).

Assumptions

Outer Space Habitat Design is concerned with space traveling[2] and space settlement of conscious humans as users, taking into account the achievement of a normal life during and after the outer space experience. In the following papers, authors from different disciplines discuss their views on increasing space habitability on the basis of the following specific assumptions.

A. Outer space habitability is of primary importance in specific context:

Long duration missions (L.D.)

Short duration missions that need quality of life (e.g., Mission tester for L.D.)

Space tourism

Outer space cultural expression

B. Habitability must be design for a conscious user (e.g., the space traveler should not be in hibernation)

The user has to experience reality and is able to express himself on it;

The user needs to have a good quality of life to be able to create and be motivated;

The user needs to know the importance of mother planet protection;

---

[2] This concept is also present in Asimov's Laws of Robotics.

C. The outer space User is natural:

   The user is not genetically manipulated;

   The user is not under drug effects;

   The user is not artificially or surgically modified.

D. The habitat should guarantee psycho-physiological well-being during and after the space mission. The following problems have to be solved in habitat design:

   Mental problems (psychological awareness and responsiveness within the space context);

   Physical problems (weightlessness and gravity difference effects, radiation influence);

   DNA modification (radiation exposure, attitudinal selection of pre-existing traits);

   Possibility of normal life on Earth after the mission;

   Outer space environmental risk for the habitat safety (micrometeorites, radiations, crew psycho-physiological reliability, etc.).

Confutation of the above mentioned assumptions could be the basis for other future and different designs. Projects not based on this assumption cannot be related to Space Habitat Design as presented in this paper. The foreseen speculations related to habitat design may appear to some extent science fictional, for example: the hypothesis of complete hibernation during space travel, so in this case there won't be need for a habitable habitat but only for a shield and medical care robotics.

Also the hypotheses (against Assumption C) of artificially creating a human being able to live in space. This has already been seriously considered; one of the most interesting suggestions supposes a brain without bone and muscle that doesn't need artificial gravity to survive in long duration mission.[3] This assumption is not against the use of artificial gravity, but rather taken as an assumption that the weightlessness effect on the human body has to be resolved in the way that the human being can be able to return to his normal terrestrial life after the space experience.

For example, the use of Virtual Reality can be matched with the habitat design concept as a support for the real habitat facilities; however it cannot be used as a total substitute for the real habitat dimension[4] because it may

---

[3] This hypothesis has been made during the Space Training meeting in Kusadasi, Turkey April 2007. As a part of the space training part of the group created a fantascientific society on the moon called "Ay xx". This invented society was guided by a brain without body, able to live without gravity on moon orbit.

[4] The use of virtual reality as a total substitute for the real habitat dimension has been used as a basis of fantascientific stories such as the Matrix. In this movie, people are living in a special amniotic liquid, and connected in a completely programmed reality.

have influence on the reality awareness (Assumption B). However, the main assumption for the space habitability is that the space habitat can be trusted to protect the human being as a natural traveler.[5]

Glossary

In this set of papers, particular considerations are given to the following specific disciplines. Short definitions of the kind of philosophical approaches are presented first:

Extreme Design: It is a branch of Design with a multidisciplinary approach that intends to increase the life quality in the extreme framework at the limits of the human survival, finding design solution and supporting cultural expression (www.extreme-design.eu, June 2007). It creates designs proficient to solve the psycho-physiological stress factors created by:

    a. physical problems (radiation, different gravity or weightlessness and body adaptation)

    b. psychological conditions: distance from mother Earth, feeling of life risk (cause: space dust, meteorites…technical break down), stress.

    c. confinement problems: dimensional (interior spatial restriction), temporal (mission time not easy to be re-schedulable), social (groups dimension limitation), biological (abstention from the natural earthly stimuli).

Outer Space Extreme Design: multidisciplinary approach intended to find design solutions for outer space extreme framework. It is the application of extreme design at the maximum level between the extreme contexts: in outer space, however it takes all the typologies of psycho-physiological pressure of the confined environment.

Outer Space Ergonomics: is the application of human factors scientific knowledge concerning human-machine interface to the design of objects, systems and environment for human usability in outer space.

Outer Space Anthropology: Space Anthropology—the study of exaptations (Latin: ex and aptus): potentialities or archetypes of the function now needed within the new environment, pre-existing in the human species, that allows the physical and cultural adaptability to the outer space (as an aspect of the ongoing human evolution and cultural development) (Ferrino M., 2004; Masali M. et al. 2005).

---

[5] Travel is defined as process or time involved in a person or object moving from a location to another.

Outer Space Psychology: as a specific topic in the area of Applied Psychology, addresses the impact of the living and working conditions in space and during space flight on human behavior, performance, mood, and behavioral health. It includes basic issues of human adaptation to the extreme conditions in space as well as operational issues of selection, training, and support of astronauts. (Manzey, 6.2007, courtesy personal communication)

Outer Space Art: "Contemporary art which relies on space activity for its implementation" (Malina, R., 2002), in the space habitat it is able to interact with the feelings and the moods of the habitants, opening their minds and increasing the well-being.

Outer Space Design: A discipline that is aimed to "contribute to the process of improving living and working conditions in space, considering - at the design stage - all the human factors that are indispensable for creating a 'personnel-friendly' environment: it must be comfortable, pleasant and efficient" (Dominoni, A., 2002, p. 65)

Outer Space Visual Design: design of visual input in the outer space context, meant to increase the well being in the user's experience utilizing human factors and cultural expressions.

Natural Design: "discipline based on natural elements" aims to create "context of art and design interaction that will increase the psychophysiological well-being." There are two main philosophies about it; one presupposes to "use the natural elements present in the local environment" and the other the recreation of natural element: evoking natural structure or stimuli (as "variety and the variability of terrestrial stimuli") (Ono et al. 2007).

References

Aguzzi, M., Häuplik, S., (2006), Design Strategies for Sustainable and Adaptable Lunar Base Architecture. IAC-06-D4.1.03 Paper presented at the meeting of the International Astrophysical Conference, Valencia, Spain.

Birren, F., (2006). Color & Human Response: Aspects of Light and Color Bearing on the Reactions of Living Things and the Welfare of Human Beings. Copyright Wiley & Sons 1978. Paperback: USA

Dominoni, A., (2002). Industrial Design for Space. Cinisello Balsamo, Italy: Silvana Editoriale.

Guilton, J., Guilton, M., (1981). The Ideas of Le Corbusier on Architecture and Urban Planning. New York, United State: George Brazillier.

Malina, R., (2002). The Definition of Space Art. The OURS Foundation. Retrieved May 24, 2007, from http://www.arsastronautica.com/definition.php

Masali, M., Schlacht, I., Fubini, E., Riccò, D., Dominoni, A., Bertagna, G., Boeri, C., Ferrino, M., Gaia, E., (2005). La percezione sensoriale in ambienre spaziale Micerogravitazionale; ricerca delle relazioni sinestetiche con il colore. ASI Proposta per la presentazione di nuovi progetti sull'abitabilità spaziale. Unpublished investigation, Dipartimento di Biologia animale e dell'Uomo, Università di Torino, Italy.

A.A, V.V, NASA (1995). Man System Integration Standards, NASA-Std-3000, revision B, Charter 8.6.2.2 Visual Design Considerations, Volume I. Houston, USA: Nasa Johnson Space Center. Retrieved May 24, 2007, from http://msis.jsc.nasa.gov/

Ono, A., Schlacht, I., Masali, M., (2007). Abstract: Lunar Zen Garden; Natural Design as a Key for Reliability in Space Extreme Environment. Hyderabad: India, 58th International Astronautical Conference.

Schlacht, I. (July 2007). Extreme Design. Retrieved July 2007 from www.extreme-design.eu

Schlacht, I., Ferrino, M., Masali, M. (2005) Interview on Color Design in Space. Turin, Italy: Unpublished investigation Archives: Thales Alenia Spazio Italia, Turin Human Factor Department".

Schlacht I. L., Masali M., Ferrino M. (2006). Interior Design and Sensorial Perception in Microgravitational Outer Space Environment. Journal of Biological Research 81(1): 112–115

Schlacht, I., (2006) Il design del colore negli habitat spaziali. Requisiti per la progettazione del colore in ambienti estremi extraterrestri. Unpublished master thesis, Politecnico di Milano, Italy.

Hermann, M., About open space – an executive summary, Chicago 2005 (www.openspaceworld.org)

Owen, H. (2003) The practice of peace, Berlin.

Additional Reading

Bond, P., (2002). The Continuing Story of the International Space Station. Springer. Chichester (England).

Masali M., Ferrino M., Schlacht I. (2005) L'uomo nello spazio: ricerche di adattabilità. La progettazione del benessere nella postura, prossemica e nel colore. Atti del XVI Congresso degli Antropologi Italiani (Genova, 29–31 ottobre 2005). Il processo di umanizzazione. Antonio Guerci, Stefania Consigliere, Simone Castagno (a cura di). Edicolors Publishing, Milano Italy 641–650.

Sloan A. W. (1979). Man in Extreme Environments. Charles C. Thomas Publisher. Springfield (USA).

13

Anthropology: Physical and Cultural Adaptation in Outer Space

Melchiorre Masali
Physical Anthropologist, Università di Torino

Marinella Ferrino
Ergonomist and Physical Anthropologist
Thales Alenia Space Italia

Monica Argenta
Cultural Anthropologist

Margherita Micheletti Cremasco
Physical Anthropologist, Università di Torino

Melchiorre Masali is a retired professor of physical anthropology at the Turin University. Starting from a technological background, his works concentrate on the analysis and interpretation of the interaction of Man with the natural, cultural, technological and outerspace environment and adaptive variations of the early and recent populations. His studies are particularly focused on the relationship of Man with the artificial environment, based on the problematic of the interpretation of body architecture, posture and shape. He is an honorary member of the Italian Society of Ergonomics (SIE), Physical Anthropologist of Universita di Torino, SIE Italian Ergonomist Society scientific board member, Human Factor Consultant of Thales Alenia Space Italia.

Marinella Ferrino, PhD in physical anthropology and Certified European Ergonomist (EurErg). She operates in the Physical Architecture–Ergonomics Department of Thales Alenia Space Italy. She is a member of the Columbus industrial team in support of the (ISS) International Space Station program performing configuration/human factors verification for payload engineering integration of U.S. payloads destined for the Columbus Module. She has been involved in several EU projects focused on virtual reality tools development for aerospace applications including habitability, space design, and human behavior in extreme environments in support of the future

Moon/Mars missions. Ergonomist and physical anthropologist, SIE Italian Ergonomist Society certificated as European Ergonomist (Eur-Erg) Thales Alenia Space Italia.

Monica Argenta earned her BSc in anthropology at Goldsmith's College, University of London, and her MA in applied social research at Westminster University. A teaching fellow at the University of Milan and Padua, she works as a consultant for several Italian governmental and local authorities. Interested in gender and intercultural studies, she is currently conducting research on Caribbean communities working in extreme environmental conditions on the Alps. Cultural anthropologist, BA anthropology, MA sociology, London University.

Margherita Micheletti Cremasco, PhD is a biologist with a PhD in physical anthropology and postgraduate Master of Ergonomics Faculty of Humanities expert in Cognitive Ergonomics. Certified as European Ergonomist (Eur-Erg by CREE), researcher of physical anthropology in the Faculty of Sciences, Turin University. Her scientific activity regards mainly the following research topics: anthropometric and biomechanical surveys on present and past Italian population, children anthropometrics regarding secular trend, and relation among obesity, lifestyle, sports, and habits about technology use. Most of her activity is in the field of posture analysis in working with VDT, biomechanical workload evaluation, and risk analysis in the industrial field. She teaches courses in anthropology, anthropometry and ergonomics at Turin University. Physical Anthropologist/Certified European Ergonomist (EUR ERG) Lab. of Anthropometry and Ergonomics, Faculty of Sciences Turin University, Italy.

ABSTRACT Man's challenge to live in Outer Space reflects on the domains of both physical and cultural anthropology: new forms of biological adaptation and new social structures may be a strong challenge for the studies of Man, insofar as Mankind faces specific environmental conditions that may affect locomotion, working capabilities, living conditions and particularly well-being. Interior design and interface usability mainly depend on human characteristics that are modified by the different environmental circumstances. Humans are the product of biological and cultural adaptation to our planet achieved over millions of years of Primates and Hominids species' evolution and of the development of intelligence, speech, and manual capability. Body shape may be different; things may not be in the expected place, but also kinship and sexual relationship terrestrial models may be difficult to transfer to extremely long space missions. This essay is intended to elucidate some aspects of the understanding of existing human traits' performance in a new uncommon environment. The process, known as "exaptation," allows to correctly infer the biological and cultural built-in mechanisms to design man-environment-machine interfaces for a correct approach to Man's trial to live in Space.

Introduction

Body shape may be different, things may not be found in the expected place, gravity may be absent, weaker, or distorted by centrifuge, squirrel cage or constant acceleration engines (for example the americium ions engine proposed by Carlo Rubbia), but also kinship and sexual relationship earthly models may be hardly transferred to extremely long space missions or settlements.

The fundamental environmental factor that makes critical the adaptation of humans is the absence of gravity or its modification. Gravitation force is the most stable environmental factor on Earth and all evolutionary history of living beings involves a positive adaptation to it: Being unsuitable for this factor means inexistence! This is quite true particularly for species that are strongly affected by gravity, like birds, tree-dwelling primates, but also upright walking Man is a typical product of a gravitational adaptation by antagonism to gravity force. Strange as it might seem, we have found a model, an archetype, an exaptation of the whole process of 1g/0g adaptation in our evolutionary background, in a Madagascar lemur, the Sifaka (*Propithecus diadema*). Among these primates, upright locomotion is characterized by the typical vertical clinging and leaping tree dwelling combined with upright land locomotion and the stunning ballistic behavior in the leap that creates instantly, but continuously repeated 0g conditions (Lombard *et al.*, 2002). Nevertheless, zero gravity, Moon gravity, Solar system planets and moons, hypo and hyper gravities are conditions never experienced by terrestrial organisms except, maybe, microorganisms dispersed in space by the solar wind. So, how may adaptation proceed according to evolutionary laws? On Earth, there have been, according to Mayr, evolutionary momentums of fundamental importance such as from aquatic life to land life, from land life to flight, that were possible only to the carrier of a truly improbable combination of traits, and this would explain why such processes have been so infrequent.

In view of an anthropological approach to Space, in which, instead of millennia, the available time for adaptation is infinitesimally short, the process can be only found in the domain of "exaptations" (from Latin *ex* = previous and *aptus* = adapted; Gould and Vrba, 1982). Within the "exaptation" framework, an organism's characteristic is not necessarily a product of natural selection as far as its current use is concerned. Instead, an organism's characteristic could come from an "archetype", already developed under specific environmental pressure, but well-designed in response to requiring new environmental conditions. A good example of human "exaptation" in Outer Space could come from our sensory organs equilibrium and vision.

Background

Body movement, orientation, and posture are strongly influenced by gravity and vision: Gravity works on the equilibrium organ, the "vestibular apparatus"[1] of the inner ear first and foremost, represented by the ampoule and semicircular canals. The system is silent under microgravity steady and uniform motion conditions, but reacts to linear and angular inertial variations. Examples include our upright forest ancestors (*Ardipithecus ramidus*, *Orrorin tuagenesis*) as well as bipedal and tree clingers.

One of the cases of deep transformation that may affect the interface relationship and may be the "0g Weltanschauung," not in the common philosophical sense, but in a real sense of world viewpoint, is the orientation or the sight line with respect to changed body shape. The sight line drops 25–30 degrees down with respect to the Ohr-Augen-Ebene (OAE) or Frankfurt horizontal Plane (Frankfurter Verständigung, 1884: Man that looks at the horizon of the classic view of the anthropological tradition). As a matter of fact, we look about 5 meters away on the ground to see way ahead and, maybe, obstacles and perils. In an evolutionary frame this means an extraordinary conflict with the rotation of the basicranium and the increasing of the occipital surface.

According to Delattre, a positive effect of this process was the "hominisation" (becoming Man) of the head, generating a bigger brain and a bent pharynx: intelligence and speech as the curved vocal tract allows the tongue to modulate the resonance cavities that produce the vowels. The adaptation is not a changing of the gravity force, but the rotation of its vector direction with respect to the body axis. This means a "gravitational revolution" for mankind.

In the framework of this research we think Man is on the edge of a second "revolution" in which microgravity is the dominant force to play a role. As a consequence, if vestibular and self perceptive systems are defective, the aptitude to know where one is in space can be taken over by auditory, visual, and tactile senses. In particular, vision becomes of paramount importance. Vision in space can be considered by different aspects: eye-body posture relationship and changing of light parameters perception mainly related to the so called neutral posture.

Human Body Shape and Environmental
Interaction in Microgravity

Body shape, according to D'Arcy Thompson (1917, 1969) depends on the variation of the scale relationship with the body itself and the surrounding physical world, insofar as Nature always operates respecting the proportions, so that all things have their correct measure. Marcus Vitruvius Pollio and Leonardo Da Vinci used geometric grids based on the ancient Greek Ὀργυιά, the measure of the extended arms divided in eight feet up to Durer and Le Corbusier to get measures and proportions for a painting sculpture and design. Nevertheless the human body, as all living beings, is not the outcome of a drawing board blueprint, but the result of a process ruled by genetic information, growth hormones, nutrition, weight and muscular forces acting on the body and its skeletal structure. In earthly gravity a healthy person will reach his predictable shape through a growth process regulated by genetic heritage, orthostatic mechanisms, bone-muscle strength interaction, self-perception, visual and vestibular reflexes. Nevertheless pathologies such as poliomyelitis show that without the correct stimuli the proper body shape is not achieved. Even though a spacecraft crew is usually composed by adults, this may not be true in future. Furthermore, keeping the healthy body shape of adults is quite a medical and kinesiological challenge even on Earth! The well known neutral posture in 0g may not be pathology and may be fully reversible, but 0g generates a distorted relationship among the bodily geometry, such as between hands and the eyesight line.

A real comprehension of such problems in the actual context of outer Space requirements calls for strong feedback from the final users, the astronauts. This would increase the knowledge of user needs and would allow us to integrate dimensional, physiological, psychological and behavioral patterns in order to understand the new "culture" of space on-orbit and to reorient the habitability requirements in the design loop of future projects.

For the crew a spacecraft, or a space settlement, is a combined Laboratory–Office–Home place and it embodies ethical, social, and cultural aspects that must be assessed to determine a viable architectural design. In other words, habitability criteria should be based on a Human Centered Design approach which consider the specific zero-gravity constrained on conditions of life in outer space. The interior design of a Habitation Module involves a rethinking of human well being aspects that include applications of crosscultural knowledge such as: standards of living, environmental aesthetics, attitudes to crowding, proxemic aspects, like the use of space and time—i.e. the sort of anthropological issues tackled by Edward T. Hall.

Physical and cultural anthropological issues may offer significant insights to interior designers creating privacy and recreational areas and, of course, sustain a workplace for long term spaceflight missions. Evaluation criteria are needed to define parameters useful to orient the layout of the habitation module interior design. For this, a general environment re-definition is necessary. Infra-cultural conditions of the crew interacting with each other and living in zero gravity conditions where restraint needs, orientation, posture, and visual perception are strongly modified with respect to Earth can not be left out. The human body and culture also have a strong impact on the perception of the working zones, therefore, time, space, social roles and task

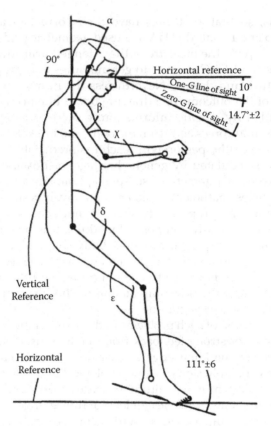

FIGURE 13.1
Weightless neutral body position.

typologies need dedicated tools to define a new index of socialization and comfort for the design process.

To define human proximity, visual interaction and comfort index, both quantitative and qualitative data have been collected in some past tests and experiments. For example, the relationships between real posture and digital mannequin simulation can be determined in order to obtain quantitative values (angles, distance, and orientation) starting from qualitative values (behavior and visual interaction). As an example of the applicability of such research, an analysis was performed to evaluate laptop computer used by different percentiles male and female in various Space Station areas and layout conditions.

In the same way, we may consider some of the results obtained using Habitability Concept Models for the ISS workplace design (Ferrino and Gaia, 2002, 2003a, 2003b). These aspects focus on the real user's needs and are considered added value to increase the quality of the habitable crew interdisciplinary ways of thinking, where engineering, proxemics, and anthropological requirements have shared goals.

FIGURE 13.2
Virtual mannequin (JACK®) analysis (Courtesy of NASA and Alcatel Alexia Space, Italy).

In particular, specific software programs such as the Digital Mannequin (JACK) together with CATIA modeling are the main tools used for accessibility, visibility, operability and human interface evaluations. A European Union (EU) Project on Virtual Reality Environment (VIEW) was developed with the aim to introduce this tool for training and body movement analysis in real time (Kalawsky, 1993; Boher et. al, 1997).

Employing virtual dimensions for workplace design implies multidisciplinary aspects, bringing together expertise and know how from the fields of architecture, psychology, ergonomics, and mental health, etc. Inevitably, this approach is concerned with individual and social issues of work organization and innovative interface such as Virtual Environment techniques. These ventures take a user centered approach, with a strong ergonomic, human-factors and psychology background.

Conclusion

Assuming its comparative perspective and conclusions as a proper point of view for our investigations (Lombard et al. 1002) we looked at Delattre and Fen's pioneering art studies that stated that the human shape should be seen as a product of basicranial remodeling, driven by the transition to an upright posture by the function of the labyrinth. We found a model, an archetype (Exaptation) of the whole process of 1g/0g adaptation in a Madagascar Lemur, in the vertical locomotion of the Propithecus characterized by the typical vertical clinging and leaping tree dwelling combined with upright land locomotion and the stunning ballistic behavior in the leap. Thus, we recognized in this species a proper comparative model to study Human adaptive behavior to terrestrial erect posture and, with a little bit of imagination, to the zero gravity on board spacecraft. Following the experiments of nature in our ancestry, this may be a model of exaptation for our future adaptation. 1g/0g adaptation could be already contempleted by our evolutionary history and should not be necessarily considered pathology. However, body shape and posture are central to our understanding of the world and are strictly linked to the symbolic meaning humans are accustomed to give to it. Anthropology, both physical and cultural could offer some interesting insights and practical applications in the attempt to forsee and design our development in Outer Space.

References

Aguzzi, M., Häuplik, S., (2006), Design Strategies for Sustainable and Adaptable Lunar Base Architecture. IAC-06-D4.1.03 Paper presented at the meeting of the International Astrophysical Conference, Valencia, Spain.

Burzio, L., (2002). Abitare lo spazio: user needs and user orientation. Unpublished master thesis, Politecnico di Torino, Italy.

Burzio L., Ferrino. M., Masali, M., (2003). Laptop Usability Analysis and Human Centered Design Working in Microgravity. HAAMAHA 8th International Conference of Human Advance Manufacturing Agility & Hybrid Automation. Conference Proceedings 26–30 May Rome pp. 361–365.

D'Arcy Wentworth T., (1969). Crescita e forma (1917 Growing and Shape), ed. Boringhieri, Torino.

Ferrino M., (2002) Home-Office nello Spazio: Ergonomia, Prossemica e Design in Microgravità (HU) Habitat Ufficio, N.114 pp. 91–93, Milano, Alberto Greco Editore.

Ferrino M., (2004). Living in Outer Space: Anthropology and Proxemic Methods for Human Adaptation Analysis in Microgravity Environment. ATTI SIBS

Ferrino M., Gaia E. (2002), Habitability Concept Models for Living in Space IAA International Astronautical Congress, 16/21 October, Houston, Texas Accepted Abstract.

Ferrino M., Gaia E. (2003a). Human Centered Design Tools and Virtual Prototyping Applications in Manned System Design: a Macroergonomic Approach. 2nd ESA Space Systems Design, Verification & AIT Workshop, Abstract 15–16 April, ESTEC, Noordwijk, The Netherlands.

Ferrino M., Gaia E., (2003b). Workplace for the Future: A Multi-Disciplined Approach for Living in Outer Space. HAAMAHA 8th, International Conference, 27–30 May, Rome.

Ferrino M., Masali M., (1997), Modelling and Analysis Tools for Ergonomics Design. Workshop on Modelling: Methods and Application IMA-CNR Genoa.

Lombard E., Masali M., Gamba G., Ferrino M., Fenoglio A, (2001), Evoluzione della postura unmana comparata a quella del Propithecus in funzione dell'adattabilità all'ambiente microgravitazionale: ipotesi interpretative. Lo studio dell'Uomo verso il Terzo Millennio. Croce di Magara. Spezzano Piccolo (CS), 25–28 Settembre.

Leoni Zambini M., (1992). Rivista I.B.M.

Additional Reading

Masali M., (1972). Postural Relevance of Sergi's Temporal Tetrahedron in Primates. Journal of Human Evolution.

Masali M., Ferrino M., Schlacht I.L. (2006). Ginnastica e Coreografia nello Spazio? (Gymnastics and choreography in outer space) in G. Trucchi; Atti; pp. 145–146.

Masali M., Schlacht I.L. (2006). Corps et mouvement dans l'Espace. (Body and movement in Outer Space). IXe Universitè Europeenne d'Eté: Anthropologie des Populations Alpines "Corpo attività fisica e patologia: un percorso dal passato al presente" Quaderni Asti Studi Superiori. Diffusione Immagine Editore ISDN 88-89277-09-2 pp 50–54) Asti, Italy.

Masali M., Ferrino M., Schlacht I. (2005). L'uomo nello spazio: ricerche di adattabilità. La progettazione del benessere nella postura, prossemica e nel colore. Atti del XVI Congresso degli Antropologi Italiani (Genova, 29–31 ottobre 2005). Il processo di umanizzazione. Antonio Guerci, Stefania Consigliere, Simone Castagno (a cura di). Edicolors Publishing, Milano Italy 641–650.

Masali M., Ferrino M., (2001) Antropologia Spaziale: uno spazio per L'Antropologia? XIV Congresso degli Antropologi Italiani. Lo studio dell'Uomo verso il Terzo Millennio. Croce di Magara. Spezzano Piccolo (CS), 25–28 Settembre.

Masali, M. (1994), Il figlio degenere della Macchina per scrivere. Ergonomia 1/4–17.

14

Visual Design: Color and Light for Well Being in Outer Space

Irene Lia Schlacht
Cand. PhD Human Machine System
Dipl. Industrial Designer
Technische Universität, Berlin

Henrik Birke
Cand. Dipl. Aerospace Engeneer
Technische Universität, Berlin and DLR

Irene Lia Schlacht is an Italian designer born in Milan. After various experiences and several publications on the outer space field, she is now studying for a doctorate at the Technische Universität Berlin, under Prof. M. Rötting. The doctorate's title is "Visual Design for Outer Space Habitability." She earned a design masters degree in 2006 from the Politecnico di Milano; her thesis was titled "Colour Requirement in Outer Space Habitats" (Prof. D. Riccò). She was also an outer space human factor researcher at Università di Torino (Prof. M. Masali). After completing a report on outer space color design during her stage at Thales Alenia Space Italy, she directed an experiment on color perception in microgravity, taking parabolic flights with the European Space Agency (ESA) in September 2006. The goal of her research is to improve the working conditions of astronauts using design and ergonomics. Today, together with experts in several fields from several countries, including the Japanese space artist Ayako Ono, the Italian space anthropologist Prof. Melchiorre Masali, and the American Psychologist Scott Bates, she has created an outerspace research group called www.Extreme-Design.eu. Cand. PhD Human Machine System, Dipl. Industrial Designer, Technische Universität Berlin, Department of Human Machine System.

Henrik Birke, Cand. Dipl. Aerospace Engineer, Technische Universität Berlin and DLR.

ABSTRACT As fundamentals for user well-being in this paper we will explain why in particular the visual stimuli like colors, light, artistic or natural visual inputs (e.g., plants) have to be considered as the most important factors in outer space design. In this paper is presented an analysis for the visual design requirements of a spacecraft design with an ergonomic approach, in particular regarding:

1. Habitat Visual Configuration
2. User Visual Conditions
3. Visual Design Requirement and Project

Introduction

How is the perception of space when you float in it? First of all, the orientation in the first 2–3 days changes physically and becomes completely visual. However, also the interaction changes and we come by another dimension of space to become explorable. In the beginning we were just walking on the 2D surface, now we can float in the 3D volume.

This changes entirely the way to make use of the space. There is no up and down and the concept of floor, wall and ceiling doesn't have any sense. We can use all parts as functional, but that may have effects on the orientation which is already unbalanced.

If we consider a space settlement on Moon or Mars, the configuration will be influenced by difference of gravity; on the Moon jumping is easy while on Mars the larger gravity makes it a bit more difficult. The solution lies in the flexibility, modularity and the variability of the configuration, for example, in case we need a habitat able to arrive on Mars in weightlessness or artificial gravity conditions that has to be useful also after the landing. These and other considerations can be the basis for Outer Space Visual Design.

Contents

As a guideline to solve the problematical question of space environment visual design and to increase the well-being of the user, the following topics are to be discussed:

1. Visual condition of the space habitat: arrangement of space, light and color.
2. Visual perception condition of the user: stress, posture and visual deficit related with the weightless environment
3. Objectives of the visual design: achieving skills, relaxation, aesthetics (open the mind), increasing the well-being.

Background

The Importance of Vision

Vision,[1] in comparison with the other sense perceptions, is the most important[2] body function that allows us to perceive the outside reality. As the president of the IACC (International Association of Color Consultants) states "Color and light are the major factors in Man-made environments; their impact influences Man's psychological reactions and physiological well-being." The author also states that "it is no longer valid to assume that the only significant role of light and color is to provide adequate illumination and a pleasant environment" (Manke, F., 1987, p.x).

In the Man–environment relationship, visual stimuli have an active role on the human organism. Anxiety, headaches, lack of concentration, inefficiency, bad moods, visual problems, nervousness, and stress can be caused by the environment (Manke, F., 1987).

Indeed, the interior colors may act on the central and vegetative neurological system. Even if this factor is minimal, in long term space missions it will generate strong effects on the psychological mood of astronauts. This appears to be demonstrated by all the psycho-physiological experiments on the influence of light and of interior color on humans, but even on apes, chiefly accomplished in America during the 1970s–1980s (Wise et al. 1988).

The visual topics approached in the context of the weightless environment are: perception, pollution, orientation, configuration, deficit, needs, and discomfort.

Visual Condition of the Space Habitat

Today the psychological well-being of the user is not the first factor of interest; discomfort and visual pollution are factors that are not considered important in present space stations. The reason is that the International Space Station (ISS) was created for short duration space missions, where human mental and physical stability is not the first issue. However, in case of any problems the astronauts can be replaced in less than a week. Considering the context of space tourism, long duration missions (and also utilizing the ISS to study the needs for long duration mission design), the visual condition of the space habitat needs to make a huge step forward.

---

[1] Visual perceptions are: movements, shapes and colors "caused by certain quality of light that the eye recognizes and the brain interprets." Therefore, light and color are inseparable elements, and, "in the design of human habitat, equal attention must be devoted to their psychological, physiological, visual, aesthetical, and technical aspects" (Manke, F., 1987, p. ix).

[2] Approximately 80% of the sensorial information about the world is the visual kind, 1/3 of the brain function is dedicated to visual information elaboration and in details 83% of our memory is visual (Romanello, I., 2002).

Visual Orientation

We need a visual design configuration that improves the orientation (Connors et al. 1985): "On Earth, the vestibular organs and eyes let people know in which way they are moving and how fast. However, visually induced feelings of self-motion are inhibited if vestibular signals fail to confirm the motion. It is hypothesized that with exposure to weightlessness, people suppress vestibular signal and become increasingly dependent on vision to perceive motion and orientation" (Mallowe 1991).

In microgravity, visual perception acquires even more importance than usual, as we were able to demonstrate by interviews carried on a sample of 10 astronauts from different nations between 2005–2006 (Schlacht et al., 2005). However, self-orientation is of primary importance in the micro gravitational space environment. Light, Color and Sound can be employed to expand one's sense of space, creating different atmospheres to give orientation in space (Aguzzi, Häuplik, 2006, p.4). Investigations related to the general improvement of orientation in space stations were done by Aoki et al. (2005).

Visual Configuration for Spaciousness

Visual design is also important for space configuration. As mentioned in the NASA Standard "As a mission duration increases, there is a greater tendency for the crew to feel confined and cramped. This can affect psychological health and crewmember performance. The feeling of spaciousness can be achieved visually through the arrangement, color, and design of the walls and position of the space module" (A.A, V.V., NASA, 1995, p. 8.6.2.2).

In the NASA Standard the factors that are mentioned as influences on the spaciousness of the visual sensation are (A.A, V.V., NASA, 1995, p. 8.6.2.2):

- Distance of the viewer, space overestimation increase with distance.
- Room Shape, volume overestimation increase in irregular room shape.
- Viewing along a surface increases the distance perception in relationship to viewing the same distance along an empty space.
- Lighting and Color, lightness and desaturation increase the volume perception.
- Items that visually detract from long view axes decrease the perceived room value.
- Windows increase the sense of spaciousness and psychological well being.

Actual Main Configurations

The interiors of the ISS have dominant characteristics in accordance with the US-European and Russian design skills that are fundamentally cultural.

European and USA typology:
- modularity
- cold
- white and light blue
- orientation through labels

Russian typology:
- space ship configuration
- familiar
- different colors: brown floor, white ceiling, green walls
- visual orientation with color and configuration of the structure

The Russian section turns out to be more pleasant to the astronauts. In the study of the interior of the Russian module, color design has been applied to the up and down orientation, also focuses the study of color perception in 0 g. All these factors have been positively used to increase the reliability, well-being and the efficiency of the astronauts. (Burzio, L. 2001)

Space Strike from Visual Monotony

The main factor that has to be taken into account in an artificial confined environment is that the crew is isolated from the natural variability of the mother earth environment. Having a walk in a garden, seeing the different colors of the season, watching the sunset, feeling the wind, are all romantic concepts that make our life beautiful, and are all natural stimuli characterized by variability.

Monotony of visual stimuli is a cause of strong discomfort. A key case of "visual monotony" is the Skylab Space Station 4, where the psychological aspects were not considered in the course of the mission that led to a hostile atmosphere and a complete insustainability of the crew. The crew judged the look of the modules as having little comfortable and emphasized the lack of chromatic variety: the interiors were mostly grey and the dominant clothing color was golden brown. The lack of colors and contrasts with the background caused difficulty in tracing objects.

The crew considered that one of the more pleasant sights was the effect of the numerous colorful lights crammed on the main dashboard in the twilight.

FIGURE 14.1
Color inside Skylab.

The only accents of color were the blue anodized handles. The astronauts suffered from insomnia, irritability and depression. As a consequence of stressing time schedule and bad environment design, Call, Gibson and Pore (Skylab 4;16.11.1973/8.2.1974) resorted to the first extraterrestrial strike, later denied from the same astronauts (Serlenga, S., 2001).

User Conditions for Visual Perception

Perception Alteration

Due to the limited environment and, above all, because of the absence of gravity in space habitats, the brain perceives reality in a totally different way. The senses of taste, olfactory, tactile, hearing and sight, turn out altered as a result of the effects of weightlessness.

The ESA astronaut Roberto Vittori described his mission on the ISS during an informal meeting with the author as follows: sensory perception has been completely overturned, tactile and olfactory and sight were less sensitive and feelings were modified regarding the perception on the Earth. Moreover, as referred to by the astronaut Umberto Guidoni, images seem to slide more slowly.

These are the main modifications that occur at sensory perception level (Schlacht, I., 2006):

- Sound: It is sharper
- Olfactory: It diminishes
- Taste: It diminishes and it changes also modifying the subjective preferences.
- Visual: It diminishes and lightly changes without compromising the general recognition of the elements.
- Tactile: It diminishes.

All these factors create a much more complex and hardly predictable sensory environment, where the importance of vision stimuli planning becomes increasingly important. "The important role of vision in a space mission was already suggested from the first researches that found that a 0 g environment can alter visual ability." In the absence of gravity, visual perception is modified at chromatic level, intensity, sharpness and in the angle of vision (Connors et al. 1985).

Visual Ergonomics

To create a visual ergonomic environment, we must consider visual alteration that is intrinsic to the space habitat:

- Angle of vision
- Myopia
- Color perception modification

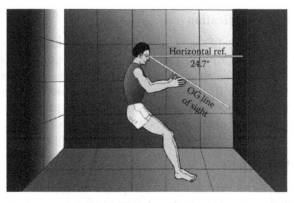

FIGURE 14.2
Angle of vision in microgravity, © Valentina Villa 2008.

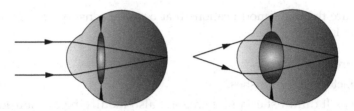

FIGURE 14.3
Effect of myopia on ocular bulbs, © Valentina Villa 2008.

In order to design the spacecraft interiors, we must consider that humans change posture from normal erected to "neutral" which is similar to the posture of a fetus. In the neutral posture under zero gravity, the visual angle is shifted by about a 24 degree slope (about the slope of the Frankfurt plane vs. the vestibular plane) as a consequence of the physiological tilting of the "vestibule plane" with respect to the "Frankfurt plane," used by anthropologists to define the conventional horizontal orientation of head and vision.

Astronaut Myopia

The astronauts suffer from myopia caused by the limited environment in which they live. The modules of the ISS are of very limited dimensions: The inside space is a rectangle parallelepiped of square section. The height and width of the parallelepiped is in general 2 meters and the length varies from 4 to 8 meters, approximately. The only possibility in which the astronauts can effectively perceive far objects is when they watch the Earth panorama from the porthole.

An effect of the limited dimensions of the modules is the lack of opportunities to focus on far objects: the continuous demand for proximal vision and the consequent constant accommodation of the crystalline lens generate the myopia.

Color Perception Modification

According to the architect Daniele Bedini, Space Architecture teacher from the International Space University in Strasbourg, the eye perceives the colors are more attenuated in the absence of gravity as a consequence of the minor oxygen contribution. For this reason colors are proposed to be more intense (Rita, L., 2000). Also color vision turns out to be modified because of the modified shape of the ocular receivers in the absence of gravity.

Color Perception Inquiry

A few experiments concerning color perception in weightlessness have been done, mainly in the 1960s (White 1965, Kravkov 1952, Kitayev-Smyk L. A. 1963, Wise J. A. 1988, White 1965). The latest experiment was made in 2006 by a

FIGURE 14.4
Astronaut reading a book in microgravity, © NASA.

European student group supported by ESA during parabolic flights. Based on the anomaloscope[3] principle they measured the range in which two color images were perceived as equal in normal and microgravity conditions. Two tests were made for the hue red and blue and two tests for the chromatic saturation of yellow and the achromatic saturation of grey, which is defined as brightness.

The data of 720 tests in normal gravity and 48 tests in microgravity recorded by six subjects showed a general changing of hue and saturation (see following image). In particular, the hue-tests demonstrated on the one hand a decreasing of the sensibility for the color red in relationship to blue and on the other hand an increase of the sensibility for red in relationship to green. Additionally, an increase of the sensibility for green in relationship to blue could be shown. In the saturation part of the tests first a general decrease of the brightness-sensibility became evident. Also the sensibility for saturation of lower saturated yellow became less. What is more, this data showed also an increase of the sensibility for saturation of higher saturated yellow. The statements about the changes are made only qualitatively due to the low number of tests with statistical analysis done in microgravity. As a consequence more investigations should be made in the future, in particular on the hue green to accomplish the three basic colors. Furthermore the direction of the color changing should be altered to the opposite to avoid influences of the driven color path [Schlacht, Birke, Masali—Astra Aeronautica IAC HIM 2007].

The causes of color perception changes is not really clear but a few hypotheses have been suggested. The first one assumes the reason in the change of position and shape of the lens, so that the focal point on the retina differs in microgravity from the one in normal gravity, which causes a changed

---

[3] An anomaloscope is an instrument used to test for color blindness. It is able to detect whether a person is a dichromat or a tritanope. The apparatus was invented by the German ophthalmologist and physiologist Willibald A. Nagel (1870–1911) who named it "anomaloskop" in 1907 (Nagel 1907).

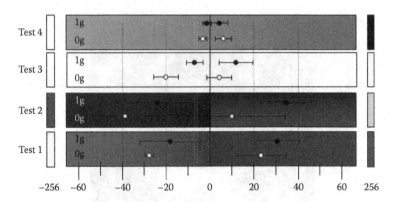

FIGURE 14.5
Mean values of the deviations from "equal color" and RMS-values of the deviations, ©
CROMOS.

use of the 3 kinds of cones. Otherwise, the reason could also be a possible change in the fraction of oxygen. As mentioned before the cause is not really understood yet, but the enormous importance of further investigations is obvious. Especially in the planned long-time missions to Moon or Mars the well-being of astronauts will be much improved by the use of an adequate color-concept for the habitats.

Figure 14.5 shows the mean values of the deviations from equal color (black dots for 1g and yellow dots for micro-g) for all subjects and the RMS values of the deviations (black horizontal lines), which indicate the fluctuation of the recorded values. On the left side one can see the starting colors, on the right side the end colors.

Visual Design Requirement and Project

Visual Design of an isolated and confined space habitat has a strong influence on the physical and psychological comfort of its users as on the worthiness of the environment. The feeling of being confined causes unavoidable changes in an astronaut's psychophysical conditions. However, the well-being, particularly in the case of long duration missions, can be considerably increased through sensorial and psychological stimulations such as light, colors, and changes in wind direction, hot and cold temperature, which are the characteristics of Earth's environment and are able to activate vital processes that are implicit in any human being (Flaborea 2002).

Color has psychologically positive effects: an attractive place reduces the hard work, improves the interpersonal rendering, productivity and relationships (Romanello 2002). Color takes part in the cortical activity, in the functions of the independent nervous system and in the hormonal activity; moreover it stimulates aesthetic and emotional associations (Manke 1996).

From this understanding, it is important to underline the importance of a structured planning of the colors, to answer to the demand of productivity and of psycho-physiological well-being correlated to it.

Color and Art

In a field such as that of space exploration, where technology is of primary importance, also artistic expression can provide answers to really felt needs.

Here follows a contribution from the space artist Liuccia Buzzoni, specialized in color applications for astronauts' creativity and well being:

> Art and especially the use of colours in a world which is deprived of them, can greatly contribute to the well-being of astronauts, in that it is reassuring while it also acts as a psychological compensation and an incentive to creativity.
>
> Each colour has its own vibratory frequency, and colours, with their positive or negative characteristics, have an effect upon our whole being. Depriving a person of colours has the effect of making it more difficult for him/her to believe in himself/herself, as a powerful source of energy is taken away. Colours in space can be a source of emotions and can arouse a person's deepest sense of life. Colours can support the astronauts' body which otherwise could tend to let itself go, thus affecting whatever activities the astronauts are carrying out.
>
> Art with its colours is a way to awaken primeval feelings, such as the instinct of survival which everybody houses in himself/herself. It is also a way to make people aware of their feelings and fit for communicating and expressing them.
>
> Art with its colours can help stimulate memory and senses. At a physical and psychological level it can make it easier for astronauts to find balance, orientation, stability and a sense of gravity. Bearing in mind the importance of a creative use of colours on board space stations, a video has been shot, "Colours in space," in cooperation with and thanks to the experience of Roberto Vittori.

The study of the effect of colors upon man in a condition of microgravity is also at the basis of a research carrying out from Liuccia for ASI (Italian Space Agency) as part of their research program. The actual experiment about the effect of colors it was project in the context of Paolo Nespoli's mission, while the Turin-based company Altec will be responsible for all backup activities.

How can you paint in space? To this purpose a series of informally painted transparent plastic plates have been devised. They can be superimposed so to obtain polychrome effects and personal creations. I have painted the plates in different nuances of the seven primary colors, bearing in mind the sensations they evoke and transmit in common experience on earth. Brush-strokes freely laid ... fit for being interpreted ... Natural colors, saturated and bold. Thanks to the colored plates, astronauts can manipulate colors without

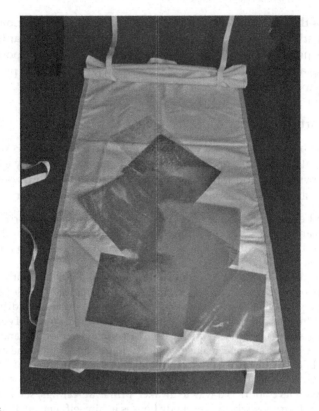

FIGURE 14.6
Color in Space, facility from Liuccia Buzzoni to paint in µg. Project "Colors in Space" is inserted by the ASI "Agenzia Spaziale Italiana" (Italian Space Agency), into its research program. The backup for the experiment is offered to the Altec (Advanced Logistic Technology Engineering Center) company in Turin and the Institute of Psychology of the Padua University of Neurosciences, in the context of the Nespoli mission.

dispersing them around. They can also have the feeling of actually painting and creating. The differently colored plates will be arranged according to the "artist's" best judgement on a panel inside a special sliding bag. The bag, made of a fabric called "Nomex," is also used to carry the color plates. It can be put away in the most suitable place in the I.S.S.

The goals of this space research are those of measuring, recording and evaluating the individuals' reactions to colors when long exposed to microgravity.

We know that on board an orbitating station spaces are restricted. Moreover, astronauts have to comply with scientific modules. But the aim of the research is just this: to make it possible to escape routine and to live as creatively as possible in conditions of microgravity.

To conclude this speech, let me express a wish: may a new place of cultural synergy come into being, a place where to express creativity, where not to feel lonely anymore.

From the Theory to the Project

In the design of long duration missions, the reliability of man becomes the successful key of the mission characterized by a confined and isolated extreme environment. Space Haven Inflatable Unit is an outer space habitat project which has been developed to test this type of reliability. Colors and interior decor influence well-being, create physiological reactions and modify the perception of reality. In a confined environment like a "Space Inflatable Unit," which is isolated from the natural environment, these reactions are amplified. In the design of colors and interior decor it is therefore necessary to recall the normal physical and psychical conditions whose characteristics are variety and variability in time (Bretania 2003).

Color Design Objectives

To create a color design of long duration space mission habitats you need to follow three objectives based on the NASA Standards Requirements (ISS Interior Color Scheme NASA 2001). These objectives are studied to create a color design with the aim of a "Human centered philosophy" and to promote the well-being and the productivity of the user.

1. Psycho-physiological well-being: The light and the colors have psycho-physiological influences on the person, therefore they must be considered in the project in order to maximize the well-being, to increase the efficiency and to guarantee the reliability of the person.
2. Orientation: The sense of direction in microgravity after 3/5 days is totally entrusted to visual perception. For this reason it is necessary to use an immediate visual configuration created according to instinctive replies to natural signals to which we are accustomed on Earth, such as "sky" up and "earth" down. According to NASA Standard 3,000, 8.4.2, 8.4.3-b: The orientation comes first of all through visual input like color.
3. Activities support: The colors and the light should respect the needs of several activities, and increase the comfort facilitating functions and needs such as privacy.

Color Design Requirements

In order to achieve the goals of well being, orientation and activity support, the following key requirements have been identified in the development of color design for a space isolated environment. These have been identified by NASA and ESA:

A. Safety: Color and decor should respect the safety requirements.
B. Visibility: Color and decor should answer to every need of vision.

C. Flexibility: Color and decor should be planned considering their physical-psychological influence and multi functionality. They should facilitate both daily life and acknowledgment of the various activities and needs (Manke 2003).

D. Variation: The environment must be variable in time, must be able to stimulate senses like in natural conditions, so as to maintain a normal status of conscience, perception, attention, concentration and intellectual activity (Deribere 1968).

E. Variety: Color and decor should create a variable environment in order to obtain a psycho-physiological positive environment.

F. Customization: Color and light should be changeable according to personal requirements both for work needs and aesthetic taste, particularly in the personal areas such as the crew quarters. The image shows an example of requirement applications through a "human centered design" realized by I-Guzzini's Italian group. It represents a habitat isolated from "terrestrial sun-light cycle" and illuminated with "Sivra" biodynamic artificial light. In particular, the project will respect the following requirements:

- Requirement E Variety: The habitat is an isolated and confined space decorated with various colors.
- Requirement G Naturalistic evocation: The "Sivra" artificial light recalls the natural sun-light effects.

FIGURE 14.7
"Sivra", © I-Guzzini.

- Requirement D Variation: The artificial light will re-create the natural color variations during the course of a whole day.
- Naturalistic evocation: decorated elements, materials, references to natural landscapes (photographs, pictures, video or colors compositions) should be arranged with the aim of providing direction references, relaxing, reducing stress, resting eyes from computer work.

Aesthetic Purposes

For a project regarding internal color and décor for a "Space Inflatable Unit" we have the following color design.

Color Selection

Colors have been chosen regarding their psycho-physiological nfluences in order also to provide orientation in space and to support the activities going on. Regarding color choices, in absence of gravity, due to a minor oxygen contribution, eyes perceive colors in a more subdued way.

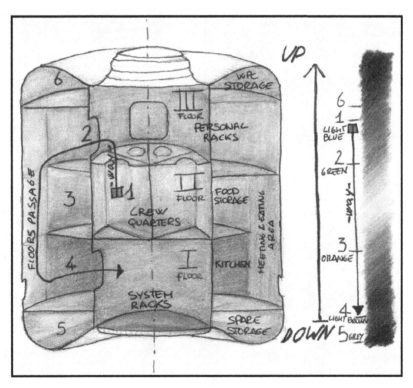

FIGURE 14.8
Color design of a "Space Inflatable Unit," © Irene Schlacht 2005.

Color in Orientation

With the aim of making orientation easier, all pavements are grey and all ceilings are white to give the sense of stability below and open space above. Every dominant color has been linked to each internal and external floor in

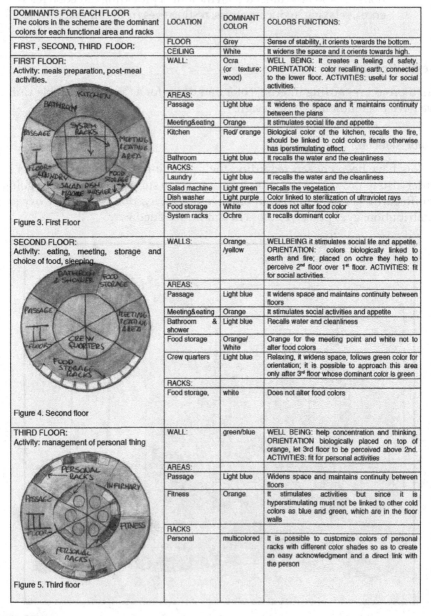

DOMINANTS FOR EACH FLOOR The colors in the scheme are the dominant colors for each functional area and racks	LOCATION	DOMINANT COLOR	COLORS FUNCTIONS:
FIRST , SECOND, THIRD FLOOR:	FLOOR	Grey	Sense of stability, it orients towards the bottom.
	CEILING	White	It widens the space and it orients towards high.
FIRST FLOOR: Activity: meals preparation, post-meal activities. Figure 3. First Floor	WALL:	Ocra (or texture: wood)	WELL BEING: It creates a feeling of safety. ORIENTATION: color recalling earth, connected to the lower floor. ACTIVITIES: useful for social activities.
	AREAS:		
	Passage	Light blue	It widens the space and it maintains continuity between the plans
	Meeting&eating	Orange	It stimulates social life and appetite
	Kitchen	Red/ orange	Biological color of the kitchen, recalls the fire, should be linked to cold colors items otherwise has iperstimulating effect.
	Bathroom	Light blue	It recalls the water and the cleanliness
	RACKS:		
	Laundry	Light blue	It recalls the water and the cleanliness
	Salad machine	Light green	Recalls the vegetation
	Dish washer	Light purple	Color linked to sterilization of ultraviolet rays
	Food storage	White	It does not alter food color
	System racks	Ochre	It recalls dominant color
SECOND FLOOR: Activity: eating, meeting, storage and choice of food, sleeping Figure 4. Second floor	WALLS:	Orange /yellow	WELLBEING it stimulates social life and appetite. ORIENTATION: colors biologically linked to earth and fire; placed on ochre they help to perceive 2nd floor over 1st floor. ACTIVITIES: fit for social activities.
	AREAS:		
	Passage	Light blue	It widens space and maintains continuity between floors
	Meeting&eating	Orange	It stimulates social activities and appetite
	Bathroom & shower	Light blue	Recalls water and cleanliness
	Food storage	Orange/ White	Orange for the meeting point and white not to alter food colors
	Crew quarters	Light blue	Relaxing, it widens space, follows green color for orientation; it is possible to approach this area only after 3rd floor whose dominant color is green
	RACKS:		
	Food storage,	white	Does not alter food colors
THIRD FLOOR: Activity: management of personal thing Figure 5. Third floor	WALL:	green/blue	WELL BEING: help concentration and thinking. ORIENTATION biologically placed on top of orange, let 3rd floor to be perceived above 2nd. ACTIVITIES: fit for personal activities
	AREAS:		
	Passage	Light blue	Widens space and maintains continuity between floors
	Fitness	Orange	It stimulates activities but since it is hyperstimulating must not be linked to other cold colors as blue and green, which are in the floor walls
	RACKS		
	Personal	multicolored	It is possible to customize colors of personal racks with different color shades so as to create an easy acknowledgment and a direct link with the person

FIGURE 14.9
Floors color scheme of a "Space Inflatable Unit," © I.Schlacht 2005.

the same order we can find in nature, so that lower floors have colors linked with earth, while top floors have sky colors.

The crew quarters, even if situated in the central part, are blue in order to evidence the chromatic sequence of the way; in fact, from the crew quarters one can only go to the 3rd floor from which it is possible to come down, chromatically as well, to the 1st floor.

Color Functions for Each Activity

Implementation

In confined space environments, the use of images or decor showing landscapes reduces stress, balances heart pulses and rests eyes.

In order to meet the above-mentioned requisitions, the following implementations have been considered:

- Decors landscape-oriented: placed in the external wall of the inner cylinder, where racks are not previewed; they increase the global orientation and maintain the chromatic dominance of the floors, while singularly they re-propose the natural chromatic scale.
  - Accomplished objectives: all
  - Accomplished requirements: flexibility, variation, variability, customization, natural recalls
- Monitors: communication, relaxation, enjoyment functions (i.e., screening water).

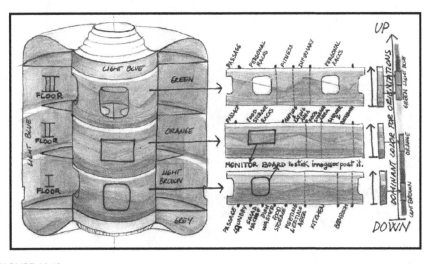

FIGURE 14.10
Implementations "Space Inflatable Unit," © I. Schlacht 2005.

- • Accomplished objectives: psycho-physiological well being, support to activities
- • Accomplished requirements: all
- • Boards: used to pin images, post-it and as a meeting point.
  - • Accomplished objectives: support to activities
  - • Accomplished requirements: visibility, flexibility, variety, variation
- • Biodynamical light: variable lighting recalling the natural light effects in the course of a whole day through automatic changes in intensity, direction, brightness and predominant wave length. It creates space, provides a sense of time passing and varies color perception. It includes the positive part of the ultraviolet beams, so as to make the crew benefit from the exposure to the solar light: in particular, the reduction of viruses in the air. It is placed "on top" increasing up and down orientation, and it has adjustable intensity.
  - • Accomplished objectives: psycho-physiological well being, orientation
  - • Accomplished requirements: all

FIGURE 14.11
Astronaut reading, © NASA.

Conclusions

In terms of human centered design logic, in long duration space missions, colors, light and interior decor must have among their purposes: psycho-physiological well-being, orientation, and supportiveness for all activities. It is therefore necessary to recall, through stimulating elements, the "normality" in confined artificial environments. Physical and psychical conditions can be improved featuring variety and natural variations occurring in time according to the principle of natural design (Romanello 2002) (Manke 1996) (Bertania 2003).

References

A.A, V.V, NASA (July 1995). Man System Integration Standards, NASA-Std-3000, revision B, Charter 8.6.2.2 Visual Design Considerations, Volume I. Houston, USA: NASA Johnson Space Center. Retrieved May 24, 2007, from http://msis.jsc.nasa.gov/

Aoki, H., Ohno, R., & Yamaguchi, T. (2005). The effect of the configuration and the interior design of a virtual weightless space station on human spatial orientation. Acta Aeronautica 56, (pp.1005–1016).

Bretania, G. (2003). Vivere nello spazio abitato, Laboratorio Colore, Politecnico di Milano, Milan, Italy. Mimeography.

Burzio, L., (2001). Abitare lo spazio: user needs and user orientation. Unpublished master thesis, Politecnico di Torino, Italy.

De Grandis, L. (1996). Teoria e uso del color. Mondadori, Milano, Italy.

Déribéré, M. (1968) Dipingere la casa, Zanichelli, Bologna, Italy.

Durao, M.J. (2002). Color in space architecture AIAA 2002-6107, Houston, Texas.

Flaborea, M. (2002–2003) Sistema a luce dinamica per l'illuminazione ambientale della stazione spaziale internazionale. Politecnico di Milano, Milan, Italy.

Itten, J. (1965). Arte del colore. Milan, Italy: Il Saggiatore.

Kitayev-Smyk L. A. (1963). Study of achromatic and chromatic visual sensitivity during short periods of weightlessness. NASA, Center of Aerospace Information, Hanover.

Kravkov (1952). Color Vision. Moscow.

Kravkov (1952). Color Vision. Moscow.

Mahnke, F., (1996). Color, Environment and Human Response. New York: Van Nostrand Reinhold.

Mallove (2001). LIFTOFF TODAY? Shuttle Mission to Explore Space Motion Sickness. MIT News Office at the Massachusetts Institute of Technology, Cambridge, Mass. Retrieved July 2007 from http://web.mit.edu/newsoffice/tt/1991/may22/24740.html

NASA, Mary Connors M., Harrison A., Akins F., (1985), Living Aloft: Human Requirements for Extended Spaceflight, cap: 2–3.

NASA (1995). International Space Station Flight Crew Integration Standard (SSP-50005 Revision B NASA-STD-3000/T). National Aeronautics and Space Administration, Space Station Program Office, Johnson Space Center, Houston, Texas.

NASA (2001). International Space Station Interior Color Scheme (SSP 50008 Revision C). National Aeronautics and Space Administration, Space Station Program Office, Johnson Space Center, Houston, Texas.

Rita, L. (2000), Ambienti confinati, Newton Artiche.

Romanello, I., (2002). Il colore: espressione e funzione, Hoepli, Milano, Italy.

Serlenga, S., (2001). Studio dell'abitabilità del modulo HAB della Stazione Spaziale internazionale. Unpublished master thesis, Politecnico di Milano, Italy.

White, W.J. (April 1965). Effect of Transient Weightlessness on Brightness Discrimination, Aerospace Medicine (page 327–331).

Wise, B., Wise, J., (August 1988). The Human Factors of Colors in Environmental Design: A Critical Review (Contractor Report 177498) Department of Psychology, University of Washington, Seattle, Washington.

Wright, F.L., (1987). Modern Architecture, being the Kahn lectures for 1930, by Frank Lloyd Wright. United States: Southern Illinois University Press. Carbondale and Edwardsville.

Wright, F.L., (1955). The future of Architecture. The Architectural Press. London. United States: Horizon Press.

Additional Reading

Brambillasca S. Schlacht I. L., Masali M. 1.2007 Esperimenti italiani volano in assenza di peso. pp.19–21, Vol. 86 Aerotecnica Missili e Spazio, ESAGRAFICA srl, Roma, Italy. (Online at: http://www.aidaa.it).

Durao J.M., Favata P., 2003, Color Considerations For The Design Of Space Habitats, AIAA 2003-6350, Long Beach, California.

Durao, M.J. (2002). Colour in space architecture AIAA 2002-6107, Houston, Texas.

Fanelli A. 2007. Zero Gravity. pp.42–4, Habitat Ufficio n.137, Alberto Greco editore, Milano, Italy.

Genko L.V., Task H.L. (1984). Testing changes in visual function due to orbital environment, Aerospace medicine research labs, Wright-Patterson AFB, Ohio.

Masali M., Schlacht I.L.; 2006. Corps et mouvement dans l'Espace. (Body and movement in Outer Space). IXe Universitè Europeenne d'Eté: Anthropologie des Populations Alpines "Corpo attività fisica e patologia: un percorso dal passato al presente" Quaderni Asti Studi Superiori. Diffusione Immagine Editore ISDN 88-89277-09-2 pp 50–54) Asti, Italy.

Masali M., Ferrino M., Schlacht I.L 2006; Ginnastica e Coreografia nello Spazio? (Gymnastics and choreography in outer space) in G. Trucchi; Atti ; pp.145–146.

Masali M., Ferrino M., Schlacht I. 2005 L'uomo nello spazio: ricerche di adattabilità. La progettazione del benessere nella postura, prossemica e nel colore. Atti del XVI Congresso degli Antropologi Italiani (Genova, 29–31 ottobre 2005) Il processo di umanizzazione. Antonio Guerci, Stefania Consigliere, Simone Castagno (a cura di). Edicolors Publishing, Milano Italy 641–650.

Popov V., Boyko N. (1967). Vision in space travel. Aviatsiya i Kosmonautika No. 3 (page 73–76).

Pugnè F., Stucchi N.(1999). Olivero A. La Percezione Visiva. UTET, Torino, Italy.

Riccò D. (1999). Sinestesie per il design, Etas, Milano, Italy.

Scharff L. (2005). Internet publication of research. Visual Perception. Department of Psychology Stephen F. Austin State University Nacogdoches (USA).

Schlacht I.L., Rötting M., Masali M.; 2008. Habitability in Extreme Environment. Visual Design for Living in Outer Space. Doctoral workshop pp. 873–876. Proceedings of 54.GfA, GfA press, Dortmund, Germany.

Schlacht I.L., Ono A.; 2007. Natural design habitat on the Moon—Lunar Zen garden. pp. 219–220, 9th ILEWG/ILC 2007, ESA Press.

Schlacht I.L. 2007. Art, Design and Human Metamorphosis in Extreme Environments MutaMorphosis Book of Abstracts (published on line http://www.mutamorphosis.org).

Schlacht I.L., Birke H., Brambillasca S., Dianiska B. 2007 Achromatic and chromatic perception in microgravity. CROMOS experiment in the ESA student parabolic flight campaign. In Monica Monici, ELGRA News. Bulletin. Vol.25, 9.2007 pp.193–195.

Schlacht I.L., Rötting M., Masali M., 2006. Color design of extreme habitats as a psychological support for the reliability (ID: A658026) ESA proceedings. Tools for Psychological Support during Exploration Missions to Mars and Moon, ESA, ESTEC, Noordwijik, The Netherlands.

Schlacht I.L., Masali M., Ferrino M. 2006 Light and color like biological stimuli for well being during long duration space missions. (published online at http://www.cosis.net: N. reference COSPAR2006-A-00692).

Schlacht I.L., Masali M. 2006 Screens as light biological variable in Microgravitational Space Environment. (published online at http://www.cosis.net: N. reference COSPAR2006-A-01114).

Schlacht I.L., Masali M., Ferrino M. 2005 Interior design and Sensorial Perception in Microgravitational Outer Space Environment. Journal of Biological Research, Rubettino Editore, Catanzaro, Italy, ISBN 88-498-0859-3.

15

Art: Art as a Psychological Support for the Outer Space Habitat

Ayako Ono

Artist in Residence at ESA/ESTEC
International Space University, SSP

Ayako Ono is a space artist and an international activist, working in the fields of art and design, specializing in the relationship between the cosmos and human beings. Her academic background is fine arts and she has job experience in computer-based graphic design. She is involved in a space art project through a joint research project with the Japanese Space Agency and Prof. Takuro Osaka. She was a member of the student team that proposed a sculpture which uses music to show different movement in Earth's gravity and microgravity. She became a leader of the team and took over its fundamental experiment and activities: art in microgravity and many art exhibitions at space related events including the first space art exhibition at the International Astronautical Congress in 2005. Thanks to a scholarship from the Japanese government she was able to study space art in Europe and the United States and became an artist in residence at the European Space Agency. She is interested in interior design including soundscapes, sculpture for microgravity, and landscape design for the Moon/Mars and their potential for use as psychological support. After writing her contribution for this book, in April 2009 she began studying for a PhD in psychiatry from the Department of Behavioral Medicine at Tohoku University Graduate School of Medicine.

ABSTRACT   While space technologies are advanced, the designs that support holistic human existence in space tend to lack stimulation for the human senses. Nowadays, "Art and Science" is commonly stated to have integral humanistic ideas and abilities. To improve the quality of life simultaneously in space and on Earth (as a spin-off of space developments), this paper demonstrates some examples and possibilities. The topics are: the artistic manipulation of exterior and interior design elements, the possibilities of Space Art, and the psychological effects of all of these.

The first focus is the role of Art, and the second is the explanation and the definition of Space Art. The third is examples and projects of Space Art. Space Art could be Art Therapy and everyone could be an artist because of the psychological benefits. Several examples of psychological support are demonstrated.

Introduction

In outer space, we face different challenges than on Earth. For instance, the length of day and night is not like on Earth, and we usually need to stay in a confined space. As another example, the sunlight in outer space is dangerous because there is no ozone layer to protect us from harmful radiation. However, the sunlight can affect feelings positively, so it will be needed for well-being.

The purpose of this paper is to change the sense of value about Art itself and to think about possibilities of art and design in the context of space. The focus of this paper is the effect on feelings, how to enjoy Art as a creator, and supporting the idea that "all humans are creators." Culture is changing, and to live in the present society, we need to focus on our own spirit.

Creation of visual arts is helpful to find our real selves. When we create, our inner feelings will appear. Visual art therapy works in this way (K. Ono, 2005). Improving quality of life through artistic expression is the objective of this work.

Background

The Definitions of Space Art

The term "Space Art" has many connotations and definitions even among the practitioners of this genre in the space community. However everybody agrees that Space Art is a new genre of contemporary—art which relies in some way on space activity. Roger F. Malina, the editor of "Leonardo: The Journal of Art, Science and Technology" has tracked its development over the years. He defines Space Art as "contemporary art which relies on space activity for its implementation" and lists seven broad categories:

1. Fine art which exploits sensory experiences generated through space exploration. New landscapes become accessible through space photography and film. Space illustrators anticipated some of these and make use of the photographic record from space exploration.

2. Art which expresses the new psychological and philosophical conceptions developed through the exploration of space. The primary example of this is the concept of the Earth as a whole system – a concept made concrete by the first views of the whole Earth seen from space.

3. Art in space, viewed from Earth.

4. Art on Earth, viewed from space.

5. Art in space, viewed in space.

6. The applied arts such as space architecture, interior design and furniture design.

7. Fine art which takes advantage of new technologies and materials created through space activities. The most important of these make use of satellite systems to create simultaneous global artworks.

Roger Malina goes on to point out that the works of some of the most important illustrators, i.e. Chesley Bonestell, David Hardy, and Ludek Pesek, not only anticipated some of the results of space exploration, but in some sense made space exploration possible by generating public interest and support as well as by helping scientists to plan and illustrate their experiments.

As to the ultimate relevance of space art to space activities, Malina (1989) states: "The creation of contemporary art is inextricably tied to the process of creating human civilization. Within this perspective, art making will occur as a part of space exploration, and in fact art making must be encouraged in space as one of the ways without which, in the long run, human use of space will be incomplete and unsuccessful."

Space artist and planetary scientist William K. Hartmann (1990) cites four roles for space art:

1. Encouraging scientific exploration

2. Recording historical evolution, planetary exploration

3. Promoting international cooperation

4. Synthesizing information to stimulate new ideas about the universe and our relationship to it

Landscape

Landscape is scenery and view. Landscape design is the art of arranging or modifying the features of a landscape, an urban area, etc., for aesthetic or practical purposes. Landscapes including architecture practice arranging or modifying the features of Space Art.

Soundscape

Soundscape means an atmosphere or environment created by or with sound, such as the raucous sounds of a city street or the relaxing sounds of ocean waves. Soundscape design is a new interdisciplinary art combining the talents of scientists, social scientists, and artists (particularly musicians).

Both landscape and soundscape design attempt to discover principles and to develop techniques by which the social, psychological and aesthetic qualities of the environment may be improved. To the extent that it attempts to understand individual, community and cultural behaviour, soundscape design takes the broad perspective of a communicational discipline and touches such other areas as sociology, anthropology, psychology and geography. For universal healing effect, the landscape in space, or spacescape, design should have virtual nature, such as the grain of wood, artificial flowers, and a natural pattern. Mimicking day and night time is an important factor to be considered when creating a comfortable environment.

Quality of Life

For astronauts/cosmonauts, isolation, fellow crew members and quality of life could be issues [Nicogossian, 1992]. Scientists are developing and testing the Crew Exploration Vehicle (CEV) and sending robotic probes to the lunar surface with the purpose of returning to the Moon by 2020. The goal is to work there for increasingly extended periods of time. Therefore, our preparations need to offer an environment for mankind to live and work in outer space. For this purpose, it is important that considering what is required in the environment is given sufficient thought early in the development process. Especially from now on, our environment should be symbiotic, collaborative, sustainable, and regenerative [Narumi, 2001].

Now, the environment in outer space is a relatively new circumstance for mankind.

- What is an ideal environment in which mankind could hope to live and work in outer space?
- What should the environment consist of?

The International Space Station (ISS), spaceship, and lunar base will be confined spaces and people may feel isolated. To improve the environment, we must focus on the mentality of daily life. The environment should consist of balance, and we need relaxation for our life. Art and design can help to make a tranquil mood.

Solutions and Recommendations

Art Therapy, Recreation and Inspiration

If we need to work in outer space for a long time without vacations back to the Earth, we may have psychological issues. We are creators of our lives. The creation of Art could be leisure and at the same time art therapy. Also, interior design and spacescape design offer nice ambient atmosphere for relaxation in place of Earth's nature.

Key elements

- Art therapy already exists, and its application for Space Art will be possible.
- Recreation: Art could be a hobby and will be connected to well being.
- Inspiring people: Art and design would be able to encourage more people, not only astronauts/cosmonauts but also space tourists, to go and stay there.

Examples of Art Therapy, Recreation and Inspiration

Spacescape (Landscape) Design

In this section, a "Lunar Zen Garden" as an example of spacescape design and art therapy is introduced (A. Ono, 2007). Zen is a way to struggle with issues, and there is an unmistakable and increasing interest in Zen Buddhism among psychoanalysts. The first approach to well-being was stated thus: well-being is being in accord with the nature of man (Fromm, 1960). Nature is the keyword of Zen. Also, it is well known that most Japanese gardens attempt harmony with nature. Spacescape design should also be in harmony with nature because this will be helpful to get calm feelings. Therefore, Zen could be basically a technique by which to overcome depression or a mental breakdown of people in isolated long-term missions.

The aim of this idea is not to promote Zen Buddhism but to propose the possibilities of Zen gardens including natural form and its therapeutic effects. This will be a kind of Occupational Therapy and Sandplay Therapy, as well. Making a garden could be both group and individual work. Rather than a passive environment, such as only rocks and sand without growing plants and animals, an interactive space that enables individual and group exploration of creative design would be a possible tourist attraction. At the Ryoanji Temple in Kyoto, one of the most famous Zen gardens, to discover why it has a calming effect on the hundreds of thousands of visitors who come every year, Kyoto University researchers found that the seemingly random collection of rocks and moss on this simple gravel rectangle formed the outline of a tree's branches (AA.VV. Kyoto University researchers, 2002). The researchers found that the unconscious perception of this pattern contributes to the enigmatic appeal of the garden. As another example, Zen gardens use rounded forms, and this makes us remember our mother's body unconsciously (Iwai, 1986).

Landscape gardener Yasuo Kitayama told that stone arrangements don't need a technique but that love is everything (Kitayama, 2007). So, we need to enhance spirituality through creation.

FIGURE 15.1
Lunar zen garden.

One Zen garden style called "karesansui" literally means dry landscape. The main elements of karesansui are rocks and sand, with the sea symbolized by sand raked in patterns that suggest rippling water (a dry stream) with minimal compositions.

These solar panels represent shapes of natural elements (flowers) mounted on rotating systems to point toward the sun. Colourful solar panels should be used. Currently, they are being developed at Gifu University in Japan

The rocks and sand engender calm. In karesansui gardens, considerable emphasis is placed on the beauty of "empty" space. We may be able to find a harmony with the Moon's natural beauty. Four major aesthetic ideals

FIGURE 15.2
Flower-shaped solar panel.

displayed by karesansui gardens are asymmetry, tranquillity, simplicity, and naturalness (Suzuki, 2002).

Keeping a Zen garden requires a lot of dedication in maintaining it, and by doing so this type of garden will benefit people both physically and spiritually. However, this does not necessarily mean that astronauts/cosmonauts have to turn the soil. They make a plan for the garden before hand and make robots to turn the soil by remote control. This is also creative work, and the idea and sense are much more important. The notable point of this lunar base is that it consists of artistic elements such as flower shaped solar panels and habitat modules; therefore, the habitat modules were designed to coordinate these elements in a favourable position such as a fractal because nature is full of fractals.

This spacescape design utilizes habitation modules in fractal shapes like snowflakes. It is an application of fractal geometries. Self-similarity will make it easy to add new modules. This area is an example of expansion and is assumed to use local materials for the buildings. Each public room (kitchen, gym, laboratory, meeting room) and private rooms (sleeping area) lie next to a plant cultivation area. Each plant cultivation box will be connected by pipes, and the length of the pipe should be minimal. Therefore, each room will have plants. The small parts depicted in black are airlocks to avoid an air leak and heavy damage. The central core coloured in yellow is regenerative Life Support Systems (A. Ono et al. 2006).

The Lunar Zen Garden contains a mountain called tsukiyama. Tsukiyama literally means constructed mountain. The older term was kasan (artificial mountain). The tsukiyama, which is modeled on Mt. Fuji, is a kind of borrowed-landscape (Schumacher, 1995–2007).

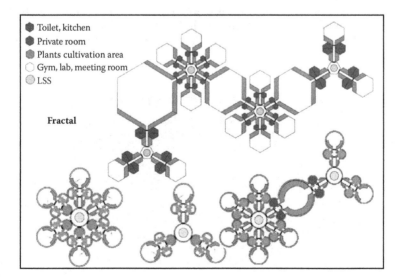

FIGURE 15.3
Module extension concept: fractal.

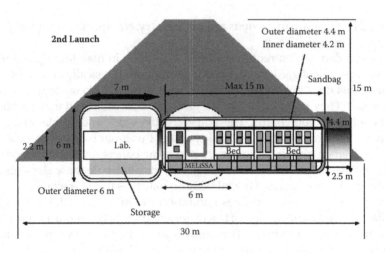

FIGURE 15.4
Concept of modules inside the lunar Zen garden.

There is another similar design: a white sand Mt. Fuji model called Kougetsudai at Ginkakuji Temple, a Zen temple which was established in the 15th century. The cone-shaped trapezoid Kougetusdai was a geometrically-designed mountain and a kind of installation. The creation date and purpose of this mountain is not certainly known, but it is considered to be an innovative design to enjoy the reflection of moonlight and to feel the pretty view with the moon placed above the mountain.

The spacescape design on the Moon will be able to use the same method to enjoy the Earth's beauty like a blue and green Moon. This installation may be a tourist attraction. The mountain should be able to contain one aluminium module and three inflatable modules for radiation shielding. At the same

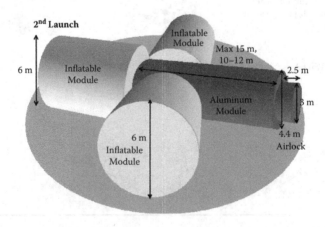

FIGURE 15.5
Another view of modules inside the lunar Zen garden.

FIGURE 15.6
Interior design.

time, this mountain will be an aesthetic spacescape design to enjoy both the Earth and the Moon's beauty. An important exemplification of these physiological principles is found in the charm of symmetry (A. Ono, et al. 2006).

Private Space

Simple is beautiful, and simplicity is a very important factor in Zen. Accordion curtains can hide a messy living area and make a private space very easily. The curtain could be a printed fabric that looks like sand raked in patterns in a Zen garden. Edible plants also become like photos or displayed art depending on the frame and simplicity of the design.

Cultivation area will be extended to use edible plants like foliage in each room. The design should be very simple. Then, the windows look like photos (refer to the plants section).

Also, music with headphones must be helpful to make privacy because it can cancel noise and improve someone's feelings with favourite music. Noise cancelling headphones will be the easiest way to improve daily life in working places such as the International Space Station (ISS).

The ISS has a high average noise of low frequency. The noise level inside of the ISS is about 60dB (see Figure 15.7), but it can be reduced to 25dB to 40dB by olefin hydrocarbon sheets, depending on the frequency difference. 35dB is the maximum noise limit at midnight. As a matter of fact, the ventilation system and many electronics make noise in a space module. To make a comfortable environment in outer space, auditory perception is a key factor deeply combined with our feelings and vision. Therefore, soundscape design and music therapy are recommended.

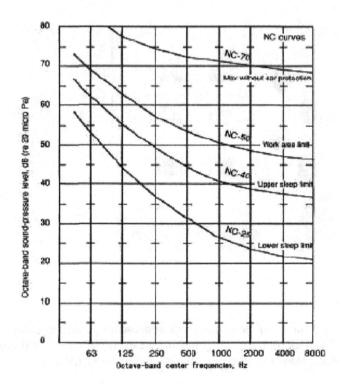

FIGURE 15.7
ISS interior noise criteria curves (ESA).

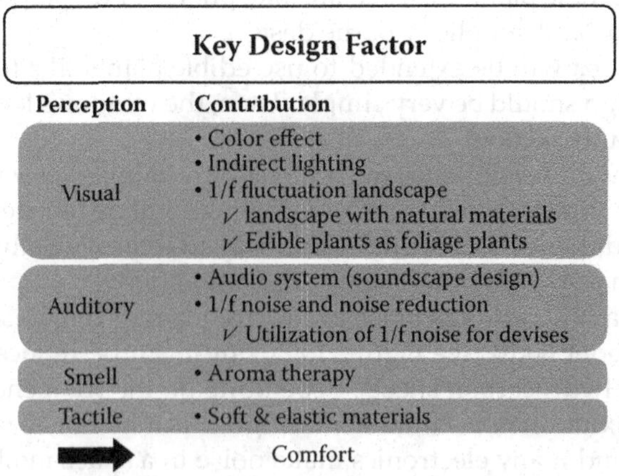

Key Design Factor		
Perception	**Contribution**	

Visual	• Color effect
	• Indirect lighting
	• 1/f fluctuation landscape
	   ν landscape with natural materials
	   ν Edible plants as foliage plants	

Auditory	• Audio system (soundscape design) • 1/f noise and noise reduction ν Utilization of 1/f noise for devises	
Smell	• Aroma therapy	
Tactile	• Soft & elastic materials	
→	Comfort	

FIGURE 15.8
Proposed solutions.

In NASA's research on human factors for the ISS, there are no strong links between environmental color and particular judgemental or emotional states (Wise, et al. 1988). Color decisions are simply considered as questions of personal taste.

Wise et al. explain "However, the influence of restful color on perception and feeling strongly relies on rather invariant patterns of simultaneous and successive contrast produced by their chroma and value dimensions. Psychological experiments aiming at studying the effects of color on time perception showed that accurate and normal time perception was observed only in the presence of green stimulation. Indeed, some color behavior effects obey a more primitive set of biological imperatives" (Wise, et al. 1988).

Therefore, bright green and several other tones of green are recommended colors in this proposal.

Common Space

Public Art

An artificial rainbow (see Figure 15.10) could also be made by a spotlight in a lunar base. This artwork uses a motor to turn the transparent ring, which transforms the rainbow as it turns.

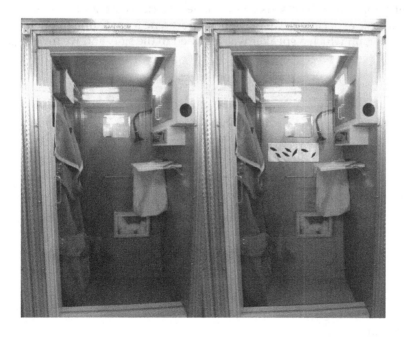

FIGURE 15.9
Effect of color.

Art Suitable for Microgravity

The purpose of introducing art suitable for microgravity is to inspire people through a cultural engagement with the experience of space. After space tourism is highly developed for us all, microgravity art will be a tourist attraction. In the near future, cultural dimensions will be important for long-term space pioneers as well.

Experiment of Space Art by C.S.A.

The "Sound Wave Sculpture" series were artistic experiments of new expression in a microgravity environment. It was created by Conference of Space Art (C.S.A.) a Japanese group interested in Space Art (A. Ono, 2005).

Since sound waves create air vibrations, micro particles can be "sculpted" in a cylindrical container and different music compositions can be used to generate different shapes and motions. Only light objects react to sounds in a 1g. environment (normal gravity on the Earth), but we noticed that the effects differed in a microgravity environment. Heavier objects can be moved by music composed with sinusoidal waves. Particles of different masses showed different motions. This series investigated the contents and the containers' shape for future artistic expressions. Music combined with sight is the point of this piece and many people said it is a sight one never tires of.

A parabolic flight was used for this experiment to make use of the free falling nature of microgravity environments. On the descent of a parabolic flight, people and objects inside the plane experience weightlessness.

As a basic experiment, we put spherical styrene foam (2 mm diameter) into a cylinder, on which a speaker was attached to the side, and we investigated

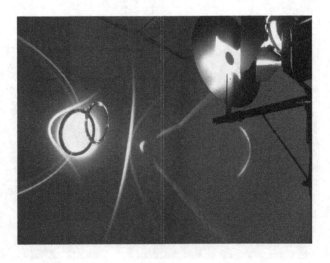

FIGURE 15.10
"Round Rainbow" by Olafur Eliasson (Hirshhorn Museum). Such expression will also be a comfortable spacescape or interior lighting design.

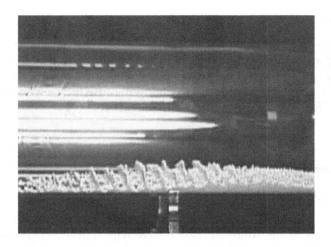

FIGURE 15.11
1 g, "Sound Wave Sculpture 2" by C.S.A.

the relationship between the formations of particles and the frequency of sinusoidal sound waves. Music composed by sinusoidal waves generated beautiful reactions in the cylinder.

Results of the Experiment

In the microgravity environment, high frequency sound made a more obvious reaction, and the movements looked like living things in water.

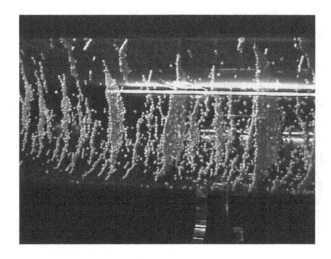

FIGURE 15.12
Microgravity, "Sound Wave Sculpture 2" by C.S.A.

Future Prospects

Adding variations of music, the movements of the objects can become more interesting and comfort inducing. As future possibilities, we can make public art in a space resort or its therapeutic effects in small spaces like space ships. The sounds and motions like water will cause the viewers to think of the earth. This will also be of interest to scientists.

Soundscape Design

Usually, daily life is full of sound combined with sights. Soundscape means an atmosphere or environment created by or with sound such as the raucous sounds of a city street or the relaxing sounds of ocean waves. Soundscape needs to be designed as a counter balance. Hence, soundproofing or anechoic rooms and soundscape design for relaxation rooms or private rooms will be recommended. In addition, if current ventilation systems can be changed, the electrical fan could be a new model of ventilation system which causes 1/f noise. National, a Japanese company, produced an electrical fan using 1/f fluctuation, and it is well known that 1/f fluctuation exists in nature. Electrical fans could reduce stress because biological rhythms have 1/f fluctuation. Noise with a 1/f spectrum has been investigated for 80 years, and there have been many theories. 1/f fluctuation also could be used for music therapy. Regarding the experimental soundscape design, multi-channel

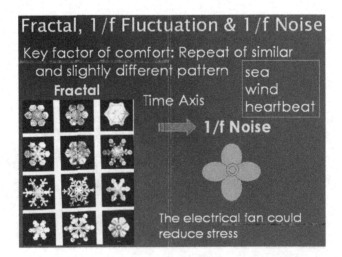

FIGURE 15.13
Fractal, 1/f fluctuation and 1/f noise.

speakers should be mounted on the walls and will offer sounds of nature and calm music. Feeling calm sounds over the full body has proven healing effects. This could be effective for human sensorial stimulation and relaxation (A. Ono, et al. 2006).

The techniques of soundscape design include the elimination or restriction of certain sounds (noise abatement), the evaluation of new sounds before they are introduced indiscriminately into the environment, the preservation of certain sounds, and above all the imaginative combination and balancing of sounds to create attractive and stimulating acoustic environments.

Conclusion

Space Art as Art Therapy may be a hobby for astronauts/cosmonauts that will help their well-being during their long-term missions. On the Moon and Mars, delay of communication is caused. To overcome such kinds of stress and solitude, art will be helpful to keep their mental balance. This section specified only visual and sound art therapy, but tactile stimulation, taste, and smell are also important for well-being. Both active and passive experiences through art help mental health. Space Art is not only for Art Therapy but also stimulation. Good Space Art makes the viewer want to "go there" and inspires many people.

Future Research Directions

For future opportunities, change of the image of art itself, in other words, change in thinking would be important. For instance, how to enjoy art as a creator is different from just a viewer's experience.

The role of art is to stimulate the imagination and to enhance the sensitivity of humans. In the near future, when the population increases outside of the Earth's atmosphere, Art in space can provide a more positive image of outer space environment for this radical transformation in terms of health and for creating new perspectives for the people living on Earth and in space. Anything could be used for expressions (see Figure 15.14), and everyone is a creator of his/her own life.

Vibration and Music Therapy

Music is a kind of vibration. More focused on the coenaesthesia is vibroacoustic therapy.

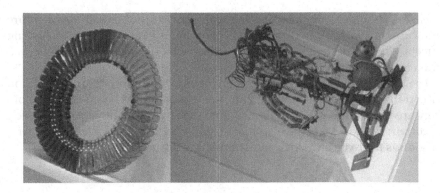

FIGURE 15.14
Examples of junk art: empty glass bottles and scrap metal.

Why has music been developed? That is because music contributes to well-being, i.e. reducing life stress through comfortable music. Music could make rhythms both mentally and physically effective. There are two types of music therapies, active and passive. Passive is just listening to music, and active music therapy is playing a musical instrument or singing a song. As music therapy, harmonious music and vibration would work for healing.

Art Gallery in Outer Space

This proposal is to build a relaxation room which can be used as an art gallery connected to the ISS or Space Hotel. Astronauts could relax even in

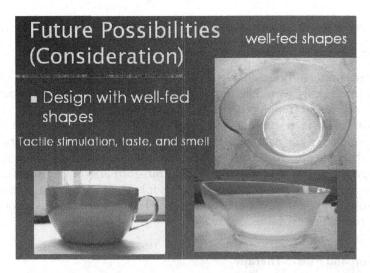

FIGURE 15.15
Examples of good design.

this confined space, and artists could use this place as a new environment for expression. For soundscape design, the relaxation room will be made as a totally silent, anechoic room which would have multi-channel speakers mounted on the wall, where astronauts can enjoy quiet sounds to relax the whole body. They could also enjoy feeling music vibrations with low-pitch beat sounds, and the lighting pattern could be varied to suit the musical atmosphere. The room could also have objects and fine art designed to enhance relaxation through tactile stimulation and sight.

Microgravity Art

Microgravity Art will be developed by many artists in the future. Some projects already in existence are continuing the investigations. The "Sound Wave Sculpture" as described above has two directions for development: One is a bigger dynamic version for a space ship or a space hotel; another is a compact version which can create visible sculpture from music.

Lunar Zen Garden

This art project will be continued from simulation to completion with the help of engineers and scientists. Some parts of a simulated Lunar Zen garden have already been built at the European Space Agency (ESA/ESTEC).

This project will be filmed in the near future and will be shown at an exhibition.

References

AA.VV., (2002). Zen garden secrets revealed. Kyoto University researchers Japan: BBC News.

Beuys, J., (statement dated 1973). First published in English in Caroline Tisdall: Art into Society, Society into Art (ICA, London, 1974), p.48. Capitals in original.

Fromm, E., (1960). Psychoanalysis and Zen Buddhism, Zen Buddhism and Psychoanalysis. New York (to be published), The USA: The Zen Studies Society, Inc.

Iwai, H., (1986). Deep Psyche of Colours and Shapes, Japan: NHK Books, p.42.

Ono, A., (2005). Environment Design in Space: Lunar Base and the ISS, Japan: IAF.

Ono, A., Hatanaka, N., Nakagawa, C., Chi, Y.K., (2006). Environmental Design: Space Art for the ISS and for a Lunar Base, The USA: AIAA.

Ono, A., (2007). Art for Psychological Support, Tools for Psychological Support during Exploration Missions to Mars and Moon, Noordwijk, The Netherlands: ESA/ESTEC.

Ono, K., (2005). Expressive Arts Therapy Drawing, Clay, Music, Drama, Dance. Japan: Seishinshobo.

Kitayama Y., (2007). Landscape gardener, Professional—Work Site, "Onore wo Dasazu ni Jibun wo Dasu ", Japan: NHK.

Narumi, K., (2001). Work of Urban Environmental Design (in Japanese), Japan: Gakugei shuppansha.

Nicogossian, A.E., (1992). "Meeting Human Needs" 29th Goddard Memorial Symposium, AAS 91-313. AAS Goddard Memorial Symposium, 29th, Washington, DC, March 14–15, 1991. American Astronautical Society (Science and Technology Series, Vol. 81) A95-87467, San Diego.

Schumacher, M., Copyright (1995–2007). Rock Gardens, Dry Landscapes, Hill Gardens. Karesansui, Kasan, Tsukiyama, Others, A to Z Photo Dictionary – Japanese Buddhist Statuary, URL: http://www.onmarkproductions.com/html/japanese-gardens.shtml [cited 1995] [viewed 1. 8. 2006].

Suzuki, D., (2002). What is Zen?, Japan: Kadokawa Sophia Bunko.

Truax, B. (1999). Handbook for acoustic ecology. Cambridge Street Publishing. England. Soundscape Design, retrieved from: http://www.sfu.ca/sonic-studio/handbook/Soundscape_Design.html (10 September 2006).

Wise, B.K.. Wise, J. A (1988). The Human Factors of Color in Environmental Design: A Critical Review, NASA Contractor Report 177498.

Additional Reading

D'Amico and P. Mazzettiome, (1986). "Noise in physical systems and 1/f noise" Proceedings of the 8th International Conference on "Noise in Physical Systems" and the 4th International Conference on "1/f Noise" (9–13 September, 1985) Rome (Italy). Editor Hardbound, Holland.

Naumburg, M., (1953). "Psychoneurotic Art: its function in psychotherapy," Grune & Stratton, New York.

Schlacht I.L., Ono A.; 2007. Natural design habitat on the Moon – Lunar Zen garden. pp. 219–220, 9th ILEWG/ILC 2007, ESA press.

Schlacht I.L. 2007. Art, Design and Human Metamorphosis in Extreme Environments. MutaMorphosis Book of Abstracts (published on line http://www.mutamor-phosis.org).

Wakao, Y., (2006). Thinking about Music Therapy. Ongakunotomo-sha. (Japan).

16

Psychology: Natural Elements as a Well-Being Stimuli in Outer Space

Scott C. Bates

Assistant Professor, Psychology
Utah State University

Joshua Marquit

Graduate Student, Psychology
Utah State University

Scott C. Bates, PhD is an associate professor in the department of psychology at Utah State University. With a background in social and environmental psychology, Dr. Bates has been pursuing the study of the non-nutritive benefits of having and/or tending plants in space in collaboration with researchers in the Space Dynamics Laboratory, National Aeronautics and Space Administration (NASA), and the Institute for Bio-Medical Problems (IBMP) in the Russian Federation. His PhD is in experimental psychology from Colorado State University.

Joshua Marquit, BS, is a graduate research assistant in the department of psychology at Utah State University. His research interests include a wide array of topics within environmental, social, and health psychology. He has worked closely with Dr. Scott C. Bates, Space Dynamics Laboratory, National Aeronautics and Space Administration (NASA), and the Institute for Bio-Medical Problems (IBMP) in the Russian Federation on a project focused on the non-nutritive benefits of viewing, growing, and/or tending plants in extreme environments. He is currently working on a PhD in experimental and applied psychological science at Utah State University.

ABSTRACT   The nature of long-duration space flight, including increased crew autonomy and isolation from standard models of psychological support, necessitates new models and modes for psychological support (Kanas & Manzey, 2003). It is our primary thesis that design decisions on the outer space habitat impact psychological support.

The presence of natural elements in long-duration space mission can be used as countermeasures; researchers have explored psychological benefit to exposure to nature.

To identifying and understanding the potential benefit of natural stimuli as plants in space, multiple perspectives, samples and variety of measures and indicators are presented.

Using Natural Elements in Habitation Design

For our purposes, "natural elements" include any design element that has a basis in nature. In some cases, they are limited to specific sensory domains such as images of nature. However, natural elements can also cross these sense-domains; plants are an example of this as they can be seen, felt, tasted, touched and even heard. The National Aeronautics and Space Administration's Man-Systems Integration Standards (1995) include recommendations regarding architecture design that suggest the use of décor that feature natural/naturalistic themes for the purpose of stress reduction.

Kaplan (2001) referred to the need to enhance "micro-restorative opportunities." This need may be particularly important in the context of space travel. One well-researched micro-restorative opportunity comes with seeing and/or interacting with nature and/or natural elements. The idea that natural environments, and natural elements in built environments, are important to people is not surprising. The varieties of humankind's relationships with nature can be easily observed across cultures. For instance, among some groups of indigenous peoples, harmony with nature is clearly emphasized (Schultz, 2002), while among other cultures nature generally, and wilderness specifically, is seen as something to be conquered (Nash, 1982). Obviously, the relationship among plants and people is an important one. Throughout history the meaning of nature to humans has gradually changed. Nature has been viewed as wild and unfinished, requiring interventions from humankind to tame and refine. Phrases such as "subdue" and "dominion over" were commonly used by politicians and natural resources managers when describing the human role in this relationship. For some, the meaning of nature was determined by its potential economic value to humans. The prevailing belief was that "nature was merely…a storehouse of raw materials: were thought to be valuable only to the extent that they could be used to create wealth" (Holling, Burke, and Folke, 1998, p. 345). Scientists, including environmental psychologists, have been grappling with this idea for some time.

The biophilia hypothesis, for instance, is the idea that humans are drawn to nature and natural elements—or at least express preferences for nature— and that this orientation is, in part, based on humankind's evolutionary development. Edward Wilson originally defined biophilia as "the innate tendency to focus on life and lifelike processes" (Wilson, 1984, p.1). Kellert (1993) extended this idea to include that "human dependence on nature…extends far beyond the simple issues of material and physical sustenance to encompass as well the human craving for aesthetic, intellectual, cognitive, and even spiritual meaning and satisfaction" (p. 20). From this perspective, given that humans developed in a natural environment and that that natural environment includes a variety of features that can hinder (e.g. poisonous plants and/or animals; great heights) or assist (e.g. open vistas for easy monitoring for predators) survival, it clearly follows that humans would likely prefer to be around particular aspects of nature.

Provided this background, there are two important theories from environmental psychology that can inform this discussion. They are Stress Reduction Theory, articulated by Ulrich and colleagues (Ulrich, et al, 1991) and the Attention Restoration Theory, described by Kaplan and Kaplan (Kaplan & Kaplan, 1989; Kaplan, 1995).

There is mounting evidence from various academic disciplines that nature, and experiences with nature, have a restorative component. In this way, restorative environments are those that foster positive emotional affect. Two theoretical supports for this idea have been generated. The first explanation was outlined by Ulrich et al. (1991), who argued that specific natural settings foster "restorative responses." Some of these "restorative responses" include stress reduction, anger reduction, and an overall restoration in energy and well-being. His theory of stress reduction was based on the concept of biophilia and the notion that humans have a biologically based affinity toward nature. This evolutionary perspective states that humans have developed a "natural" response to natural landscapes and are attracted to and ascribe specific restorative components to specific natural settings. Ulrich et al. argued that humans have spent many years co-evolving with nature and have developed through generations of experiences, a predisposition to affiliate with nature, developing a biologically-based affinity for settings that are natural as opposed to built or urban environments. Additionally, Ulrich et al. explains that humans experience setting-related stress due to the fact that many of us spend much of our lives in urban settings. He contends that we have not been able to adapt to the urban landscape because it has only affected a few generations of humans.

The second explanation was outlined by R. and S. Kaplan (1989), as well as Kaplan (1995). The couple argued that the restorative psychological benefits of nature is directly associated with attention restoration. Attention Restoration Theory (ART) states that human fascination of restorative environments has the potential to restore our ability to direct attention to challenges in our lives. In 1995, Steve Kaplan described the process by which humans restore their ability to concentrate to challenges. He explained that many of life's challenges require a direct concentration or attention to complete. To accomplish the task one must exert effort to maintain focus by ignoring distractions and gratification from emotions or other actions. Directed attention requires a tremendous amount of effort to maintain, sensitive to interruption and distractions, and difficult to recapture once lost. Following the completion of challenges, humans suffer from what is described by Kaplan as "directed attention fatigue" or mental exhaustion. Directed attention fatigue is a direct by-product of prolonged, focused attention to the challenge. As result of this fatigue, humans must find a way to relax and re-energize. Kaplan argues that recovery from this fatigue requires sleep, which by itself is insufficient, and involuntary and effortless attention to fascinating activities. One such activity is viewing natural settings. Natural settings are an important source of fascination for humans. These settings provide both "soft fascinations," such as scenic vistas and sun rises, and an escape from day-to-day challenges. The act of observing natural settings provides the mechanism for which stress recovery can occur.

Each of these theories is more similar than different. Both theories find their origins in landscape preference research and support the notion that experiences with nature have restorative psychological components. Additionally, both Ulrich and the Kaplans agree that nature is innately important to humans because of its role in maintaining psychological health. Hartig and Evan (1993) and Kaplan (1995) argue that the two theories will eventually evolve into one all-encompassing theory.

There are two important issues that frame our discussion of the impact that plants and other natural elements can have on people (Bates, Gushin, Marquit, Bingham & Sychev, 2007). First is the notion of benefit domains. A variety of benefit domains for plants and other natural elements have been identified. They are: a) physiological; b) psychological; and c) social. In each case, studies have been identified that support the notion that the relationship between plants and other natural elements, and people, can be observably positive. The second important issue has to do with interaction style—passive or active. Passive interaction includes looking at other natural elements without direct interaction. This may include staring out a window, or at a computer screen, showing images of nature. In addition, this can include simply being in a space that includes a natural element: having a houseplant, or hearing trickling water. The active style of interaction includes activities as simple as care-taking for a plant, as complex as gardening, or as immersive as hiking in nature.

These two dimensions benefit domain and interaction-style, and form an important diagram that represents the varieties of ways in which the literature on the plant/human interaction can be framed.

Passive Interaction with Natural Elements

There is evidence that simply having natural elements in our environment simply to look at may be beneficial (Ulrich and Parsons, 1990). In this passive manner of interacting with nature, a variety of potential positive benefits become apparent.

It seems that humans will, at times, create artificial contact with nature when real contact is not possible. In a study conducted by Heerwagen and Orians (1986), the ways in which office workers decorated their work-spaces were analyzed in an effort to detect if there was a discernible difference between those workers who had consistent access to a natural view—in this case, a window—and workers who did not. They found that while all workers used more nature-oriented material for decoration as compared to non-nature oriented material, workers with windowless offices used three times more nature-oriented material than did workers with windowed offices.

One particularly relevant study was conducted by Wise and Rosenberg (1988), who measured both physiological responses and preferences to natural elements in a simulated space-station. In this case, the bulkhead was designed to simulate one of four scenes: a savannah, a mountain scene, modern abstract art, or blank (which served as a control). The most preferred scene, based on self-report, was the mountain scene. Additionally, both nature scenes (savannah and mountain) were more effective at impacting physiology than either the control or the abstract art.

These two studies support a basic point: people prefer to look at nature-based stimuli. These results connect well to studies of basic preference for visual scenes. It is well established in laboratory experiments that people prefer to view landscapes that include plant-life as opposed to built environments (Ulrich, 1979). For instance, Kaplan, Kaplan and Wendt (1972) found that slides with images of nature were greatly preferred over those with images of urban settings. The difference was not accounted for by complexity of the images.

Beyond these preferences for natural stimuli, however, are studies that were designed to detect potential benefits, across all three domains (physiological, psychological, social) for simply seeing nature.

Three important studies by Roger Ulrich are keys to this idea. In 1979, Ulrich conducted a study using college students to examine the effects of natural scenery on stress related to course examination. He found that by viewing a series of scenes of natural settings, college students could reduce the intensity of the stress induced by course examination. In 1984, Ulrich studied the impact that natural scenery had on health-recovery by evaluating and comparing the outcomes for 46 patients recovering from gallbladder surgery. All were exposed to one of two room-conditions: a window view with trees, or a window view of another building. Ulrich found that patients in the recovery room overlooking the trees recovered more quickly, had fewer post-surgical complications, received fewer negative evaluative comments from members of the hospital staff, and used fewer pain-reducing medications. Finally, in a study conducted in 1991, Ulrich et al. conducted an investigation of the effects of viewing videos of natural and urban settings following a stressful video. The stress-inducing video was a 10-minute, black-and-white video used by the industrial industry to reduce on-the-job accidents. The film portrayed a variety of graphic accidents including mutilations and bloody situations. Immediately following this video, participants watched one of six

10-minute color videos of urban or natural scenes. During viewing sessions, participants' blood pressure, skin conductance, and muscle tension were measured to determine stress arousal. Additionally, participants were asked to verbally express their feelings while watching the videos. Ulrich et al. found that the rate of stress recovery was more rapid for those that viewed the video of natural settings as opposed to urban settings. Interestingly, they also found that the participants' heart rate slowed in both the stressful and nature videos, but not in the urban video. They hypothesize that the slowing of the heart rate is a product of heightened attention required by both the stressful and nature videos but not the urban videos.

Beyond the preferences, physiological impact of seeing nature is the idea that there are psychological benefits to having nature around us. Researchers have investigated the impact that nature or natural elements can have on mood, specifically on negative mood (e.g. depression, anger).

One relatively recent advancement in the study of mood is the use of cognitive priming. Priming is an experimental technique in which participants are exposed to a stimulus briefly—some at, and sometimes just below, perceptual awareness. Two experiments have used this priming technique to assess mood reactions to natural scenes. In both Korpela, Klemettila and Hietanen (2002) and Kietanen and Korpela (2004), this priming technique was used to present nature—or urban—scenes to participants. This priming was followed up with human vocal expressions of emotion. Participants were then asked to make a quick-judgement regarding the affective valence of the vocal expression. Korpela et al. (2002) found that participants reacted more quickly and accurately to "anger" stimuli when primed with urban scenes while reacting more quickly to "joy" when primed with natural stimuli. Thus, they concluded that nature generates positive affect while urban scenes are more likely to generate negative affect. In a second study (Hietanen and Korpela, 2002), faces were used as the follow-up stimuli and natural scenes that varied on their restorativeness and preferences were presented as primes. Again, positive emotion was associated with restorative nature and negative emotion was more closely associated with non-restorative nature.

Another interesting study of the impact that plants can have on mood was conducted by Cackowski and Nasar (2003). In this study, participants were randomly assigned to one of three videotaped highway drives that varied in their level of vegetation. State/trait anger was assessed, as was frustration tolerance (time spent on unsolvable anagrams). They found that while anger did not differ across the conditions, frustration tolerance did: a higher level of present vegetation was met with more frustration tolerance.

The impact that natural elements have on other aspects of psychological functioning have also been studied. Examples include Kuo (2001) and Kuo and Sullivan (2001), who studied the impact of having nature (e.g., nearby trees and grass) in an urban setting on a sample of urban poor. Kuo (2001) assessed attentional functioning, and effectiveness in managing major life issues and found higher levels of functioning among those participants who had been randomly assigned to public housing buildings with nearby nature (Kuo, 2001). In Kuo and Sullivan (2001), levels of mental fatigue, levels of aggression and violence were assessed in these groups. There, too, so-called nearby nature seemed to impact these variables positively.

Humans, of course, are not exclusively passive. While the evidence presented thus far is in support of the idea that minimally having nature, or elements of design that invoke nature, can create positive psychological support, the other important aspect is that of more active involvement with nature or natural elements.

Active Interaction with Natural Elements

While much of what we have summarized has been in the realm of passive-interaction, design can also include more active elements. More active interaction with plants has also been the target of investigation and may inform design decisions in the context of long-term space missions or other extreme environments.

One of the primary ways in which one can interact with nature is by simply walking through it. While there are many studies that examine and describe this type of interaction, for our purpose we will focus on the type of interaction that is possible in space: gardening.

Researchers have been actively looking at the impacts of gardening for decades. In this context, several studies emerge as particularly important. For instance, R. Kaplan (1973) collected self-report data from a variety of types of gardeners (e.g. community gardeners, home gardeners, plot gardeners) and found three categories of benefits. These included: tangible outcomes such as fresh vegetables or flowers; satisfaction outcomes such as desire to work in soil, or see things grow; and aesthetic pleasure, seeing plants grow or diversion. Each sub-type of gardeners, and a variety of background variables, predicted these particular outcomes.

The impact that tending to houseplants has on perceived control has also been studied. In this case, elderly nursing home residents were assigned to experimental (high-control) and control conditions (low-control; Langer & Rodin, 1976). In the experimental condition, plants were placed in common areas and the nursing home residents were told that the maintenance of the plants was their responsibility. In the low-control condition, plants were included in the common areas but were cared for by staff. Behavior and self-report measures showed benefits to participants in terms of alertness, active participation and well-being. Indeed, these benefits seemed to be sustained over time as nurses-ratings, as well as medical and mortality data showed (Rodin & Langer, 1977). In short, perceived control—implemented by using plants—was associated with better outcomes.

A variety of positive therapeutic effects of working with plants have been explored. Horticulture-therapy is defined as "the engagement of a person in gardening-related activities, facilitated by a trained therapist, to achieve specific treatment goals" (American Horticultural Therapy Association, 2007). The field of horticulture therapy was established in the 1970s (Relf, 1973) and continues to flourish across multiple domains (Relf, 1992). The *Journal of Therapeutic Horticulture* is published annually and summarizes the use of horticulture for a variety of therapeutic purposes.

Another example of the potential therapeutic impact nature or natural elements is found in research by Fabor Taylor, Kuo and Sullivan (2004), who studied "green activities" and their impact on children with attention-deficit disorder. In this study, parents of 96 children diagnosed with ADD were surveyed and asked about weekend and after-school activities that assisted functioning, and general surroundings. They found that so-called "green" activities were identified as "best" for symptom management in that they were related to better post-activity attentional functioning. That is, children with more green-exposed play settings were rated as higher functioning.

Considering that the current and future psychological dynamics of space travel (Kanas & Manzey, 2003) impact perceived control, attention and mood positively, as well as providing interesting and compelling leisure activity may justify the inclusion of plants and other natural elements in future design considerations.

Conclusion

Kanas and Manzey's book, *Space Psychology and Psychiatry* (2003) includes a summary of the current state of the art in understanding psychological processes, impacts and consequences of space travel. Given the psychological impacts including the variety of stressors including isolation, confinement, danger, and monotony since the start of human space travel, psychological countermeasures have been a topic of exploration. Habitability factors, such as those outlined herein, provide a crucial support for humans living in these extreme environments.

Natural Element in Habitation Design: Remarks

Much of the evidence presented in this section has come from studies of the impacts of people's interaction with, or viewing of, nature. We believe that what these studies point toward is the necessity of incorporation of natural elements into the design of habitats for space travel. Previously, we have presented options and ideas for habitation design in terms of visual design, ergonomics and art. In each of these, it is likely justifiable to include natural elements as a way to express what E.O. Wilson identified as humankind's biophiliac tendencies.

Again, humankind's connection to nature is not a surprise, nor is it new. In *Defiant Gardens*, Helphand (2006) reveals the consistency that humans have created gardens under the extreme circumstances of war or imprisonment. He concludes that the creation and maintenance of gardens—that most basic of natural elements—arise from basic human needs for life, connection to "home," for hope, for action, and for peace. Ultimately, he noted, "As human beings we identify with nature's vitality; it is alive, like us." (p. 212).

References

American Horticultural Therapy Association. (2007, May 29). Retrieved May 29, 2007, from http://www.ahta.org/information/faq.cfm.

Bates, S.C., Gushin, V.I., Marquit, J.D., Bingham, G., & Sychev, V.V. (2007, March). Plants as countermeasures in long-duration space missions: A review of the literature and research strategy. Presented at the European Space Agency Workshop: Tools for Psychological Support.

Cackowski, J.M, & Nasar, J.L., (2003). The restorative effects of roadside vegetation. Implications for automobile driver anger and frustration. Environment and Behavior, 35(6), 736–751.

Fabor Taylor, A., Kuo F.E., & Sullivan, W.C., (2004). Coping with ADD. The surprising connection to green play settings. Environment and Behavior, 33(1), 54–77.

Heerwagen, J.H. and G. Orians. 1986. Adaptations to windowlessness: A study of the use of visual decor in windowed and windowless offices. Environment and Behavior 18(5):623–639.

Hietanen, J.K. and Korpela, K.M. (2004). Do both negative and positive environmental scenes elicit rapid affective processing? Environment and Behavior, Vol. 36, No. 4, pp. 558–577.

Kaplan, R. (1985). Nature at the doorstep: Residential satisfaction and the nearby environment. Journal of Architectural and Planning Research, Vol. 2(2), pp. 115–127.

Kaplan, R. and Kaplan, S. (1989). The experience of natures: A psychological perspective. New York: Cambridge University Press.

Kaplan, S. (1987). Aesthetics, affect, and cognition: Environmental preference from an evolutionary perspective. Environment and Behavior, Vol. 19, No. 1, pp. 3–32.

Kaplan, S. (1995). The restorative benefit of nature: Toward an integrated framework. Journal of Environmental Psychology, 15, 169–182.

Kellert, S.R. (1993). The biological basis for human values in nature. In S. R. Kellert and E.O. Wilson (Eds.), The Biophilia Hypothesis. Washington, DC: Island Press.

Korpela, K.M., Klemettila, T., and Hietanen, J.K. (2002). "Evidence for rapid affective evaluation of environmental scenes." Environment and Behavior, Vol. 34, No. 5, pp. 634–650.

Kuo, F. E. (2001). Coping with poverty: Impacts of environment and attention in the inner city. Environment and Behavior, Vol. 33, No.1, pp. 5–34.

Kuo, F.E. and Sullivan, W.C. (2001). Aggression and violence in the inner city: Effects of environment via mental fatigue. Environment and Behavior, Vol. 33, No. 4, pp. 543–571.

Nash, B.C. (1982). Wilderness and the American mind (3rd ed.). New Haven, CT: Yale University Press.

National Aeronautics and Space Administration. (1995). Man-Systems Integration Standards, NASA-STD-3000.

Relf, D. (1992). The Role of Horticulture in Human Well-Being and Social Development. Timber Press.

Relf, P.D. (1973). Horticulture—a therapeutic tool. Journal of Rehabilitation. 39(1), 27–29.

Schultz, P.W. (2002). Environmental attitudes and behaviours across cultures. In W.J. Lonner, D.L. Dinnel, S A. Hayes, & D.N. Sattler (Eds.), Online Readings in Psychology and Culture (Unit 8, Chapter 4), (http://www.wwu.edu/~culture), Center for Cross-Cultural Research, Western Washington University, Bellingham, Washington USA.

Taylor, A.F., Kuo, F.E. and Sullivan, W.C. (2001). Coping with ADD: The surprising connection to green play settings. Environment and Behavior, Vol. 33, No. 2, pp. 54–77.

Ulrich, R.S. (1984). View through a window may influence recovery from surgery. Science, 224, 420–421.

Ulrich, R.S. (1999). Effects of gardens on health outcomes: Theory and research. Chapter in C.C. Marcus and M. Barnes (Eds.), Healing Gardens: Therapeutic Benefits and Design Recommendations. New York: John Wiley, 27–86.

Ulrich, R.S., Simons, R.F., Losito, B.D., Fiorito, E., Miles, M.A., & Zelson, M. (1991). Stress recovery during exposure to natural and urban environments. Journal of Environmental Psychology, 11, 201–230.

Ulrich, R.S. and R. Parsons. (1992). Influences of passive experiences with plants on individual well-being and health. In: D. Relf (ed.). The Role of Horticulture in Human Well-Being and Social Development: A National Symposium. Timber Press, Portland, OR.

Wilson, E.O. Biophilia. Cambridge, MA: Harvard University Press.

Wise, J.A. and E. Rosenberg. 1988. The effects of interior treatments on performance stress in three types of mental tasks. CIFR Technical Report No. 002-02-1988. Grand Valley State University, Grand Rapids, MI.

Additional Reading

Ono, A., (March 2007). Art for Psychological Support, Tools for Psychological Support during Exploration Missions to Mars and Moon, ESA, ESTEC, Noordwijik, The Netherlands.

Schlacht I.L., Rötting M., Masali M., (March 2007). Color design of extreme habitats as a psychological support for the reliability. Tools for Psychological Support during Exploration Missions to Mars and Moon, ESA, ESTEC, Noordwijik, The Netherlands.

17

Perspectives: Multidisciplinary Approach for User Well-Being

Irene Lia Schlacht

Cand. PhD Human Machine System
Dipl. Industrial Designer
Technische Universität, Berlin

Irene Lia Schlacht is an Italian designer born in Milan. After various experiences and several publications on the outer space field, she is now studying for a doctorate at the Technische Universität Berlin, under Prof. M. Rötting. The doctorate's title is "Visual Design for Outer Space Habitability." She earned a design masters degree in 2006 from the Politecnico di Milano; her thesis was titled "Colour Requirement in Outer Space Habitats" (Prof. D. Riccò). She was also an outer space human factor researcher at Università di Torino (Prof. M. Masali). After completing a report on outer space color design during her stage at Thales Alenia Space Italy, she directed an experiment on color perception in microgravity, taking parabolic flights with the European Space Agency (ESA) in September 2006. The goal of her research is to improve the working conditions of astronauts using design and ergonomics. Today, together with experts in several fields from several countries, including the Japanese space artist Ayako Ono, the Italian space anthropologist Prof. Melchiorre Masali, and the American Psychologist Scott Bates, she has created an outerspace research group called www.Extreme-Design.eu. Cand. PhD Human Machine System, Dipl. Industrial Designer, Technische Universität Berlin, Department of Human Machine System.

ABSTRACT To create a living quarter in space, of course we must reduce the cost and increase the productivity, but we should also take art and culture expression as a part of the basis for the humanity of the human being to be different from a machine.

As stated in the nine principles of the modern architecture of Frank Lloyd Wright, we should use "ornament that came out of the nature of the materials, to make the whole building clearer and more expressive as a place to live in" (Wright, F. L., 1987, p. 74).

Introduction

This group of papers has been written by specialists in different fields coming from different cultures and countries; the main objective is to give specific solutions and different ideas on how to increase the habitability in space.

Different approaches of the Outer Space Habitat Design have been presented to point out that to improve habitability we need to make the most from the start of the project with a multidisciplinary understanding.

Today Art, Psychology, Anthropology and Visual Design are not taken as an important functional part of the design of space habitats. It is rather a question of building a system which works well from the standpoint of the technical aspects.

Human protection and the provision of labor and research areas stand in the foreground, which is comprehensible, since it is absolutely necessary as a part of the human habitat design also for short duration missions. However in the context of long time duration missions like space travelling and space settlement, this way of thinking has to be widened.

The importance of human well being has increased and psychological aspects are now of particular relevance to the health of the user. The question of spacecraft habitability has to be taken into consideration.

Proposing the theory of an American psychologist, two Italian anthropologists, a Japanese artist and German engineer and pedagogue, we discovered a common point of view: the relationship with the Natural Design Philosophy. Natural Design implies the well being of humans utilizing natural local elements or imported biological stimuli:

- use of natural elements present in the Space local environment (e.g., Moon stone)
- evoking natural earth stimuli as structures and inputs to guarantee the well being of humans as earthly animals (e.g., day–night solar circadian cycle, seasonal climate and light variation)

In conclusion, regarding flights to Mars or the creation of a habitat on the Moon, the engineering elements of the system like energy support, radiation shielding and propulsion have to be combined from the start of the design with physical and psychological disciplines to build up a proper human environment.

Remarks

Outer Space Habitat Design is focused on the creation of the conditions in which the user can experience space travel and also the importance of the protection of the Mother Planet.

Outer Space is today a place present in the mind's eyes of everybody! Common thinking of human beings is that after we caused the destruction of the Earth we will migrate to another planet to colonize it as common parasites do.

Nevertheless not all of us know that another Earth to colonize has not yet been discovered; Mars or the Moon are not places that can guarantee human life, Venus is the extreme of greenhouse effect! There is yet not a reliable shield to protect humans from space radiation and a solution for weightlessness is still remote.

Star Trek's Enterprise, Kubrick's Spacecraft, O'Neil-type Space Colonies, are quite utopist. Still Space may soon become an "Everyone Experience," as the Astronauts say[1]; with the philosophy to get us to discover "how fragile and precious our planet is."

Future Possibilities

One discipline itself cannot find the solution for an entire complex system. However, space-based habitats pose major challenges regarding communication between varied disciplines. Future research to design the journey and space habitats needs to focus more on the multidisciplinary demands and appropriate ways of communication and interaction.

Here Yaari Pannwitz (designer, open space facilitator) and Dominik Ringler (social scientist, open space facilitator) present the "open space technology" as a tool, a method as well as a process that enables groups to deal with and solve complex issues in this context.

Open Space Technology vs. Outer Space Habitat Design

How do we meet the need for an appropriate way of communication in the multidisciplinary approach? How do we take advantage of chaos to come up with order? Open Space Technology is a technique invented in USA that allows the creation of projects with multidisciplinary methodologies; this technique is applied here to design Outer Space Habitats. Open space is a proven "large group intervention" which allows us to achieve results in days for what other approaches require months and years. The open space setting works with 5 up to 2000 people and takes 1.5 to 3 days.

---

[1] This concept was expressed particularly by the Canadian Robert Thirsk and also by the Japanese Chiaki Mukai at ISU lecture (Strasbourg November 2006), the Dutch Andres Kuipers during the success meeting (Köln April 2007), and other astronauts.

By now there are quite a few examples where open space worked successfully. *"Opening space for outer space"* can be applied, for example, in international projects such as on the "Moon Mars Workshop,"[2] and the "Space Generation internet community annual meeting."[3]

Developing highly complex adaptive (flexible and able to learn) systems in a chaotic and mostly unknown environment—outer space—open and self-organising systems in which chaos is appreciated are needed.

> Chaos appears in multiple forms. It is always painful if you happen to be caught in the path, but for all that pain there appears to be a purpose—opening space in the old order so that the new may appear. It might just be that this life we hold so dear is less about the established forms, and existing order, than the journey itself. In which case the chaos we experience is by no means just a painful incidental, but rather an essential component, for the journey would clearly cease without open space in which to move forward.
>
> **(Owen 2003, p. 6).**

Identifying givens and adapting a minimal level of formal structure is required to ensure the flexibility to cope with the constant process of formation and adjusting to a constantly changing and highly challenging environment.

The idea of open space is to get the "whole system" together—opening space for outer space. No agenda is predetermined, no speeches are held, and there is actually no plan! There is just a common theme which leads into a direction. The subject must be complex and urgent, the answer must be unknown, and multidisciplinary is a must. Self-organization is supported, communication is highly effective since small groups of people only work on the issues they are qualified for and capable of getting them done, issues they have a passion for and are willing to take responsibility for. The participants create and manage their own agenda of parallel working sessions around the central theme of strategic importance, interconnecting, e.g., interior design, ethics, ergonomics, anthropology, psychology, art, design, engineering, communication, philosophy, biology, architecture, medicine, mathematics....

> Open Space works best when the work to be done is complex, the people and ideas involved are diverse, the passion for resolution (and potential for conflict) are high, and the time to get it done was yesterday. It has been called passion bounded by responsibility, the energy of a good coffee break, intentional self-organization, spirit at work, chaos and creativity,

---

[2] "Moon Mars Workshop" (MMW) is an annual meeting organised with the collaboration of ESA, it consists of a group of young students from different fields and countries that meet for one week to create new ideas and projects on the theme of Space Habitats.

[3] "Space Generation internet community annual meeting" at the MMW, young students from different fields and countries that meet for one week to create new ideas and projects on the theme of Outer Space Design.

evolution in organization, and a simple, powerful way to get people and organizations moving—when and where it is needed most.

(Hermann 2005, p. 1).

Creating adaptive complex systems for outer space is like sitting on a fence...leaving the well-known behind and jumping into the unknown (giving up control). The same principles which are used for any open space setting should also work here: whoever comes is the right person, whatever happens is the only thing that could have, whenever it starts is the right time, when it's over it's over/when it's not over it's not over.

Remarks

As a final future input we propose a project where we can apply the "open space technology:"

- Space tourism: space as "everyone's experience"
- Natural design: human factors and natural stimuli for well-being
- Space education and communication: space as an experience to learn respect for the mother planet.
- Cultural utilization of space: space can be used for artistic expression and to enlarge the human wisdom, not only for economic goals (the cultural application of ISS is a department of ESA).

References

Aguzzi, M., Häuplik, S., (2006), Design Strategies for Sustainable and Adaptable Lunar Base Architecture. IAC-06-D4.1.03 Paper presented at the meeting of the International Astrophysical Conference, Valencia, Spain.

Guilton, J., Guilton, M., (1981). The Ideas of Le Corbusier on Architecture and Urban Planning. New York, United States: George Brazillier.

Hermann, M., About open space—an executive summary, Chicago 2005 (www.openspaceworld.org)

Owen, H. (2003), The Practice of Peace, Berlin.

Holmann, P. /Devane, T., (2007). The Change Handbook (second revised edition), San Francisco.

Owen, H., (1997). Open Space Technology: A User's Guide, Second Edition, San Francisco.

Owen, H., (1999). The Spirit of Leadership: Liberating the Leader in Each of Us, San Francisco.

Owen, H., (2000). Power of Spirit: How Organizations Transform, San Francisco.

Chapter Acknowledgments

This work has been possible thanks to the passion and the effort of all the coauthors whose teamwork bridged deeply differing cultures, time zones, disciplines, ages and approaches. To Valentina Villa, the other illustrators a special thanks for the images, to iGuzzini illuminazione SpA and NASA a special thanks for the pictures. Thanks to the space artist Liuccia Buzzoni for the contribution in the Space Design section, Dominik Ringler and Yaari Pannwitz for the contribution with Open Space Technology section in the Perspectives (final part).

We would like to thank the Chair of Human-Machine Systems, Prof. of the Technische Universität Berlin for its support, in effort, time, and facilities. Politecnico di Milano: Prof. Dina Ricco, team of "Laboratorio Colore," Prof. Giulio Bretagna, Prof. Annalisa Dominoni, Prof. Bandini Buti, from the Design Faculty and from the Aerospace Faculty: Prof. Amalia Finzi, Prof. Cesare Cardani, Ing. Stefano Brambillasca. Thales Alenia Space Italia; team of Human Factor and their collaborators. Arch. Giorgio Musso, Ing. Enrico Gaia, Dr. Vincenzo Guarnirei, and the Thales collaborators: Ing. Blaine Sessions (Sofiter), Jaime Forero (NASA), the anonymous astronauts who participated at the interview on color habitat design. Technische Universität Berlin: Prof. Matthias Rötthing, Prof. Dietrich Manzey. Cospar and IAA association. The reviewer and friends Melih Bakirtas and Prof. Baerg from the TU-Berlin. Special thanks to Ms. Naoko Hatanaka, a student of architecture at The University of Tokyo in 2006; Ms. Chitose Nakagawa, a student of architecture at Tokyo National University of Fine Arts and Music in 2006; and Mr. Yong Kyu Chi, product designer; and COLOSSO Co. Ltd in Seoul, for their generous contribution of images and ideas.Special thanks as well to the European Space Agency (ESA) including Mr. Dieter Isakeit and Dr. Bernard H. Foing for the invitation to ESA/ESTEC, ESA/ESRIN, and the 9th ILEWG International Conference on Exploration and Utilisation of the Moon, and also for all their cordial support and assistance. Also special thanks to Mr. Gianfranco Visentin, Mr. Pantelis Poulakis, Mr. Willem Van Hoogstraten and other specialists for technical support. All the persons who supported the authors at Noorwijk ESA center with effort and friendliness. Arch. Alessandra Fenoglio and Dott. Marco Moietta of the Dept. of Human and Animal Biology of the Turin University for collaboration in graphics and library search to part of Anthropology. Thanks as well to The Art Catalyst including Ms. Nicola Triscott, who organized and supported the exhibition at ISEC. Also thanks to Arch. Jun Okushi, Space Projects Group; Arch. Mark M Kohen, and NASA-Ames for their generous support and encouragement to pursue this analysis. Prof. Takuro Osaka at the University of Tsukuba for the first opportunity to participate in a Space Art project in Japan. Leonardo Olats: Prof. Annick Bureaud, Prof. Roger Malina. Arch. Sandra Häuplik Techical University Vienna and Liquifer team Wien. Cospar and IAA association. We would like also to thank everyone else not mentioned but involved for their consultation, time, and support.

Section IV

The Human Condition

18

An Analysis of the Interface between Lunar Habitat Conditions and an Acclimatized Human Physiology as Defined by the Digital Astronaut Project

Richard L. Summers

Department of Emergency Medicine and Department of Physiology and Biophysics, University of Mississippi Medical Center, Jackson, Mississippi

Thomas G. Coleman

Department of Emergency Medicine and Department of Physiology and Biophysics, University of Mississippi Medical Center, Jackson, Mississippi

Robert L. Hester

Department of Emergency Medicine and Department of Physiology and Biophysics, University of Mississippi Medical Center, Jackson, Mississippi

Richard L. Summers, MD is a professor of emergency medicine and an assistant professor of physiology and biophysics at the University of Mississippi Medical School, where he serves as director of emergency medicine research. He received his MD degree from the University of Mississippi Medical Center in 1981. After clinical training in internal medicine from 1981–1982 at the University of Mississippi Medical Center he entered advanced doctoral graduate studies in the Department of Physiology and Biophysics under the tutelage of Dr. Arthur C. Guyton. In 1989 he joined the faculty of the University of Mississippi Medical Center and became an assistant professor in the Department of Physiology and Biophysics. After becoming board certified in emergency medicine in 1993 he switched his emphasis to his present clinical position. Through his collaborations with NASA, he has investigated the mechanisms of physiologic adaptations to microgravity. In addition to his clinical responsibilities, Dr. Summers currently oversees the development of the NASA Digital Astronaut Project.

ABSTRACT   Background: The physiologic acclimatization of humans to the lunar environment is complex and requires an integrative perspective to fully understand the requirements for settlement habitat conditions. A large computer model of human systems physiology (Guyton/Coleman/Summers Model) provides the framework for the development of the Digital Astronaut used by NASA in the analysis of biologic adaptive mechanisms. The model offers a means for the examination of the interface between a lunar adapted human physiology and potential habitat environments.

Methods: The current Digital Astronaut model contains over 4000 equations/ variables of biologic interactions and encompasses a variety of physiologic processes of interest to humans during spaceflight. The model is constructed on a foundation of basic physical principles in a mathematical scheme of interactions with a hierarchy of control that forms the overall model structure. Physiologic relationships derived from the evidence-based literature are represented as function curves within this structure. Different physiologic systems and body organs are connected through feedback and feedforward loops in the form of algebraic and differential equations to create a global homeostatic system. The model also contains a biologic-environment interface with external conditions such as temperature, barometric pressure, atmospheric gas content and gravity. During computer simulation studies, the predicted physiologic responses to a habitat environmental change are tracked over time.

Results: Computer simulations using the model have been found to accurately predict the physiologic transients seen during entry into, prolonged exposure to, and return from the microgravity and bed rest environments. Computer simulation studies suggest that humans with a lunar adapted physiology would be more vulnerable and less tolerant to extreme changes in habitat temperature, humidity and atmospheric oxygen content as compared to an equivalent earth-based setting.

Conclusions: An analysis of the interface between proposed lunar habitat conditions and an acclimatized human physiology as defined by the Digital Astronaut Project may be important to reduce potential health risks. This system can be used as a tool in the technical planning and design of lunar settlements.

Introduction

The potential risks for human health to be considered during extraterrestrial habitat design are difficult to assess. Predicting the impact of changes in environmental conditions on human physiology and the allowable variations in that environment will be an important consideration in the planning of lunar habitats [1,2]. However, it is evident that there is a certain degree of human physiologic adaptation to conditions of microgravity exposure during spaceflight and probably also in partial gravity environments. If lunar habitat environmental structures are developed around established earth-based human physiology, it is uncertain that these criteria will provide a safe dwelling for the 1/6 g adapted individual.

Since there is very little data concerning human physiologic adaptations to the lunar gravity, computer simulations using a mathematical model of human physiology may provide the only means for assessing the impact of these habitat conditions on human health. A large integrative computer model of human systems physiology (Guyton/Coleman/Summers Model) provides the framework for the development of the Digital Astronaut used by NASA in the analysis of biologic adaptive mechanisms [3,4,5]. The model provides a means for the theoretical examination of the interface between a lunar adapted human physiology and potential habitat environments. A systems analysis approach using the Digital Astronaut computer model was employed to examine the possible impact of different habitat environments on a lunar adapted human physiology. In this paper, an example of the physiologic effects of an extreme change in habitat temperature is presented.

Methods

Digital Astronaut

The current Digital Astronaut model is a special adaptation of an existing benchmark computer model (Guyton/Coleman/Summers model) developed by the investigators over the past 30 years [3,4,5,6]. The benchmark model contains over 4000 variables of biologic interactions and encompasses a variety of physiologic processes of interest to humans during spaceflight including cardiovascular functioning and adaptation to microgravity, bone metabolism, neurohormonal adaptations to weightlessness, and general nutritional and metabolic mass balance. The process of model building is centered around the concept of a hierarchy of control in which relationships are constructed primarily on a foundation of first principles (i.e., mass balances, physical forces). The current Digital Astronaut model will serve as the framework for continued future model expansion to include a greater detail of many of the existing systems as well as the addition of other systems of interest. The model can be solved using common numerical methods on a variety of computing systems. The software interface supporting the model is designed to provide for simple interaction of the user through a desktop platform with current personal computing technology or with a mainframe. The model and software support system allows scientists to perform complex systems studies and theoretical hypothesis testing on specific research questions surrounding human exposure to microgravity [7]. The model structure is presently specified in compiled C++ code but is being translated into XML in a component-based format (kidney, liver, circulation, etc) with a top down profile (molecular to cellular to organ to system to whole body) and extensive documentation as a part of the model description. An interface between environmental and biologic conditions is being developed within the Digital Astronaut Project. The current model platform allows for analysis of the impact of changes in altitude, barometric pressure, atmospheric humidity, external temperature and gravity on physiologic responses (Figure 18.1). Also possible is a prediction of the influence of changing available water and salt conditions as well as variations in the atmospheric gases (oxygen, nitrogen, hydrogen, carbon monoxide) and nutritional metabolites.

In order to build confidence in the integrity of the predictions by the Digital Astronaut, it is necessary that the model undergo a rigorous validation process as each new system completes a phase of development. The most important part of this process is the comparison of physiologic endpoints that typify and define the adaptation to microgravity to those predicted by the model. The model predictions are validated against experimental findings by demonstrating that the predicted values are within the 95% confidence interval of the established target value. The current model has been validated for general cardiovascular and metabolic functioning [4].

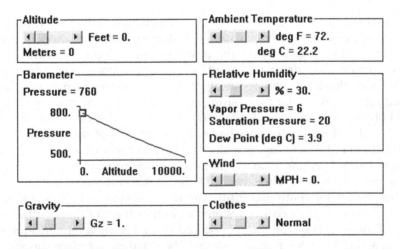

FIGURE 18.1
This figure shows the variables that interface between the Digital Astronaut and the external environment of the habitat in the model. These variables, currently at normal values, can be modified over a wide range of inputs.

Computer Simulation Studies

As an example of the methodology, the theoretical effect of changes in habitat temperature on a human whose physiology is adapted to the lunar environment is compared to that predicted for a human on earth in similar condition. The analytic procedure involves recreating the scenario of exposure to spaceflight and subsequent acclimation to 1/6 g for a virtual astronaut in a simulation environment. The impact of the variation in habitat temperature, barometric and oxygen conditions on human physiologic variables was analyzed and the sequential changes were recorded during the time course of the simulated protocol. Utilizing the rich details of the model, the particulars of the physiologic adaptations to these perturbations were examined including an examination of the relative changes in the blood chemistry constituents. This methodology allows us to develop a sophisticated approach to habitat analysis from the perspective of the very complex physiologic adaptive process. A similar technique has been used successfully to understand mechanisms pertaining to biologic adaptation to spaceflight and the microgravity environment that were not intuitively obvious otherwise [5,8,9]. By this method we can support the human safety predictions of our habitat design ideas through an exacting and quantitative approach.

Results

A computer simulation study was performed in which a wide range of human physiology was monitored for an earth bound astronaut. Simultaneously an identical astronaut adapted to the Moon's gravity was observed for comparison. After a week of stabilization of physiologic parameters at a habitat temperature of 72° F, the computer simulation was changed to model an abrupt change in the environmental temperature to 100° F for both astronauts. The allowable drinking water intake was limited to 2 liters/day to simulate the limitations in available resources on the Moon. After the change in habitat temperature, there was a rapid and progressive loss of plasma volume for both astronauts while the extracellular plasma sodium concentration rose precipitously. Changes in these and other physiologic parameters were reflective of the development of a profound dehydration which was more pronounced in the astronaut on the Moon (Figure 18.2). The time to development of a clinical condition of heat stroke in the astronauts (as determined by loss of consciousness) was found to be 47 hours for the Moon based astronaut as compared to 55 hours for his earth bound counterpart (Figure 18.3).

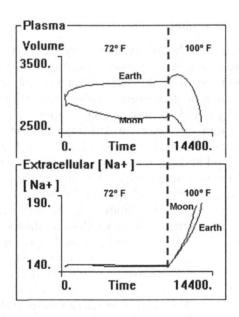

FIGURE 18.2
This figure shows the output from a computer simulation using the Digital Astronaut in which temperature of the habitat is abruptly changed from 72 to 100 degrees Fahrenheit while the drinking water allowance remains constant. The changes in plasma volume and extracellular sodium concentration [Na+] over time (minutes) are followed for an astronaut on earth and compared to changes predicted for an astronaut adapted to the lunar gravity.

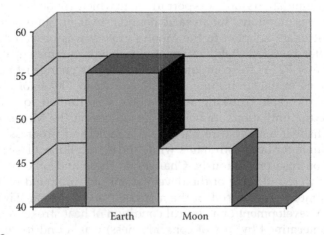

FIGURE 18.3
Comparison between a Moon-adapted astronaut and an astronaut on Earth in the time to development of heat stroke after a change in habitat temperature to 100 degrees Fahrenheit and a fixed drinking water allowance.

Conclusions

Any design of a lunar habitat should also consider the impact of environmental conditions on human health and safety. The physiologic acclimatization of humans to the lunar environment is complex and requires a sophisticated approach to predict the requirements for settlement habitat conditions. An advanced analysis of interactions between any proposed limits for lunar habitat conditions and a human physiology acclimated to the Moon's gravity may be important to reduce potential health risks. Computer simulation studies using NASA's Digital Astronaut suggest that humans with a lunar adapted physiology would be more vulnerable and less tolerant to extreme changes in habitat conditions as compared to an equivalent earth-based setting. The example provided in this study also suggests that this system can be used as a tool in the technical planning and design of lunar settlements.

References

1. Litton CE. Lessons learned studying design issues for lunar and Mars settlements. Life Support Biosph Sci. 1997;4:127–44.
2. Ashida A. Considerations of human's long stay in closed systems. Acta Astronaut. 1994;33:149–53.
3. Montani JP, Adair TH, Summers RL, Coleman TG, Guyton AC. Physiological Modeling and Simulation Methodology: From the Mainframe to the Microcomputer. J of the Miss Acad of Sci 1989;24:41–54.
4. Summers RL, Coleman TG, Meck JV. Development of the digital astronaut project for the analysis of the mechanisms of physiologic adaptation to microgravity. Acta Astronautica 2008; in press.
5. Summers RL, Coleman TG. Computer Systems Analysis of the Cardiovascular Mechanisms of Reentry Orthostasis in Astronauts. Comput Cardiol. 2002;29:521–525.
6. Coleman TG. HUMAN: mathematical model of the human body in health, disease, and during treatment. ISA Trans 1979;18:65–73.
7. Summers RL. Computer simulation studies and the scientific method. J. of Applied Animal Welfare Sci. 1998 1: 119–131.
8. Summers RL, Martin DS, Meck JV, Coleman TG. Computer Systems Analysis of Spaceflight Induced Changes in Left Ventricular Mass. Comput Biol Med 2007; 37:358–363.
9. Summers RL, Coleman TG, Platts S. Systems Analysis of the Mechanisms of Cardiac Diastolic Function Changes After Microgravity Exposure. Acta Astronautica 2008; in press.

19

Mental Health Implications of Working in a Lunar Settlement

Chester S. Spell

School of Business–Camden
Rutgers University, Camden, New Jersey

Chester S. Spell, PhD is an associate professor of management at the School of Business–Camden, Rutgers University. He received his PhD from the Georgia Institute of Technology. In addition to his interest in extreme work environments, his current research interests focus on the relationship between working conditions, employee mental health, and substance abuse in the workplace. A current project also investigates how employee behavior, both in and away from the workplace, has been monitored and evaluated by organizations and how monitoring techniques have changed over time. His research has appeared in the *Academy of Management Journal,* the *Journal of Management,* the *Journal of Organizational Behavior,* and the *Sloan Review of Management.*

ABSTRACT Technology makes it possible for human beings to live and work in places that in the past we could neither reach nor have survived in. Extreme environments present unique opportunities for psychological research and applications. At the same time, research on the effect of working in extreme, isolated environments is relevant for planning and staffing a potential lunar base.

This paper's purpose is to review what we know about the effect on humans of working in extreme environments, as well as recommendations of how to minimize the deleterious effects on crews working under such conditions. In reviewing what we know to date, we will point to the implications for planning and staffing a lunar base. The paper will also report recent research in the relationship between group dynamics on psychological distress and discuss why this recent research is relevant for lunar base planning. Finally the paper concludes with recommendations drawn from current research.

Review of the Literature on Extreme Environments

The majority of psychological research involving the effects of working in Extreme and Unusual Environments (EUEs) has been done in polar stations, space vehicles, and undersea habitats. The people living and working under these conditions have been studied through a variety of methods, including psychometric and projective tests, interviews, participant observations, field studies, and simulations. In drawing conclusions about the results of psychological studies on EUEs it is important to remember that how people experience an environment is more important than the objective characteristics of the environment [1]. So, most of the following descriptions of EUE effects on people will be seen in terms of the perceptions of the environment rather than the objective characteristics. The effects on people who work for extended periods in extreme and isolated environments can be grouped in four categories. The categories are feelings of social isolation, confinement, restriction of sensory functions, and disruption of circadian rhythms.

While some of the research on working in EUEs suggests that there are significant challenges in creating and maintaining the psychological health of the workforce, the research is not unequivocal that all people working in EUEs experience psychological strain. For example, autobiographical accounts of astronauts [2] claim on balance very positive reactions to space travel (albeit of a duration of only a few days). There are studies of people working at Antarctic stations that report many more positive than negative experiences. The same is true of individuals working for periods of time in undersea laboratories. Twenty-six of 28 participants in an undersea habitat study were willing to do it again. The long-term after effects of such experiences are also strikingly positive according to some reports. Both self-reports and other data show that people who have come through a demanding capsule mission are mentally and physically healthier, more successful, and more insightful than they had been or when compared to matched control groups [3]. Therefore, in a wide variety of EUEs, some people working for short or even extended periods report no psychological problems arising out of isolation or confinement.

The perception of confinement and the monotony of seeing the same small set of people day after day have been identified as key sources of psychological distress connected with EUEs. Thus social factors have been seen as a key source of mental health. Smith [5] concluded that after 2 weeks or more of confinement, the foremost psychological irritants were inadequate leadership and the behavior of others. Interpersonal problems have been identified in a wide variety of EUEs including Antarctic station [5], in submarines [6], and space [7]. In fact, interpersonal relations were soon identified as the biggest issue when analyzing diaries of Antarctic personnel [8].

Conflict between inhabitants of EUEs due to interpersonal issues has been identified in many studies. These conflicts have sometimes arisen from seemingly trivial matters, such as a murder over a chess game reported by Palinkas [9], as well as assaults with a hammer [10]. Schwetje [11] even speculated that crimes committed during space missions could be defended on grounds of irritability due to working conditions. There have been widely documented reports that Russian and American astronauts sometimes have problems, especially if there is only one Russian among several Americans, or vice versa.

Some of the problems connected to conflict among people in EUEs may be due to the fact that individuals are removed from their typical work routines and roles (before living in the EUE) and so their role expectations may be unclear (what do they need to do or how should they behave as a member of the group).

In addition to social problems, environmental characteristics play a role in mental health issues for at least some inhabitants of EUEs. Again, much of this evidence comes from studies conducted at polar stations. Social conditions are made challenging because individuals may feel isolation yet paradoxically they may feel crowded because of restricted living space [12]. Living at the pole has been connected to seasonal changes in mood and behavior due to long nights. This effect in connection to the polar environments has often been referred to as the "winter-over syndrome." Winter-over syndrome has been associated with depression, irritability, insomnia and reduced cognitive abilities [13]. "Big eye" (insomnia) and "long eye" (losing touch with reality) are other terms popular among polar crews to refer to different winter over effects.

Group Dynamics and Mental Health

While psychological research done using people in EUEs is central to understanding the mental health implications of living and working on the Moon, other research is also relevant. The research literature on people working in groups in "regular" jobs tells us that there is a connection between how fair people think they are being treated in their job and mental health. Also, recently research has found that shared perceptions of fairness of work assignments, responsibility, and other job-related factors, a construct called fairness climate, may be relevant for the psychological well-being of individuals in the group.

Spell and Arnold [14], in a study of 483 employees in 56 work groups, found the work group's "fairness climate" can affect individual levels of anxiety and depression. They examined mental health (anxiety and depression) as related to both procedural justice (the perceived fairness of work procedures, e.g., how work assignments and schedules were made) and distributive justice (the fairness of the actual rewards and other compensation distributed to each individual in the group). These effects were explained using a social contagion framework, which describes the creation of group effects that would occur in addition to influences on individuals. Spell and Arnold found that the interactive effects of distributive and procedural justice climates significantly influence individual feelings of both anxiety and depression. An implication of this study is that in cohesive groups (especially those groups with limited interaction outside of group) establishing and maintaining fair procedures is especially significant since perceptions of fairness will be solely formed by the individual and their group members.

Another recent study with implications for potential lunar base workers examined faultlines, which are subgroups or divisions within work groups [15]. For example, a group with three men and three women would have a gender faultline; one with three people in their 20s and two in their 50s would be said to have a faultline with respect to age. In addition to demographics, faultlines can exist with respect to experience levels and other worker characteristics. The Bezrukova and Spell study found that groups with strong faultlines and unsupportive supervisors were less anxious and depressed than groups with weak faultlines. Also, groups with strong faultlines and perceived unfairness were less anxious and depressed than groups with weak faultlines. Thus, subdivisions within groups seemed to have a moderating effect on psychological distress and were in that sense beneficial.

While counterintuitive at first glance, the groups with strong rifts or factions may provide individuals with emotional and other support in the face of unsupportive supervisors, poor working conditions, or other stressors. More generally, this research points to the role that group composition plays in individual mental health and complements studies on EUEs that also suggest that beyond selecting people for individual attributes, the ability to get along with co-workers is important. One speculation from these results is that the effects of group composition may be exacerbated in the extreme and isolated work environment of a lunar base.

Recommendations

While many recommendations may be made from the research existing on EUEs, one theme from this review is that social interactions and the composition of the work group staffing a potential lunar base is a significant consideration in individual mental health of base workers. Therefore, there should be some considerations given to selecting people to work on a lunar base on the basis of group compatibility rather than selecting based on just individual factors (technical expertise, psychological fitness, etc.). Selecting teams of people to work in EUEs is nothing new; the Russian space program used a test requiring a group of candidates to drive a small car for long distances to see how the team interacts. The importance of psychological tests for future space missions to Mars has also been recognized by NASA [16].

While the need to consider group composition and its effect on mental health is clear, we still need to know more about the personality traits that are important in working on a lunar base. Although the personality factors relevant to particular habitats are not fully identified, we may tentatively group them along two dimensions.

One way that personality traits can impact mental health is due to the tendency of many volunteers for work in EUEs to have a personality associated with risk-taking and adventure, and openness to new experiences. Yet, many of these volunteers discover the reality of working in EUEs is made up of monotonous, routine, boring tasks in an environment where they work daily with the same small group. Also, workers in EUEs that value autonomy may instead find that such environments are very controlling due to the conditions and organizational demands to follow rigid procedure [17]. While simulators have been used to try to get individuals accustomed to the controlled environment, at every opportunity the individuals living at the station should be given a say in schedules and other decisions rather than only to managers or others not living at the base.

In addition to social factors, other studies have suggested that psychological well being is enhanced by attention to the physical environment. In general, this means, again, extending to the crew of a lunar base the autonomy to personalize and customize their working and living space. Kwalleck & Lewis [18] suggest that worker performance may be affected by the type of wall color chosen. Providing different colors throughout the base living area may alleviate boredom [17]. These results suggest that the physical environment is an important factor in shaping psychological well-being and that more research is needed on the connection between environmental characteristics and perception, performance, and psychological well-being.

Overall, prior research suggests that environmental characteristics, group composition and social factors are key factors that need to be considered in selecting and managing individuals to staff a lunar base [19]. Yet, most of this research is either based on data from space missions of just a few days, space station missions of 6–12 months, or polar station missions of a few months. There is more research done on group composition and mental health but it is done on settings that are not in EUEs. None of these environments are exactly like a lunar base and it is unlikely therefore that we can extrapolate the findings here directly to what working and living at a lunar base would be like. Yet, the research does tell us those social and environmental factors that are critical concerns in affecting the psychological state of people working on a base. Therefore, such factors should be considered in both selecting people for the staff of the base, training the staff, and designing the base itself.

References

1. Suedfeld P. 1991. Groups in isolation and confinement: environments and experiences. In Harrison AA, Summit J. How "Third Force" psychology might view humans in space. *Space Power*, 10, 135–46.
2. Collins M. 1974. *Carrying the Fire: An Astronaut's Journeys*. New York: Farrar, Straus, Giroux.
3. Burr R., Palinkas L. 1987. Health risks among submarine personnel in the U.S. Navy, 1974–1979. *Undersea Biomed. Research*, 14,535–44.
4. Palinkas L. 2003. The psychology of isolated and confined environments. *American Psychologist*. 58, 353–363.
5. Smith S. 1969. Studies of small groups in confinement. In *Sensory Deprivation: Fifteen Years of Research*, (ed. JP Zubek) pp. 374– 403. New York: Appleton-Century-Crofts.
6. Sandal G., Endresen I., Vaernes R., Ursin, H. 1999. Personality and coping strategies during submarine missions. *Mil. Psychology*.
7. Bluth B., Helppie M. 1986. Russian Space Stations as Analogs, NASA Grant NAGW-659.Washington, DC: NASA. 2nd ed.
8. Stuster J, Bachelard C, Suedfeld P. 1999. In the Wake of the Astrolabe: Review and Analysis of Diaries Maintained by the Leaders and Physicians of French Remote Duty Stations. Santa Barbara, CA: ANACAPA Sci.
9. Palinkas L. 1990. Psychosocial effects of adjustment in Antarctica: Lessons for long duration spaceflight. *J. Spacecr.* 27, 471–77.
10. *New Zealand Herald*. 1996. FBI sent in to Antarctic base. *Herald*, Oct. 14, p. 83.
11. Schwetje F. 1991. Justice in the Antarctic, space, and the military. In Harrison A., Summit J. *How "Third Force" Psychology Might View humans in Space. Space Power*, 10, 383–94.
12. Weiss, K., Gaud, R. 2004. Formation and Transformation of Relational Networks During an Antarctic Winter-Over. *Journal of Applied Social Psychology*, 34, 1563–1586.
13. Palinkas LA, Browner D. 1995. Effects of prolonged isolation in extreme environments on stress, coping, and depression. *Journal of Applied Social Psychology*, 25, 557–76.
14. Spell, C., Arnold, T. 2007. A multi-level analysis of organizational justice climate, structure and employee mental health. *Journal of Management*, 33, 724–751.
15. Bezrukova, K., Spell, C. 2006, August. Faultlines in Diverse Workgroups: Reconsidering the Justice-Psychological Distress Relationship. Academy of Management Meeting, Atlanta, GA.
16. Behavioral Health and Performance Team. 1999. Behavioral health and performance program plan: Definition and implementation guide. Houston, TX: Medical Operations Branch, NASA Johnson Space Center.
17. Stuster J. 1996. Bold Endeavors: Lessons from Space and Polar Exploration. Annapolis, MD: Nav. Inst.
18. Kwallek N, Lewis CM. 1993. The impact of interior colors on the crew in the habitation space module. In *Proc. IDEEA Two Conf.*, p. 213. Montreal: Cent. North. Stud. Res., McGill Univ.
19. Suedfeld, P., Steel, D. 2000. The environmental psychology of capsule habitats. *Annual Review of Psychology*. 51, 227–253.

20

Humans: The Strongest and the Weakest Joint in the Chain

Jesper Jorgensen

SpaceArch
Copenhagen, Denmark

Jesper Jorgensen, MSc is a space psychologist employed in the private space industry and a researcher on the coming Mars500 experiment in Moscow. His special research interests are in the effects of confined environments on the sensory system and in development of technological advanced countermeasures to psychological stressors in coming long-term manned missions. He has extensive experience in development of models and theories in cooperation between psychologists, artists, architects and engineers in space design studies. Member of AIAA space architecture technical committee and IAA study group on Space Architecture.

ABSTRACT   Human factors will be one of the most important factors for a coming Moon base mission. Human factors understood as the way the individual and the group of crewmembers cope and interact with each other, with the technology and the environment on the Moon and inside the habitat. A permanent presence on the Moon will challenge the present concepts of human spacefaring, and add many new dimensions to our knowledge on problems and methods in what will be the long term agenda: permanent space colonization.

New components concerning design and architecture of habitats, composition of crews in size, gender, qualifications and the understanding of sociological problems in the crew, in the leadership and in the management of the mission from Earth, will probably be more prioritized in the coming decade as the space agencies get a growing understanding of the need to do more research in human factors and in the relation between humans, technology and the environment. The human factor could be the single most important factor, which can change a mission from a success to a catastrophic failure, because humans unlike many mechanical systems are very dynamic and usually experience a great deal of change in attitude, behaviour, performance, and health over the course of a mission (Eckard, 1999). This despite the tight selection procedures prior to a mission.

Human Factors: A Brief Definition

Human factors must be defined as the relation between: humans, technology and the environment, understood in the sociological and socio-economical context in which the project is performed. In the past, human factors were mostly understood as the adaptation of astronauts to the technology onboard and to the extreme environment in space, but as missions extend in time and distance, and technology develops to more advanced stages, this simplistic approach has shown to be obsolete.

Until ISS was built, space projects were mostly national prestige projects, but turned international and larger, with a demand for both the agencies and the space crews to collaborate across cultures and political systems. Projects turned into patchwork projects with many different contributors and players, and often with a more complicated organisational and political structure. A coming Moon Base will probably, in relation to the high costs and needed technological development, be another international project. Political control, media interest and involvement of the public will therefore presumably get a high priority in coming projects with a direct focus on the single crewmember in the coming mission. As the Moon Base will be built in a future and much more technologically advanced time, with a direct and continuous surveillance of the projects in the media or over Internet (in a much more advanced version than today) more direct public and political influence must be foreseen. The public and the politicians must be allowed direct observation and influence on priorities and decisions, and probably turning the design, building and execution processes of the project into a much more fluctuating system than current large scale space projects. This situation where there is much involvement from all fronts must be integrated into the planning. There is therefore an argument for including the contextual situation around the project in the human factors.

Natural Extreme Environments

Mankind has a long history of living in extreme environments, from the millenniums wide experiences of people surviving and developing civilizations in polar regions, in high mountain areas, deserts, and on isolated islands, to our newer history of early polar expeditions and more than 45 years of presence in space. In the natural settlements in extreme environments on Earth, we do have a normal diversity in age, gender, skills, personalities and patterns of diseases and pathological behavior in the population, but in the expedition crews, the members are selected under tight conditions to exclude any foreseen psychological and health problems. Seen in the long term perspective, the Moon missions will probably in the beginning follow a traditional and strict selection strategy, but as the pace of space colonization grows, diversity in generations, gender and qualifications will change to a more settlement or family orientated approach, which at present is quite provocative to most of the space society.

There is no doubt that, despite the fact that the members in families are non-selected, valuable information can be gathered from the human experience in living in extreme environments. Given that eventually family members will need to survive in a confined habitat, it is important and interesting to develop methods to train such groups together.

Seen from an anthropological point of view, the long history of surviving in a polar region in a remote habitat must contain information about traditional methods in conflict resolution, lowering social stress, handling long period of isolation and darkness, for example. Very little of the large anthropological knowledge from studies in these areas has been connected to future spaceflight, but could it be one of the future research directions in space psychology and architecture, especially with long term missions to planets in outer solar system, and with the perspective on space colonization in mind. Some designers have used these principles in obtaining basic information on core elements in "the meal situation" in different locations and cultures worldwide. What defines a cozy meal and how is social coherence created in the common meal however it is eaten, outside or inside the house or on the ground or by a table? Can this knowledge be transferred into design of the place for meals in a space habitat? (Nyström, Reuterswärd 2003) In the coming years more focus will be put on design processes with a shift from the concept of "surviving" to "living" and "feel at home" will be facilitated by new research and be a key component in simulations and design development for future Moon missions. The growing number of female space architects and designers will probably support this development.

Artificial Extreme Environments

In modern extreme environments, such as nuclear submarines, polar research stations and space stations, effects of these confined and extreme environments are quite well documented. These three different extreme environments are often used as reference for each other, because there is some difference in crew composition, culture and the general background of crewmembers. Research from these areas provides some of the background needed for the design of simulation studies done as preparation for long-term interplanetary space missions. Analogue environments on

Earth might supply larger sample sizes than possible amongst space crews, due to the very limited number of persons with longer stays onboard the space stations (Sandal, 2001). The limited number of people who actually have been in space, creates a methodological problem, because information from the spacefarers is not uniform, with different viewpoints and reactions, which in combination with the few number of people included, make it difficult to generalize knowledge by scientific methods. The information must therefore in many cases be categorized as anecdotal (Kanas, Manzey, 2003). Information from analogue environments and simulations are therefore important sources of knowledge on problems in future spaceflight.

Simulations of Space Flight

A number of simulations have been done of space missions, from early spaceflight where isolation and sensory deprivation was the main concern for single astronauts and cosmonauts to simulations of longer missions and larger crews, with focus on psychological factors, group relations and psychological effects on the health of isolation. Simulations can focus on some of the components of the travel ("partial" or "single event" simulators), or be a full scale simulation of a longer interplanetary mission. In general simulations are limited concerning the effects of microgravity and the restrictions laid out in the ethics of scientific experiments. As a result of these limits any test participant can at any time leave the experiment, and the management of the simulation does have the responsibility not to bring any danger to the participants. This suggests restrictions on how the most dangerous situations can be simulated, for example, whether a seriously ill test person will be taken out of the experiment, and not treated inside, as could be the situation in an interplanetary mission. Similarly, a simulation will not be able to simulate the "point of no return" situation a coming Mars mission will be in, most of the time after launch, or a situation where a Moon base crew cannot leave the base for evacuation.

Experiences from one of the long term international isolation experiments (>300 days) showed that interpersonal conflict can result in test persons leaving the experiment. The experiment therefore failed in one of the key elements which will be of great importance in coming long-term missions with gender and cultural mixed crews: how to handle conflicts.

Simulations can yield a large amount of knowledge on human factors, how technology, interior design, and human psychology and social systems cooperate with each other in a safe and controlled environment with room for mistakes, correction and repetition of components or actions. But it does have the limitation as a first aid course: In real life situations conditions and actions can be different or the human mind can mobilize conditions not seen in simulations, where an unconscious element of play often will be present.

Actual Simulation Studies

One of the actual "single event" simulations is the NEEMO project run onboard the Aquarius underwater habitat outside the coast of the Florida Keys. The teams, often crew members from the astronaut corps, train and simulate missions to ISS. Anther single event simulation is done by the Mars Society in 2 simulators in the Utah desert and in the Alaska tundra. Normal lengths of the crews' stays are 14 days, but some longer experiment periods are planned. The public relation effects of these missions are probably higher than the scientific output, but they can maybe pave the way for larger and more relevant simulation facilities run by non-governmental partners.

Alternative single factor simulations are done or planned to be done, to simulate physiological effects of microgravity due to prolonged permanent bedrest (approximately 3 months) in a position where the head of the test person's bed is tilted to −6°. This gives some of the effects of microgravity on the cardio-vascular system and the immobilisation in such a lying position has effects on the muscle–skeletal system. Psychological researches, focusing on the single test person are done in parallel.

Concerning the coming Moon–Mars missions, full scale missions are more relevant. One of the only international space mission simulations done until now was the 1999 SFINCSS (Simulation of Flight of International Crew on Space Station) isolation study at the Institute for Biomedical Problems (IMBP) in Moscow. This simulation raised a number of problems in cooperation across cultures and in conflict management in international crews, besides a large amount of knowledge on the effects of isolation on the human mind and body.

Another of the very few full scale simulations relevant for Moon–Mars settlements was the BIOSPHERE 2 project in Arizona, which showed both technological and sociological problems related to a closed environment, and the lack of full knowledge on both management of ecosystems and human relations in a confined environment.

The next planned experiment is the "Mars500" project, scheduled to start in spring 2009 with an initial experiment where a crew of six members is isolated for 105 days, followed by a full simulation over 500 days by another test crew with six persons simulating a mission to Mars. The experiment is based on a simulated journey to the planet, a landing on Mars with a part of the crew, the rest in orbit, before the whole crew returns to Earth. The Mars500 project is planned to include both a small greenhouse and a Mars surface mock up.

Human Factor Problems in Space Flight or Analogue Environments

Research from both analogue situations and space flight has shown a number of general psychological problems, which will be relevant for a coming Moon settlement too, as they are related to the circumstances of isolation more than the location. Problems arise from living in a confined space, the limited individual manoeuvring room and from living in a tight and closed social system over extended time. Amongst significant general problems observed are:

Chronic Fatigue

Fatigue is a well known and documented event, which occur in all analogue and space settings, and chronic fatigue or "asthenia" (term used by Russian researchers) is well documented as a result of high workload, isolation, tensions and the lack of real recreational facilities in a situation where a prolonged time span worsens these conditions. Often a pattern is seen, where exhaustion starts from the middle of the mission, independent of the actual length and grows towards the end of the mission. There are of course very individual responses to this situation and a tendency to worsen if co-stressors are added. There is a lowering of symptoms if an individual is active or collective countermeasures are put into action.

The effect of chronic fatigue results in sleeping disorders, impaired performance, internal conflicts and depressive mood conditions. The effect of mission related exhaustion is observed in space, with a tendency to worsening after more than 4 months onboard (Kanas et al. 2001). An investigation into cultural factors has showed that the level of fatigue is related to the level of depression amongst Russian crew members, but not the Americans. And in an opposite way, anxiety is related to depression in Americans but not in Russians (Boyd et al. 2007). It is evident that depression and asthenia reactions can be resolved with increased audio-visual contact with family and friends on Earth before full-scale psychiatric problems develop (Gushin, 2003).

For the design of coming permanent human missions to the Moon, it will be important to develop a model for recreation onboard, both in the architecture and in the psychological countermeasures. Further research into chronic fatigue, its individual, social and cultural applications is needed, and models for psychological countermeasures must be developed.

Disturbed Circadian Rhythm

Lunar days vary in duration, and can be as long as a few minutes to over 24 hours. Light intensity and darkness are stronger than on Earth and the day/night cycle more varied than here, depending on the location on the lunar surface. We know that humans can survive even crude disturbance in day night shifts, from the permanent winter dark or 24 hours midnight sun in Polar Regions, to 90-minutes day/night shift onboard space stations in orbit around Earth. Subjective reports from space state that the most important factors contributing to impaired well-being and performance during flight are sleep disturbance and fatigue. Objective studies show a disturbance of sleep length and quality in space on shorter missions and at the beginning of longer missions, with a tendency to adapt over time (Kanas, Manzey, 2003).

Seen in a habitat perspective, it has been observed in crews on Antarctic research stations and in populations living north of the polar circle, that alertness is higher, the sleep cycle is much shorter and people more socially active in the summertime. In the darker periods people tend to be more introverted, less interested in social contacts and more at risk of mild depression like symptoms. Intensity of light and its variation over time have a direct hormonal response to the brain, signalling to be awake or go to sleep. Future research is needed to get knowledge of the physiological effects of the light/ darkness characteristics on the Moon and the influence on the affected crewmembers' behaviour and mental state. Effective countermeasures must be developed, both on the physiological and psychological side, and the design of the habitat must include this knowledge to compensate for or to protect against the sunlight to give maximal comfort to daily activity and to protect, structure and secure the quality of sleep and avoid a "free run" of the circadian system in the crewmembers (Kanas, Manzey, 2003).

Leadership Roles

As space missions onboard MIR and ISS have extended the time in space up to more than a year, a shift is observed from short missions done in high efficient mode in a military-like command structure (for example, shuttle missions in 14 days) to less intensive operations with a more flat or soft leadership modus, with a higher independence of the crew. Roles soften up over time, and accordingly open up for rivalry of leadership roles. Dominant members of the crew challenge the leader and his ability to manage the often strong personalities in the crew who are highly educated specialists in their respective fields. A situation quite comparable to the conductor of a symphony orchestra, with both a partiture, lots of soloists and a common sound as the task. On the longer missions the leader must manage to create fruitful alliances with the dominant members, avoid scapegoating or other problems with members at risk of being expelled from the group, create a reliable and well functioning working and living environment and must be prepared to handle being the responsible manager of unforeseen or dangerous situations. The crew leader is often in a vulnerable situation: being part of the crew, but must be able to be alone outside the crew when an unpopular decision must be taken. Some have suggested the crew medical doctor to be outside the command system, so he could act as an independent advisor or discussion partner for the leader.

For a coming Moon mission more knowledge on management and development of management roles in small groups is needed. Experiences from analogue environments must be tested in a space habitat concept and new models of leadership of groups with specialists with both high individualism and self-esteem must be developed. As preparation for further interplanetary missions, with full crew autonomy, the Moon Base is a relevant test bed.

Intercultural Conflicts

Coming Moon crews will consist of a group of mixed races, gender and cultures, even if the mission will be pure national. Experiences from recent spaceflight have shown the importance of cultural factors, which may be hidden under short missions but will surface over time. Behaviour, the invisible personal "distance" between individuals, mood and conflict are related to cultural factors, besides all the visible symbols of culture: food, art, music (Boyd et al., 2007). Daily situations which on a short mission don't raise conflicts, can over time be important factors of irritation and have major impact on cooperation and social life. An example could be the cultural differences on the view of the importance of the evening dinner as a social meeting point, demanding conversation, time and cosiness in opposition to an effective individual eating "on the run" in solving the daily tasks. National red-letter days have different rituals and meanings, which can be overseen by other crewmembers. In a more globalized world years ahead, cultural differences will still exist, but may be more complicated and related to new subtle stratification in societies, and not ethnicity or nationality known today. As astronomer Carl Sagan forecasted, mankind will over time see cultural diversity as strength and create a new difference between "earth kind" and "space kind" as humans adapt more to life in space and in permanent space bases (Harris, 1996).

Before a Moon base for an international crew is designed, cultural differences must be considered, so the base will have symbols and places for what is important to the members' culture, and room for the crew to make their own hybrid culture as a Moon base crew.

Ground Control/Space Communication Disturbances

Conflicts between ground crews and space crews are well documented, and the role of conflicts as a system to create coherence inside the space crew in stressed situations, is well described. If control or crew loses confidence in the other, it raises a number of important problems from: "are messages reliable," "are we fully orientated about the situation onboard" to open and hostile conflicts and mutiny. One of the present research strategies investigates methods to make voice and semantic analysis of communication with crewmembers to detect stress, depression or anger in the verbal communication. A more reliable method could be to train crews to be more independent in coming simulations, acting as a ship's crew, by leaving behind the concept of strict ground control. As the mission leaves the low orbit around Earth, nature will limit the ability of ground control to directly control missions.

Environmental Problems Related to the Moon

On the Moon reduced gravitation and direct radiation from the Cosmos and the Sun will be the two major environmental factors having impact on human factors. Microgravity does have a major influence on the human body, and needed are a number of physiological countermeasures to minimize the constant destruction of muscle and bone tissue due to the lack of action from gravitational forces on the muscle-skeletal system. Tissue in both muscles and bones in the mobile system is reduced by approximately 1% each month in microgravity. A heavy daily training programme over several hours in space does not exclude this, but can only lower the effects of the atrophy. Other known problems in microgravity are disturbances in the balance system, body fluid shift and metabolic problems.

On the lunar surface, base members will live in 1/6 of normal gravity on Earth. We have until now only few days of experience with this condition from the former Apollo missions and further research in the special gravitational conditions is needed. The gravitational situation on the Moon may influence human factors, both in the planning of daily activities in the habitat and the design of the base system.

A future Moon base should handle the heavy radiation on the surface of the planet, due to the lack of both a protective atmosphere and a magnetic field. Radiation consists of both particles from the Cosmos and the Sun. The shielding design will have influence on the interior of the Moon Habitat regarding the volume of the interior, choice of materials and how the interior living quarters are arranged. One design model has, for example, humans inside the habitat core, with equipment and protective materials outside the core (Eckart, 1999). Other ideas are an underground base on the Moon, with the overlying rock as shielding. Radiation gives rise to similar considerations in relation to the opportunities for the crew to do EVAs, and the related risk of quickly obtaining the maximum allowable radiation dose and therefore being "grounded" inside the base or returned prematurely to Earth.

Another environmentally related human factor problem, which will affect the base crew, will be the reaction to solar particle events, coming from plasma ejections from the Sun, with faster less damaging particles arriving first on the Moon followed by slower more damaging particles, all happening in few minutes. A satellite based warning system close to the Sun can give some minutes of warning time. From a human factors perspective, both the constant alertness and short reaction time needed when sun particles are underway must be foreseen to have a negative influence on mood and performance of the affected crewmembers. For crews on EVAs, constant alertness must be higher and more stressful. This crew may, if being more than minutes from the shelter in the base, bring Moon rovers with the necessary shielding along.

A Permanent Moon Mission: New Human Factor Problems?

The previously described general psychological problems related to being in an extreme environment may be extended by specific problems related to the characteristics of the Moon habitat. It has been discussed if the problems foreseen for a human mission to Mars will be comparable to problems on the Moon. A short distance to Earth, the possibility for normal duplex radio and TV communication and the subjective emotional experience of being close to Earth, will probably not be similar to the isolation from Earth on Mars. Here the situation will be a severe disturbance in communication to ground control and the absence of a return opportunity to Earth in case of an emergency. Until now we do not know the psychological and emotional effects of seeing Earth as a pale blue dot in the sky, as will be the situation on Mars. A Moon base will therefore still be part of the concept of near Earth space exploration as former Apollo Moon landings, station MIR and ISS, but will be able to simulate further problems related to the next step: Mars.

Moon Missions as Part of the International Society: Human Factors

How the political and societal environment affects both the individual and the teams in space flight projects is not fully researched. There seem to be a theoretical and scientific vacuum around the important questions on how the societal response to missions, both planned and activated projects, influence the psychology of the individual astronaut and the team, both on Earth and in space. Until now manned space missions have been a solely governmental responsibility, performed in a national framework or as an international cooperation between national agencies. The private initiative has not reached that stage at present, where it can compete with public space flight, and will still be far away from realizing a Moon mission or a base habitat.

In the history of manned space flight, the direct influence from political power has been supportive or destructive dependent of the present political agendas. There is no doubt that working as an astronaut or space scientist in a political environment where space science is a high priority in societal life is much more motivating than being in a constant spotlight as a user of taxpayer money and haunted by nervous management or press departments demanding for constant legitimating of the financial spending of the programme.

Future Studies

Research today in human factors of coming interplanetary missions have the obvious problems that the research is based on present knowledge and technology, but the missions will presumably be at least 10 years ahead, where knowledge from research and the technological development will be very much ahead of the present status (Krishen, 2008). We must therefore both lay out tracks for future development and research, and be open for knowledge we at present do not know the importance of or knowledge we at present "do not know we don't know."

Seen on a crosscut of the presentations on astronautical conferences in recent years, there is a wish for inclusion of modern technology as a diagnostical tool for social and psychological problems in the crew, and as an effective tool for individual psychological countermeasures. PDA based monitoring systems, "digital friends" (Hoermann et al., 2008) and psychoeducative programmes are some of the ideas, but at present the lack of sufficient virtual technologies is the main problem in realizing the models in an operative mode. Future technological development will probably solve these problems, and maybe introduce advanced technologies which can only be dreamed of today. On the non-technical frontier, both extended simulation programmes, more qualitative and anthropologically founded and neuropsychological research will give further knowledge on human coping, interpersonal cooperation and new models for conflict resolution.

And because we are working with the future, we have to reconsider whole our concept of mission structure, composition and the single crewmembers personality and qualifications. Space mission designs, both on the technological and the human factors side are bound in a very conservative tradition, based on high skill engineering and a personnel structure taken from the air force due to historical reasons as astronauts and cosmonauts were recruited from the corps of military test pilots.

Principles for a good design, giving respect to the comfort of the crew, without compromising on mission agenda or safety have been done before. Human spaceflight until now has been a general success with focus on safety and technical compliance. Now the next step will be to develop both the human capability to handle the new missions and to create habitats for living more than surviving. And we do not need to reinvent these principles. It has been done before in our history at the time of early polar exploration. Leadership by men such as Shackleton has shown that the democratic leader can function even in dangerous situations and can combine his own leadership with the knowledge of the crew in a mutual responsibility.

Another example is the principles in designing the schooner *Fram* in late 1800s by Fridjof Nansen and Roald Amundsen. With a strong focus on both the capability to perform the polar trips in heavy ice and weather, and to manage the 2-year-long overwintering in the ice, they gave maximal comfort to the crew aboard. *Fram* had room for social and cultural life, work, and voluntary solitude aboard the ship. The crew consisted of persons with multiple, both formal and informal, skills and personalities. And in opposition to space crews, an artist was seen as important for the expedition for documentation of all the non-technical and emotional matters.

Let this be an argument for including a diversity of professionals in designing the Moon base and investigating human factors in a broad frame. We need historians, sociologists, psychologists, artists, doctors, engineers, information technology specialists and all the others who can give their learned views to the mission design procedure.

Seen from a space psychological and sociological perspective, future long term missions to the Moon will be very important as an understanding of the human factor problems related to longer interplanetary missions out in the solar system. If a permanent base can be run on the Moon with a low ratio of psychological and physiological problems, the understanding of interaction between human, environment and technology in space will extend the borders of human expansion in an unlimited way. This will be an experience that will allow the human race to expand in outer space and build new space colonies.

Literature

Ball J.R. et al. 2001. *Safe Passage. Astronaut Care for Exploration Missions*. National Academy Press, Washington D.C.

Boyd J.E., Kanas N., Gushin V., Saylor S., 2007, Cultural differences in patterns of mood states on board the International Space Station. *Acta Astronautica* (61) 668–671.

Eckard P., 1999, *The Lunar Base Handbook. An Introduction to Lunar Base Design, Development and Operations*. Space Technologies Series, The McGraw-Hill, New York.

Gushin V.I., 2003, Problems of distant communication of isolated small groups. *Human Physiology*, 29, 5, 39–46.

Harris P.R., 1996, *Living and Working in Space. Human Behaviour, Culture and Organisation*. John Wiley & Sons, New York.

Harrison A.A., Clearwater Y.A., McKay C.P., 1991 *From Antarctica to Outer Space. Life in Isolation and Confinement*. Springer Verlag, New York.

Hoermann H-J., Johannes B., Salnitski V.P., 2008 The "Digital Friend" A knowledge-based decision support system for space crews. *Acta Astronautica* (63) 848–854.

Human Health and Performance for Long-Duration Spaceflight Position paper by Space Medicine Association & NASA. *Aviation, Space and Environmental Medicine* 24 (6) June 2008.

Imhof B. et al. 2004. *Musings towards a new genre in [Space] architecture.* Liquifier, Vienna.

Kanas N., Salnitsky V., Gushin V. et al., 2001, Asthenia—Does it exist in Space? *Psychosomatic Medicine*, 63, 874–880.

Kanas N., Manzey D., 2003, *Space Psychology and Psychiatry.* Kluwer Academic Publishers, Dordrecht, Holland.

Krishen K., 2008, New technology innovations with potential for space applications. *Acta Astronautica* (63) 324–333.

Lineger J.M., 2000. *Off the Planet.* McGraw Hill, New York.

Nyström M., Reuterswärd L., 2003. *Meeting Mars, Recycling Earth.* Svensk Byggtjänst, Lund, Sweden.

Palinkas L.A., 2003, The psychology of isolated and confined environments. Understanding human behaviour in Antarctica. *American Psychologist* 58 (5) 353–363.

Sandal G.M., 1996. *Coping in extreme environments: The role of personality.* Doctoral thesis, University of Bergen, Norway.

Sandal G.M., 2001. Psychosocial issues in Space: Future Challenges. *Gravitational and Space Biology Bulletin* 14 (2) June 2001.

21

Here to Stay: Designing for Psychological Well-Being for Long Duration Stays on Moon and Mars

Sheryl L. Bishop

University of Texas Medical Branch
Galveston, Texas

ABSTRACT Current psychological and sociocultural considerations for Moon and Mars bases are embedded within a complex matrix of long duration issues that have been demonstrated to significantly impact on human behavior and performance in challenging environments characterized by isolation and confinement. Planning for psychological health for long duration inhabitation missions rather than short duration task accomplishment profiles is further complicated by the need to address both flight issues as well as station issues. The fundamental problem today in developing effective countermeasures for long duration missions is the fact that space missions are not truly psychologically comparable to any other undertaking humans have ever attempted (e.g., long-duration stays on orbital space stations, historical expeditions to unknown parts of the Earth, wintering-over in Antarctica, long-term submergence in submarines) differing most notably in the enormous distance to travel, the unique separation from the rest of humanity and the extraordinary demands of an environment characterized by the lack of any fundamental environmental factor (e.g., air, water, gravity) heretofore required for survival. Mission scenarios call for crewmembers to endure extraordinary long periods of extreme confinement and isolation compounded with the absence of normal sensory input that would regulate psychological and physiological functioning. Increased known psychological risks related to individual performance, behavioral health and crew interactions will be magnified by completely new unpredictable psychological challenges under these extraordinary conditions of isolation and confinement from the rest of humanity.

A broad spectrum of social psychological and behavioral research has contributed to the emerging realization that many of the negative psychological factors of long duration Moon/Mars missions (e.g., prolonged isolation, confinement, exposure to unpredictable and unknown extreme environment, reliance on closed loop environmental system) could be mitigated by designing habitats that thoughtfully countered some of these impacts. The intersection of psychology and psychosocial factors with habitat design allows us to implement psychological support in non-intrusive, holistic environmental modalities that are preventive in orientation rather than palliative.

The present paper discusses those psychological factors with mission structure or design implications that can be addressed by a thoughtful and proactive implementation of psychological and psychosocial support systems and countermeasure environmental elements, including enabling technologies on the horizon that would significantly contribute to the successful psychological adaptation of long duration space inhabitants.

Introduction

Speculation, investigation and discussion about human missions to the Moon and Earth's planets has been ongoing as far back in human literature as Lucian of Samosata (c. A.D. 125–after A.D. 180), one of the first to write of voyages to the Moon and Venus, extraterrestrial life and wars between planets centuries before Jules Verne and H. G. Wells. Scientific challenges to Percival Lowell's 1895 explanation of the "canals" on Mars were issued by A.R. Wallace in 1907 [1, 2]. While the predominant foci of these early discussions were largely technological, there has always been a sub-rosa theme centered on the human element of space exploration.

In the last 20 years, the need for a systematic evidence-based science of human performance in space has necessitated the utilization of isolated, confined environments as analogs for space to investigate individual and team functioning and performance. Various characteristics, both environmental and situational, of the space environment can be found among a variety of terrestrial extreme environments and simulation facilities scattered around the world. Although specific conditions of the settings vary, extreme environments share common characteristics in their reliance on technology, physical and social isolation and confinement, the inherent high risk and associated cost of failure, presence of high physical, psychological, and cognitive demands, the need for human-human, human-technology, and human-environment interfaces, and the demand for team coordination, cooperation, and communication [3]. Those factors that will significantly affect the physiological and psychosocial well-being of long duration space crews have slowly moved to the forefront of importance as the focus has moved from mere survivability to thriving in space and extra-terrestrial environments.

Over this period, a number of periodic serial white paper publications have drawn together experts from a wide spectrum of domains to assess the state of knowledge about the myriad factors that are proposed to contribute to a successful mission to Moon, Mars and beyond. The recitation of the known and unknown across the years has repeatedly underscored one notable conclusion: The human element is both the greatest strength of a mission and the greatest weakness. As early as the late 1960s, lunar design projects were grudgingly acknowledging that group interaction would be the primary limiting factor in long duration missions [4]. Improvements and advances in technology, engineering and technological based human factors have not been mirrored by similar advances in human factor issues dealing with human health and well-being. While medical issues have become more specifically defined and identified, psychological and psychosocial considerations are still grappling with questions that were raised from the earliest days of space exploration: What is the best fit individual for a mission? How do we compose the best group? How do we select, train and support the best group for a mission?

Background

The impact of challenging environments on individual functioning and performance must be examined from the multiple and interacting perspectives of physiological, medical, and psychosocial domains. Psychological and sociocultural issues are critical components to mission success and have clearly been shown to significantly impact human behavior and performance in most challenging environments, especially those characterized by isolation and confinement [5-8]. A broad spectrum of social psychological and behavioral research contributed to the emerging realization that many of the negative psychological factors of long duration missions to the Moon or Mars (e.g., prolonged isolation, confinement, exposure to an unpredictable and unknown extreme environment, reliance on a closed loop environmental system) could be mitigated by habitability designs that thoughtfully countered these impacts. The intersection of psychology and psychosocial factors with habitat design is proving to be a very productive approach to improving understanding of individual and group factors by allowing us to implement psychological support in a non-intrusive, holistic environmental modality that is preventive in orientation rather than palliative. It has facilitated progress in separating the pure interpersonal factors from those that are embedded in responses to the situation/environment in which individuals and groups find themselves. This has enabled investigators to address the "best fit" issue from the perspective of matching individual capabilities (e.g., adaptation, tolerance for ambiguity, stress hardiness, interpersonal orientation, needs for autonomy) with mission characteristics (e.g., duration, crew size, mission goals, authority hierarchy, habitat).

Within psychology, the areas of specific focus have also been varied (see Table 21.1). With only a small number of researchers, little funding, few opportunities for accessing teams in analog environments and fewer still in accessing actual space crews, progress has been understandably slow. The greatest attention to date has been a focus on individual factors since selection has dominated the concern of all space programs. However, group functioning and performance are rapidly becoming recognized as equally

TABLE 21.1

Areas of Space Psychosocial Research

Individual Characteristics
Personality
Hardiness
Stress/Coping
Leadership Style
Interpersonal Orientation
Performance
Group Characteristics
Group Dynamics (relationships, conflict, cooperation, compatibility)
Group Composition (Gender, Skills, Nationality)
Group Performance
Group Identity
Group Fission and Fusion Factors

critical to mission success, especially when addressing factors that impact long duration mission profiles.

Space is, arguably, the most extreme environment humans have faced to date. Not only are we exposed to the presence of unique environmental and physiological extremes (e.g., temperature, radiation, absence of gravity) but we are challenged to adapt and perform in the ultimate isolated confined environment in which known parameters are few and are characterized by catastrophic outcomes for mistakes. Regardless of whether we are focusing on life support concerns, medical concerns, or psychosocial concerns, the key questions are the same:

1. What are the effects?
2. Do they impair functioning?
3. Are they self-limiting or progressive?
4. Are they reversible and when?
5. What are the countermeasures? [3]

The impact of isolation and confinement is repeatedly underestimated by teams and mission control groups. Isolated confined environments (ICE) are characterized by serious stressors from both the immediate environment and from external factors that span communication, resources, inescapable environmental conditions, unpredictable, unfamiliar and potentially dangerous contingencies, and reductions in the gratification of basic human needs for affection, security and personal significance.

A summary of the most commonly reported psychological problems in these environments exemplifies how pervasive physiological disturbances are as well as psychological reports of boredom, loss of motivation, homesickness, irritability and difficulties in concentration (Table 21.2) [8]. In addition, anecdotal evidence has repeatedly indicated problematic areas in which even short-term group functioning has been compromised by the presence of communication breakdowns, interpersonal conflict, individualized responses to environmental stressors and conflicts over authority and control [9].

A quick summary of findings from my own studies underscores the growing evidence for high impact factors across different environments including space (Table 21.3) [8, 11–15]. For instance, there is repeated evidence that certain personality profiles show a greater or lesser hardiness or robustness to deal with confinement, isolation, and the physiological as well as psychological stresses inherent in such extreme environments. Preliminary evidence seems to suggest that different personality profiles are optimal for different environments, e.g., the trekking expedition compared to the station mission. Similarly, there are parallel issues regarding gender, professional, cultural and age configurations that impact on selection and group functioning issues as well.

TABLE 21.2

Reported Problems in Confined Environments

Reported Problems	MIR	US Space	SUBS	POLAR	SIMS
Interpersonal conflict	X	X	X	X	X
Somatic complaints	X	X	X	X	X
Sleep disturbances	X	X	X	X	X
Homesickness	X	X	X	X	X
Boredom, restlessness	X	X	X	X	X
Decrements in performance	X	X	X	X	X
Decline in Grp compatibility	X	X	X	X	X
Substance abuse	-	-	?	X	-
Communication breakdowns	X	X	X	X	X
Conflicts with mission control	X	X	?	X	X

The persistent reoccurrence of mismatches between subjective and objective measures of stress across these teams provides for a potential multimodal methodology to monitor stress and coping as well as a monitoring capability to assess perceptual accuracy which may be severely challenged in extreme environments. Inaccurate self-monitoring represents potentially serious challenges to safety, health, well-being, performance and mission success. Not only are individuals at risk for not recognizing dysfunctional responses, but inability to identify dysfunctions would prevent or delay timely intervention and successful long-term adaptation. What threatens the individual ultimately threatens the team and the mission.

Designing for Psychological Well-Being

Since the early 1980s, 14 behavioral issues with mission and habitability design implications have been repeatedly identified as critical issues related to space and have been reaffirmed by subsequent reviews, research, panels and anecdotal team reports (Table 21.4) [16]. Understanding the psychological drivers for these elements promotes greater insight into design improvements for future Lunar and Mars missions. I will briefly cover those that lend themselves to psychosocial preventive measures which can be incorporated into habitat design. A more thorough discussion of these features can be found in any number of papers including my own [8, 17].

TABLE 21.3

Summary of Teams Researched

	Cave Divers	Antarctica	Australia	Polar Expeditions	Simulation Short Dur.	Simulation Long Dur.
Duration	6+ w	11 m	10 d	2–4 w	2 w	4 m
Group Composition	Mixed Gender, International	Males Only, Monoculture	Mixed Gender, International	Males Only, Monoculture	Mixed Gender, International	Mixed Gender, International
Distinct Personality Profiles	X	X	X	X	X	X
Mismatches b/w Stress	?	X	X	X	X	?
Differential Group Morale	X	X	X	X	X	X
Performance Decrements	X	X	X	X	X	X

TABLE 21.4

Behavioral Health Issues Affecting Design

	Physiological	Psychological
Food	✔	✔
Sleep	✔	✔
Clothing	✔	✔
Exercise	✔	✔
Medical support	✔	
Personal hygiene	✔	✔
Waste disposal and management	✔	
Onboard training, simulation and task preparation	✔	✔
Leisure activities		✔
Group interaction		✔
Habitat aesthetics		✔
Outside communication	✔	✔
Privacy and personal space		✔
Behavioral issues related to microgravity environments	✔	✔

Structural layout and habitat design can address a number of these behavioral issues by:

- Maximizing habitable volume with configurations that are perceived as more spacious (e.g., narrow spaces perceived to be unpleasant and cramped).
- Utilizing multiple compartments to provide for variety and segregated use.
- Using color and lighting to enhance desirable moods, reduce feelings of crowding and promote physiological synchronization
- Using windows, digital displays, or art to counter feelings of confinement and monotony and provide visual depth.
- And designing for multiple uses of greenhouses for food production, leisure activities, stress reduction, crafts, gardening, small group interaction, exposure to full spectrum lighting and natural fractals.

Many features can promote positive group interaction and reduce negative interaction. Desirable factors will promote group fusion, i.e., social bonding, group identification, social support. Negative factors to avoid are those that produce group fission, i.e., tension, conflict, discord, social isolation, scapegoating (blaming a convenient person), and miscommunication.

Many factors can potentially contribute to either fission or fusion goals in multiple ways.

For instance, separation of private functions from public is critical. Such separate spaces meet needs for solitude and privacy, allow individuals to regulate social interaction and provide for individual control over the amount of contact with others. It provides for individual personalization and individual differences in activities while moderating feelings of crowding and confinement. Yet there also needs to be spaces that *require* and facilitate group interaction. One of the inherent dangers in isolated groups is the general drift of certain individuals into complete isolation from his crewmates. Spaces that enforce human contact and interaction serve to preserve bonds between crewmembers as well as close contact needed for accurate group monitoring and intervention when needed. To achieve these dual needs within the very limited space available in confined habitats, we need flexible, definable and re-definable interior environments that provide for group as well as individual activity.

Sleep disruption is another persistent and high impact issue. Closed loop environments are inherently noisy due to life support machinery and restrictions on structure design. Some one HAS to have the cabin next to the toilet. The ambient noise of the environment itself has shown significant impacts to sleep for crews on Mir, Shuttle, and ISS. When you add the noise of human activity in a closed environment that lacks sensory day/night cues, there are serious impediments to normal sleep. Countermeasures for circadian disruptions can involve use of phase shifting to reset synchronicity or prevention through entrainment of sleep with artificial light/ dark cues, scheduled meals and other activities.

In ICEs, it is often the mundane that plays the largest role in interpersonal difficulties. Santy et al. (Table 21.5) conducted a survey of American astronauts in 1993 and found the highest rates of miscommunication, misunderstandings and interpersonal conflicts were in the moderate to high mission impact category and occurred during the inflight operations over matters of payload/experiment resources, housekeeping tasks and personal hygiene issues [18]. Little things become big things in isolated, confined environments. Adherence to hygiene standards have been shown to be important for health and self-esteem as well as enhanced morale and standards of personal decorum and respect for others. Almost all previous habitats and space facilities have been characterized by hygiene facilities that were laborious to use, lacked privacy, and were cumbersome, and cramped, all of which produce barriers to maintenance of personal hygiene and group standards, setting the stage for conflict and dissent.

In addition, there are significant differences between individuals, cultures and the genders concerning matters of housekeeping and hygiene. Early attempts at designing ISS were stymied by national differences in bathing. Americans wanted showers and the Russians wanted a sauna…they ended up with wetwipes. The very fact that these disagreements have occurred in

TABLE 21.5

Miscommunication, Misunderstandings, and Interpersonal Conflicts

Mission Phase	Number of Incidents (% of total)	IMPACT (% of row category)		
		Low	Moderate	High
Preflight training	9 (21)	6 (67)	3 (33)	0 (0)
Inflight operations	26 (62)	9 (35)	12 (46)	5 (19)
• Payload/ experiments	4 (15*)	1 (25)	3 (75)	0 (0)
• Housekeeping	5 (19*)	1 (20)	3 (60)	1 (20)
• Personal hygiene	5 (19*)	1 (20)	3 (60)	1 (20)
Postflight activities	7 (17)	3 (43)	4 (57)	0
Total	42	18 (43)	19 (45)	5 (12)

percent of Inflight Operations Mission Phase

almost all ICE groups, regardless of specific environment, and to varying extent to most individuals involved, strongly suggests that future groups will be at risk for these experiences also.

Confined environments suffer from a limited range of visual variety and stimulation. Lack of visual variety has been clearly associated with decrements in concentration, performance, mood, increases in boredom, restlessness and risky behaviors designed to fill the gap. "Personal décor," in the form of clothing, is an efficient way to provide variety, stimulation and personal expression. There should be a variety of colors for clothing and the garments should be adjustable so that they can be easily configured from pants to shorts, long sleeves to short sleeves or from warmer to cooler to allow individual adjustment to the ambient interior temperature. Although a critical factor in the plans for use of power and water, laundry facilities not only improve opportunities for clothing variability but will reinforce attention to hygiene.

Currently, exercise is the primary countermeasure for bone decalcification and muscle atrophy. Crews will not only exercise routinely as part of medical countermeasures but will likely seek out some form of exercise as recreation as well. The amount of exercise required to counter physiological deconditioning in lunar gravity is unknown. However, the expected need for substantive exercise, requires exercise systems and equipment that are intrinsically motivating and inherently fun to keep crewmembers attentive to physiological and psychological health. Equipment should be located close to hygiene facilities and related work but far enough away from sleep facilities and labs concerned with biological or vibration contamination.

The importance of food to individual and group psychological well-being for long duration space crews cannot be overstated. Meals are central events and food and eating are major sources of stimulation and variety for all ICE groups, including space. Meals promote orientation to group welfare, provide opportunities for face to face social exchange and associate satiation of physiological needs with positive group interaction. Provisions should be in place as much as possible for individual selection, snacking, and protocols for the sharing of food preparation tasks, i.e., kitchen duties. Provisions for contributing to food resources through individual gardening activities provide sources of intrinsic and extrinsic rewards and group recognition. Combining station keeping greenhouse duties with small personal allowances for individual gardens has been found to be an effective motivational approach of combining work with pleasurable recreation.

Interpersonal competitiveness has repeatedly been demonstrated to contribute to intergroup strife and discord [19–22]. Therefore, the emphasis should be on beating personal bests rather than besting others, and activities that offer opportunities for growth or testing and improving mission related skills are highly preferable. Use of immersive virtual reality systems could provide myriad venues for such challenges as well as address dysphoria (negative or aversive moods) by enriching expeditionary environments with games or pleasurable activities. Similarly, preparing media and professional presentations will be strong work-leisure activities. Opportunities to deliver these presentations via video conferencing will be a significant positive countermeasure to isolation.

Events and holidays should be scheduled as well as spontaneous. Experience in Antarctica and other long duration polar environments guarantees that Lunar and Mars crews will develop their own variety of "sports," games and holidays as well. Creative outlets will take many forms but provisions for crafting and gardening can serve dual purposes by resulting in products that will enhance the circumstances of the group, e.g., additional furniture or musical instruments, as well as provide relaxing and pleasurable hobbies.

Ideally, communication capabilities will include private send and receive, full fidelity audio/video/3D immersion via virtual reality and/or telepresence technology. Minimally, there should be personal encryption codes to ensure confidentiality of communication, open channels to Earthside for personal, professional and mission communication. Communication issues are not only hardware and technology related. Careful consideration needs to be given to protocols addressing private consults with medical or psychological personnel, families and, perhaps, scientific or proprietary investigators. These protocols need to be in place and understood by all. Sensitive issues that have potential consequences for crew well-being need to be explicitly agreed upon and followed by crews and mission control, e.g., how negative news involving family or friends should be handled. Similarly, one persistent challenge that has routinely emerged across all remote missions is the maintenance of remote command and control relationships. Perspective sharing and careful attention to clear communication is critical and requires sustained priority and training on both sides.

EVAs will be major sources of stimulation, anticipation, planning, and activation. All members should participate in EVAs as they provide opportunities for habitat egress and relief from confinement. All members should be involved in projects that afford them opportunities to engage in scientific discovery and the creation of new knowledge. There should be no one that does not have a meaningful exploratory project whether it be improved power generation or the search for life. Technological capability will be required to support personal growth opportunities that will be critical for learning new skills relevant to the mission and personal interests. Facilities on-site should provide computer accessibility from labs, workspaces and private quarters.

Schedules should be largely crew controlled. Persistent evidence suggests that micromanagement by Mission Control not only produces feelings of intrusion, work demand, stress and resentment among remote crews but interferes with performance and production by increasing incidences of conflict and inefficient use of time. A clear commitment to protection and separation of non-work and leisure time must be in place and enforced. Use of production/science space and performance of maintenance tasks must be shared equitably among crew. Imposition of external reprioritization of projects, tasks, or production schedules must be in consultation with the crew and not simply imposed.

Distance plays a key role in the amount of on-board monitoring and intervention capability that is required. Mars will require a greater degree of self-support and monitoring capabilities than Moon. However, the Moon will still require a larger degree of self-support and crew capability than previous stations.

Monitoring of cognitive, psychological, stress, group assessments should be both passive and active. We will need real-time feedback assessment capability for crew self-monitoring to enable first line crew control and response for adaptation and implementation of countermeasures.

Project Boreas: An Applied Example

Like the purpose of this conference, there have been ongoing efforts to bring all these elements together within design projects for both Moon and Mars. As an example I would like to quickly feature one such project on which I participated. Project Boreas was a 2-year international endeavor for the British Interplanetary Society to design a Mars polar base [23]. The polar location presented several unique features in extremes of Martian weather and prolonged day/night cycles similar to Earth's polar regions. The resulting Boreas Polar Base consisted of five modules arranged around a sixth central growth module (Figure 21.1) [24]. This configuration provides maximum

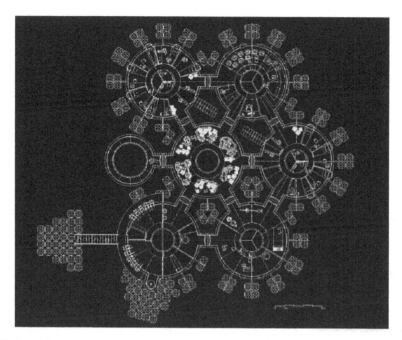

FIGURE 21.1
Boreas layout.

accessibility and multiple choices for traffic flow, reducing potential conflicts over intrusions and accessibility restrictions.

Presence of plant and greenery in the Growth Module provides ecological fractal properties, terrestrially familiar and restful surroundings, colors associated with stress reduction, exposure to full spectrum Earth normal lighting, water features, a workbench for creative construction and crafting activities, and spaces between plantings for semi-private gatherings.

Each module core was to be decorated in a different environmental theme. These themes would generate perceptions of changing environments as one traveled from one module to another. Use of immersive and virtual reality systems was incorporated by a proposed Biotrope system comprised of digital reality themed media that could be linked to real environments on Earth and updated periodically or in continuous real-time transmission to provide greater variety and stimulus.

Windows were positioned in both side walls and ceilings in most chambers at different heights to provide visual variety (Figure 21.2) [24]. Exterior lighting outside the habitat would provide visual detail and wayfinding elements as well as illuminate features during the polar night. There were many more features that cannot be adequately covered in the time here. The link to the publication table of contents is provided for those interested (http://www.bis-spaceflight.com/sitesia.aspx/page/170/id/980/l/en).

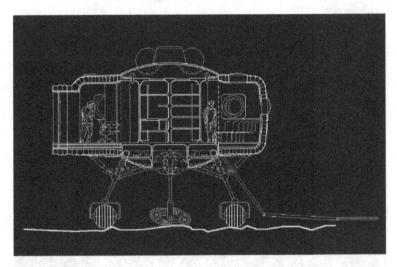

FIGURE 21.2
Views from multiple windows.

Summary

These are some of the major issues we have to consider for long duration teams in order to select, train and support the teams under the extraordinary circumstances inherent in a long duration space mission. A number of countermeasures are being designed into proposed habitats to address factors of isolation and confinement. Still, we have many more questions than answers. Earthside analogs will continue to help us grapple with defining our levels of "adequate preparation" in the face of ideally pre-defined levels of "acceptable risk" or even "acceptable losses" as we apply lessons learned to future space expeditions and missions. The next steps will be to bring together all these domains to inform and contribute to countermeasures that are inherent in the individuals we select, the training, mission and scheduling protocols we develop, and the habitat that will shelter and nurture humans as they advance the boundaries of new "extreme environments" in which humans are here to stay.

References

1. Wallace, A.R. "Is Mars Habitable? A critical examination of Professor Percival Lowell's book 'Mars and Its Canals' with an alternative explanation," Macmillan, London, 1907.
2. Lowell, P. "Mars," Houghton, Mifflin and Company, Boston, 1895.

3. La Patra, J.W. "Moon lab: Preliminary design of a manned lunar laboratory." A Stanford/Ames summer faculty workshop study, NASA, 1968.

4. Herring, L. Astronaut draws attention to psychology, communication. *Journal of Human Performance in Extreme Environments*, 1997; 2(1), 42–47. (Reprinted from APS Observer, 1995, September.)

5. Manzey, D., Lorenz, B. Human performance during prolonged space flight. *Journal of Human Performance in Extreme Environments*, 1997;1 (2), 68.

6. Manzey, D., Albrecht, S., Fassbender, C. Psychological countermeasures for extended manned space flights. *Journal of Human Performance in Extreme Environments*, 1995; 1 (2), 66–84. (Reprinted from *Acta Astronautica*, 35, (4/5)).

7. Morphew, M.E., Maclaren, S. Blaha suggests need for future research on the effects of isolation and confinement. *Journal of Human Performance in Extreme Environments*, 1997; 2 (1), 52–53.

8. Bishop, S. Evaluating teams in extreme environments: From issues to answers, International Space Life Sciences Working Group (ISLSWG) Group Interactions Workshop, *Journal of Aviation, Space and Environmental Medicine*, July, Vol. 75, no. supplement 1, pp. C14–C21(1), 2004.

9. Morphew, M.E. and Maclaren, S. Blaha suggests need for future research on the effects of isolation and confinement, *Journal of Human Performance in Extreme Environments*, 2 (1), pp. 52–53, 1997.

10. Bishop, S. Evaluating teams in extreme Environments: Deep caving, polar and desert expeditions, *Society of Automotive Engineers*, 32nd International Conference on Environmental Systems (ICES), San Antonio, Texas, July 2002.

11. Bishop, S.L., Dawson, S., Rawat, N., Reynolds, K., Eggins, R., and Bunzelek, K. Assessing teams in Mars simulation habitats: Lessons learned from 2002–2004, Jonathan D. Clarke (Ed), *American Astronautical Society's Science and Technology* series, Invited Papers, Vol. 111, pp. 177–196, 2006.

12. Bishop, S. L. and Primeau, L. Assessment of group dynamics, psychological and physiological parameters during polar winter-over, Proceedings from the Human Systems Conference, Nassau Bay, Texas, June 20–22, 2001.

13. Bishop, S. L. A comparison of male and female teams in surviving the Australian outback, Proceedings from the Human Systems Conference, Nassau Bay, Texas, June 20–22, 2001.

14. Bishop, S. L., Dawson, S., Reynolds, K., Eggins, R., Rawat, N. Bunzelek, K. Integrating psychosocial results from small group analogue studies: Three studies from the Mars desert research station simulation habitat, Proceedings from the 15th Humans in Space, Gratz, Austria, May 15–22, 2005.

15. Bishop, S. L., Sundaresan, A., Pacros, A. Patricio, R. Annes, R. A comparison of homogeneous male and female teams in a Mars simulation, Proceedings from the 56th International Astronautical Congress, Fukuoka, Japan, October 17–21, 2005.

16. Stuster, J. Strategies for exploration habitability during long-duration space missions: Key issues associated with a mission to Mars, in C. R. Stoker (ed), *The Case for Mars III, Science and Technology Series*, 74, pp. 181–191, 1989.

17. Bishop, S. L. Psychological and psychosocial health and well-being at Pole Station, British Interplanetary Society, *Project Boreas: A Station for the Martian Geographic North Pole*, 2006, pp 160–171.

18. Santy, P.A., Holland, A.W., Looper, L., Marcondes-North, R. Multicultural factors in the space environment: Results of an international shuttle crew debrief. *Aviation, Space and Environmental Medicine*. 1993; 64:3, 196–200.

19. Palinkas, L.A., Gunderson, E.K. and Holland, A.W. Predictors of behavior and performance in extreme environments: The Antarctic space analogue program, *Aviation Space and Environmental Medicine*, 71, pp. 619–625, 2000.
20. Rose, R. M., Helmreich, R.L., Fogg, L.F. McFadden, T. Psychological predictors of astronaut effectiveness, *Aviation, Space, and Environmental Medicine*, 64, pp. 910–5, 1994.
21. Kozerenko, O.P., Gushin, V.I., Sled, A.D., Efimov V.A. and Pystinnikova, J.M. Some problems of group interaction in prolonged space flights, *Human Performance in Extreme Environments*, 4, pp.123–127, 1999.
22. Chidester, T.R., Helmreich, R.L., Gregorich, S. Geis, C.E. Pilot personality and crew coordination: Implications for training and selection, *International Journal of Aviation Psychology*, 1, pp.25–44, 1991.
23. British Interplanetary Society, *Project Boreas: A Station for the Martian Geographic North Pole*, 2006.
24. Green, M. *Project Boreas*, ANY Limited Design Studios, University of the Arts, London, England.

22

Indoor-Air Quality Implications of ^{222}RN from Lunar Regolith

François Lévy
synthesis international, USA
Houston, Texas

John Patrick Fardal

François Lévy, M. Arch, MSE born in Paris, France and raised in the United States, Lévy received his bachelor of arts in liberal arts (philosophy, mathematics, and classics) from St. John's College in Santa Fe, New Mexico and received his Master of Architecture degree from the School of Architecture at The University of Texas at Austin. He has subsequently been an adjunct lecturer at the School of Architecture. He was also a board member of American Youthworks (a service learning organization serving at-risk youth) for five years, where he was involved in affordable housing programs. In architectural practice, he has worked on a variety of projects, including single-family residences, a new subway line for the Paris Métro, and an industrial plant in the U.K. In 2006 he enrolled at UT's College of Engineering, where he is pursuing an MS in architectural engineering, with an emphasis on sustainability and (informally) design for extreme environments. He has a lifelong passion for space exploration, and is honored to be able to contribute to the field, however modestly.

John Patrick Fardal, MSE in environmental and water resources engineering from the University of Texas, and earned two BS degrees in civil engineering and computer science from UT Austin and UT San Antonio, respectively. His professional experience includes computer programming, civil engineering permitting, and pipeline design. He will enter the Texas College of Osteopathic Medicine for his MD in July of 2009.

ABSTRACT Recently interest has grown in resuming lunar exploration with the possible establishment of long-term bases. Due to payload costs and the scale of permanent bases, there is a compelling need to employ *in situ* resources, leading to extensive use of lunar soil, or regolith, in such bases. Regolith is prone to ^{222}Rn exhalation. We modeled two scenarios for radon exhalation in regolith–based lunar construction. We examined the potential for human health risks due to ^{222}Rn decay-product exposure in such closed, hermetically sealed environments where lunar inhabitants would potentially have direct, long-term contact with regolith. We found the potential for significant health concerns, but more detailed data on the physical properties of regolith-aggregate concrete and its ^{222}Rn exhalation rates is required to accurately determine radon emanation and diffusion coefficients.

Introduction

After the late 1960s and early 70s when there was an abundance of research, interest in lunar missions and exploration declined. More recently, however, scientific and engineering research on lunar characteristics and interest in the establishment of lunar bases has increased. Such bases would have numerous benefits, including astronomical platforms, staging areas for crewed expeditions to Mars, and facilitating exploitation of lunar resources themselves (*Schmitt 2003, Duke 1998*). However, transportation of construction materials to lunar sites is extremely expensive. For example, delivery costs for Apollo 17 were approximately $70,000 kg^{-1} (*Schmitt 2003*). Even assuming drastic reductions in launch costs, there is a clear advantage to maximizing ISRU (*in-situ resource utilization*) (i.e., *Lin et al. 1997, Benaroya et al. 2002, Schmitt 2003*). Several engineering approaches have thus been suggested which make use of lunar soil, or regolith, as a significant building material (*Benaroya et al. 2002*). It is reasonable to assume the possibility of lunar bases with very long-term inhabitants, given transportation costs and the unlikelihood of frequent returns to Earth (*Schmitt 2003*). While early habitats will certainly include living modules transported from Earth, one can therefore foresee long-term and large-scale habitats whose construction employ ISRU. Researchers have proposed construction methods that potentially place lunar inhabitants in close proximity to regolith. Benaroya *et al.* (*2002*) provide a review and summary of a range of proposed lunar habitat designs and construction techniques. Several of these proposed lunar habitats employ regolith berms, lunar concrete, or fused regolith.

The density of lunar regolith is approximately 2,000 kg m^{-3} according to Boles *et al.* (*1997*) and simulated versions have been measured at 1,910 kg m^{-3} per Klosky *et al.* (*2000*). Keihm and Langseth (*1975*) estimate its depth as being from 10 to 30 m. The surface layer is a very fine, electrically charged, adhesive, abrasive dust arising from the pulverization of the lunar regolith by meteorite and micrometeorite impacts (*Benaroya et al. 2002*), and under certain circumstances it remains resuspended (*Kolesnikov and Yakovlev 2002*). It is inorganic and contains no water with the possible exception of ice at polar cold traps (*Duke 1998*).

Other researchers have established the lunar presence of ^{222}Rn, a natural by-product of ^{238}U decay (i.e., *Gorenstein and Bjorkholm 1977, Wieler and Heber 2003*). More recently, higher ^{222}Rn concentrations have been supposed to correlate to greater tidal stresses in equatorial regions and other areas of crustal weakness where gases can more easily escape (*Lawson et al. 2005*). Yaniv and Heymann (*1972*), Friesen and Heymann (*1972*), Lambert *et al.* (*1975*), and Friesen and Adams (*1976*) further establish ^{222}Rn levels and exhalation rates in lunar regolith. The human health risks of ^{222}Rn are well documented (*Nazaroff and Nero 1988, Krewski et. al 2005*). With a 3.7 day half-life, inhaled ^{222}Rn may readily decay to ^{214}Po and ^{218}Po in the lung tissue, and these elements are known carcinogens. There are several plausible lunar base configurations which would avoid direct human contact with lunar regolith. However, ISRU implies use of lunar regolith as a construction material. Any ISRU deployment of regolith must provide the following: radiation protection; shielding from meteorites, micrometeorites, and ejecta; and thermal protection. In many scenarios, such deployment of regolith might be external to a pressurized, habitable environment. Would applications bringing inhabitants of lunar bases in near or direct contact with lunar regolith over long durations pose health risks to inhabitants?

Many factors must be considered in determining the risk to lunar inhabitants due to ^{222}Rn exposure. In a habitat in direct contact with regolith-based concrete, the actual concentration of radon in the habitat depends only on the exhalation rate from the regolith and the decay rate of radon, as habitat atmosphere is continuously being recirculated. The radon exhalation rate for regolith-aggregate concrete depends on the radon exhalation rate of the regolith used in the concrete and the radon exhalation rate of the cement. We found no sources that measured exhalation rates, but information is available about radon emissions from both regolith (*Yaniv and Heymann 1972*) and typical cementitious materials, like Portland cement (*Kovler et al. 2005a*). Useful models for radon diffusion rates are suggested in Nazaroff and Nero (*1988*); Rogers *et al.* (*1994*); Anderson (*2001*); Kovler *et al.* (*2005*); Krewski *et al.* (*2005*); de Jong and Dijk (*1996*).

We investigate two scenarios in which regolith would be introduced within a pressurized lunar habitat. We then calculate the amount of ^{222}Rn which would be diffused into the habitable space in both cases, using both assumed exhalation rates and numerical modeling. We compare our results to accepted ^{222}Rn exposure levels, and draw preliminary conclusions of health risks.

Model Parameters

In order to begin to ascertain potential risks of ^{222}Rn exhalation for hypothetical lunar bases, a series of logical assumptions based on existing research were made. The primary supposition was that regolith will feature prominently as a construction material in long-term lunar bases. Two plausible scenarios were investigated in which regolith and/or regolith-aggregate lunar concrete would be deployed within a habitable space. In order to determine plausible quantities for regolith concrete in such structures, three motivating factors justifying regolith or regolith concrete use were considered.

Lunar radiation is experienced as GCR, galactic cosmic radiation, general background radiation from which Earth's atmosphere protects terrestrial life. In addition, SPEs, solar proton events are brief, uncommon but intense solar flares which bombard the Moon with protons ejected during solar flares. Silberberg *et al.* (1985) suggest that at minimum a 2 m regolith cover would provide adequate radiation shielding assuming human occupation of a base 80% of the time. Benaroya *et al.* (2002) suggested having at least 2.5 m of regolith cover in order to keep radiation to 0.5 REM. A thicker cover might provide opportunities for secondary radiation due to radiation hitting atoms in the regolith cover and scattering electrons and other radiation (*Benaroya et al.* 2002), but this is unquantified.

Meteorite and micrometeorite impacts and secondary ejecta due to near-misses are a significant long-term hazard to lunar structures. In order to reduce impact risk to generally acceptable ranges, Jolly *et al.* (1994) recommend regolith shielding to depths of 3 to 4 m.

Diurnal swing on the Moon is considerable, and indeed the whole environment is one of temperature extremes, with daytime means around 380°K ranging to around 400°K, and nighttime temperatures approximately 90°K to 120°K, dropping to 40°K in the perpetual darkness of shaded polar craters (so-called cold traps). At a depth of 1 m, however, temperatures stabilize to approximately 238°K (*Heiken et al.* 1991) or 253°K (*De Angelis et al.* 2002). Furthermore, high-mass lunar structures can be used for thermal storage (*Marra 2006*), suggesting a 2–3 meter regolith shielding to maintain a habitat at ambient temperatures.

These considerations suggested construction techniques producing buried or covered installations, rather than lighter structures arrayed on the lunar surface. Some of the former solutions might involve direct human contact with regolith. In two plausible such scenarios lunar regolith would be integrated into habitable spaces such that inhabitants might come into direct contact with it. Compacted fill under a concrete slab could be installed in an inflated structure, as proposed by Chow and Lin (1988) and reported by Benaroya *et al.* (2002). Alternately, cast lunar concrete could be deployed in exposed surfaces, constructed using a dry-mix/steam-injection method (*DMSI*) (*Lin et al.* 1997).

In the first ISRU scenario, an inflated structure's floor is ballasted with compacted regolith, and a topping slab is poured in order to produce a smooth, level floor. At 1/6-g soils are less resistant to displacement because surrounding soils exert a smaller confining stress (*Benaroya et al.* 2002). It is not therefore a foregone conclusion that a lunar foundation would be 1/6 the

FIGURE 22.1
Artist's rendering of a vehicle-based lunar installation. Given radiation and meteorite hazards for long-term habitat, such a base would be most appropriate for short-duration exploration. Image credit: NASA/JSC image #S93-45585.

thickness of its terrestrial counterpart. While Chow and Lin propose that the compacted fill lie on top of the bottom of the inflated membrane, which would likely seal off ^{222}Rn outgassing from lunar fissures below the habitat, radon might still be outgassed by the fill or slab itself. For computational purposes, a slab thickness of 0.1 m over a 2 m pad of compacted regolith was assumed, and the structure's dimensions were assumed to be 10 m by 10 m by an average height of 3 m, for an overall regolith-aggregate concrete surface area of 100 m^2 and volume of 300 m^3.

In the second ISRU scenario, lunar concrete with a regolith aggregate is deployed either as a surface assembly or as a retaining structure, or both. A dry-mix/steam injection (DMSI) method of concrete casting has been proposed for the waterless, near-total lunar vacuum. Lin *et al.* (1997) reported experimental test samples of such a concrete with regolith aggregate having a compressive strength exceeding 70 MPa. Given the above discussion of the shielding requirements of regolith, for all but floor conditions a preliminary assumption of 0.5 m for a concrete wall thickness was made, as a substitution for the required 4 m of protective regolith. This second habitat is a concrete structure with regolith-aggregate concrete walls, ceiling, and floors, all 0.5 m thick. The dimensions of the second habitat are identical to the dimensions of the first habitat for computational and comparison purposes, although it is likely that the inflatable structure would ultimately be significantly larger than the concrete structure. Terrestrial concrete is highly variable in its permeability due to a variety of factors such as aggregate type and size, cement paste to aggregate ratios, and water content of the mix (*Dinku and Reinhardt 1996*). While it is difficult to assess the permeability of lunar concrete without further research, an oxygen permeability, kO, of 10^{-16} m² (*Romer 2005*) was assumed, based on a water to cement ratio for DMSI concrete of 0.50 (*Lin et al. 1997*). This might suggest that regolith concrete thicknesses necessary for radiation, impact and thermal shielding might yield a fairly airtight structure without recourse to an internal membrane. Such a benefit could, however, potentially expose inhabitants to ^{222}Rn.

The most significant parameter in determining the radon concentration in a lunar habitat is the exhalation rate of radon from the regolith concrete section that contacts the habitable space. In order to ascertain the exhalation rate, two scenarios were examined. In the first, the exhalation rate was determined from the work of Lambert et al. (*1972*) with lunar fines, or finely crushed lunar soil. In the second scenario, an exhalation rate using the emanation rate of radon from lunar regolith fines (*Yaniv and Heymann 1972*), diffusion coefficients for radon through terrestrial concrete (*Rogers et al. 1994*), and numerical modeling of the regolith-aggregate concrete sections using Fick's First Law, were determined, details of which are found in the following sections.

Scenario One: Assumed Radon Exhalation Rate

A radon exhalation rate of three atoms cm^{-2} min^{-1} of regolith-aggregate concrete surface area was used (*Lambert et al. 1975*), although this value was based on emissions from lunar fines, not regolith-concrete. It is unknown whether radon is contained only in the fines, or permeates lunar regolith.

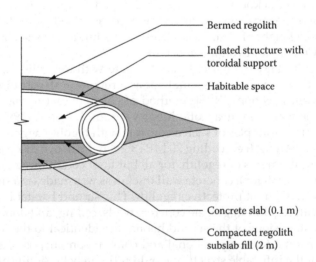

Bermed regolith

Inflated structure with toroidal support

Habitable space

Concrete slab (0.1 m)

Compacted regolith subslab fill (2 m)

FIGURE 22.2
Partial section through an inflatable structure, based on a design proposed by Chow and Lin (1988).

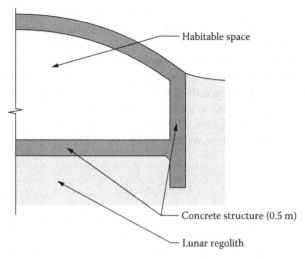

FIGURE 22.3
Diagrammatic partial section through a dry-mix/steam-injection method (DMSI) concrete structure.

Since ^{222}Rn is a decay product of uranium, however, our supposition is that radon is found throughout potential aggregates, not just the finest fractions. This supposition is supported by research (*Lawson et al. 2005*). This exhalation rate, along with the surface areas for both habitats, 320 m^2 and 100 m^2, respectively, were used to calculate the total emission rate of radon into the airspace of each habitat. With no external ventilation in this structure, the model for determining steady state concentration was taken from the work of Yaniv and Heymann (*1972*) as:

$$C_\infty = ES/V\lambda \qquad\qquad \text{[Eq 22.1]}$$

where E is the radon exhalation rate (Bq m^{-2} s^{-1}), S is the surface area of the source, in this case the regolith-aggregate concrete (m^2), V is the volume occupied by habitat atmosphere (m^3), and λ is the decay constant of radon (2.06 x 10^{-6} s^{-1}) (*Kovler et al. 2005, Part 1*).

Adsorption to surfaces and aerosols was not considered in this model, although the expectation is that radon would not strongly adsorb to such surfaces and that the steady state concentration in the habitat airspace would not be significantly altered due to such adsorption. We also did not model thermal and pressure-driven transport through the regolith-concrete layers.

Scenario Two: Calculated Radon Exhalation Rate

Determining the radon exhalation rate for regolith-aggregate concrete involved several steps. First, the emanation rate of radon was determined from the work of Yaniv and Heymann (*1972*). The maximum emanation rate of 0.9 atoms of ^{222}Rn g^{-1} hr^{-1} was used to simulate a worst-case scenario. The emanation rate was used to determine the number of atoms of radon that escaped the solid matrix of regolith-aggregate concrete and entered the pore habitat atmosphere space, which made those atoms available for diffusion through the regolith-aggregate concrete. Next, the density for regolith-aggregate concrete was determined to be 2 g cm^{-3}, based on the density of regolith. A diffusion coefficient for radon through terrestrial concrete of 10^{-8} m^2 s^{-1} was determined from the work of Rogers et al. (*1994*). Fick's First Law:

$$J = -D(\partial C / \partial x) \qquad\qquad \text{[Eq 22.2]}$$

was then used to numerically model the flux of radon through small slices of the regolith-aggregate concrete, where J is the flux, D is the diffusion coefficient of radon, and $\partial C/\partial x$ is the change in pore space radon concentration with respect to a distance.

Numerical Modeling

Each section of regolith-aggregate concrete was divided into 50 slices. Time intervals of 60 seconds were used for calculation of radon emission, diffusion, and decay in each slice for each time interval. Time intervals were allowed to progress until a steady state condition existed in the surface layer. The flux from the surface layer to the habitable space was then used to determine the emission rate of radon into that space.

From this point, the equations used to determine the steady state concentration in the habitable space were identical to those used in scenario one. The concentration of radon on the regolith-aggregate concrete surfaces was assumed to be zero to provide a worst-case scenario for exhalation rate (i.e., to create the largest possible driving force for diffusion from that slice to the habitat airspace). The model was run twice: first for the wall and ceiling sections, then for the regolith-aggregate concrete slab top. Fick's first law was used to develop a finite difference solution for diffusion through the different materials. The concentration of radon in all *i* layers of regolith-aggregate concrete and regolith, other than the top and bottom layers, was predicted as:

$$C_i^{n+1} = \frac{C_i^n + \dfrac{D\Delta t}{\Delta x^2}\left[C_{i-1}^{n+1} + C_{i+1}^n\right]}{1 + \dfrac{2D\Delta t}{\Delta x^2}} + C_{emitted} - C_{decayed} \qquad \text{[Eq 22.3]}$$

TABLE 22.1

Summary of Parameters for Assumed and Modeled ^{222}Rn Exhalation
Calculations for Both Lunar Habitat Scenarios

Parameter	Inflated Regolith-Ballasted Structure		Regolith-Concrete Structure	
S, regolith concrete surface area	100	m^2	320	m^2
V, volume of habitable space	300	m^3	300	m^3
λ, decay constant of ^{222}Rn	2.06	$x\ 10^{-6}$ s^{-1}	2.06	$x\ 10^{-6}$ s^{-1}
E, exhalation rate (assumed)	500	Bq m^{-2} s^{-1}	500	Bq m^{-2} s^{-1}
D, diffusion coefficient through concrete	10^{-8}	m^2 s^{-1}	10^{-8}	m^2 s^{-1}
Δt	60	s	60	s
Δx	4.2	cm	1	cm

where C_i is the concentration at the midpoint of the slice of material (atoms m^{-3} $_{material}$), $C_{emitted}$ is the concentration increase due to emission of radon (atoms m^{-3} $_{material}$), $C_{decayed}$ is the concentration change due to decay of radon (atoms m^{-3} $_{material}$), D is the diffusion coefficient of radon through the material in question (m^2 s^{-1}), Δt is the time interval of the time steps used in the numerical solution (s), Δx is the thickness of each slice of material (m), n is the time step number, and i is the number of the slice, with values increasing through the material.

The concentration of radon in the outer layers of the concrete structure required a slightly different formula due to an assumed concentration of zero at the surface next to the habitable airspace and a slightly different thickness, since the outermost layer is measured from its center to the surface of the concrete, rather than to the center of the adjacent layer.

The compacted fill was assumed to create a flux of 3 atoms cm^{-2} min^{-1} into the lowest layer of the concrete slab, which was assumed to have flux only into the adjacent concrete slice. This was a conservative assumption, leading to the largest possible flux into the lowest concrete layer. Once this flux was calculated, it was used as the radon exhalation rate in the equation from scenario one to determine the steady state concentration in the habitat airspace, which assumed no external ventilation. Table 22.1 summarizes our parameters for each habitat scenario.

Results and Discussion

The calculated assumed and modeled ^{222}Rn concentrations for both structures are shown in Table 22.2.

TABLE 22.2

Lunar Habitat ^{222}Rn Concentrations for Assumed and Modeled ^{222}Rn Exhalation Rates

	Radon Concentration Concrete Structure (Bq m^{-3})	Radon Concentration Inflatable Habitat (Bq m^{-3})
Assumed Exhalation Rate	530	170
Modeled Exhalation Rate	110	84

The higher porosity of regolith allows radon to diffuse through it faster than through regolith-aggregate concrete. Since the only factor that lowers radon concentration is radon decay, the longer it takes for a radon atom to diffuse through a material, the greater the chance that it will decay before entering the habitat airspace. A dense material with low porosity and permeability will have a lower radon diffusion coefficient, and thus a lower radon exhalation rate, than a more porous and permeable material. The results clearly show such a trend, with the regolith-based exhalation rate leading to a significantly greater radon concentration than the modeled terrestrial concrete-based exhalation rate. This trend supports the idea that regolith-aggregate based concrete would be a safer material to have within the habitat airspace than compacted regolith.

Our results are also based on a variety of assumptions and estimates, beginning with lunar concrete characteristics for ^{222}Rn content, density, porosity, and thickness. As on Earth, lunar ^{222}Rn content is site-dependent. Its presence has been detected at a variable number of sites over time from lunar missions (*Yaniv and Heymann, 1972; Gorenstein and Bjorkholm, 1977; Lawson et al., 2005*). For example, the Apollo 16 CSM detected the presence of ^{222}Rn at Grimaldi crater in 1972; the Lunar Prospector mission in 1998 detected none at Grimaldi (*Lawson et al. 2005*). Proposed lunar base locations should therefore be locally investigated for local ^{222}Rn, preferably from the ground.

Different concretes have a great variety of diffusion-related characteristics, such as density, porosity and permeability, depending on a variety of factors, including water content and fabrication techniques. Lin *et al.* (*1997*) were only able to produce 1 cm^3 of dry mix steam regolith concrete due to the very limited availability of lunar aggregate. More research is therefore needed to confirm the actual properties of regolith-aggregate concrete in particular. The exhalation rate of regolith-aggregate concrete is unknown without such research, although it should prove to be lower than the exhalation rate of pure regolith, but without reliable data on lunar concrete this cannot be confirmed. Future lunar missions should endeavor to produce lunar regolith-aggregate concrete samples in situ so that lunar concrete characteristics, including porosity and diffusion, could be experimentally determined, in addition to the structural characteristics of such lunar concrete which would be of interest to lunar base designers and civil engineers.

EPA (*Environmental Protection Agency*) guidelines recommend taking remedial action if the radon concentration in a habitat exceeds 148 Bq m^{-3}, and that action be considered if levels exceed 74 Bq m^{-3}. OSHA (*Occupational Safety & Health Administration*) and the NRC (*Nuclear Regulatory Commission*) have established a standard maximum of 3700 Bq m^{-3} averaged over a 40-hour work week for workplaces (*Nuclear Regulatory Commission, 2005*). Given that regolith-aggregate concrete lunar base inhabitants would be exposed to ^{222}Rn nearly continuously, the EPA standard is more appropriate. Both habitats exceeded the EPA value for the assumed radon exhalation rate. The above modeled exhalation rates achieve levels well within 50% of EPA "remediate" standards, and exceed the "consider remediation" standard. While lung cancer risks to lunar inhabitants are perhaps less immediate or dramatic than GCR and SPEs, meteorite impacts, or temperature extremes, our calculations nevertheless suggest that ISRU application of regolith in an indoor environment is a credible risk.

In cases where ^{222}Rn exists at a dangerous level, Gao *et al.* (2002) suggest a way to reduce risk through the use of a polymerized plaster coat over concrete, leading to significantly lower radon exhalation rates. Transportation of such additives could prove to be costly for the protection derived from their use, however, even if ultimately less costly than a non-ISRU habitat. Alternately, the work of Daoud and Renken (2001) shows a promising method for reducing the diffusion coefficient of radon by using flexible thin-film membranes. In their tests, Daoud and Renken (2001) demonstrated a decrease in radon concentration of up to 95% when flexible thin-film membranes were placed adjacent to concrete samples. Such a decrease would bring the radon concentrations for all scenarios to acceptable levels. Another possibility for reducing radon concentration in the habitats would be to use habitat atmosphere filters containing activated carbon beds, as studied by Bocanegra and Hopke (1989). This method could pose difficulties in a lunar environment with regeneration of the activated carbon. All of these remedial measures entail additional transportation and deployment costs, and therefore should only be undertaken once further research confirms our findings.

Conclusions

Radon concentrations in hypothetical lunar habitats are a potentially significant health concern if lunar regolith is used as a component of an internalized construction material. Given the current data, it appears that radon concentrations might reach dangerous levels in certain construction types. With a complete lack of habitat atmosphere replacement in a lunar habitat, it will be of the utmost importance that all cost-effective means of reducing radon concentrations are employed. The use of thin film membranes may be an especially effective way to reduce radon concentrations for a relatively low cost. However, without detailed data on the physical properties of regolith-aggregate concrete, it isn't possible to accurately determine radon emanation and diffusion coefficients from regolith-aggregate concrete, and thus exhalation rates. More research is required to provide such data.

Acknowledgments

The authors would like to thank Dr. Jeffrey Siegel of The University of Texas at Austin, Department of Civil, Architectural and Environmental Engineering for providing the initial impetus for this paper and for his guidance. We also thank Dr. Mike Duke of the Center for Commercial Applications of Combustion in Space at the Colorado School of Mines for his very generous and thoughtful critiques and comments.

References

Anderson, C.E., *Numerical Modeling of Radon 222 Entry into Houses - An Outline of Techniques and Results*, The Science of the Total Environment Vol. 272, May 14, 2001, pp. 33–42.

Benaroya, H., Bernold, L., and Chua, K.M., *Engineering, Design and Construction of Lunar Bases*, Journal of Aerospace Engineering, Vol. 15, No. 2, April 1, 2002, pp. 33–45.

Bocanegra, R., Hopke, P.K., *Theoretical Evaluation of Indoor Radon Control Using a Carbon Adsorption System*, JAPCA-The Journal of the Air & Waste Management Association, Vol. 39, No. 3, March 1989, pp. 305–309.

Boles, W.W., Scott, W.D., and Connolly, J.F., *Excavation Forces in Reduced Gravity Environment*, Journal of Aerospace Engineering, Vol 10. No. 2, April 1997, pp. 99–103.

Daoud, W.Z., and Renken, K.J., Laboratory Assessment of Flexible Thin-Film Membranes as a Passive Barrier to Radon Gas Diffusion, The Science of the Total Environment, 272, 2001, pp. 127–135.

De Angelis, G., Wilson, J.W., Clowdsley, M.S., Nealy, J.E., Humes, D.H., and Clem, J.M., Lunar Lava Tube Radiation Safety Analysis, Journal of Radiation Research, 43: Suppl. December 2002, pp. S41–S45.

De Jong, P.E., van Dijk, W., van Hulst, J.G.A., van Heijningen, R.J.J., *The Effect of the Composition and Production Process of Concrete on the ^{222}RN Exhalation Rate*, Environmental International 22 Supplement 1 (1996), pp. S287–S293.

Duke M., *Lunar Polar Ice: Implications for Lunar Development*, Journal of Aerospace Engineering, Vol. 11, No. 4, October 1998, pp. 124–128.

Friesen, L.J., and Adams, J.A.S., *Low Pressure Radon diffusion - a laboratory study and its implications for lunar venting*, Geochimica et Cosmochimica Acta, Vol. 40, 1976, pp. 375–380.

Friesen, L.J., and Heymann, D., *Model for Radon Diffusion Through the Lunar Regolith*, Earth, Moon and Planets, Vol. 3, No. 4, 1972, pp. 461–471.

Gao X.M., Tam, C.F., and Gao, W.Z., Polymer cement plaster to prevent Radon gas contamination within concrete building structures, Building and Environment, Vol. 37, 2002, pp. 357–361.

Gorenstein, P., and Bjorkholm, P.J., Radon Emanation as an Indicator of Current Activity of the Moon, Physics of the Earth and Planetary Interiors, Vol. 14, 1977, pp. 289–292.

Heiken, G.H., Vaniman, D.T., and French, B.M., editors, Lunar Sourcebook: A User's Guide to the Moon. Cambridge: University of Cambridge Press, 1991.

Hodges, R.R., Ice in the Lunar Polar Regions Revisited, Journal of Geophysical Research, Vol. 107, No. E2, 5011, 2002, pp. 6–1 – 6–7.

Hodges, R.R., Reanalysis of Lunar Prospector Neutron Spectrometer Observations over the Lunar Poles, Journal of Geophysical Research, Vol. 107, No. E12, 5125, 2002, pp. 8–1 – 8–5.

Howell, J., Lecture delivered by former director of Johnson Spaceflight Center to UT Space Society, Austin, Feb. 15, 2006 and private conversations.

Jolly, S.D., Happel, J., and Sture, S., Design and Construction of Shielded Lunar Outpost, Journal of Aerospace Engineering, Vol. 7, No. 4, October 1994, pp. 417–434.

Keihm, S.J., and Langseth, M.G., Microwave Emission Spectrum of the Moon: Mean Global Heat Flow and Average Depth of Regolith, Science, New Series, Vol. 187, No. 4171, Jan. 10, 1975, pp. 64–66.

Klosky, J.L., Sture, S., Ko, H., and Barnes, F., Geotechnical Behavior of JSC-1 Lunar Soil Simulant, Journal of Aerospace Engineering, Vol. 13, Issue 4, October 2000, pp. 133–138.

Kolesnikov, E.K., and Yakovlev, A.B., Vertical Dynamics and Horizontal Transfer of Submicron-Sized Lunar-Regolith Microparticles Levitating in the Electrostatic Field of the Near-Surface Photoelectron Layer, Planetary and Space Science, Vol. 51, Issue 13, 2003, pp. 879–885.

Kovler, K., Perevalov, A., Levit., A., Steiner, V., and Metzger, L.A., *Radon exhalation of cementitious materials made with coal fly ash Parts 1 and 2,* Journal of Environmental Radioactivity, Vol. 82, N° 3, 2005, pp. 321–334 and pp. 335–350.

Krewski, D., Lubin, J.H., Zielinski, J.M., Alavanja, M., Catalan, V., Field, R.W., Klotz, J.B., Letourneau, E.G., Lynch, C.F., Lyon, J.I., Sandler, D.P., Schoenberg, J.B., Steck, D.J., Stolwijk, J.A., Weinberg, C., and Wilcox, H.B., *Residential Radon and Risk of Lung Cancer: A Combined Analysis of 7 North American Case-Controlled Studies,* Epidemiology, Vol. 16, 2005, pp. 137–145.

Lambert, G., le Roulley J.C., and Bristeau, P., *Accumulation and Circulation of Gaseous Radon Between Lunar Fines,* Philosophical Transactions of the Royal Society of London Series A-Mathematical Physical and Engineering Sciences 285 (1327), 1975, pp. 331–336.

Lawson, S.L., Feldman, W.C., Lawrence, D.J., Moore, K.R., Elphic, R.C., and Belian, R.D., *Recent Outgassing from the Lunar Surface: The Lunar Prospector Alpha Particle Spectrometer,* Journal of Geophysical Research, Vol. 110, E09009, 2005.

Lin, T.D., Skaar, S.B., and O'Gallagher, J.J., *Proposed Remote-Control, Solar-Powered Concrete Production Experiment on the Moon*, Journal of Aerospace Engineering, Vol. 10, Issue 2, April 1997, pp. 104–109.

Marra, W.A., *Heat Flow through Soils and Effects on Thermal Storage Cycle in High-Mass Structures*, Journal Of Aerospace Engineering, Vol. 19, No. 1, January 2006, pp. 55–58.

Nazaroff, W.W., and Nero, A.V., editors, *Radon and its Decay Products in Indoor Air*, 1988.

Nuclear Regulatory Commission, NRC Regulations (10 CFR), Radionuclides (Radon-222), http://www.nrc.gov/reading-rm/doc-collections/cfr/part020/appb/Radon-222.html, updated May 27, 2005.

Rogers, V.C., Nielson, K.K., Holt, R.B., and Snoddy, R., *Radon Diffusion Coefficients for Residential Concretes*, Health Physics, Vol. 67, No. 3, September 1994, pp. 261–265.

Schmitt, H., *Private Enterprise Approach To Lunar Base Activation*, Advances in Space Research, Vol. 31, Issue 11, 2003, pp. 2441–2447.

Silberberg, R., Tsao, C.H., Adams, J.H., and Letaw, J.R., *Radiation transport of cosmic ray nuclei in lunar material and radiation doses*, Lunar Bases and Space Activities of the 21st Century (W.W. Mendell, Ed.), Lunar and Planetary Institute, Houston TX (1985), pp. 663–669.

Wieler, R., and Heber, V.S., *Noble Gas Isotopes on the Moon*, Space Science Reviews 106, 2002, pp. 197–210.

Yaniv, A., and Heymann, D., *Measurement of Radon Emission from Apollo 11, 12, and 14 Fines*, Earth and Planetary Science Letters, Vol. 15, 1972, pp. 95–100.

23

PAC: Protected Antipode Circle at the Center of the Farside of the Moon for the Benefit of All Humankind

Claudio Maccone

Member of the International Academy of Astronautics (IAA)
Torino, Italy

Claudio Maccone, PhD was born in Torino (Turin), Italy. In 1972 he obtained his first degree (Laurea) in physics at the University of Turin with the top mark of 110 points out of 110 and "praise" ("lode"). In 1974 he obtained a second degree (Laurea) in mathematics at the University of Turin, again with the top mark of 110 points out of 110 and "praise" ("lode"). In 1974 he was awarded a Council of Europe Higher Education Scholarship by the British Council, enabling him to read for a PhD at the Department of Mathematics of the University of London King's College. There he obtained his PhD in 1980. His thesis embodied the Karhunen-Loève eigenfunctions of the power-like time-rescaled Brownian motion, later published in several papers. In 1977 he was awarded a Fulbright scholarship enabling him to study and reside in New York City. There he researched the theory of stochastic processes at the Department of Electrical Engineering of the Polytechnic Institute (now Polytechnic University) of New York. He joined the Space Systems Group of Aeritalia (now Alenia Spazio) in Turin in 1985 as a technical expert for the design of artificial satellites. At Alenia, he is involved in the design of space missions such as the Quasat satellite for radio astronomy, the tethered satellite flown by the U.S. Space Shuttle in 1992 and 1996, and the design of a solar sail to reach Mars while being pushed by sunlight. In 1993, he submitted a formal M3 Proposal to ESA for the design, construction, and launch of the "FOCAL" space mission. This spacecraft/antenna is intended to be launched outside the solar system to the distance of 550 Astronomical Units (3.17 light days) to exploit the huge radio magnification provided by the gravitational lens of the Sun, as predicted by general relativity. In 1994 his first book was *Telecommunications, KLT and Relativity*. In 1997 he was elected corresponding member of the International Academy of Astronautics (IAA) in the Class of the Engineering Sciences. In 1998 he published his second book, *The Sun as a Gravitational Lens: Proposed Space Missions*. This book was awarded the 1999 Book Award for the Engineering Sciences by the International Academy of Astronautics. In 2000 he was elected co-vice chair of the SETI Committee of the International Academy of Astronautics and was appointed coordinator of the IAA Cosmic Study on the Lunar Farside Radio Lab. In 2001, Asteroid 11264 was named "Claudiomaccone" in his honor by the International Astronomical Union (IAU): "Claudio Maccone (b. 1948), an Italian scientific researcher and technical expert at the Alenia Spazio in Turin." In 2001 he was elected full member of the International Academy of Astronautics. In 2002 he was awarded the Giordano Bruno Award by the SETI League "for technical excellence in the service of SETI," as described at the web site http://www.setileague.org/awards/brunowin.htm. In 2004 he took early retirement from Alenia Spazio, which gave him full time and freedom for research. He has published over 70 scientific and technical papers, most of them in *Acta Astronautica*.

ABSTRACT The international scientific community, and especially the IAA (International Academy of Astronautics), have long been discussing the need to keep the Farside of the Moon free from man-made RFI (Radio Frequency Interference). Consider the center of the farside and specifically crater Daedalus, located very close to the Antipode of the Earth, i.e., on the equator and at 180 deg in longitude. Daedalus is ideal to set up a future radio telescope (or phased array) to detect radio waves of all kinds that it is impossible to detect on Earth because of the ever-growing RFI. In this paper we propose the creation of PAC (Protected Antipode Circle), a circular piece of land on the Farside centered at the Antipode and spanning an angle of 30 deg in longitude, latitude and all radial directions from the Antipode.

Defining PAC, the "Protected Antipode Circle"

The need to keep the Farside of the Moon free from man-made RFI (Radio Frequency Interference) has long been discussed by the international scientific community. In particular, in 2005 this author reported to the IAA (International Academy of Astronautics) the results of an IAA "Cosmic Study" that had been started back in 1994 by the late French radio astronomer Jean Heidmann (1920–2000) and had been completed by this author after Heidmann's death (see, for instance, [1] and [2]).

The center of the Farside, specifically crater Daedalus, is ideal to set up a future radio telescope (or phased array) to detect radio waves of all kinds that are impossible to detect on Earth because of the ever-growing RFI.

Nobody, however, seems to have established a precise border for the circular region around the Antipode of the Earth (i.e., zero latitude and 180 deg longitude both East and West) that should be PROTECTED from wild human exploitation when several nations will have reached the capability of easy travel to the Moon.

In this paper we propose the creation of PAC, the Protected Antipode Circle. This is a large circular piece of land of about 1820 km in diameter, centered

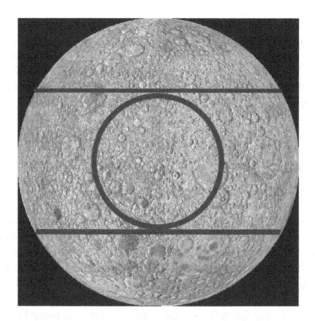

FIGURE 23.1
PAC, the Protected Antipode Circle, is the circular piece of land (1820 km in diameter along the Moon surface) that we propose to be reserved for scientific purposes only on the Farside of the Moon. At the center of PAC is the Antipode of the Earth (on the equator and at 180 deg in longitude) and, near to the Antipode, is crater Daedalus, an 80 km crater proposed by the author in 2005 as the best location for the future Lunar Farside Radio Lab. Inside Daedalus, the expected attenuation of the man-made RFI (Radio Frequency Interference) coming from the Earth is of the order of 100 dB or higher.

around the Antipode on the Farside and spanning an angle of 30 deg in longitude, in latitude and in all radial directions from the Antipode, i.e., a total angle of 60 deg at the cone vertex right at the center of the Moon.

There are three sound scientific reasons for defining PAC this way:

1. PAC is the only area of the Farside that will never be reached by the radiation emitted by future human space bases located at the L4 and L5 Lagrangian points of the Earth-Moon system (the geometric proof of this fact is trivial);

2. PAC is the most shielded area of the Farside, with an expected attenuation of man-made RFI ranging from 15 to 100 dB or higher;

3. PAC does not overlap with other areas of interest to human activity except for a minor common area with the Aitken Basin, the southern depression supposed to have been created 3.8 billion years ago during the "big wham" between the Earth and the Moon.

Figure 23.1 shows a photo of the Farside of the Moon, the two parallels at plus and minus 30 deg drawn by solid lines, and PAC, the Protected Antipode

Circle, a solid circle centered at the Antipode and tangent to the above two parallels at plus and minus 30 deg.

In view of these unique features, we propose PAC to be officially recognized by the United Nations as an INTERNATIONAL PROTECTED AREA, where no radio contamination by humans will possibly take place now and in the future. This will be for the benefit of all Humankind.

Urgent Need for RFI-Free Radio Astronomy

In order to detect radio signals of all kinds, as radio astronomers do, it is mandatory to firstly reject all RFI (Radio Frequency Interference). But RFI is produced in ever increasing amounts by the technological growth of civilization on Earth, and has now reached the point where large bands of the spectrum are blinded by legal or illegal transmitters of all kinds.

Since 1994, the late French radio astronomer Jean Heidmann pointed out that Radio astronomy from the surface of the Earth is doomed to die in a few decades if uncontrolled growth of RFI continues. Heidmann also made it clear, however, that advances in modern space technology could bring Radio astronomy to a new life, if Radio astronomy is done from the Farside of the Moon, obviously shielded by the Moon's spherical body from all RFI produced on Earth.

In view of the following developments in this paper, we present now a short review about the five Lagrangian points of the Earth-Moon system, shown in Figure 23.2.

Terminal Longitude λ on the Moon Farside for Radio Waves Emitted by Telecommunication Satellites in Orbit around the Earth

In this section we prove an important mathematical formula, vital to select any RFI-free Moon Farside Base. We want to compute the small angle l beyond the limb (the limb is the meridian having longitude 90° E on the Moon) where the radio waves coming from telecommunications satellites in circular orbit around the Earth still reach, i.e., they become tangent to the Moon's spherical body. The new angle $\lambda = \alpha + 90°$ we shall call "terminal longitude" of these radio waves. In practice, no radio wave from telecom satellites can hit the Moon surface at longitudes higher than this terminal longitude λ.

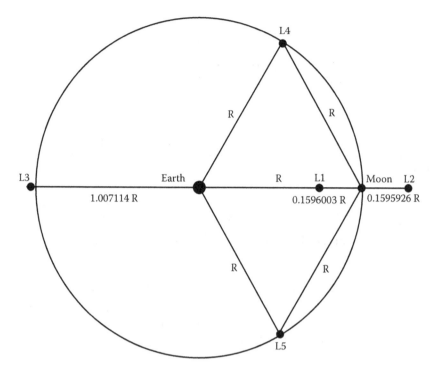

FIGURE 23.2
The five Earth-Moon Lagrangian Points (i.e., the points where the Earth and Moon gravitational pulls on a spacecraft cancel out!): Let R denote the Earth-Moon distance that is 384,400 km. Then, the distance between the Moon and the Lagrangian point L1 equals 0.1596003*R, that is 61350 km. Consequently the Earth-to-L1 distance equals 0.8403997*R, that is 323050 km; The distance between the Moon and the Lagrangian point L2 equals 0.1595926*R, that is 61347 km; The distance between the Earth and the Lagrangian point L3 equals 1.007114*R, that is 387135 km; The two "triangular" Lagrangian Points L4 and L5 are just at same distance R from Earth and Moon.

To find α (see Figure 23.3) we draw the straight line tangent to the Moon's sphere from G, the point tangent to the circular orbit having radius R. This straight line forms a right-angled triangle with the Earth-Moon axis, EM, with right angle at G. Next, consider the straight line parallel to the one above but from the Moon center M, intersecting the EG segment at a point P. Once again, the triangle EPM is right-angled in P, and it is similar to the previous triangle. So, the angle α is now equal to the EMP angle. The latter can be found, since:

1. The Earth-Moon distance $\overline{EM} = D_{Earth-Moon}$ is known and we assume its worst case (Moon at perigee): Earth-Moon distance equal to 356410 km.

2. The $\overline{EP}$ segment equals the $\overline{EG} = R$ segment minus the Moon radius, R_{Moon}.

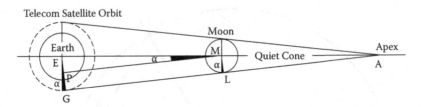

FIGURE 23.3
The simple geometry defining the "Terminal Longitude, λ" on the Farside of the Moon, where radio waves emitted by telecom satellites circling the Earth at a radius R are grazing the Moon surface.

3. Using Pythagoras' theorem one finds $\overline{PM} = \sqrt{(\overline{EM})^2 - (\overline{EP})^2}$.

4. The tangent of the requested angle α is then given by $\tan\alpha = \frac{\overline{EP}}{\overline{PM}} = \frac{\overline{EP}}{\sqrt{(\overline{EM})^2 - (\overline{EP})^2}}$.

Inverting the last equation and making the substitutions described in points 1, 2, and 4, one gets the terminal longitude λ of radio waves on the Moon Farside (between 90° E and 180° E) emitted by a telecom satellite circling around the Earth at a distance R:

$$\lambda = \mathrm{atan}\left(\frac{R - R_{Moon}}{\sqrt{D^2_{Earth-Moon} - (R - R_{Moon})^2}} \right) + \frac{\pi}{2}.$$

Here the independent variable R can range only between 0 and the maximum value that does not make the above radical become negative, that is,

$$0 \le R \le D_{Earth\text{-}Moon} + R_{Moon}.$$

The equation above for λ shows that the $\lambda(R)$ curve becomes vertical for $R \rightarrow \left(D_{Earth-Moon} + R_{Moon}\right)$ and $\lambda = 180°$.

Selecting Crater Daedalus Near the Farside Center

This author claims that the time will come when commercial wars among the big industrial trusts running the telecommunications business by satellites will lead them to grab more and more space around the Earth, pushing their satellites into orbits with apogee much higher than the geostationary one. A *"safe"* crater *must be selected East along the Moon equator. How much further East?* The answer is given by the diagram in Figure 23.4, based on the above equation for λ.

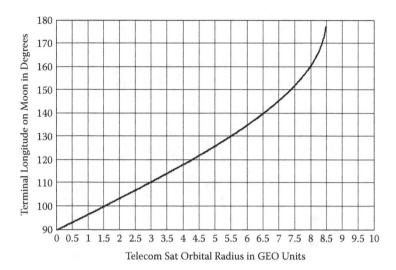

FIGURE 23.4

Terminal longitude λ (vertical axis) on the Moon Farside versus the telecom satellites orbital radius R around the Earth (horizontal axis) expressed in units of the Earth's geostationary radius (42241.096 km).

The vertical trait predicted by our equation for λ shows up in Figure 23.4 as the "upgoing right branch." This shows that, if we only keep the equation for λ into account, the maximum distance from the Earth's center for these telecom satellites is about 8.479 times the geostationary radius, corresponding to a circular orbital radius of 358148 km. Was a telecom satellite put in such a circular orbit around the Earth, its radio waves would flood Moon longitudes as high as about ~175° or more. However we did not consider the Lagrangian points yet!

So, it will never be possible to put a satellite into a circular orbit around the Earth at a distance of 358148 km, simply because this distance already lies beyond the distance of the Lagrangian point L1 nearest to the Earth, that is located at 323050 km (Lagrangian points are, by definition, the points of zero orbital velocity in the two-body problem!).

So we are now led to wonder, what is the Moon Farside terminal longitude corresponding to the distance of the nearest Lagrangian point, L1? The answer is given by the above equation for λ upon replacing R = 323050 km, and the result is λ = 154.359°. In words, this means the following: *the Moon Farside Sector in between 154.359 E and 154.359 W will never be blinded by RFI coming from satellites orbiting the Earth alone.*

In other words, the *limit* of the blinded longitude as a function of the satellite's orbital radius around the Earth is 180° (E and W longitudes just coincide at this meridian, corresponding to the "change-of-date line" on Earth). But this is the *Antipode* to Earth on the Moon surface that is the point exactly opposite to the Earth direction on the other side of the Moon. And our theorem simply

FIGURE 23.5
AS11-44-6609 (July 1969)—An oblique view of the Crater Daedalus on the Lunar Farside as seen from the Apollo 11 spacecraft in lunar orbit. The view looks southwest. Daedalus (formerly referred to as I.A.U. Crater No. 308) is located at 179 degrees east longitude and 5.5 degrees south latitude. Daedalus has a diameter of about 50 statute miles (~ 80 km). This is a typical scene showing the rugged terrain on the Farside of the Moon, downloaded from the web site: http://spaceflight.nasa.gov/gallery/images/apollo/apollo11/html/as11_44_6609.html

proves that the antipode is the most shielded point on the Moon surface from radio waves coming from the Earth. An intuitive and obvious result, really.

So, where are we going to locate our SETI Farside Moon base? Just take a map of the Moon Farside and look. One notices that the antipode's region (at the crossing of the central meridian and of the top parallel in the figure) is too rugged a region to establish a Moon base. Just about 5° South along the 180° meridian, however, one finds a large crater about 80 km in diameter, just like Saha. This crater is called Daedalus. *So, this author proposes to establish the first RFI-free base on the Moon just inside crater Daedalus, the most shielded crater of all on the Moon from Earth-made radio pollution!*

Our Vision of the Moon Farside for RFI-Free Science

Let us replace the value of $\lambda = 154.359°$ with the simpler value of $\lambda = 150°$. This matches perfectly with the need for having the borders of the Pristine Sector making angles orthogonal to the directions of L4 and L5. The result is this author's vision of the Farside of the Moon, shown in Figure 23.6.

Figure 23.6 shows a diagram of the Moon as seen from above its North Pole with the different "colonization regimes" proposed by this author. One sees that:

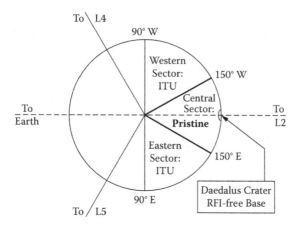

FIGURE 23.6
Our vision of the Moon Farside with the Daedalus Crater Base for RFI-free Radio astronomy, Bioastronomy and SETI science. Future International Space Stations (ISS) might be located at both the L4 and L5 Earth-Moon Points in the decades to come. Only Point L2 will have to be kept free at all times.

1. The near side of the Moon is left totally free to activities of all kinds: scientific, commercial and industrial.

2. The Farside of the Moon is divided into three thirds, namely three sectors covering 60° in longitude each, out of which:

   a. The Eastern Sector, in between 90° E and 150° E, can be used for installation of radio devices, but only under the control of the International Telecommunications Union (ITU-regime).

   b. The Central Sector, in between 150° E and 150° W, must be kept totally free from human exploitation, namely it is kept in its "pristine" radio environment totally free from man-made RFI. This Sector is where crater Daedalus is, a ~ 100 km crater located in between 177° E and 179° W and around 5° of latitude South. At the moment, this author is not aware of how high is the circular rim surrounding Daedalus.

   c. The Western Sector, in between 90° W and 150° W, can be used for installation of radio devices, but only under the control of the International Telecommunications Union (ITU-regime).

Also:

1. The Eastern Sector is exactly opposite to the direction of the Lagrangian point L4, and so the body of the Moon completely shields the Eastern Sector from RFI produced at L4. Thus, L4 is fully "colonizable."

2. The Western Sector is exactly opposite to the direction of the Lagrangian point L5, and so the body of the Moon completely shields

the Western Sector from RFI produced at L5. Thus, L5 is fully "colo-nizable" in this author's vision. In other words, this author's vision achieves the *full bilateral* symmetry around the plane passing through the Earth-Moon axis and orthogonal to the Moon's orbital plane.

3. Of course, L2 may not be utilized at all, since it faces crater Daedalus just at the latter's zenith. Any RFI-producing device located at L2 would flood the whole of the Farside, and must be ruled out. L2, however, is the only Lagrangian point to be kept free, out of the five located in the Earth-Moon system. Finally, L2 is not directly visible from the Earth since it is shielded by the Moon's body, and just fur-ther supporting the case for "leaving L2 alone"!

The Further Two Lagrangian Points L1 and L2 of the Sun-Earth System: Their "Polluting" Action on the Farside of the Moon

There still is an unavoidable drawback, though. This is coming from the *further two Lagrangian points L1 and L2 of the Sun-Earth system*, located along the Sun-Earth axis and outside the sphere of influence of the Earth, with a radius of about 924646 km around the Earth. Precisely, the Sun-Earth L1 point is located at a distance of 1496557.035 km from the Earth towards the Sun, and the L2 point at the (virtually identical) distance of 1496557.034 km from the Earth in the direction away from the Sun, that is toward the outer solar system. These two points have the "nice" property of moving around the Sun just with the same angular velocity as the Earth does, while keeping also at the same distance from the Earth at all times. Thus, they are *ideal places for scientific satellites*.

Actually, the Sun-Earth L1 Point has already been in use for a scientific sat-ellite location since the NASA ISEE III spacecraft was launched on 12 August 1978 and reached the Sun-Earth L1 region in about a month.

On December 2, 1995, the ESA-NASA "Soho" spacecraft for the exploration of the Solar Corona was launched. On February 14, 1996, Soho was inserted into a halo orbit around the Sun-Earth L1 point, where it is still librating now (2007).

As for the Sun-Earth L2 point, there are plans to let the NASA's SIM (Space Interferometry Mission) satellite be placed there, as will be ESA's GAIA astrometric satellite as well.

So, all these satellites do "POLLUTE" the otherwise RFI-free Farside of the Moon when the Farside is facing them. Unfortunately, the Moon Farside is facing the Sun-Earth L1 point for half of the Moon's synodic period, about 14.75 days, and it is facing the Sun-Earth L2 point for the next 14.75 days. Really all the time!

This radio pollution of the Moon Farside by scientific satellites located at the Lagrangian Points L1 and L2 of the Sun-Earth system is, unfortunately,

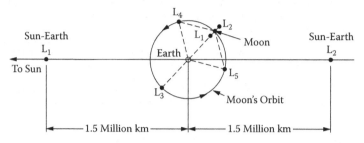

**Five Lagrangian Points
of the Earth-Moon System**

FIGURE 23.7
(Courtesy of Dr. Robert "Bob" Farquhar, Johns Hopkins University Applied Physics Laboratory, Laurel, MD, USA) In addition to the five Lagrangian Points of the Earth-Moon system (already described in Figure 1) the next two closest Lagrangian Points to the Earth are the Lagrangian Points L1 and L2 of the Sun-Earth system. These are located along the Sun-Earth axis at distances of about 1.5 million kilometers from the Earth toward the Sun (L1) and outward (L2). Unfortunately, spacecraft located in the neighborhood of these L1 and L2 Sun-Earth Points do send electromagnetic waves to the Farside of the Moon. Examples are the ISEE-III and Soho spacecraft, already orbiting around L1, and more spacecraft will do so in the future around both L1 and L2.

UNAVOIDABLE. We can only hope that telecom satellites will never be put there. As for the scientific satellites already there or on the way, the radio frequencies they use are well known and usually narrow-band. This should help the Fourier transform of the future spectrum analyzers to be located on the Moon Farside to get rid of these transmissions completely.

Attenuation of Man-Made RFI on the Moon Farside

In a recent paper presented by this author at the International Astronautical Congress held in Valencia in October 2006, his co-worker Salvo Pluchino succeeded in computing the RFI attenuation on the Farside of the Moon [3]. A basic result proven there are the RFI attenuation values shown in Table 23.1.

Perhaps even more important than the "generic" frequency values listed in Table 23.1 are the following precise line frequencies of high scientific importance again taken from the paper [3]. In practice, these are the attenuations of man-made RFI to be expected at crater Daedalus and within the PAC. It should also be stated that these are the attenuation values assuming that the Moon is *not* surrounded by a very thin ionosphere. Since a very tiny Lunar Ionosphere might possibly exist, however, the values below might be slightly incorrect.

TABLE 23.1

Radio Wave Attenuation in the Lunar Equatorial Plane and at Lunar Longitude
$\lambda = 180°$ (i.e., Near the Daedalus Crater) for Radio Sources Emitting at 100 kHz,
100 MHz and 100 GHz, respectively. All attenuation values are in dB.

Frequency or radio waves	$f = 100$ kHz	$f = 100$ MHz	$f = 100$ GHz
Source in GEO	−42.62 dB	−72.62 dB	−102.62 dB
Source in an orbit passing through the L1 point	−30.32 dB	−60.32 dB	−90.32 dB
Source still at			
L4 or L5 Lagrangian points	−29.15 dB	−59.15 dB	−89.15 dB

Conclusions

The goal of this paper was to make the readers sensitive to the importance
of protecting the Central Farside of the Moon from any future wild, anti-
scientific exploitation. In particular, we gave sound scientific reasons why
the PAC, Protected Antipode Circle, should be declared an international
land under the Protection of the United Nations, or, in absence of that institution, by direct agreement among the space-faring nations. The Farside of
the Moon is a unique place for us in the whole universe: it is close to the
Earth, but protected from the radio garbage that we ourselves are creating
in ever increasing amounts that is making our radio telescopes blinder and
blinder. The Farside cannot be left in the realtors' hands! And this is an
urgent matter! Some international agreement must be taken for the benefit
of all Humankind.

Acknowledgments

The author would like to thank the International Academy of Astronautics
(IAA) for allowing him to serve as Coordinator of the "Lunar Farside Radio
Lab" IAA Cosmic Study in the years 2000–2005. Interest in the international
role that the United Nations might play in the creation of the PAC also came
from several members of the International Institute of Space Law (IISL), in
particular Prof. Vladimir Kopal, Chair of the IAA "Scientific-Legal Liaison
Committee." Hopefully, these and other far-sighted minds will unite their
efforts to save the Farside of the Moon from new man-made RFI pollution

TABLE 23.2

Attenuation in the Lunar Equatorial Plane and at Lunar Lat $\lambda = 180°$ (Near the Daedalus Crater) for Radio Waves Having Some of the Most Important Frequencies used by Radio Astronomers to Explore the Universe.

Origin of Radio Waves	Radio Frequency f	Source in GEO	Source in Orbit at L1 Distance	Source Still at L4 or L5
ELF	0.003 MHz	−27.39 dB	−15.57 dB	−14.61 dB
VLF	0.030 MHz	−37.39 dB	−25.10 dB	−23.94 dB
Jupiter's storm	20 MHz	−65.63 dB	−53.33 dB	−52.16 dB
Deuterium	327.384 MHz	−77.77 dB	−65.48 dB	−64.30 dB
Hydrogen	1420.406 MHz	−84.14 dB	−71.85 dB	−70.68 dB
Hydroxyl radical	1612.231 MHz	−84.69 dB	−72.40 dB	−71.23 dB
Formaldehyde	4829.660 MHz	−89.46 dB	−77.17 dB	−75.99 dB
Methanol	6668.518 MHz	−90.86 dB	−78.56 dB	−77.39 dB
Water vapor	22.235 GHz	−96.09 dB	−83.79 dB	−82.62 dB
Silicon monoxide	42.519 GHz	−98.90 dB	−86.61 dB	−85.44 dB
Carbon monoxide	109.782 GHz	−103.02 dB	−90.73 dB	−89.56 dB
Water vapor	183.310 GHz	−105.25 dB	−92.95 dB	−91.78 dB

right over there! Finally, the support of the author's co-workers Salvo Pluchino and Nicolò Antonietti is gratefully acknowledged in the study of the mathematical problems to quantify the RFI on the Moon Farside.

References

1. C. Maccone, "The Quiet Cone Above the Farside of the Moon," Acta Astronautica, Vol. 53, 2003, pp. 65–70.
2. C. Maccone, "Moon Farside Radio Lab," Acta Astronautica, Vol. 56, 2005, pp. 629–639.
3. S. Pluchino, N. Antonietti and C. Maccone, "Protecting the Moon Farside Radiotelescopes from RFI produced at future Lagrangian-Points Space Stations," paper IAC-06-D4.1.01 presented at the International Astronautical Congress held in Valencia (Spain), October 2–6, 2006.

24

Developing the Moon with Ethics and Reality

David M. Livingston

The Space Show ®

David M. Livingston, DBA is the founder and host of The Space Show®, the nation's only talk radio show focusing on increasing space commerce, developing space tourism, and facilitating our move to a space-faring economy and culture. In addition, Dr. Livingston is an adjunct professor at the University of North Dakota Graduate School of Space Studies. He earned his BA from the University of Arizona, his MBA in International Business Management from Golden Gate University in San Francisco, and his doctorate in Business Administration (DBA) also at Golden Gate University. His doctoral dissertation was titled "Outer Space Commerce: Its History and Prospects." Livingston is a business consultant, financial advisor, and strategic planner.

ABSTRACT   On January 14, 2004, President Bush announced the Vision for Space Exploration (VSE), a new public space program for the United States, which calls for the United States to not only return to the Moon, but also to explore and visit Mars and other destinations in the solar system. While the Moon is considerably closer to Earth than Mars, we have already been there, and the costs and technology needed to get to the Moon are much more available and affordable than those needed to get to Mars. It is not surprising that the announcement of the VSE has prompted many in the entrepreneurial NewSpace community to make bold plans for commercial lunar visits and establishing businesses and operations on the Moon. While there is no doubt about the seriousness of companies making such plans, one must question the potential effect of these announcements on the emerging commercial space industry, an industry which already has a difficult time being taken seriously by investors, the financial community, the media, and many other important segments of our population. This paper examines some of the issues facing future commercial lunar enterprises and business plans.

Much of the information used in this paper comes from the author's more than 8 years of hosting The Space Show (www.thespaceshow.com), a radio talk show devoted to expanding space development and commercialization.

Over 1100 interviews are currently available on archives, many of which are on this and related topics.

Introduction

Some of the most common issues discussed on The Space Show are the issues of plausibility and credibility regarding commercial space ventures, particularly those that have been announced, planned, or promoted as eventually taking place on the Moon. As The Space Show is broadcast to all seven continents, listeners from around the world contact the show regarding lunar business plausibility and credibility, questioning what is being said by the guests, telling us what they hear in news reports, and generally wondering how any of these businesses could be taken seriously by investors when, more often than not, the businesses that are making such grandiose announcements have no real business plans to speak of. This listener feedback comes from a friendly community that would like to see these ideas become reality and would like to help, not harm, the industry. However, not everyone is going to be so friendly towards the claims made by NewSpace communities. What happens if, and when, critically-thinking members of the general business community, general media, or policymakers hear the claims? Can it discredit or damage the industry as a whole? Are there inherent ethical concerns and, if so, what are they and how do they affect the industry? Or are these claims just foolish and laughable?

The prevailing thought seems to be that such business promotional talk is really nothing more than unrealistic rhetoric which is unlikely to damage the commercial space industry's reputation in the financial and other important communities, but people hardly ever explain why this is so. Based on the feedback from The Space Show listeners and conversations at space-related conferences, it is necessary to explain why many of the claims are, or could be, damaging and also why there is an ethical element to this discussion.

Getting Back to Business Basics

There are some business facts that, although basic, always apply whether the business is terrestrial, on the Moon, or going to and from the Moon. For example, the business must have capital and continued access to working capital, especially in its early days as a young business since start-ups consume capital for growth and unexpected developments. A business must have a market for its products or services and the market must be easily economically exploited. The business also needs to make a profit servicing its market. Traditional business valuation and measurement tools are just as applicable for future lunar businesses as they are for businesses here on Earth. These tools include those such as risk and reward analysis, return-on-investment (ROI) studies and determinations, due diligence, and assumption building. Lunar-related businesses would not get a free pass on any of these business tools or measuring devices if they would like to attract investor money and become successful.

In addition, there are some unique factors that would apply to the lunar business. For example, public support can be highly critical due to issues surrounding and impacting the environment of cislunar space.[1] These issues include, but are not limited to, environmental issues, the U.N. treaties regarding the use of space, developed- versus undeveloped-nation competition, how best to invest/spend public funds, and more.

Business plans are as important for a lunar business as they are for any business on Earth. As to the contents of the business plan, there are no special considerations for lunar businesses despite that it is more complex to write given that neither lunar businesses, nor most of the elements which would make them up, do not yet exist and may not exist for a few more years. When the elements do not exist, it can be very difficult if not virtually impossible to "cost out" the business. For example, if there is no way to get to the Moon to carry out the intended business, how can the real business costs be determined if we cannot arrive at the job site? While it may be possible to extrapolate from earlier missions to the Moon or something else that is deemed similar, the resulting conclusions would be likely inaccurate and important disclosures regarding the assumptions and projections used in the plan would need to be made. If such disclosures are not made, the business would be discredited, and it would also reflect a serious breach of business ethics or, at best, incompetence if the party did not know such information was essential.

Even given these considerations which need to be made, there are a number of lunar business propositions and development plans in the commercial space world. Without being specific to any one cislunar business plan, examining some of the issues common to such business planning is important.

Lunar Business Reality Check

The emerging space industry is fragile in terms of financial investment, market development, and public support. Even the government program, the Vision for Space Exploration (VSE) that NASA is working to implement, is having more than its share of problems relating to public support and the need for public financing spanning multiple presidential and congressional elections. One major reason for the lack of public support relates to credibility in the mind of the public. For example, does the public believe that NASA can put together a program to take us back to the Moon and on to Mars? Will Congress support the mission from start to finish over the coming years or will they change it or even cancel it, especially in light of our present day economic problems? Regarding the private sector, could businesses demonstrate competence at all levels in order to bring in significant investment to do this type of business? This means management must have realistic business and marketing plans that show reasonable and achievable goals. Make no mistake about it, the public and private sectors influence each other regarding interest in space development.

With this in mind, we look at what would be involved in promoting a private company's lunar business venture. Management knows that the business will not immediately unfold. On the other hand, promoters often treat the proposed lunar business as being not just realistic and plausible, but ripe for exploitation. In order to create a business of any type, even on Earth, one must be able to get to the job site. Imagine any business, except perhaps an online business, where one could not physically get to the business or the center of operations. This is true for the Moon. There is no way to get to the Moon at this time. How credible can one be in promoting a business when nobody can get to the job site?

If we look at how NASA is planning to take us back to the Moon, we see it taking years, perhaps a decade or more and the cost is in the billions. The hardware and architecture needed, as controversial as it is, is much beyond the scope of private sector companies, especially those promoting or discussing a lunar business or commercial lunar opportunity. While it is probably true that some private company could get to the Moon in a less costly way than NASA, maybe even a year or two earlier, at present and into the foreseeable future there simply is no transportation to the Moon. Even getting there with a robotic mission, a feat that is easily doable today, requires costly rockets and carefully sized small payloads since most of the mass of the rocket will be the fuel necessary to take the payload to the Moon. To talk about a lunar business today as if it is just around the corner, viable, or even real, is a disservice to the space community because it flies in the face of reality and sophisticated investment capital knows reality from fantasy.

Due diligence is the name of the game when studying most investment opportunities, particularly high-risk ventures. Certainly, a lunar business is considered to be high risk. Those entrusted to protect investment dollars and performing due diligence will not invest much in the project once they discover there is no current path to the Moon. This will also cast a shadow of doubt over the management team which may not have disclosed this prior to accepting investment money. As there is always a competing use for investment dollars, known as the opportunity cost, funds will flow to projects that are credible, can produce a current or future acceptable return, and to a management team that instills confidence and has proven expertise. Unfortunately, promoting the lunar business does just the opposite for the management team. With no way to get to the Moon, a major cost factor in working the spreadsheet to see if the business can be profitable, and under what circumstances, cannot be accurately determined or even estimated.

Markets are another key issue in this discussion. A business must have a product or service to sell and there must be buyers for these products or services. The business needs to be able to successfully exploit the market it has chosen to engage in, and produce a profit. What are the markets for products coming from a lunar business?

One commonly mentioned product is Helium-3 (He-3), to be mined on the Moon and transported back to be used for energy on Earth. A simple investigation of this possibility results in finding out that He-3 can produce energy as a fuel for a fusion process/plant. However, we do not have fusion capability. Fusion is theoretically possible, but has not been successfully developed despite billions of dollars invested by governments and private businesses. Even with these billions of investment dollars, we are not even close to an engineering breakeven point and that status is still years from anything commercial. So to talk about mining He-3 at this time is to talk about something that might be possible in the intermediate or even more distant future; it is not something that can be realistically approached.

In addition to the major problem of the need for fusion development prior to need for He-3, there is the issue of actually mining something on the Moon. Even if a company intended to do it robotically, at this time no mining equipment exists that can work on the Moon. Even if we assume that we can readily make such equipment, we have no transportation for it to be taken to the Moon. Even if we assume we can design it to go to the Moon using one of our heavy lift vehicles, Atlas V, Delta IV, and the under construction Falcon 9, or a foreign heavy launch vehicle, how many launches would it take, given the relatively small payloads, to get the equipment to the Moon with our existing rockets? Larger vehicles are currently dependent on the NASA program for a successful VSE and that is at its earliest, around 2017-2020, if nothing happens to the program. Further, if the robotic mining equipment were on the surface of the Moon, how would it be maintained, repaired, or even replaced given today's limitations on transportation to the Moon, technology, engineering, and manufacturing?

Mining on the Moon must also be looked at to see if it is reasonable in terms of technology, engineering, and mining capability. Reasonable, that is, in terms of promoting the business today as if it is doable, not something that might be doable in a decade or more. According to Dr. John Lewis, "recovering a single tonne of Helium-3 requires perfect extraction and recovery of all the gas from 100 million tonnes of regolith, a seemingly implausible amount."[2] Again, the issue here is really one of timing and present-day capabilities if one is to be taken seriously in the lunar business promotion. Mining

He-3 may, in fact, prove economically attractive and fusion may become a reality, so this potential process cannot and should not be discarded. But it is not ready for prime time just yet.

Human Factors

For there to be a successful lunar business involving humans, the situation becomes far more complex than robotic missions. Not only is the transportation problem a major obstacle to returning to the Moon anytime soon, humans are faced with a host of human factors issues that must be mitigated if we are to settle on the Moon for longer than a short visit and to do anything useful while there. The human factors issues briefly mentioned in this section include microgravity, radiation, medical issues, and the toxicity of lunar dust.

Microgravity

According to human factors medical experts and astronaut flight surgeons, no one knows the minimum amount of gravity needed for healthy human life.[3] The Moon has one-sixth the gravity of Earth and we simply do not know how this low level of gravity will affect humans over the long run. Researchers know that there are serious medical consequences from low gravity including bone loss. Physical exercise and weight-bearing activities in space help to mitigate the problem, but astronauts suffer from not living in the Earth gravitational force. In the future, we may find out that lunar gravity is insufficient for humans to live and work on the Moon while remaining healthy. In any event, this is an issue which needs further research. While we will likely find ways to effectively counter the negative effects of low or zero gravity, we do not know the costs of countermeasures, the timing for their development, or the possible ways such measures may influence other human factor space issues.

Radiation

Simply stated, there is no safe level of ionized radiation and, without an atmosphere to shield the lunar settler/worker, the radiation exposure on the Moon is going to be significant and dangerous. While lunar settlers might live underground or in some sort of shielded structure, working on the surface of the Moon will expose the astronaut to significant radiation. At this time, there is no effective shielding that is cost effective and not mass heavy.

Medical Issues

While the Moon is only a few days from Earth, there is no 911 to call for a medical emergency. Lunar workers will need to be self-sufficient for their medical care, but our ability to use medicine as we know it on Earth is quite limited in space and certainly untested on the Moon. Many pharmaceuticals do not work as well in space and have a shorter shelf life. Surgery is untested and would certainly be a challenge. How reasonable is it to think that we can put humans on the Moon to work in a business, to be miners, or to do something else without appropriate emergency medical care?

Lunar Dust

From what we know from the Apollo missions, lunar dust is quite toxic in addition to being a serious problem for spacesuit and equipment maintenance. NASA is looking at ways of containing lunar dust outside a lunar habitat, but even if that can be done effectively, we still have the issue of how the dust impacts the equipment and its reliability. There is no doubt that this issue will have a resolution over time, but if we are talking about a lunar business today, one can only raise this issue as an additional red flag to the many that have already been discussed.

Ethical Issues

Applying ethics to a space-based business takes us into a new and untested area because space-based businesses do not yet exist. For purposes of this paper, the definition of a space-based business is one that is operated independently of the Earth, although it will clearly be in communication with Earth command and control centers. One can apply terrestrial ethics to a space-based business, but it is unclear how appropriate the ethics can or will be. One area that is open to ethical consideration is the conduct of the businessmen and women working and promoting the space businesses. Using our lunar business example, consider the ethics of promoting the lunar business as reality or even doable now or in a few years. This ethical discussion has to be considered as there are consequences for such promotions.

As mentioned earlier, promoting the lunar business as if it can be conducted in a short time raises all sorts of red flags with investors, regulators, and other space-aware people. Such actions can and do trash credibility, poisoning the well or polluting the pond so to speak, making it harder if not impossible for legitimate space businesses to be funded or even taken seriously. Not only is this a business issue, but it goes right to the heart of the ethics of those involved in these activities. As the issues of credibility and polluting the pond are well known within the space community, one has to question the ethical behavior of those insisting on such promotions.

Ethical Journalism

One can easily argue that news reporting and journalism are not always ethical regarding terrestrial stories, so why expect it to be different with space stories and news? The big difference is the vulnerability of the developing space industry and the formidable challenges regarding funding, a regulatory environment, safety, and public support. A space journalist writing for a space publication or an online service will often report on a company or individuals as if their work and their plans are serious, just a step away from completion, and on track to be highly successful. Sometimes these stories are even used by the businessmen and women to help raise capital from space-interested investors. The problem is that the journalist knows better and perhaps even knows that the company or the people that he or she is writing about are not credible. There are many journalists who frequently do this sort of reporting because if they do a truth or hit piece, they won't get invited back for another interview. Further, the word will spread with other companies to avoid this or that journalist. So they usually write a positive story or give a company a positive spin even if the facts don't warrant such a story. Is this ethical? Probably not, and it is a problem for the industry because such articles do attract capital to various ventures. When the ventures eventually tank, not only have investors lost their investment, but also the developing industry gets another black eye and the professional capital markets are even more convinced that space is too hard for private companies and is only going to unfold under the control of NASA.

Ethical journalism is an issue in the space community. Writing about lunar businesses as if they are real today, just around the corner, and operating off a first-rate business plan has unfortunate consequences as they contribute to poisoning the well for financing. Is there a solution? There is, but since journalists do not want to damage their future opportunities to get stories and pursue their careers, expecting a change toward maintaining strict integrity and adopting ethics in the process seems slim. I believe it would be as unrealistic to expect significant journalistic improvements as it would be to expect major improvements in self-policing this developing futuristic industry.

To be fair to journalists, they often say that it is not easy to know if a business or plan is flawed and will have the impact suggested. Often, they state that they are not businessmen and women, they are simply journalists and it is not really their job to critique the business and its promotional aspects. Their job is to get the story out to the public. Having talked to so many people as a result of The Space Show and learning the degree to which people look to media and journalism of all kinds for accuracy, the side effect of this type of reporting can be damaging to more than just the business. Journalists should engage in at least a minimum of due diligence before undertaking the interview or the story. While it might be too much to hope for that the journalist would write a realistic article that was factual even if negative, were ethics on the reporter's mind, we would see a huge improvement in reportage. This improvement would minimize many of the detrimental consequences and unfulfilled expectations of many people who buy into what is written without their own due diligence or questioning.

It is also important to note that ethics for lunar and space development champion reality over perception, deflects opposition, enables those taking an ethical approach to their business and conduct to take and remain on the high road, and drives critical thinking and planning. In addition, it fosters a high level of due diligence with redundancy. While many may think an ethical approach to the business may prove more costly, the opposite is true in that it greatly facilitates and shortens the time in the decision-making process, lowers costs over time, and it enables a larger percentage of company funds to go to needed operations rather than fighting battles or putting out fires. Ethics in business, and this is true with the lunar and space business, helps to attract and retain quality employees. This is an extremely important issue for any business, especially one operating in the cislunar environment.

Additional Ethical Issues

Not only are there business ethics that need to be considered in promoting a lunar enterprise, there are some important, large-scale ethical issues that also must be considered for lunar and space business development. The fact that these issues remain in somewhat of a state of limbo does not help the promoter or business considering lunar and space commerce now or in the near future.

Space and lunar property rights are huge issues for space development and are also significant ethical issues. Do we have property rights like we have in the United States or in other western countries? What about those nations that do not have a space program and cannot go to the Moon or anywhere else in space? Do they get omitted from the equation?

This brings us to the very controversial issue of benefit sharing. U.N. treaties call for benefit sharing, but businesses that can and are planning to operate in space do not consider sharing their revenues with other nations, especially since they incur all the risks and the costs. This is a significant ethical issue which is often under discussion within the space community and with space policy leaders in the United States, the United Nations, and in other countries. It is doubtful that this issue will be resolved until a company actually gets to the Moon or a celestial body and starts doing something commercial, even making a profit. Such an act is likely to be a catalyst to legal or regulatory activity leading to a resolution of this issue. How ethical the resolution will be is an unknown item. Since this is a risk factor for a lunar business, it is recommended that benefit sharing become a cost disclosure item since it has the potential to be a cost burden for any lunar business.

There are also ethical concerns about strip mining on the surface of the Moon by many people. Even the launching of rockets has ethical overtones because rockets damage the ozone layer on a short-term basis and they discharge pollutants in the atmosphere that might adversely impact climate management. At the present time, there are so few rocket launches on a global basis that this is not very significant. However, if the thousands of launches a year envisioned by the space community are ever realized, this might very well become an important issue with potential to alter or damage Earth's atmosphere. Finally, since space development is international, involving numerous countries with a variety of cultures and ethical considerations, whose ethics will be applied to space development? Will space ethics represent our ethical standards in the United States? What about ESA, Russia, China, India, Israel, Iran, or countries in South America? As we continue our drive towards becoming spacefaring, we have the potential to complicate our ethical concerns ranging from lunar and space businesses, to space activities conducted from Earth and across many countries and cultures.

Promoting A Lunar Business

Lunar businesses will continue to be promoted as if they are realistic, ready to go, and totally plausible. Understanding this suggests that there should be some guidelines for how such a lunar business should be discussed and promoted to avoid the problems that have been discussed. The suggestions offered below reflect the ethical high road for the businessmen and women involved in these futuristic businesses. The suggested disclosures, if followed, will minimize the issues detracting from credibility reflecting on the industry and other business promoters engaged in more realistic ventures with interested financial partners.

First, entrepreneurs must use full, reality-based disclosure for key elements of the lunar business. Businessmen must disclose the transportation problem: there is no way to get to the Moon at this time, there is an absence of operating hardware to make the lunar trip, and that there is no reasonable projection of lunar transportation costs. Thus, all projections reflecting business costs and possible returns on investments are to be considered as highly speculative and likely inaccurate even though they reflect management's best ability to estimate such costs. This disclosure must be comprehensive and written in a way that is easy and clear to understand, not in legalese or in manner that might obscure the actual facts being disclosed.

Another disclosure regarding markets for the goods and services for the lunar business would be necessary. The disclosure should state that there is no factual way to assess the market for the goods and services produced on or initiated on the Moon. Thus any projected ROI must be considered highly speculative despite management's best ability to gauge the potential market. The same type of disclosure needs to be made for most of the entries that will be found on pro forma representing possible financial outcomes for the business.

Given that the regulatory environment for lunar businesses is currently in flux, a disclosure needs to be inserted in the plan and in the promotional efforts stating that uncertainties exist in the regulatory environment and that the business could be adversely impacted by regulations from the United States, the United Nations, or even other nations.

Another disclosure needs to be made regarding maintenance and repair of equipment once the equipment is gotten to the Moon. As mentioned earlier, lunar dust is a problem. How does one cost out lunar dust repair and maintenance issues? Do they omit the disclosure? Or do they deny it is a problem or possibly shift the problem forward to be the responsibility of a business or management team off into the future? Lunar business plans do not usually address this issue.

If one is planning to conduct a business where the needed support technology is not in existence, such as fusion needed for He-3 usage, this requires full disclosure of the facts, such as Helium-3 fusion is not yet available and there is no estimate of when fusion will be commercially available, if ever.

The lack of property rights must be disclosed as well as the fact that U.N. treaties call for benefit sharing. While the United States is not a signatory to the U.N. Moon Treaty calling for benefit sharing, it is a signatory to a major U.N. treaty, the Outer Space Treaty, which also calls for benefit sharing. The potential exists for benefit sharing to adversely impact the revenue stream of a lunar or any space business and investors and interested parties need to be informed of this fact.

In addition to the disclosures mentioned above, the lunar business must be projected as a potential future business with some sort of reasonable timeline supported by actual data on how long it might take to reach needed business development milestones. For example, what is the milestone for when lunar transportation might be available and will it be government-supported transportation or affordable private lunar transportation? One could produce reasonable estimated timelines given what NASA is experiencing in working toward accomplishing the VSE, then extrapolating to a private sector timeline based on past experience with private-sector development versus public-sector development. This would be a speculative estimate, so obviously the assumptions underlying the results would need to be included and would also require disclosures.

When presenting these lunar business ventures at conferences or public proceedings, they must be presented as futuristic followed by an explanation of why they are being promoted today, why funding is currently being sought, and why an investor should consider such a futuristic investment.

If this approach to discussing and promoting the lunar business today were carried out, there would be far fewer issues to be critical of. So far, we have not seen this happen, and thus space advocates must continue to address this issue at conferences or when a guest on The Space Show presents a business plan that is neither viable nor plausible now or in the near term, but might be doable on a longer-term basis.

La La Land vs. Reality

One of the most persistent problems in discussing the subject of lunar development and lunar commerce centers on what is reality today, what is plausible for the future, and what people perceive, believe in, or wish to be real, thus it must be real. While one may find it hard to understand the wish list or a "true believer" approach to space as being real, in many ways and arenas, it is the predominant approach.

One example common in this approach is the claim that present day businesses can mine He-3 on the Moon and return it to Earth. The reality is that, at the present, there is no commercial way to get to the Moon let alone a reliable and routine way of getting to the Moon, forgetting for the moment if it is economical to do so. There is no way to mine He-3 on the Moon. There is no equipment to do it; there is no way to maintain the equipment which would be operating in a very harsh environment, and there is no way to return the product to Earth. Even if all of the above were resolved, He-3 requires fusion energy and despite billions of dollars having been spent on fusion energy research, we do not have fusion and we are not even close to achieving engineering break even let alone commercial break even. That said, the He-3 venture is consistently discussed and put forth as reality today. This is not reality, and is unlikely to even be plausible and for anyone to promote it as reality suggests that they live in la la land rather than the present.

Another example of this, though not related to the Moon, is the Space Solar Power (SSP) cause. Did you know we already have $100/lb to LEO launch costs and around $200/lb to GEO, so that launching massive space solar power satellites can be done so economically? When space solar power supporters are asked to identify the rockets that can do so at this cost, there is often a gap of dead silence or the declaration that the rockets are being built, but don't worry, they are just around the corner. In addition, SSP true believers ignore the sad status of our electrical grid which cannot sustain new power inputs, the fact that we have no heavy lift rockets, and that we have no potential to lift the mass needed to GEO for SSP satellites. Other issues prevail, such as beaming technology and the efficiency of the beam, political resistance to beaming energy through the atmosphere, the fact that the satellites for use in beaming power to Earth do not yet exist, and more. The point is, SSP is not reality based today but like lunar activities, it is seen as reality by large numbers of promoters and advocates for the cause.

Countless other examples of this exist, but as suggested earlier in this paper, one might wonder: what does it matter? Such advocates and promoters are not really hurting anything, are they? However, they are hurting legitimate efforts to develop lunar commercial efforts and more. Such unrealistic ideas, ventures, and efforts are seen exactly as that by those who know the industry, financial people doing due diligence into an investment or business opportunity, and policy makers who understand what is reality today, plausible tomorrow, and fantasy. The purveyors of fantasy as reality cause serious players to be taken less than seriously and this has the potential to interfere with the legitimate emerging entrepreneurial and commercial space industry.

Many have asked why people subscribe to fantasy as if it's real. Many guests and listeners have suggested answers, mostly thinking that space enthusiasts have a sense of entitlement about them regarding space and space expectations because of the role science fiction film and literature has played in our culture. Furthermore, many who grew up in the Apollo years expected we would be on Mars by now, we would be routinely visiting lunar settlements, and that we would be truly space faring. People thought this to be a sacred promise of the space program, NASA, the government, or even Robert Heinlein, Sir Arthur Clarke, and others. Over the years, confusing what should have been for what is, wishing for what should have been, and having all this reaffirmed culturally through the space enthusiasts community, space advocate organizations, and like minded advocates promoting a cause, we instead get a blurring between fantasy and reality.

Many say that the fantasy is the dream and everything stems from a dream. But a fantasy is not a dream. A dream has a chance to become real depending on the success of those pursuing the dream. A fantasy can never be real and, in fact, endlessly loops over and over again as a fantasy. In 2006, I wrote a piece for the Space Cynics blog comparing a fantasy and a dream. This article can be read at: http://spacecynic.wordpress.com/2006/05/23/dreams-fantasy-and-kool-aid-exploring-the-meani/. There must be no mistaking a fantasy for a dream. Yes, dreamers can turn their dreams into a rocket, a space venture, someday even a lunar or Martian settlement or mining operation. A fantasy will always loop around as a fantasy. It is tragic when a person is trapped by their fantasy and does not see that, but this happens all too often. All too often, a person steps up to the plate only to claim the status of a true believer, a wish list thinker, or a person who confuses reality with la la land.

In terms of building infrastructure and a successful commercial environment for engaging in lunar commerce for the future, we must be grounded
in reality and we must claim reality as our place of residence. It behooves us
all to be grounded in reality, to know and understand sufficient engineering, physics, business, financing, marketing, and human factors to be able to
discern reality from Dusty Springfield's 1964 hit, "Wishin' and Hopin'."

Conclusion

Space business venture discussions, business plans, and promotions will
become more common as interest grows in space development from both the
public and the private sectors. If the assumption is made that the VSE will be
successful and will actually look like what we think it will look like today,
assuming no congressional or future administration changes to the plan, there
will be an increasing interest in lunar commerce. To avoid discrediting the
developing industry, damaging its potential in financial markets and investor
communities, and increasing the chances of less than favorable regulation, we
need to make sure that, when these businesses are discussed, promoted, and
business plans circulated, not only are they ethical in their disclosures, but they
are based on reality in order to demonstrate management competence. The
industry is now in its infant stage of development and it is highly vulnerable to
being considered to be lacking credibility. This would turn investors and Wall
Street away, and would cause alarm or fear resulting in an overly aggressive
regulatory regime. This downside can be avoided by following the suggestions
outlined in this paper and by being grounded in reality, not fantasy.

References

1. Cislunar space refers to the volume of space within the moon's orbit or a sphere
   formed by rotating that orbit. *Cislunar* is Latin for "on this side of the moon."
2. John S. Lewis, "A Proposed International Legal Regime for the Era of Private
   Commercial Utilization of Space," *The George Washington University Law Review*
   37 (2005): 748.
3. Jim Logan, The Space Show, http://archived.thespaceshow.com/shows/783-
   BWB-2007-07-03.mp3; 7-7-07.

Section V

Planning and Analogues

25

Lunar Base Living: Beyond the Pioneering Stage

James D. Burke

The Planetary Society
Sierra Madre California

James D. Burke is an alumnus of Webb Institute and Caltech and a former U.S. naval aviator, employed at the Jet Propulsion Laboratory from 1949 to his retirement in 2001. In retirement he is active in The Planetary Society and the International Space University, where he has been a faculty member in each ISU summer session since 1989. Burke's main professional interest is in the exploration and settlement of the Moon. He is a Fellow of the British Interplanetary Society and a member of the American Geophysical Union and the American Institute of Aeronautics and Astronautics. Married since 1950 to his wife, Caroline, with five children and three grandchildren, he is a current pilot, lifelong yachtsman, and determined advocate of educating young people toward enjoying learning and achievement.

Introduction

This chapter addresses some lunar subjects mostly unexamined outside the realm of science fiction. If the large investments now proposed, leading to permanent human residence on the Moon, do actually occur, it will be logical to consider how best to exploit the results. At present a stated goal is to prepare for human missions to Mars. But once the Moon has served as a stepping stone, what then? Will established lunar base facilities and operations simply waste away like the detritus of an ended war, or will other uses be found for them in a longer-term scenario? If so, how will those other uses be paid for and with what ultimate intentions? Here we approach these questions from the standpoint of the discipline of Futures Studies, a method that considers alternate possibilities without prediction, but with intent to highlight preferred futures that could be advocated and realized.

Initial Assumptions

To enable a specific discussion we make the following assumptions at the outset:

(a) An international lunar program with sustained human presence at a south pole base will become reality.

(b) Over time, increasing dependence on local natural resources will greatly reduce the essential logistics flow from Earth.

(c) Technical and managerial advances will lead to vastly improved performance of all systems, including Earth-Moon-Earth transport, in terms of cost, safety and ability to evolve into more advanced versions.

(d) Information systems will evolve to provide essentially unlimited data flow, at a cost trivial in comparison to that of transporting physical objects, between Earth and Moon as well as from point to point on the Moon. Navigation, selenodesy and timing on the lunar surface and in cislunar space will be as accurate and convenient as the analogous functions are now on Earth with GPS.

Terrestrial Settings

With these assumptions we are now ready to consider what might be going on here on Earth that would provide an environment for the long-term success of lunar settlements. Clearly none of the above happy outcomes is possible without some improvement relative to present-day economic and political conditions here. NASA's lunar program, with its Constellation transport system and associated proposed facilities on Earth and Moon, will continue to be starved for funding and public support unless there is some policy breakthrough resulting in a justified and broadly agreed shift of national priorities. No other nation or group of nations has committed to a real program for sustained human presence on the Moon. So long as lunar settlement remains merely a distant hypothetical goal, lunar programs will limp along at their present pace.

This does not mean that there can be no progress. Over time, in many parts of the world, capabilities are being built, both technical and managerial, that could be exploited in a real lunar settlement program if the political environment were to become supportive. What might cause this change to occur? The central need is for a peaceful and abundant society, the kind that has existed from time to time in at least some parts of human civilization.

Today, doom appears inevitable as the world economy encounters resource exhaustion, wars, hunger, fear and despair. But long-term solutions exist, and already some movement toward applying them can be observed.

Of course there is always the possibility of a collapse due to any of several quick or slow catastrophes–an asteroid impact, a rapid global climate shift, or simply a failure of people and their governments to act before a descent becomes irreversible (Ref. 1). In such an instance there would be no point in imagining a large, near-term lunar settlement program. In any event we should by then have implanted an archive in the Moon (Ref. 2) to aid immediate recovery and to assist in civilization's long climb back.

Let us then examine only those scenarios that include success in present lunar programs, followed by the appearance of practical questions about what to do next on the Moon.

Program Options

With an established infrastructure near the lunar south pole, and with the main exploratory goal having moved on to Mars, what are the realistic possibilities beyond that period? One choice would be simply to abandon the Moon base, perhaps leaving behind some automated scientific equipment as was done with Apollo's ALSEP instruments. A more ambitious program could include continued occupation of the base, primarily for further development of techniques for the extraction and use of lunar resources such as helium-3. In this option the lunar effort would be directly competing with the Mars program, so probably the only way to sustain it would be to have it in the hands of national or non-governmental agencies not mainly committed to Mars. Some people think tourism could contribute to this solution, but at the moment the only serious lunar tourism initiative is for circumlunar flights (Ref. 3).

Beyond just continuing occupation of an existing base lies the territory we wish to examine here; namely, an endlessly expanding human lunar presence leading onward toward a two-planet civilization.

A Preferred Future

With the assumptions listed earlier, plus terrestrial progress toward an equitable and sustainable world society, it can readily be imagined that human life on the Moon, while severely constrained by the lunar radiation environment and lack of atmosphere, may still evolve into forms containing all of the good features of life on Earth, plus new features enabled, for example, by the low gravity of the Moon. In serious science fiction lunar versions of all of the fine and lively arts are to be found; human-powered flight is easy; new kinds of sports and diversions exist; the possibilities are limitless. All this is interesting, but the more important prospect is truly unpredictable; namely, a development of new human ideas, means of government, biological technologies, perhaps even philosophy and religion, among a population in continuous information contact with, but physically removed from, Earth.

Underground cities have existed in Anatolia (Refs. 4 and 5), but they are an imperfect analog because their inhabitants' habits and arts apparently did not diverge from the existing above-ground culture. In the Moon, particularly as settlements become large and diversified as to cultural norms, terrestrial history suggests that we may expect truly new social developments–perhaps the most important long-term consequence of human settlement of the Moon.

References

1. Diamond, Jared M. (2005), Collapse: How Societies Choose to Fail or Succeed. Viking Press.
2. Students of ISU (2007 summer session, Beijing) Phoenix Team Project Report. Available at www.isunet.edu
3. www.spaceadventures.com: Lunar Mission
4. www.tourismturkey.org: Central Anatolia
5. www.hitit.co.uk/tosee/cappy/ucities.html

26

Assessment of Lunar Exploration Objectives

Marc M. Cohen

Human System Integration Lead for Space Systems
Northrop Grumman Integrated Systems
El Segundo, California

Marc M. Cohen, ArchD, is a space architect who took early retirement from NASA–Ames Research Center in 2005 where he served in the Space Human Factors Office, the Advanced Space Technologies Office, and the Advanced Space Projects Branch. He currently serves as the Human Systems Integration Lead for Space Systems in the Advanced Programs and Technologies Division, Aerospace Systems Sector of the Northrop Grumman Corporation. Marc's current projects are the Altair Lunar Lander and Lunar Surface Systems. Marc has collected an AB in Architecture and Urban Planning from Princeton, an M.Arch from Columbia, and an Arch.D from Michigan. He is a licensed architect. He was a founder of the AIAA Space Architecture Technical Committee, http://www.spacearchitect.org.

ABSTRACT In 2006, NASA compiled a large collection of over one hundred candidate Lunar Exploration Objectives from across the aerospace community. NASA organized these candidate objectives into 18 "families" of related activities. NASA then asked the aerospace industry—through the Aerospace Enterprise Council—to help assess these objectives. On August 29 and 30, the Enterprise Council held a workshop at which industry representatives evaluated the objectives. The workshop produced half dozen sets of numerical ratings and text summaries covering 61 objectives that scored above zero; covering about 1/3 of the total furnished by NASA.

This paper describes and assesses the products of the workshop. The Workshop generated two types of metrics: prioritized rankings of the objectives and time phasing of the objectives. This paper applies a simple model to combine the disparate prioritization and time phasing results into a single ranking scale.

This analysis yielded several important findings. It found a natural grouping of the 61 objectives into five overarching objectives (OAOs): Engineering, Engineering Operations/Crew Activity Support, Habitation, Operations, and Science Operations. The data analysis identifies the percentage of variance from the Mars factor and the lunar time phasing for each of the OAOs.

The assessment presents observations and findings about the results from the Workshop that provides insight into the properties and interactions of the exploration objectives.

Introduction

On August 29–30, 2006, the Space Enterprise Council of the U.S. Chamber of Commerce in Washington, DC hosted a workshop to which they invited the aerospace industry to evaluate NASA's candidate lunar objective. These objectives originated in April of that year with the Exploration of the Moon and Beyond RFI that produced 800 responses and the International Lunar Exploration Workshop and other sources that produced 400 responses. From these 1200 responses, the Exploration Systems Mission Directorate at NASA HQ distilled 180 candidate lunar exploration objectives that they provided to the Space Enterprise Council. By reducing duplication and clarifying themes, the Space Enterprise Council team consolidated these candidates to 150 objectives that the workshop participants considered, rating and prioritizing them on a variety of criteria that they conceived and applied at the workshop. The outputs from the workshop consisted of "Prepare for Mars" priority votes and lunar time phase estimates for 104 objectives arranged in 18 families of related objectives ranging in size from one to 14 topics. Appendix A reproduces the transmittal letter from the Space Enterprise Council to NASA.

The analysis that this paper describes determined that of these 104 candidate objectives, some 87 were sufficiently well characterized to allow assessment to a consistent and hopefully rigorous evaluation standard. Table 26.1 presents a descriptive overview of these 87 objectives that comprise 18 scientific and technical families.

In addition to the quantitative point system of rating, the Workshop also produced a set of qualitative assessments that provided background and depth to their effort. These qualitative outputs included descriptive summaries of their results on six topics: Commercial Space, Global Partnerships, Lunar Habitation, Prepare for Mars, Public Engagement, and Science. It soon became evident that the documentation provided for Commercial Space, Global Partnerships, and Public Engagement did not really address putting payloads on the moon, so the analysis focused on the other three topics.

Purposes of This Assessment:

Create a set of metrics to interpret and measure these Data.

Develop a method of analyzing, comparing, and evaluating these survey data on a systematic and objective basis.

TABLE 26.1

Summary of Lunar Exploration Objectives by Family Group

Prefix	Family of Scientific and Technical Objectives	No. Objectives in Family	Family Mars. Factor Avg. Score	Overarching Objective (OAO)
mA	Astronomy	7	1.00	Science
mCAS	Crew Activity Support	5	14.45	Operations
mCOM	Communications	3	10.13	Operations
mEHM	Environmental Hazard Mitigation	5	9.95	Engineering
mENVCH	Environmental Characterization	11	1.44	Science
mENVMON	Environmental Monitoring	2	1.21	Operations
mGEO	Geology	14	2.57	Science
mGINF	Global Infrastructure	2	1.21	Operations
mHEO	Heliophysics	3	1.33	Science
mHH	Human Health	5	15.54	Habitation
mLRU	Lunar Resource Utilization	9	2.91	Engineering
mLSH	Life Support & Habitation	4	15.71	Habitation
mMAT	Materials	2	1.00	Engineering
mNAV	Navigation	1	12.45	Operations
mOPS	Operations	6	15.50	Operations
mPWR	Power	3	6.40	Engineering
mSM	Surface Mobility	2	13.21	Operations
mTRANS	Transportation	3	9.78	Operations
TOTAL	18 Families	87	Avg. = 7.54	

Develop a set of constructs to allow comparison of the lunar objectives across the variables of team size and representation, groupings of objectives into families, and families into "overarching objectives, " and the family size of the objectives,

Make findings about the most important parameters of the Workshop results.

Identify the sources of variance in the data.

Understand how the lunar time phasing will change with the shift from "Sortie-first" to "Outpost-first" exploration strategy.

Understand the relationship between the initial emphasis on "Prepare for Mars" versus the more pragmatic imperative—and outcome—of "Prepare for the Moon."

Approach and Overview of Method

This assessment followed this approach:

Synthesize and plot the raw data by "Prepare for Mars" priority. Assess the raw data for the lunar time phasing and organize it in a useful way.

Combine the lunar time phasing and Mars priority into a single Mars/Lunar value for each objective.

Recognizing that the 18 families were too many to manage for this evaluation, combine them into five Overarching Objectives (OAOs): Engineering, Engineering Operations (Crew Activity Support), Habitation, Operations, and Science Operations.

Translate the Mars/Lunar values to the OAO rubric, which indicated that the scores tended to be skewed by family size and by voting block effects.

To compensate for these effects, the last step was to adjust the OAOs to the same mean on the Mars/Lunar value scale, which enables a direct comparison of their point-spreads.

Scope

The scope of this evaluation depends almost entirely upon the results from the Industry Workshop. APPENDIX B shows the summary table of the objectives passed and rated non-de minimus by the Workshop with short technical descriptions. Given that output data, the challenge was to develop a method that first could produce a useful synthesis, and then analyze what it means. Figure 26.1 shows the scope of this assessment within the larger output from the Workshop. The goal of impartial evaluation required the team to act as an honest broker and to avoid imposing any of his or her biases or preferences on the assessment process or outcome. Only after completion of the analysis, does the team discuss the specific objectives, offer observations on the implications of the ratings, and state conclusions.

The "raw data" takes the form of ratings of the priority to "Prepare for Mars" exploration on three prioritization scales and on the time phasing of each objectives for lunar exploration. The lunar time phasing derived from an assessment of when NASA would need to initiate or achieve the objective, the technology readiness level, and the degree of technical difficulty to achieve it. Please note that the Workshop did not include a comparable "Prepare for the Moon" criterion. The lunar time phasing serves this purpose. The general criteria that the Workshop used to evaluate each objective were:

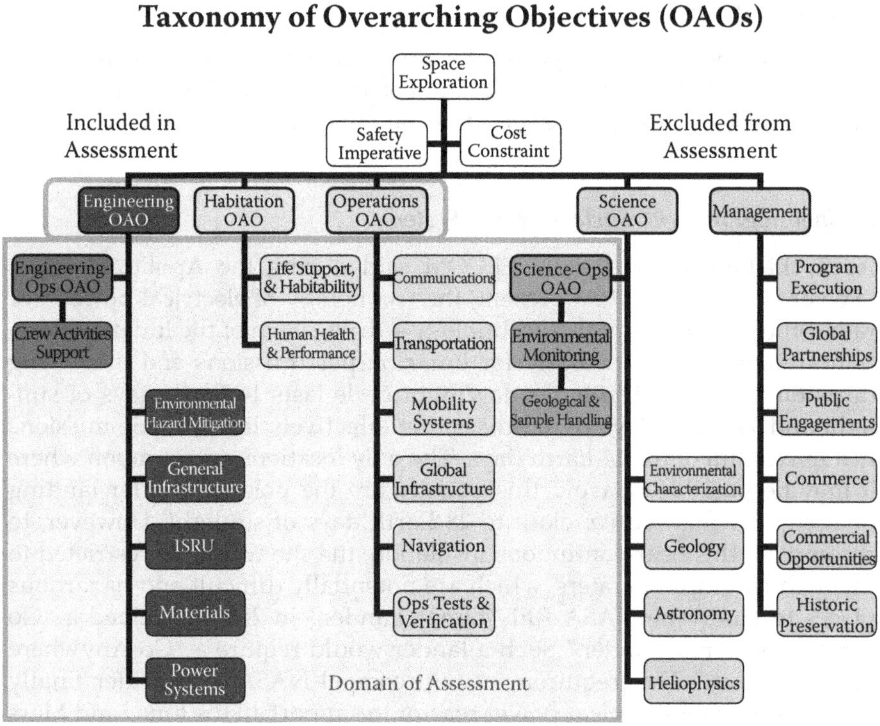

Taxonomy of Overarching Objectives (OAOs)

FIGURE 26.1
Overview of the lunar exploration families of objectives evaluated within the overarching objectives framework.

Whether objective was:
Enabling to Exploration, [or]
Enabled by Exploration, [or]
Enabling to and Enabled by Exploration (designated "hand-in-hand")

Science Objectives Anomaly

Science objectives did not score well in the Workshop proceedings. There were two reasons for this low voting result. First, there were few if any scientists present who might advocate for science objectives or vote for them. Second, the evaluation criteria given in the charter to the Workshop participants spoke only of enabling exploration of being "enabled" by exploration. It did not address—directly or indirectly—the development of new knowledge, scientific or otherwise. This near-exclusion of science objectives from Workshop results posed a challenge to the assessment team. The few science-related objectives that scored above de minimus were largely operational, "enabling," in the Workshop's terminology. The Science Operations OAO serves to recognize the role of science at least as enabling and being enabled by exploration.

Examples of Top Objectives

This section provides examples of the top-rated objectives to provide insight into what constitutes a lunar exploration objective. These five exemplars are selected from each of the five sets of overarching objectives (OAOs).

Engineering: mPWR1 Surface Power System

As Sy Liebergot, the Apollo EECOM said during the Apollo 13 crisis, "Power is everything." At present, the technology of electrical power—or rather the limitations of this technology—drive much of the lunar exploration architecture, particularly for lunar outpost missions and other long duration payloads. The lunar day/night cycle lasts 14 Earth days of sunlight and 14 Earth days of darkness that effectively limits sortie missions to a maximum of the 14 Earth days. The only locations on the moon where it may be possible to avoid this penalty are the poles that offer landing or base sites that receive close to 28 Earth days of sunlight. However, to accomplish this near-continuous insolation, the site would be restricted to the high rim-tops of craters, which are potentially difficult and hazardous places to land. The NASA RFI "Lunar Lander" in 2006 described a "Go Anywhere Lunar Lander." Such a lander would require a "Go Anywhere Power System." This requirement will compel NASA to consider finally developing a space nuclear power reactor to support all the lunar and Mars exploration requirements.

The promise of automation has always been to "free people to be more creative and productive," but often this promise has failed in various blue collar and white collar applications. However, if we want moon and Mars exploration crews to do better than the Mir and ISS crews that have devoted most of their efforts to just keeping the station alive, we must push Automation to where it can become a true partner to the crew. The critical question to answer for every mission and every implementation is how to allocate tasks the most effectively and economically between people and machines. Automation and robotics will play a front line role in making space exploration affordable, reliable, and safe. This role requires advanced research and development in human-machine interaction, human factors, cognition, perception, and mental workload. There is a teleoperation objective that rated well but not as highly as Human-Machine Partnership. We interpret this result to mean that NASA should resist the temptation to rely upon the crutch of teleoperation from the Earth to the moon because it appears easier or less expensive. This is a classic case when—as President Kennedy said—we should choose not the thing that is easy but the thing that is hard. Human-Machine Partnership will be harder to achieve but it is the way of the future in space exploration as it is in many other domains.

This definition of Closed Loop consists of physical/chemical regenerative or recycling systems and is an absolute requirement for the lunar outpost missions of six months or more. The ISS life support system is "scarred" to accommodate a Sabatier reactor to begin closing the water loop, but overall, the ISS system was not designed to be fully regenerative. Most of these technologies exist at Technology Readiness Level (TRL) 4 or 5, and the Workshop score indicates that NASA should proceed to bring them up to flight rating over the next decade. Once the development phase is complete for physical/chemical regenerative systems, it will become possible to apply them economically to shorter duration vehicles and missions such as a pressurized rover to explore the lunar surface.

This objective rates very highly because of the severe limitations in EVA astronauts walking great distances. Unpressurized and pressurized rovers will prove a vital part of exploration, and have already demonstrated their value on the Mars Pathfinder Sojourner and the Mars Exploration Rovers Spirit and Opportunity. Our inference from this maximum is that it will be highly valuable and cost-effective for robotic precursors to traverse large areas to perform surveys to identify and examine sites of scientific interest in advance of crew exploration.

There were actually two objectives to characterize resources, one from Geology, and one from Lunar Resource Utilization families. In both cases the description clearly presents an engineering objective, and there was a recommendation to combine the two objectives. In addition, there were objectives to demonstrate ISRU processes, to utilize resources, to create ISRU products, and to use native materials for construction. All of these objectives received substantial scores, although we cannot explain the variance among them. We did not have any guidelines or criteria for combining objectives, however closely linked, so we left these objectives as rated.

Methodology

Given the raw data output from the Industry Workshop, the task was to evaluate the results in an impartial and useful way. In developing this methodology, they confronted several potential difficulties: objectives time-slip, too many families of objectives, effect of family size, effect of participant voting blocks, and statistical validity. This story of developing the methodology tells the essential themes of the evaluation itself.

The methodology and findings of the evaluation are presented together because the methodology is evidence-based. The findings of one step help determine the next step in the methodology. This evaluation involved not just method, but truly the study of method to find the best approaches, testing numerous options for each step to pull together this story in as coherent a fashion as possible.

Prepare for Mars "Mars.Factor"

The first step was to develop the metric for the Prepare for Mars priority, the "Mars.Factor." The Workshop used three scoring systems for the Prepare for Mars goal. Equation 1 shows metric for this Mars.Factor that appears in Figure 26.3 and 26.4, based on Pythagoras' geometric mean. The reason for using the geometric mean instead of the arithmetic mean (average) is that it allows the addition of 1 into the summation. That way, if all the other values total zero, there is still the square root of 1, so it is still possible to use the Mars.Factor in calculations (not be possible for the root of zero).

EQUATION 1. $Mars.Factor = \sqrt{1 + MarsScore1 + MarsScore2 + MarsScore3}$

Figure 26.2 presents the legend for all the exploration objectives as they appear in Figure 26.3 and 26.4. Figure 26.3 presents the raw data from the Industry Workshop results. The "Prepare for Mars" raw prioritization appears on the Y-axis and the raw lunar time phasing appears on the X-axis. The Y-axis is a logarithmic scale to best display the closely prioritized objectives at the bottom of the scale. Figure 26.2 shows the legend for the charts in Figures 26.3, 26.4, and 26.5.

■mA1 Radio Astronomy	■mA10 Survey Science Sites	■mA2 Interferometry
■mA3 Near Infrared Astronomy	■mA6 Measure Radiation	■mA7 Exotic Matter
■mA9 NEOs	◆mCAS1 Robotic Construction	◆mCAS2 Lunar EVA Suit
◆mCAS3 Human-Machine Partnership	◆mCAS4 EVA Robots	◆mCAS5 Teleoperation
■mCOM1 Telecommunications	■mCOM2 Comm Network	■mCOM3 Commercial Info Services
◆mEHM1 Radiation Shielding	◆mEHM2 Dust Mitigation	◆mEHM3 Micrometeoroid Protection
◆mEHM4 Thermal Protection	◆mEHM5 Exhaust Blast Protection	▢mENVCH1 Thermal Environment
▢mENVCH10 Seismic Activity	▢mENVCH11 Electrical Field	▢mENVCH2 Geotechnical Properties
▢mENVCH3 Radiation Environment	▢mENVCH4 Micrometeoroid Environment	▢mENVCH5 Dust Environment
▢mENVCH6 Topography	▢mENVCH7 Exhaust Cratering	▢mENVCH8 Magnetic Field
▢mENVCH9 Gravity Field	▢mENVMON1 Space Weather Prediction	▢mENVMON2 Environmental Safety
■mGEO10 Subsurface Structure	■mGEO11 Curation & Contamination Control	■mGEO1-1 Origin of the Moon
■mGEO12 In-Situ Analysis	■mGEO1-2 Crust & Mantle	■mGEO1-4 Structure of the Moon
■mGEO2 Geology	■mGEO3-2 Cratering Flux	■mGEO5 Cosmic Radiation
■mGEO6-1 Regolith	■mGEO6-2 Regolith	■mGEO7-1 Volatiles
■mGEO8 Characterize Resources	■mGEO9 Impacts	◇mGINF5 Rescue & Recovery
◇mGINF6 Global Coordinates	▢mHEO11 Solar Observatory	▢mHEO2 Solar Radio Astronomy
▢mHEO3 Crustal Magnetic Fields	mHH1 Biological Effects	mHH2 Human Performance
mHH3 Health Care	mHH4 Habitat Effect on Health	mHH5 Environmental Health
◆mLRU1 Quantify Resources	◆mLRU10 Commercial Prospecting	◆mLRU2 Utilize Resources
◆mLRU3 Demo ISRU	◆mLRU5 Excavation	◆mLRU6 Process Resources
◆mLRU7 Produce Consummables	◆mLRU8 Construction with ISRU Materials	◆mLRU9 Lunar Products
mLSH1 Safe Habitation	mLSH2 Bioregenerative Life Support	mLSH3 'Closed Loop' P/C Life Support
mLSH8 Fire Protection	◆mMAT Effects on Materials	◆mMAT New Materials
■mNAV Guidance, Navigation & Control	○mOPS1 Crew/Surface Ops	○mOPS10 Repair Equipment
○mOPS11 Work Ops Testing	○mOPS3 Mars Analog	○mOPS7 Biological Contamination Control
○mOPS9 Crew-Centered Control	○mPWR1 Surface Power Systems	○mPWR2 Earth-Generated Beamed Power
○mPWR3 In-Space Beamed Power	■mSM1 Surface Mobility Traverse	■mSM2 Mobility/Construction Ops
■mTRANS1 Redundant Transportation	■mTRANS2 Autonomous Lander	■mTRANS3 In-Space Cryo Management

FIGURE 26.2
Legend for the lunar exploration objectives in Figures 26.3, 26.4, and 26.5.

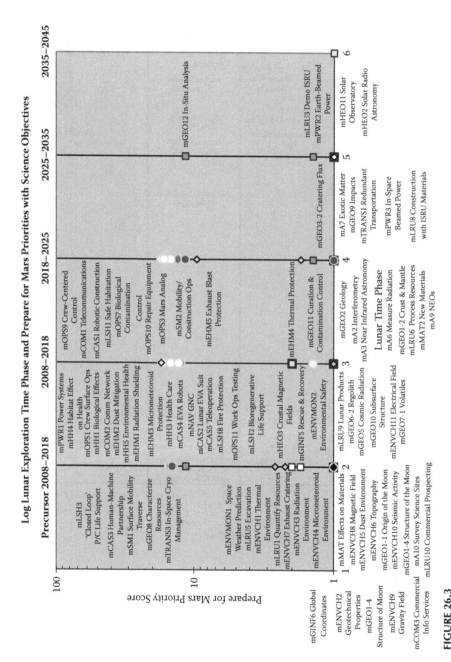

FIGURE 26.3
Logarithmic plot of the lunar objectives by prepare for Mars priorities (Mars.Factor) and lunar time phasing

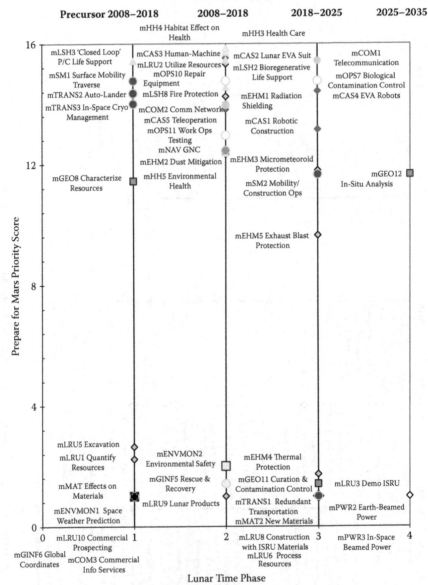

FIGURE 26.4
Non-log plot of lunar objectives.

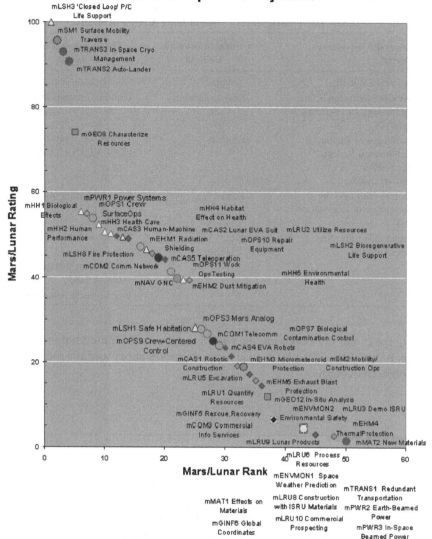

FIGURE 26.5
Mars/lunar value rating of lunar exploration objectives.

A further problem with the crowded lower tier in Figure 26.3 is that the many low-rated science objectives obscure the non-science objectives. After removing the science objectives, the engineering, habitation, operations, and science operations objectives remain. Figure 26.4 attempts to make the plot more legible by showing the objectives with the de minimus-scoring science objectives deleted. The Y-axis is an integer scale that allows better resolution

TABLE 26.2

Lunar Exploration Time-Phasing Team's Analytical Matrix

Lunar Time-Phasing Team's Time-Phasing Analytical Matrix		
"General Assessment Criteria"	Critical Subsystems Criteria	Development Phase Criteria
Technical Difficulty,	Navigation,	Objective implemented,
Provides Significant Science, Technology, or Economic Return,	Communication,	Ground test and demos,
Encourages or Facilitates Commercial Benefits, Applications, or Opportunities,	Control,	Launch required (ELV, CLV/COTS, HLV) [sic],
Involves the Public in Adventure of Exploration, and	Power,	Lunar experiments,
Emphasizes Education Opportunities.	Etc.	ISS test and demos,
		Non-human surface support, and
		Human surface support.

among the upper tier of objectives. This chart shows the bimodal distribution between the almost entirely non-science objectives above and the mainly science objectives below.

Lunar Exploration Time Phasing

The Lunar Time-Phasing team in the workshop developed a set of "general assessments" to consider *when* the mission timeline and mission architecture might be best served by accomplishing or implementing the objective. They combined these two perspectives to recommend a time phase for each objective.

Lunar Time Phase Estimation

The next step was to find a way to express the Lunar Time Phase quantitatively to incorporate into this single scale. Table 26.3 displays the values into which the Lunar Time Phase translated. Taking the mean shows when the preponderance of the preparation work must occur, so it does not hinge on when the capability will be delivered and operational on the moon, but rather it indicates the fiscal years in which to prepare it.

Objectives Time-Slip

This time-slip of objectives is quite common in the spaceflight business, and it is a sure bet that some lunar exploration objectives will slip to a later phase. The Workshop did not provide guidance about what happens to an objective that may slip in time: Does it move to the head of the next column? Does it slide sideways with the same Mars.Factor priority? Or, does it go to the bottom of the next column? It is possible to make sensible arguments for each of these dispositions. This dilemma made it obvious that the evaluation needs a way to put all the objectives across time phases. The next step was to find a way to compare all the objectives on a single, aggregate scale.

Mars/Lunar Value

In constructing the evaluation formula, the first step was to combine the three "relevance to Mars" scores for each objective as the geometric mean as the Mars Factor. The second step was to divide the Mars Factor by the Lunar Phasing to generate the Mars/Lunar score, showing the combination of Mars and lunar priorities. EQUATION 2 shows this algorithm given in units of Mars Priority/lunar exploration years, where the lunar time phase value is given in Table 26.3 below.

EQUATION 2. $Mars / Lunar\_Value = \dfrac{Mars.Factor}{Lunar\_Time\_Phase\_Factor}$

TABLE 26.3

Lunar Time Phase Equations and Values

Phase Name	Phase Period	Equation 2	Lunar Time Phase Factor (Years)
Precursor	2008–2018	$= .25\,(2018–2007)$	3.63
Phase 2 Development	2008–2018	$= .75\,(2018–2007)$	7.37
Phase 3 Development	2018–2025	$= \dfrac{2025 + 2018}{2} - 2007$	14.5
Phase 4 Development	2026–2035	$= \dfrac{2025 + 2035}{2} - 2007$	23.0
Phase 5 Development	2036–2045	$= \dfrac{2035 + 2045}{2} - 2007$	33.0

Figure 26.5 presents the Mars/Lunar Value plotted against the Excel Rank function. The value of using the Rank on the x-axis is that a horizontal gap shows where there are multiple points overlaid on the left side of the gap. This chart shows the disciplinary families of objectives and the comparative ratings of each individual objective. However, this chart is much too busy to draw inferences.

Overarching Objectives

The 18 families of the lunar objectives proved too many to keep track in terms of all the characteristics and qualities that they entailed, especially when the differences between these families were often quite subtle. A higher-level consolidation would be helpful, which is where the "Overarching Objective" concept arose. Table 26.4 shows assignment of each of the families to one of five Overarching Objectives (OAOs). Figure 26.1 shows the domain of the five OAOs within the larger universe of objectives that NASA gave to the Space Enterprise Council.

Engineering OAO

Engineering is the largest and in some respects the most obvious OAO. It concerns the development of products – mainly hardware and commodities for the Exploration Economy. The Engineering technical families include: Environmental Hazard Mitigation, Lunar Resource Utilization, Materials, and Power.

Engineering Operations OAO/Crew Activities Support

At first we considered the Crew Activity Support as part of Engineering. However, as we understood the content better, it became apparent that the Crew Activities Support objectives were more active and more complex than the other engineering objectives. Still, they were not purely operational objectives. Therefore, we established an Engineering Operations as a hybrid OAO subset.

Habitation OAO

Habitation grows from two families of objectives: Human Health and Life Support and Habitation. These two families combine to define the requirements and solutions for the crews to live and work in a healthy, reliable, and safe environment on the Moon and Mars.

Operations OAO

Operations concern a wide range of subsystems, systems, and integrated systems that support the full menu of mission activities, both human and robotic.

TABLE 26.4

Allocation of Lunar Exploration Objectives to Five Overarching Objectives, Using the Color Coding Applied to the Charts in Later Sections

Objective ID	ENGINEERING	Objective ID	ENGINEERING OPERATIONS/ Crew Activity Support	Objective ID	OPERATIONS
mEHM1	Radiation Shielding	mCAS1	Robotic Construction	mCOM1	Telecomm
mEHM2	Dust Mitigation	mCAS2	Lunar EVA Suit	mCOM2	Comm. Network
mEHM3	Micrometeoroid Protection	mCAS3	Human-Machine	mCOM3	Commercial Info Services
mEHM4	Thermal Protection	mCAS4	EVA Robots	mGINF5	Rescue, Recovery Systems
mEHM5	Exhaust Blast Protection	mCAS5	Teleoperation	mGINF6	Global Coordinates
mLRU1 Resources	Quantify		HABITATION	mNAV1	GNC
mLRU10	Commercial Prospecting	mHH1	Biological Effects	mOPS1	Crew Surface Ops
mLRU2	Utilize Resources	mHH2	Human Performance	mOPS7	Biological Contamination
mLRU3	Demo ISRU	mHH3	Health Care	mOPS10	Repair Equipment
mLRU5	Excavation	mHH4	Habitat Effect on Health	mOPS11	Work Ops Testing
mLRU6	Process Resources	mHH5	Environmental Health	mOPS3	Mars Analog
mLRU7	Produce Consumables	mLSH1	Safe Habitation	mOPS9`	Crew-Centered Control
mLRU8	ISRU Construction	mLSH2	Bioregenerative Life Support	mSM1	Surface Mobility Traverse
mLRU9	Lunar Products	mLSH3	"Closed Loop" Life Support	mSM2	Mobility Ops/ Construction
mMAT1	Effects on Materials	mLSH8	Fire Protection	mTRANS1	Redundant Transport
mMAT2	New Materials		SCIENCE OPERATIONS	mTRANS2	Autonomous Lander
mPWR1	Surface Power System	mENVMON1	Space Weather Prediction	mTRANS3	In Space Cryo Management
mPWR2	Earth Beamed Power	mENVMON2	Environmental Safety		
mPWR3	In-Space Beamed Power	mGEO11	Curation & Contamination Control		
		mGEO8	Characterize Resources		
		mGEO12	In-situ analysis		

Science Operations OAO

Just as CAS constitutes a hybrid subset of Engineering, Science Ops emerged as a subset of the Science objectives as described above. As the smallest OAO with only three objectives it makes the only direct science contribution to this evaluation. The Science Ops should be reviewed by the Science Community.

Observations and Findings

The observations describe the properties of the data that were susceptible to numerical and statistical inquiry and analysis.

Observation 1: Prioritization vs. Time Phasing as A Source of Variance

The Mars/Lunar value is essentially an interpretive measure of the interaction between the Mars.Factor and the Lunar Time Phase. To answer this question with regard to the Mars/Lunar Value, the OAOs provide the must useful unit of analysis. Figure 26.6 presents the coefficient of determination

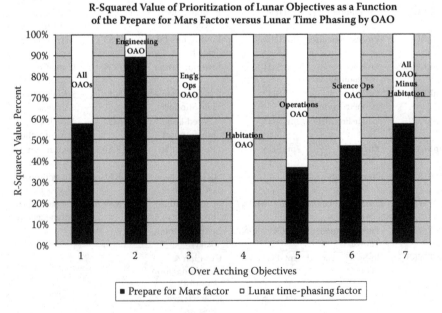

FIGURE 26.6
Coefficient of determination for prioritization vs. time-phasing

(R-squared value) results for each of the five OAOs. The R-squared value give the percentage of the variance in Y that is a function of the variance in X, where X is the Mars.Factor and Y is the Mars/Lunar Value. A complementary definition is that R^2 is the fraction of the total squared error that the model explains. This approach affords the ability to identify the remainder of the variance in M/L Value as a function of the Lunar Time Phase.

The R^2 value for all the OAOs is 57.3 percent, which means that overall, the variance in M/L is 57.3 percent a function of the Mars.Factor and 42.7 percent a function of the lunar time phase. Although these values seem like middle-of-the road results, the variability among the OAOs is much greater, representing their differing unique characteristics. For example, the Engineering OAO correlates most highly with the Mars Factor at $R^2 = 89$ percent, meaning that about 89 percent of the variance is a function of the Prepare for Mars factor, suggesting the Engineering Objectives are almost independent of their time phasing. In contrast, the Habitation OAO correlates the least with the Mars factor, at $R^2 = 0$ percent, suggesting that 0 percent of its variance is a function of the Mars factor, meaning that variance in Habitation is purely a function of the lunar time phasing. This result makes sense insofar as habitation does not come into play until the crew arrives.

Observation 2: Family Size Effect

The larger families tend to score lower than the smaller families, suggesting that there is an effect arising from different family (sample) sizes. Figure 26.7 portrays this family size effect. The concern stemming from this effect is that where the initial objective identification process deposited a larger number of objectives into a family, it had the effect of diluting the vote by the workshop participants. There are three low-scoring families among the smaller families, so the effect is not clear-cut. The families with the mid-range sizes from four to six generally score the highest. As part of the methodology, we looked at ways to adjust for effects due to family size.

The size of the families of objectives varied so considerably that a small family with very high scores would overwhelm a larger family with equal or lower scores. Therefore, the next step was to devise a means to control for the family size in a composite score. While this approach resolved the family size issue to a certain extent, the unexpected result was to display the OAOs clarified into distinct "voting blocks."

Observation 3: Voting Block Effect

While a few of the families of objectives have a wide point spread, most of the families show a pattern of clustering that makes the raw data subject to question without some handle on possible discipline-specific biases. For example, in Figures 26.3 and 26.4, the Habitation-related objectives with the mHH and

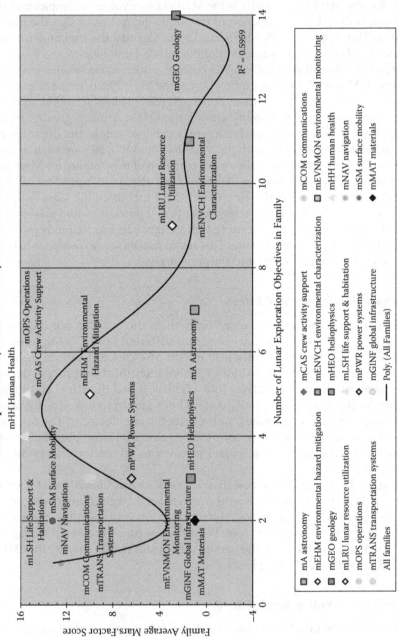

FIGURE 26.7
Chart of objectives family size effect

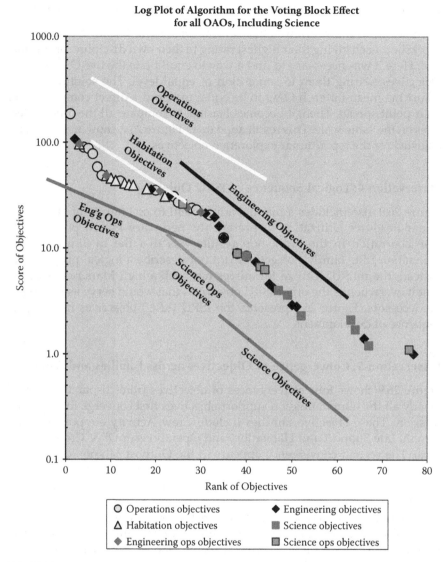

FIGURE 26.8
View of voting block effect for the overarching objectives among the workshop participants.

mLSH prefixes all appear near the top of their Lunar Time Phases. However all the Operations mOPS objectives appear only in the second and third phases. Nearly all the Environmental concerns with mENVCH, mENVMON, and mEHM prefixes appear in the bottom tier or at the bottom of the upper tier.

Solution for the Voting-Block Effect: The Mean-Adjusted Mars/Lunar Value

The voting block concern was that several constituencies participated in the Workshop, each giving their highest rating to their own discipline or product line. Here it was necessary to find a way to equilibrate the five Overarching Objectives, setting them to some kind of equal level. The solution was to adjust the mean for each OAO to be equal to afford a direct comparison of their point spread. Then, it becomes feasible to compare all the other objectives on the same scale. This result appears in Figure 26.9, showing the mean-adjusted for the top 57 lunar exploration objectives.

Observation 4: Logical Sequence of ISRU Objectives

Figure 26.9 also includes a manual adjustment to one family of objectives: Lunar Resource Utilization to present the necessary logical progression. One aberration in the Workshop results was that the In Situ Resource Utilization LRU family's objectives did not describe a logical progression. Instead, the mLRU2 Utilize Resources has such a high Mars.Factor rating that it overwhelms the other ISRU objectives that would serve as necessary predecessors. Figure 26.9 presents the ISRU (LRU) objectives in a logical sequence of development.

Observation 5: Convergence of Objectives across Families and OAOs

Figure 26.9 shows four convergences of objectives from disparate families. Nearly all the human mission support objectives first converge in the Sortie Mission. These objective families include Crew Activity support, Human Health, Life Support and Habitation, and Operations (mOPS1 Crew Surface Ops). The second convergence appears in the Outpost Mission concerning construction: mLRU8 ISRU Construction using native materials, mLRU9 Lunar Products, mCAS1 Robotic Ops/Construction, and mMS2 Mobility Ops/Construction. The third convergence involves the long-term habitation that the Outpost would serve in the context of long duration environmental exposures: mOPS9 Crew-Centered Control, mHH5 Lunar Environmental Health, mENVMON2 Environmental Safety, and mLSH1 Safe Habitation. Finally, while only the Outpost Phase can serve as a dress rehearsal for Mars, a cluster of the relevant objectives occur in that phase: mOPS3 Mars Analog, mOPS9 Crew-Centered Control, mLRU7 Produce Consumables, mLRU2 Utilize Resources, mGEO11 Curation and Contamination Control, and mOPS7 Biological Contamination Control.

Observation 6: Sortie-First Lunar Time Phasing

The analysis of Lunar Time Phasing, combined with prioritization yields a plot showing the sequence of implementation of the lunar exploration

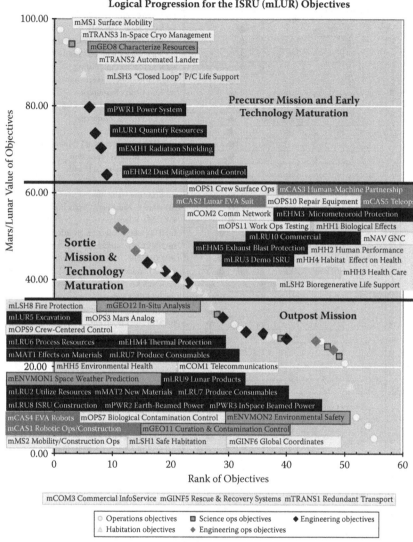

FIGURE 26.9
Mean-adjusted Mars/Lunar values for lunar exploration objectives indicating the three major mission phases.

objectives, as shown in Figure 26.10. This chart places the timeline on the vertical axis and the approximate sequence of objectives on the horizontal axis. It plots three key dates: Start, Implementation, and Deployment on the moon. One striking feature of this plot is that development of about half of these objectives must start before NASA releases its big Lunar Lander Request for Proposals, currently planned for January, 2012.

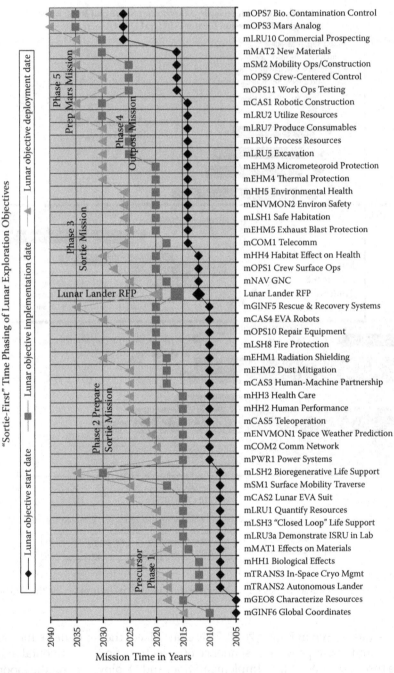

FIGURE 26.10
"Sortie–First" lunar exploration objective time phasing.

While these results may not prove dispositive for making policy recommendations, they do portray many of the central issues, conflicts, and trade-offs that NASA must address both in formulating lunar exploration policy and in implementing it.

This Aerospace Industry evaluation derives from an engineering, operations, and production perspective. It gives valuable insights into how this industry view assesses and compares these candidate objectives. In this respect, we consider four of the top five objectives as being fundamental for an engineering and production strategy for direct application to lunar lander and lunar outpost missions: Closed Loop Physical/Chemical Life Support, Surface Mobility Traverse, Human Machine Partnership, and Characterize Resources.

Recommendations

Include Science Objectives

The omission of science objectives is a very serious problem that is potentially fatal to any effort to persuade Congress to fund new human lunar exploration. Therefore, NASA should make an equivalent effort to identify, include, and support lunar science that humans will conduct and enable on the moon.

Align the Closely Related Objectives Across Disciplines

The first step to implement recommendation one might be to align science objectives with corresponding objectives from other families. For Example: The Objectives contain a parallel track of Environmental Characterization (Science) and Environmental Hazard Mitigation (Engineering).

Combine Concurrent, Cross-Disciplinary Development Objectives

The analysis shows that there is a strong affinity among closely related and simultaneous objectives. The Observation of the coefficient of determination column chart showed that the human health and habitation objectives all depend upon the timing—the same timing—of when humans arrive on the moon. Therefore, NASA should combine them into an allied effort. These objectives include:

- mLSH1 Safe Habitation,
- mENVMON2 Environmental Safety,
- mHH4 Habitat Effect on Health, and
- mHH5 Environmental Health.

TABLE 26.5

Example of the Potential Alignment of Corresponding Environmental Science and Environmental Engineering Objectives

Science Objectives: Environmental Characterization	Engineering Objectives: Environmental Hazard Mitigation
mENVCH1 Characterize Thermal Environment	mEHM4 Thermal Protection
mENVCH3 Characterize Radiation Environment	mEHM1 Radiation Protection
mENVCH4 Characterize Micrometeoroid Environment	mEHM3 Micrometeoroid Protection
mENVCH5 Characterize Dust Environment	mEHM2 Dust Protection

Unify Related Technologies That Serve Common Goals

Objectives that share related technologies and applications can integrate around their common functionalities to achieve diverse goals. For example, consider all the elements of construction on the lunar surface that may serve a common set of goals and milestones, despite deriving from several families of objectives:

- mCAS1 Robotic Construction
- mCAS3 Teleoperation
- mCAS4 EVA Robots
- mLRU5 Excavation
- mLRU8 ISRU Construction
- mSM2 Mobility Ops/Construction

Split Some Objectives in Two or More Parts

Some objectives need to be split into major component parts as separate objectives. For example, Demonstrate ISRU as stated combined both lab and lunar surface testing. Although there would be a relationship, the lab testing would come at such a lower TRL level and so many years earlier, that it was necessary to treat it as a separate objective; otherwise the first time the prioritization recognizes this objective is when it becomes operational on the moon.

- Split on the basis of time-phasing differences.
  - Example: mLRU3a Demo ISRU in Lab *versus*
  - mLRU3b Demo ISRU on Lunar Surface.
- Split on the basis of mission differences.
  - Example: mSM1a Surface Mobility Traverse—Unpressurized *versus*
  - mSM1b—Pressurized.

Restructure the ISRU Objectives

The ISRU objective as presented to the Workshop and to the Space Community at large was broken down into too many smaller objectives that made interpretation of voting impossible. The result of having too many objectives was that "Utilize Resources" was not only much higher priority than all the other objectives that would lead incrementally up to it, but it also came out earliest in the time phasing votes.

Conclusion

The most significant aspect of this assessment is that it reveals that way people in the aerospace community, predominantly engineers, think about NASA's lunar exploration objectives. This assessment shows the deep structure and complexity of this set of ideas and their interactions. It demonstrates how it is not possible to treat such objectives individually in isolation from the larger universe of exploration activities, tasks, technologies, and operational capabilities. It shows especially how limiting and even self-defeating it is to posit exploration objectives absent a core science mission. What the assessment conveys is that the constituent elements of an exploration program involve profound linkages within specific families of objectives, between families, across disciplines, and among the overarching objectives.

Appendix A: Letter to NASA

UNITED STATES CHAMBER OF COMMERCE • INTERNATIONAL DIVISION
1615 H STREET, N.W. • WASHINGTON, D.C. 20062-2000 • 202/463-5479 • 202/463-3114 FAX

September 20, 2006

Mr. Doug Cooke
Deputy Associate Administrator
Exploration Systems Mission Directorate
NASA Headquarters
Washington, D.C. 20546

Dear Doug:

On August 29-30, the U.S. Chamber of Commerce Space Enterprise Council hosted a workshop in support of NASA's development of the Integrated Global Exploration Strategy. Per your guidance, we convened the workshop to generate an industry view and assessment of mission sequencing of the objectives and how they might be arranged to fulfill potential exploration scenarios.

As a starting point for our deliberations we referred to the six exploration themes, multiple categories and corresponding objectives that were identified to date through ESMD-sponsored venues. Participants were asked to consider the time-phasing of missions, precursor requirements, and to identify issues, conflicts, and policy considerations for the exploration objectives.

Workshop participants represented member-companies of the Council, the Coalition for Space Exploration, small businesses, and members of the Space Alliance Technology Outreach Program (SATOP). Each organization was invited to send up to three representatives who were familiar with technical and operational issues that will be required to implement the goals of the Vision for Space Exploration.

While we attempted to institute and common format for the discussion of the objectives for each theme, modifications were made in real time based on input and comments from the session facilitators and the participants. The revised approaches are described in the individual reports for each of the themes.

As a matter of record, we did find the format that was provided to be somewhat inefficient for a productive assessment. There seemed to be duplication between "categories" and "themes" as well as an inconsistent level of description of each objective. Nonetheless, we gave all the provided information considerable consideration and attempted to provide information that will prove useful to your future deliberations. The participants universally appreciated the invitation from NASA to provide input to the strategy and we are prepared to collaborate through future venues sponsored by ESMD.

Sincerely,

David Logsdon
Executive Director
Space Enterprise Council

Appendix B: Lunar Exploration Objectives

Objective ID and Short Title	Objective Description	Overarching Objective
mCAS1 Robotic Construction	Develop robots which support the construction and or assembly of human Moon base and support astronaut's activities.	EO
mCAS2 Lunar EVA Suit	Develop a high performance EVA suit	EO
mCAS3 Human-Machine Partnership	Human-machine partnership	EO
mCAS4 EVA Robots	Objective: Develop autonomous robotic support for EVA and long-range exploration	EO
mCAS5 Teleoperation	Objective: Develop teleoperation capabilities to support human operation of equipment on the lunar surface. Implement human interaction systems (telepresence) to support automation technologies required for lunar operations.	EO
mCOM1 Telecomm	Objective: Implement a secure reliable and scalable telecommunications capability to support expanding telecom needs.	O
mCOM2 Comm Network	Objective: Establish a communications network that can provide high-bandwidth support for public engagement	O
mCOM3 Commercial Info Services	Objective: Provide commercial information services to the greatest extent possible.	O
mEHM1 Radiation Shielding	Objective: Provide radiation shielding for surface operation to protect crews, materials, and instruments.	E
mEHM2 Dust Mitigation	Objective: Evaluate and employ dust mitigation techniques to protect crews, materials and instruments during extended lunar stays.	E
mEHM3 Micrometeoroid Protection	Objective: Provide protection from micrometeorite bombardment during surface operations.	E
mEHM4 Thermal Protection	Objective: Provide thermal protection from the lunar day/night extremes.	E
mEHM5 Exhaust Blast Protection	Objective: Provide protection to surface infrastructure from rocket exhaust blast ejects.	E
mENVMON1 Space Weather	Objective: Monitor space weather and provide forecasting to determine risks to lunar inhabitants.	SO
mENVMON2 Environmental Safety	Objective: Monitor real-time environmental variables affecting safe operations.	SO
mGEO 11 Curation	Objective: Curation and contamination control.	SO
mGEO12 In-situ analysis	Objective: Analyze lunar samples in-situ.	SO
mGEO8 Characterize Resources	Objective: Characterize potential resources	SO

(Continued)

(Continued)

Objective ID and Short Title	Objective Description	Overarching Objective
mGINF5 Rescue & Recovery Systems	Objective: Develop lunar rescue systems with the maximum extensibility to Mars. Note: Assuming this is lunar surface rescue...Not moon to earth rescue. Need to add recovery element to objective title.	O
mGINF6 Global Coordinates	Objective: Establish a globally accepted lunar reference coordinate system to utilize in planning and executing lunar missions (standards).	O
mHH1 Biological Effects	Objective: Study the fundamental biological and physiological effects of the integrated lunar environment on human health and the fundamental biological processes and subsystems upon which health depends.	H
mHH2 Human Performance	Objective: Understand the effects of the integrated lunar environment, in particular partial gravity, on human performance and human factors.	H
mHH3 Health Care	Objective: Improve health care on the Moon by creating remote medical practice infrastructure and understanding the effect of the lunar environment on medical treatments and procedures.	H
mHH4 Habitat Effect on Health	Objective: Understand the effects of vehicle habitat, and EVA suit pressures and oxygen concentrations on human health so as to design mitigation strategies for extended stays. Recommend: Delete and Merge in to mHH1.	H
mHH5 Environmental Health	Objective: Understand the impact of Lunar environments on multiple generations of terrestrial life forms that impact human health.	H
mLRU1 Quantify Resources	Objective: Characterize and quantify the resource potential of the Moon.	E
mLRU10 Commercial Prospecting	Objective: Utilize the commercial sector to perform resource prospecting and provide mining materials processing, manufacturing construction and other services/products based on lunar resources.	E
mLRU2 Utilize Resources	Objective: Use lunar resources to enable and support future exploration missions and destinations.	E
mLRU3 Demo ISRU	Objective: Demonstrate ISRU technologies and systems to reduce risk for mission integration and commercial development.	E
mLRU5 Excavation	Objective: Perform lunar resource excavation.	E
mLRU6 Process Resources	Objective: Develop and validate tools, technologies and systems to extract and process resources on the Moon with the extension to other exploration destinations.	E

Objective ID and Short Title	Objective Description	Overarching Objective
mLRU7 Produce Consumables	Objective: Produce propellants, life support and other consumables from lunar resources.	E
mLRU8 ISRU Construction	Objective: Construct facilities, manufacture hardware, materials, chemicals and other products on the Moon using lunar resources.	E
mLRU9 Lunar Products	Objective: Investigate and develop technologies and systems that effectively utilize lunar resources and products.	E
mLSH1 Safe Habitation	Objective: Provide safe and enduring habitation systems to protect individuals, equipment and associated infrastructure.	H
mLSH2 Bioregenerative Life Support	Objective: Develop biologically based life support system components to support long duration human exploration missions.	H
mLSH3 "Closed Loop" Life Support	Objective: Develop and deploy closed loop physiochemical based life support systems to increase self sufficiency of future long duration human exploration missions.	H
mLSH8 Fire Protection	Objective: Develop and implement fire detection and suppression strategies for 1/6 G environment.	H
mMAT1 Effects on Materials	Objective: Study the effects of the lunar environment on materials so as to design mitigation strategies for extended stays.	E
mMAT2 New Materials	Objective: Investigate the development of new materials under fractional Earth gravity and high vacuum.	E
mNAV1 GNC	Objective: Establish GNC capabilities to support human missions.	O
mOPS1 Crew Surface Ops	Objective: Demonstrate human surface operations capability.	O
mOPS10 Repair	Objective: Develop repair techniques for equipment operating on the Moon.	O
mOPS11 Work Ops Testing	Objective: Engage in operation testing to understand the effect of the lunar environment on basic working tasks (with timescales applicable to early crewed Mars missions).	O
mOPS3 Mars Analog	Objective: Conduct Mars Analog Tests on the Lunar Surface.	O
mOPS7 Biological Contamination Control	Objective: Evaluate biological contamination control protocols and astrobiology measurement technologies that will be used to search for life on other planets.	O
mOPS9 Crew-Centered Control	Objective: Establish crew-centered control as the norm for lunar operations with real-time and off-line planning and scheduling.	O
mPWR1 Power System	Objective: Develop power generation, storage, and distribution systems required to facilitate increasing surface and subsurface durations.	E

(Continued)

(Continued)

Objective ID and Short Title	Objective Description	Overarching Objective
mPWR2 Earth-Generated Beamed Power	Objective: Establish a power architecture where Earth-generated power is transmitted to the lunar surface and to cis-lunar transportation assets.	E
mPWR3 In-Space Beamed Power	Objective: Provide space power for both in-space and Earth applications.	E
mSM1 Surface Mobility Traverse	Objective: Implement surface mobility systems to support both crew and cargo traverses over increasing distances.	O
mSM2 Mobility Ops/ Construction Ops	Objective: Provide surface mobility capabilities for the purpose of constructing and operating a permanent lunar outpost.	O
mTRANS1 Redundant Transport	Objective: Provide redundant transportation services on the Moon and to and from the Moon to increase access to the Moon and traversing the Moon.	O
mTRANS2 Autonomous Lander	Objective: Demonstrate Autonomous Lander	O
mTRANS3 Cryo Management	Objective: On-orbit cryogenic fluid management technologies.	O

27

A Self-Sufficient Moon-Base Analogue

Niklas Järvstråt

Division of Production Engineering University West
Trollhättan, Sweden

Niklas Järvstråt has been active in international research projects since 1993, as participant, project leader, and coordinator, and has led several national fatigue research projects. In 1994, he commenced research and networking directed towards the establishment of a self-sufficient human settlement on the moon. In the meantime, he has also developed and implemented methods for technology management and quality control at Volvo Aero. He was one of the driving forces in forming UTMIS—the Swedish fatigue network, serving as vice chairman 2000–2002. Dr. Järvstråt has established a major national conference on life prediction procedure and theory, arranged for the tenth consecutive year in October 2004. He is associate professor and head of the Division of Production Engineering at University West in Trollhättan, Sweden, and leader for the welding research group at the Swedish Production Technology Centre (PTC). Publications include topics in high temperature material modeling, technology management, quality control, fatigue and residual stress calculation. Currently (2008), he is in the process of starting up activities at the "Moon-Mine," Storgruvan in Pershyttan, Sweden, with the aim of creating a completely self-sufficient underground moonbase analogue before production starts on the moon in 2020, and then to run operations in parallel with the base on the moon, anticipating and solving problems without the dangers of external vacuum and isolations of the moon.

ABSTRACT Building the first base on the Moon without first thoroughly testing all components and the system performance in a full-scale facility on Earth would not only be dangerous but also a waste of money. How to achieve self-sufficiency at a low cost by drawing upon previous and on-going research and using facilities similar to those on the Moon will be shown in this paper.

Why Self-Sufficient?

The cost of establishing and operating a base on the Moon is evidently monumental, especially the early building phase. However, *if* the Moon base could be made self-sufficient, operating it would, by definition, be possible without additional costs. This rather self-evident reason for having a self-sufficient Moon base can be further broken down into two main reasons:

A Self-Sufficient Colony Would Serve as an Infrastructure Backbone

This is the "external" reason—it is good for those remaining on Earth that the Moon base is self-sufficient because:

- It will be easier to establish utilitarian functions on the Moon, as any added commercial or scientific activities would be at "additional cost" only, being able to capitalize on the existing infrastructure. This is very much in analogy with establishing a casino in the middle of a desert—nobody would even consider that. But a casino in Las Vegas is quite a different thing, because the infrastructure is in place with roads, shops, food and water supply. Adding a casino with employees would in principle need additional houses and extension of supply flows and Las Vegas *is* in the middle of a desert but expanding existing infrastructure is so much easier than creating one from scratch. Thus, new casinos are opened quite frequently in Las Vegas.

- There will, by definition, be no recurring cost for keeping the Moon base operational, once it has achieved self-sufficiency. Depending on the reasons behind establishing a Moon base in the first place, this may be a good, a neutral or even a bad thing. It is good if you want to keep using the Moon base, almost regardless of purpose; it will not have much of an impact if the plans are meant to be short-term, and if the aim is making a profit on Earth by shipping supplies, it is certainly not a profitable idea making the Moon base self-sufficient.

Self-Contained Society

This is the "internal" reason—it is good for colonists, lunatics, inhabitants, staff, and long-term settlers on the Moon.

- Psychological security is perhaps the most important reason and will also have an economic impact for terrestrial utilitarian purposes. This is important because it will help the Moon settlers feel at home, secure and rooted. It is not quite realistic to expect settlers to feel secure if they have to depend on a regular supply of vital goods.

However, once the settlers know that they can survive and develop their surroundings through their own efforts, the colony will start growing and attracting families; eventually children will be born and raised on the Moon. Over time, the process of making the Moon base habitable will be replaced by making it cosy, continuously improving the standard of living and maybe one day even surpassing current life on Earth.

- Further, the colonists of a self-sufficient Moon base have the advantage of living in an environment virtually immune to terrestrial budget cuts since budget cuts on Earth would essentially only affect the flow of luxury goods, not those easily manufactured on-site.

Survival Requirements

What is needed for survival? Imagine placing a group of people on the Moon, and then imagine what they need to stay alive. They obviously need oxygen to breathe, but oxygen needs to be contained in something so it can be used for breathing. An airtight shelter or spacesuits would be a minimum requirement, but shelters should also be meteorite-proof since the lack of protective atmosphere will allow radiation, small meteorites and micrometeorites to impact. A micrometeorite which would not even be a visible flash of light entering Earth's atmosphere could still puncture a spacesuit. After having secured air for breathing and shelter to protect against the dangers of space, this imaginary group of people would soon be hungry and thirsty, so food and water (or beer) is definitely on the list. A natural result of eating and drinking would be the need for the people to relieve themselves, and subsequently hygiene facilities including sewage recycling will be required. Hopefully, diseases and accidents will not be the first problem, but pretty soon some emergency ward with at least basic medical supplies will be required to keep people functional. Of course, energy to run all this needs to be provided, as well as heat and light for work and rest. This is represented by the survival requirements chart in Figure 27.1, which also takes the requirements breakdown to the second level, through needs to a set of processes thus found sufficient and necessary for human survival.

The second order process activities: Agriculture; Mining and mineral beneficiation; manufacture; energy generation, storage and distribution; logistics and life support systems; and Construction can of course be broken down further into requirements and activities. However, as shown for agriculture in Figure 27.2, there are no new activities added—the same five production process activities reappear again, except one extra requirement, namely the workforce. Since survival of the inhabitants will provide the needed working

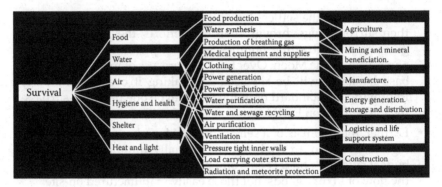

FIGURE 27.1
Survival requirements breakdown.

crew, we have, by adding survival as a primary process, seven main process groups that will need to be fulfilled for a self-sufficient Moon settlement, or, indeed for any human settlement regardless of location. Thus, it may be instructive to use this process breakdown also when studying terrestrial settlements, be it isolated outposts, small villages or major mega-cities.

For a Moon base, the basic material human needs can be quantified according to Table 27.1, which can form a good basis for quantifying the material needs and supply breakdown of Figure 27.1.

Available Resources on the Moon

According to Apollo samples, see Table 27.2, the lunar soil is rich in most resources, including metals and other minerals important for manufacturing.

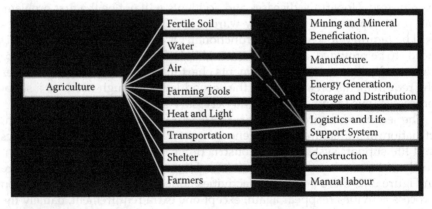

FIGURE 27.2
Agriculture requirements breakdown.

TABLE 27.1

Quantified Survival Needs (from O'Handley 2000)

	In system (total)	Need per day Person	Need per day 500 p.	unit
Breathing air	60000	1,5	750	m³
Water	28000	7	3500	litres
Food	7000	1,0	500	kg
Pressurized volume	30000			m³
Heat and light	21000	3	1250	kW

Oxygen for breathing on the Moon will never be a problem for serious colonists although the Moon lacks an atmosphere, because regolith contains about 44% oxygen, more in fact than our atmosphere! There have been numerous studies about how to extract oxygen from the regolith, the simplest, although not the most economical is probably to heat the regolith until oxygen is released. However, when considering a self-sufficient settlement, a lot of other minerals need to be extracted and in most cases this extraction will include removing the oxygen from, for example, metal oxides.

Metals, for example, compose approximately 26% of the average Apollo samples. There is also plenty of iron (10%), aluminium (9%), magnesium

TABLE 27.2

Lunar Soil Composition (Weight %, from Blair 1998 and Prado 1998)

Element	Highland	Mare	Earth	Earth rank	Applications
Oxygen[1]	45	42	47	1	Fuel, essential air constituent
Silicon[1]	21	21	28	2	Glasses, ceramics, etc. Solar
Aluminium[1]	13	7.0	8.1	3	Electric wire, structures, mirrors
Calcium[1]	11	7.9	3.6	5	Ceramics, electrical conductor
Iron[1]	4.9	13	5.0	4	Structural steel
Magnesium[1]	4.6	5.8	2.1	8	Metal alloying element
Sodium[1]	0.31	0.29	2.8	6	Chemical processing, Plant
Titanium[1]	0.31	3.1	0.44	9	High strength metal
Chromium[1]	0.085	0.26	0.01	21	Metal alloying element
Potassium[1]	0.08	0.11	2.6	7	Chemical processing, Plant
Manganese[1]	0.068	0.17	0.095	12	Metal alloying element
Phosphorus[1]	0.05	0.066	0.11	11	Plant nutrient
Sulphur[2]		0.12			Chemical processing
Carbon[2]		0.015		17	Life, chemical processing
Nitrogen[2]		0.008			Plant nutrient, air constituent
Hydrogen[2]		0.006		10	Fuel, water, chemical processing
Helium[2]		0.005			Inert gas

(5%) and titanium (2%), all important construction materials. A deficit that appears somewhat strange is that there was very little copper in the Apollo samples – whether that is due to a general copper deficit or because copper only appears in highly localised areas, has yet to be found. Although copper is the best known material apart from gold from which to make electrical wires, aluminium is quite feasible and, as seen in Table 27.2, it is abundant on the Moon.

Of non-metallic materials, silicon is 21%, calcium 9% and the plant nutrient elements nitrogen, phosphorous and potassium make up a total of about half a percent. Half a percent may not appear to be much, but it would actually represent a fertile soil, except for an unfortunate lack of nitrogen, hydrogen and carbon, all necessary for plant growth, but each only present in trace amounts, less than 0.1%. However, plants are able to make do with surprisingly lean soils and by concentrating carbon and nitrogen by a moderate factor ten or so for the soil and recycling biological matter, it should be possible to achieve good yield on most crops.

It can be concluded (O'Handley 2000) that nitrogen is the limiting factor. Assuming 0.5 %/ day atmospheric leakage (according to measured data at ISS), more than 400 metric tons of regolith of Apollo sample composition would need to be processed—per person and day—in order to replenish such losses only. This may seem very disheartening, as it would probably never be considered reasonable to process such a large amount of regolith, just to cover nitrogen needs.

Using the probable polar ice, instead of the dry equatorial regolith sampled by the Apollo missions, would help a lot, reducing the processing needs to about 100 kg, per person and day. Here, a bit of guesswork is needed, as the composition of soil at the lunar poles is not known, and a processing need of 100 kg per person and day was obtained by using the rather well-known composition of volatile-rich comets.

On the other hand, this amount of nitrogen is not really needed and it should be possible to reduce nitrogen losses. Plants use nitrogen, but most plants extract nitrogen from the soil rather than from the atmosphere. Thus, it would be possible to use air with somewhat less nitrogen than terrestrial. Unfortunately, not so much research is available on the long-term effect in humans of nitrogen-free gas mixture breathing. Nitrogen does have the beneficial effect of reducing the risk of fire, so a one-third bar pressure, pure oxygen atmosphere might not be a good choice. However, it seems probable that the leakage rate could be reduced in a permanent installation where there are no mass constraints except what escapes, and it might also be possible to create a "shell" around the base, containing a less valuable gas mixture at slightly higher pressure than the habitat. Thus, gas escaping out into vacuum would have the composition of that intermediate shell, and that might be oxygen or other easily produced yet harmless gasses. Thus reducing the losses to cheaper gasses than nitrogen would save a factor 20, and by also improving air tightness to below 0.1% atmospheric losses per day, it would be possible to reduce the processing to just 1 kg per person and day of the nitrogen rich polar ice. This seems quite feasible—one kg of ice is shovelled into the processing plant before going off to work—or, with a settlement of 1000 persons, one person is working full time loading a metric ton of ice per day.

The Polar Regions, apart from the cometary ice stored in cold traps, have one additional great advantage, and that is the so called "peaks of eternal light": crater rims and mountain peaks receiving sunlight for a very large fraction of the lunar day-night cycle. It is possible, by careful placement of photovoltaic solar panels at three locations within a radius of about 30 km close to the South Pole, to receive virtually uninterrupted electric energy supply. Thus, and because the craters on the south pole are a bit wider and deeper than at the north pole, the lunar south pole seems today to be the preferred location for a first Moon base.

Production System

After having connected available resources with the survival requirements, a production system is needed. Bearing in mind that the system should be completely self-sufficient, capable of surviving without any supplies from off-Luna, not only all resources, but also any equipment and consumables used directly by humans need to be produced in-system. Also the production system itself should be designed "producible" from in-situ resources using tools that can also be produced by the system.

Conditions

There are some crucial differences between the conditions prevailing on the Moon and the standard terrestrial conditions:

Low Gravity (1/6 of That on Earth)

The low gravity can be fun, and will facilitate carrying heavy objects, but there are also some less evident effects that need to be considered. Liquids will flow slower downhill, which would have a dramatic effect on, for example, the filling of casting moulds, unless measures are taken to counter this effect, such as using pressure to squeeze in the liquid material in the mould. There is also less convection—the flow of air and other gases and liquids caused by temperature driven density variations, cold air is heavier than hot air; thus the cold air will flow downwards replacing hotter air going up, mixing the air. Similar effects are present in any liquid or gas, but would be six times smaller on the Moon. This is sometimes a drawback, as in the habitat where it is important that stale air is circulated away to the plant areas for re-oxygenating—the lower circulation caused by lower gravity would have to be assisted by a higher degree of forced ventilation than is common on Earth. Some consideration might also be needed for liquid surfaces, which would get a slightly different geometry as the liquid is pushed down with lesser force, allowing surface tension to have larger impact. An interesting piece of information is that this could influence production features such as weld integrity, if the transition between weld metal and the pieces to be joined becomes less smooth. The notch formed would weaken the joint compared to one formed under terrestrial conditions.

Free High Quality Vacuum

There is no atmosphere on the Moon, though there are some atoms drifting about which might vaguely be described as an atmosphere. From a terrestrial industry perspective, however, the Moon "air" is high quality vacuum. Vacuum is a risk and a nuisance, of course, should you need to do excursions to the surface. But for the purpose of an underground base, surface excursions, although necessary for example when maintaining photovoltaic collectors or establishing transport connections to remote mining sites of concentrated minerals, will probably not be a main activity. However, underground mining would almost certainly take place in vacuum to eliminate the decompression risk otherwise associated with encountering a crack penetrating to the surface.

On the positive side, vacuum is beneficial to many manufacturing processes, in particular those involving molten metal alloys because oxidation and contamination problems are eliminated by the lack of foreign atoms. Indeed, high purity melting and new process steps including outgassing to vacuum may one day lead to new materials for extraterrestrial use being invented on the Moon.

Lack of Some Minerals

As mentioned earlier, the Moon has a very poor copper, hydrogen, carbon and nitrogen content, making any compound including these elements very expensive and awkward to make. Thus, replacing these materials as far as possible and pursuing as high a degree of recycling as possible in all applications containing these elements would be important.

Small Scale Solutions

The terrestrial production network is very much dependent on large-scale multi-tier sequences of producers of even the most insignificant product. This kind of large-scale production, however, would not be feasible on the Moon, and a Moon settlement will in the foreseeable future remain very small in comparison to the terrestrial scale of cities. Thus, the economy of scale employed on Earth when making millions of identical items, e.g., toothbrushes, at very small individual cost will not be possible to apply. Instead, small-scale, very flexible production methods would be required, and the machine or set of tools used for manufacturing toothbrushes will only be busy with toothbrushes for a very small part of the year. The rest of the time it must be used for producing something else needed in the settlement, perhaps there would be a brush-making tool in the settlement, but then it would have to be one capable of producing a large range of different brushes, brooms and other utensils incorporating a large number of bristles attached to a handle. And the same reasoning can of course be applied to all steps in the production process, maybe for a start there will not be different metal alloys, but just "metal" of the composition extracted from the regolith at hand using an economical process.

System Design

Keeping in mind the conditions outlined above, providing the production system necessary for fulfilling the survival requirements will involve selecting candidates from amongst existing production technologies and then replacing any overly complex components of each piece of equipment, as well as reducing or eliminating any use of resources that are difficult to produce in sufficient quantity. Without going into details about specific manufacturing technologies, a flowchart of the material needs for human survival is shown in Figure 27.3. Note that although designed for a Moon settlement, this flowchart can be applied to the Human society on Earth, a village, or a Mars base.

Habitat Design Loads

Compared to terrestrial structures, the conditions prevailing on the Moon will result in important differences in loads acting on structures. The lower gravity (1/6 of that on Earth) will drastically reduce the effect of structural weight as a design consideration, while other safety critical design issues will assume unparalleled importance in the harsh lunar environment.

Gravitation

Perhaps the most obvious difference compared to terrestrial conditions is the lower gravitational loads. In fact, this reduction will somehow be

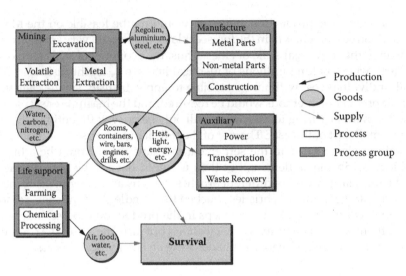

FIGURE 27.3
Manufacturing chain for material survival needs.

compensated for with the additional weight of the shielding covering structures. Consider a roof structure at 500 kg/m². Weighing 0.5*9.8 = 4.9 kN/ m² on the Earth, the same structure with 2 meter thick regolith shielding of density 1500 kg/m³ would on the Moon give a force of (0.5+2*1.5)* 1/6*9.8 = 5.7 kN/m². Thicker shielding would increase this downward load correspondingly. This means that, for surface structures in general, there will not be a reduction in the vertical loads to parallel the reduction in gravitation.

Internal Pressure

The biggest difference will come from the application of inside pressure. In a structure with inside pressure of one atmosphere, the uniformly distributed outward load will be very close to 100 kN/m², which corresponds to the gravitational load of almost 40 meters of regolith. Thus a structure has almost the same gravitational load as on the Earth, but in addition a 20 times greater upward force will create tension in the vertical members. This will change the well-understood concept of column and result in the adoption of a new concept like the "tension-compression-column" (Benaroya, 1993). Correspondingly, habitats will from a structural-integrity point of view be pressure vessels more than "buildings".

Radiation

The Moon does not have a protective magnetic field like the van Allen belts that shield the Earth from the bombardment of charged particles from the sun. Neither is there an atmosphere to keep down electromagnetic radiation. Thus, if genetic damage due to radiation is to be avoided, heavy shielding will be necessary. According to common estimates, about 2 meters of rock would provide sufficient shielding, also for rare events such as solar flares. If shelters can be provided and used during solar flares, the main body of the habitat may have significantly thinner shielding. However, such a design solution would force activities and/or comfort to be considerably reduced over extended periods of solar flares.

Meteorite Impact and Moonquakes

The Moon is much less geologically active than Earth, being smaller and thus more solidified. Thus, Moonquakes due to seismic activity will be a minor problem. The corresponding "load case" necessary to consider on the Moon is the possible impact of large meteorites. It is important to note, however, that although the Moon is pockmarked by an incredible number of meteorite impacts, large meteorites are very rare. Further, with proper radar surveillance, an advance warning would arrive with quite sufficient time to evacuate sensitive equipment and all personnel from threatened areas near the impact. Perhaps even to reduce air pressure or take other active damage attenuating actions.

Air Tightness

As discussed in the above section on survival requirements, the level of achieved air tightness will be a key parameter in determining the material supply necessary for sustaining the lunar colony. Thus, the treatment of seals and the choice of an airtight structure or an inner or outer sealing skin must be given careful consideration, and may well be one of the more demanding material challenges for the ISRU production engineers. Airlocks and seals between habitat sections and the external vacuum must also be designed to minimize air losses.

Habitat Approaches

Living on the Moon will require quite different considerations from terrestrial housing projects, and may well lead to exciting new breakthroughs in interior design and the use of living and communal spaces. Not only are external loads and design considerations drastically different, but it will be much more important to design the habitats for comfort as there is no exterior "nature" with greenery and recreation possibilities. Thus, all this must be "built-in" into the design of an underground habitat. This is a challenge that, if successfully met, may well produce solutions to improve the standard of living in many terrestrial cities as well.

Glassed Domes

The most commonly depicted "Moon colony" construction is undoubtedly the glass dome—a giant dome of glass containing the entire colony in one dome, or a few domes. This is not only in science fiction, but unfortunately also in scientific and engineering articles on Moon bases. The popularity is due to the fact that it is an aesthetically pleasing solution, which also provides a convenient view of the interior. Regrettably, the disadvantages of this design far outweigh the advantages. The well-known fact that glass has poor tensile properties, especially when somewhat scratched or damaged, is not in itself a sufficient argument against this design, since pre-tensioned tendons could be used to put the glass in compression. However, glass by its very transparency provides poor protection against radiation. Glass would let out infrared heat radiation during the night—putting shutters on an entire dome could of course be done, but is an awkward chore.

Worse, in order to protect against the dangerous components of sunlight during the day—the shutters would have to be very heavy and stay on during the day, completely destroying the glassed beauty of the dome. Thus, sadly, there will be no glassed domes on the Moon—at least not for permanent habitation. They could have uses for short-term visits such as tourist scenery spots or for housing sporting events—where the aesthetics may drive economics and stays are short enough for radiation levels to stay acceptable. Even then, measures for improving radiation protection would surely be needed, such as special glass compositions with higher opacity for dangerous radiation, heavy shielding giving shadow for spectators, or simply not using the domes in daytime or during solar flares.

Buried "Tin Cans"

The Moon is covered with several meters of fine dust, regolith. Thus, just piling enough regolith over the "glass dome" is a straightforward means to provide a quick and highly accessible radiation protection. This has probably been the most commonly suggested type of habitat in more construction-oriented papers concerning Moon bases. Of course, the dome then has no need for transparency and other materials than glass will be more appropriate. For a quick first shelter, the landing craft or other Earth imported rigid or inflatable "pressure vessels" will be a much healthier and safer habitat after radiation and meteorite shielding has been added by piling regolith dust on top.

For long-term use, concrete, aluminium, cast basalt and other in-situ produced materials have been suggested. An important design consideration for production of these kinds of habitats is to avoid structural loads during construction that are radically different than the structural loads in use. For example, it may be necessary to gradually pressurise the habitat very carefully and evenly when covering with regolith. One disadvantage, besides the need to move and secure large volumes of regolith would be the need for long entrance structures in order to reach surface without risk of regolith refill.

Sealed Lava Tubes and a Mine Shaft Society

Having technically disqualified the pretty science fiction glass dome shelter and proven that a hole in the dirt would be more feasible, what would be more natural than going back to cavemen for inspiration? But are there caves on the Moon? Surely no limestone caves such as the more common caves on Earth, since these are formed in limestone deposited by ancient coral reefs. However, although not much geological activity remains, the Moon once had only a thin solid skin over a molten core and prehistoric volcanism is known to have been considerable.

There are strong indications for the existence of large lava tunnels on the lunar surface. These natural cavities could be cleared, furnished and sealed to provide readymade shelters with a proven track record of resisting meteorite bombardment. The advantage of this solution is of course that the structure is pre-existing, while a significant disadvantage is that gaining access and clearing them from rubble might be more effort than drilling a fresh tunnel where one is needed. A significant advantage of lunar rock dwellings is that the load situation with regards to structural integrity is very similar to underground constructions on Earth—because the main load is the pressure of surrounding rock. As a matter of fact, even the airtight sealing problem has been successfully addressed in terrestrial underwater tunnels (although the acceptable leakage is much lower in a Moon colony). Another idea for subterranean (or sublunarian, rather) dwellings would be to spend just a little more effort in the mining operation to leave a sealed and furnishable void when extracting the richest mineral veins for satisfying the material needs of the colony.

Regardless of how the sublunarian cavities have been formed, structural integrity as well as meteorite and radiation protection is provided by the surrounding rock. Habitat design and construction is then left with the smaller task of designing cavity shape and size to avoid any risk of cave-ins, sealing the cavities to keep the air in, and furnishing for comfortable human habitation. Living in caves might not seem an attractive option, but there is no need at all for the "caves" to look like caves (Järvstråt, 2001). Think of large shopping arcades—under a roof but feeling a lot like outdoors. On the Moon, the need for food production would make all open areas much greener than an ordinary shopping malls, with "sunlight lighting" needed for optimum plant growth. Still, care must be taken to enhance the feeling of nature or at least cultivated farmland.

It will be an architectural and logistic challenge to arrange agricultural areas to provide recreation opportunities in as natural a setting as possible. The estimated need for 80 m² agriculture and common areas per person may not seem sufficient for a "forest walk". But, in a colony with 1000 persons and 20% of the space taken by rock support pillars, this corresponds to 100,000 m². This could be arranged as four 5-meter wide strips of greenery stretching 5 kilometres. Each a quite nice walk or running track if planned properly and probably more "nature" than used by most people today. And if the layout or design gets monotonous before the growth season is over, it is always possible to visit a neighbouring colony section.

Active Structures and Passive Safety

As outlined in (Järvstråt & Toklu 2004), for terrestrial construction, the designers are accustomed to conceiving civil engineering products of a certain capacity, beyond which the product cannot be used due to failure or passage to a non-functional stage. Examples can be buildings that collapse when a level of earthquake magnitude is exceeded, or dams capable of resisting a predetermined level of water, or highways that become blocked whenever the traffic exceeds the design value. In the first example there is failure, in the second example there is passage to a non-functional stage perhaps coupled with failure, in the third example there is non-functionality. Although these are all undesirable effects, designers and engineers have used such ultimate limit load criteria through the millennia since Hammurabi, a Babylonian king who established the first official written rules about collapsing houses, walls, dams and ditches around 1800 B.C. Only very recently, with the advent of technology and accumulation of knowledge, have the designers and constructors gained the capacity and consciousness necessary for distinguishing between the old way, passive design and passive structure, and the new way, active design and active structure.

When building for extraterrestrial conditions, the main difference compared to terrestrial conditions is perhaps that structures are much more safety critical. A structural collapse that would be damaging on Earth could well be fatal on the Moon, and any crack may become fatal if left unattended, because atmosphere will leak, bleeding vital and expensive constituent gasses (nitrogen) into vacuum, possibly cutting off communication between the areas of the settlement connected by the leaking section. That means that for these structures new safety definitions and new ultimate load limits have to be defined. Further, the structure must be monitored in such a way that danger should be detected well before the design limits are reached for any vital component of the habitat. Lastly, should dangerous values be reached, the structure must behave in such a way that failure is prevented or reduced as much as possible, i.e., the structure must be "intelligent" enough to sense the danger and must act to prevent the damage, thus the term "intelligent" or "smart" or "active" structures. Such a scheme can be made possible by satisfying the following capabilities:

- Active Design: The structure must be designed as an active structure, with well-defined danger levels to trigger corrective actions (Utku 2002).
- Monitoring System: There must be sensors placed in the structure to measure critical stresses, loads, strains, and/or displacements.
- Control Units: There must be a redundant set of control units that automatically analyze and evaluate the measurements, decide on

levels and types of actions to be taken, and command the actuators to respond accordingly.

- Active Structure: There must be actuators embedded in the structure commanded by the control units. These actuators will make the structure behave in such a way that temporary or permanent measures will be taken to prevent unacceptable damage.

An example can be the reduction of inside pressure of a structure to lower but still acceptable levels should stresses in the tendons approach prescribed limits. Active structures should thus respond automatically to any changes, especially if they could have dangerous consequences. Such measures will result in more economical construction, a decreasing of the safety factors that should otherwise be applied.

Although the use of active structures and automated control and regulation systems as described above will be necessary for emergency amelioration, it will, however, also be necessary to design with highly stable self-controlling processes whenever possible. Homeostasis is the process of environment auto-regulation encountered in many natural settings, such as societal insects. Self-convection driven by natural day-night temperature cycles to circulate the habitat air is an important homeostasis candidate on Earth, though more difficult on the Moon because of the lower gravity driving force for convection.

Air leakage could be reduced with a passive solution, such as providing a "buffer gas layer" between the habitat and the external vacuum, pressurised to higher levels than the habitat and containing gasses cheaper-than-air, safe for human and plant exposure, and easily detectable, e.g., by smell or color.

Whenever the internal pressure is the dominant loading for a structure, it will be more economical to choose pressure-efficient structural types, such as spheres, cylinders, tori, spheroids or similar forms. For flexibility, hexagonal modules could be employed with advantage. It seems to be more appropriate to reinforce the walls of these structures by pre or post tensioned tendons. The tension in these tendons can be so arranged that the walls themselves can be under tension or compression. The latter option seems to be preferable on the Moon since it is much easier to find or produce compression resistant materials than tensile resistant materials.

Choice of Site for a Terrestrial Analogue of a Large Scale Moon-Base

A site for creating a Moon-base analogue must meet several criteria: first, the environment should be as close as possible to the lunar environment. Unfortunately, it is impossible to mimic the two most obvious features of the lunar environment, as 1/6 g gravity cannot be found on Earth, and vacuum can only be created and maintained at high cost, although the vacuum pumps needed for that could be installed at any location.

The only distinctly Moon-like feature possible to find on Earth is the habitat environment, but it isn't a grey dusty desert one should look for although that may give the closest lunar surface analogue. In order to model the *indoor* conditions of a Moon-base, the building type foreseen needs to be considered: for a glassed dome habitat the grey desert would be the best choice, as may well be the case also for the "buried tin-can" habitat, though the colour of the dust would then be less important than the depth. If, on the other hand, the interest is in a "mine-tunnel" or "shopping center" habitat, existing underground tunnels are to be preferred.

Apart from being "as Moon-like as possible", it is naturally important for a Moon-base analogue to be accessible, e.g., situated reasonably close to main communications and a community of reasonable size. (Compare the discussion about Las Vegas in the section "Why Self-Sufficient.") Some possibilities could be

- An existing village or town would be ideal, if all inhabitants could be convinced to support the project by gradually reducing all imports to the village aiming for complete self-sufficiency. But however attractive the solution may seem, upholding democratic ideals, it seems impossible to find such a village of dedicated Moon fans.

- An industrial area or research park with sufficient area of indoor space could easily be converted to double as a Moon-base production system demonstrator—in particular if the research is production oriented or the industry in the area is oriented towards manufacturing of not-too-high-tech goods. However, it would be somewhat difficult to mimic the habitat spaces within a building, as the shape is square in contrast to the more cylindrical shape that mined or naturally occurring tunnels would prefer. Appearances could be changed, of course, but it would be hard to avoid a lingering feeling of fabrication.

- An analogue base in the desert or arctic is attractive from the point of view of the barren lands resembling the lunar surface. Indeed, the Mars Society is running several such habitat analogues (see www.marssociety.org/portal/groups/AnalogsTF). A disadvantage of such a location is that it is hard to find one that is also easily accessible.

- Subterranean locations are commonplace on Earth, including caves, mines and decommissioned military facilities to name a few. Caves, and in particular lava tunnels, would be ideal to mimic lava tunnel habitats on the Moon, but it might be argued that terrestrial lava tunnels should be protected as they are rare natural phenomena and some are also homes for some rare species of bats, fish and insects. Excursions and study trips are of course in order, and indeed, the Oregon L5 Society has performed a series of preliminary studies (see www.oregonl5.org/lbrt/l5ombrr1.html) in a lava tunnel complex in central Oregon. However, although a mine seems a prime candidate for underground habitat design, no such studies seem to have been performed to date. A terrestrial mine, though containing tunnels and caverns of different sizes, would be rather different in layout from a lava tunnel suitable for habitat establishment, because all connecting tunnels would by economical reasons be rather small—man sized—while the caverns created from removing ore would be large, but generally not arranged horizontally and linear in the way of a lava tunnel. However, it will be possible to identify and use both big hallways and narrow corridors, and with any mining taking place on the Moon, connection tunnels and caverns very similar to those in a terrestrial mine would soon be created, that could be conveniently converted to living space. In addition, an added advantage of a terrestrial mine, especially one containing some residual metal ore is that the mining and beneficiation part of the Moon-base analogue can be naturally incorporated and also used for demonstrating means for expanding the habitable volume by extracting useful minerals.

Activities for a Terrestrial Moon-Mine Analogue

Ideally, a complete system should be constructed in as complete detail as possible, including all the activities of Figure 27.1, or equivalently all the process steps in Figure 27.3. Naturally, it will not initially be feasible to create a full-scale society fulfilling all needs "in-house." However, it should not be difficult to include typical manufacturing components in each of the process categories, and gradually connect the systems adding more and more pieces into the process, eventually connecting the flow of materials and products into a fully interconnected network. Some early activities that don't require extensive equipment but would still provide useful insights and ideas are illustrated schematically in Figure 27.4 and include:

- **"Outdoor" mine-tunnel design,** using creative design with edible plant beds to give the underground habitat an outdoor feeling to compensate for the lack of forest, beaches and other natural recreation areas enjoyed on Earth.
- **Sunlight transmission through fibers,** collecting sunlight on the surface and leading it through cables down into the safety of the underground. Such systems are commercially available today, but

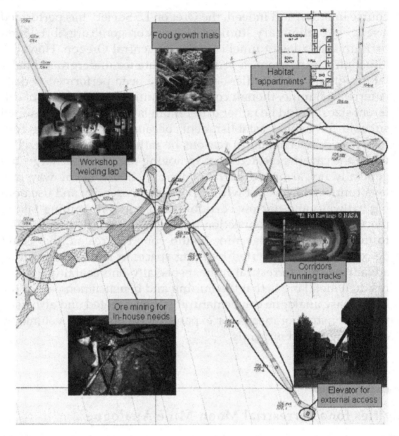

FIGURE 27.4
The Pershyttan Moon-Mine concept draft.

> may need some adaptation for transmitting light in a beneficial way
> for plant growth.
>
> - **Artificial light food-growth demo,** growing vegetables in appropri-
>   ately fertilized lunar regolith simulant under artificial light or light
>   transmitted through fibers to the greenhouse.
> - **Low-volume high-flexibility production facilities,** such as Weld
>   Metal Deposition, building metal components layer by layer with
>   each layer depositing a fraction of a millimetre of metal onto the
>   growing detail producing virtually any shape.
> - **Habitat design,** transforming parts of mine tunnel or rooms into
>   apartments with modern facilities—while reducing as far as possible
>   any "imported" goods and making as much as possible in-situ man-
>   ufactured or possible to manufacture using minimal equipment.

- **Minimal-equipment mining and ore processing,** applying manual mining techniques or small-scale semi-automated ore breaking, as well as other processing steps that can be manufactured from the metal extracted from the ore mined.

Current Status and Near Term Plans of Pershyttan Moon-Mine

Although it seems like the next logical step in space exploration and settlement, the creation of large scale system demonstration facilities unfortunately seems to be very low on the agenda of all organizations active in space research. However, such a project has recently been started close to the small Swedish town of Nora.

The old iron mine (see Figure 27.5) abandoned in 1967 is now being converted into a demonstration facility for the technologies and systems needed in a self-contained lunar settlement. Before starting the activities listed

FIGURE 27.5
Storgruvan in Pershyttan, above and below ground.

above as suitable for a Moon-mine analogue, however, the mine needs to be emptied of water and the old hoist restored and certified for carrying people down to the extensive network of tunnels at the 110 meter level. Here, demonstrator projects will be performed, by selected production industry partners, while partners in hotel and tourist industry will ascertain that the specially designed environment and the holistic view on the material needs of human society are appropriately showcased and made accessible to the broader public.

The Moon-mine analogue (www.Moon-mine.com) will be established as a think tank and demonstrator facility, not only for space exploration and settlement, but also for research on contained spaces and environmentally friendly, low resource impact, production supply chain management. The vision is to bring together old traditions from mining and handicraft with space age flexible production methods, as well as fusing the common interests of the space exploration and environmentalist movements. At the time of print of this book, the Moon-mine is at an early stage of creation, but current information regarding opportunities to collaborate and contribute, as well as reports on ongoing work and research results will become available at the www.Moon-mine.com.

References

Benaroya, H. (1993) *"Tensile-Integrity Structures for the Moon"* Applied Mechanics of a Lunar Base, Applied Mechanics Reviews, Vol.46, No:6, pp. 326–335.

Blair, B. (1998). *"Use of Space Resources - A Literature Survey"* SPACE 98 - Engineering, Construction, and Operations in Space VI, pp. 651–665, American Society of Civil Engineers, New York, NY.

Järvstråt, N. (2000). *"What Do You Mean by "Outside" in a Moon Town?"* Moon Miners Manifesto, #141, December 2000.

Järvstråt, N. & Toklu, C. (2004), *"Design and Construction for Self-sufficiency in a Lunar Colony"*, Proceedings of the Third International Conference on Advances in Structural Engineering and Mechanics (Invited lecture), pp. 110–122, Techno-Press, 2004.

O'Handley, D. (2000) *"Final Report On System Architecture Development For A Self-Sustaining Lunar Colony"*, (http://www.niac.usra.edu/studies).

Prado, M. (1998) PERMANENT - *Projects to Employ Resources of the Moon and Asteroids Near Earth in the Near Term*, Ladprao, Bangkok: Fong Tong Enterprise Co. Ltd.

Utku, Senol (2002) *"Civil Engineering"*, ACE 2002, 5th International Congress on Advances in Civil Engineering, Istanbul, Vol.1, pp. 1–17.

28

Terrestrial Analogs Selection Considerations for Planetary Surface Facility Planning and Operations

Olga Bannova

University of Houston
Sasakawa International Center for Space Architecture
Houston, Texas

Olga Bannova is a research assistant professor at the Sasakawa International Center for Space Architecture (SICSA) of the University of Houston. She received a Master of Architecture degree from the University of Houston in 2001 and a Master of Science in Space Architecture in 2005. She brings several years of professional architectural practice with Houston and Moscow firms as well as research and teaching experience at the University of Houston. Olga conducts research and design studies that address a variety of topics such as planning analyses for a broad range of space vehicles, habitats, and systems, including advanced spacecraft propulsion options, inflatable hydroponics laboratory and logistic modules, special design influences and requirements for different gravity conditions in space, and habitat concepts for extreme environments on Earth. She has given a number of presentations and publications at different international technical conferences and media. Olga is a corresponding member of the International Academy of Astronautics (IAA) and a member of the American Institute of Aeronautics and Astronautics (AIAA) Space Architecture Technical Committee. She leads the IAA Commission VI Space Architecture Study Group in development of a white paper report on Space Architecture for IAA. The study group addresses such aspects as human factors, multidisciplinary education, space tourism and commercialization, art in space, and space missions' simulators.

ABSTRACT   This paper will draw parallels and define differences between factors that drive the planning and design of human surface facilities in space and in extreme environments on Earth. Primary emphases will highlight influences upon general habitat requirements, constraints upon delivery and construction, and special provisions for safety and hazard interventions. The overall intent is to identify important lessons that can be applied across different settings which present common priorities, issues and challenges. Such environments include future bases on the Moon and Mars, offshore surface and

FIGURE 28.1
IDEEA-One conference proceedings cover.

submersible facilities, polar research and oil/natural gas exploration stations, military desert operations, and natural and man-made emergency shelters.

Important topics of emphasis include the following considerations:

Design influences driven by transport to remote sites

Environmental influences upon facilities and construction

Influences of crew sizes, types of activities and occupancy durations

Influences of construction methods and support infrastructures

Special safety and emergency response requirements

This presentation will draw upon research and design activities at the Sasakawa International Center for Space Architecture (SICSA). Information is also taken from a SICSA-sponsored conference "International Design for Extreme Environments One" (IDEEA-One) at the University of Houston which attracted more than 400 interdisciplinary participants from 12 countries representing diverse professions and environmental settings. (Figure 28.1).

Background and History

Extreme environments on Earth provide analog experience to support planning of extraterrestrial facilities and operations. Each environment presents special lessons regarding habitat design, crew operations and training, and equipment and logistical requirements for space exploration.

SICSA has extensive experience in research and design for extreme environments, including orbital and lunar planetary facilities, disaster shelters, polar stations and offshore surface and submersible habitats. Investigations have addressed such issues as hardships and challenges posed by harsh climate conditions, remoteness with restricted access and return opportunities, limitations on available equipment and support services, and ever-present safety risks. All of these environments share many kinds of technical and operational priorities. Key among these are needs for appropriate transportation and construction systems, efficient energy, effective and environmentally-responsive waste management and life support systems, maintenance and repair provisions, and emergency accommodations.

It is important to note that needs and priorities in extreme environments also represent some of the most pressing challenges and issues that face our entire planet. Increased difficulties and urgency in addressing human need and requirements in extreme environments often motivates efforts to find new and better solutions. Useful program advancements related to the extreme environment of space, for example, include important contributions to fields associated with computing and information management, material sciences, energy technologies, environmental monitoring and life sciences.

Experiences on US and Russian spacecraft, underwater vessels, and polar stations have revealed a variety of common issues (Figure 28.2):

- Cut off from "the outside," crews must learn to be resourceful, and to depend upon one another:
  - They must work to help crewmates deal with psychological and physical stresses.
  - They are required to adapt to limited comfort and recreational amenities.
  - They must be prepared for fatiguing work overloads and stimuli deprivations.
- They must be trained and equipped to deal with equipment malfunctions.
- Common types of constraints place stringent requirements and severe restrictions on habitat design and operations:
  - Limited internal volumes constrain storage and human activities.
  - Limitations on equipment, labor and processes constrain structure assembly/deployment procedures.
  - Limitations on maintenance and repairs (people, tools/spares and methods) constrain maintenance and repair options.
- Safety and operations under harsh environmental conditions and demanding mission schedules pose safety and operational challenges.

(a)

(b)

FIGURE 28.2
Destiny module interior (a), submarine interior (b), Antarctic station (c).

(c)

Human and Environmental Planning Influences

Human requirements and environmental factors specific to each different type of environment, operation and facility must be correlated with resulting planning needs. Some general considerations are listed below in Table 28.1.

Structure types and architectural forms are typically influenced by similar considerations which include the following:

- Site/environment influences
- Transportation modes
  - Capacity (volume, mass, size)
  - Delivery method

TABLE 28.1

Planning Considerations

Human Requirements	Environmental Influences
Number of occupants	Structure selection and construction options
Social/cultural influences	Climate/thermal characteristics of the site
Time frame/mission duration	Logistical requirements and scheduling
Special safety hazards	Types and levels of danger
Emergency escape means	Proximity to major transportation modes
Recycling of expendables	Type of surface transportation
Primary mission objectives/purposes	In-situ resource utilization possibilities

- Mission timeline and crew work schedules
- Site/infrastructure preparation requirements
- Facility evolution/growth projections and requirements
- Special assembly/deployment accommodations and problems

Transportation, Safety, and Emergency Response Requirements

As in space, high transportation costs, restrictions on cargo payload volumes, and limitations on periods of site accessibility pose serious constraints for extreme environments on Earth, such as polar and emergency response operations. These constraints impact the design of facilities, applicability and use of large equipment systems, re-supply of consumables, and crew rotation cycles.

For example, the only transportation available for most Greenland science facilities is LC-130 heavy-lift aircraft during the summer. The dimensions of its payload cannot exceed the size of 2.4 m × 2.4 m × 10.9 m (8 × 8 × 36 ft) and 11340 kg (25000 pounds) in weight. A short Greenland summer and therefore a short period of time when flights are available place additional restrictions on payload mass and size, which can significantly extend any construction period. To simplify construction and to make the most components of the structure exchangeable, all members of the trusses, floor and walls details, and utilities runs must fit the allowable payload size; therefore in this case all dimensions of the elements should be divisible to 2.4 meters (8 feet) (Bannova, O., Smith, I. F. C., 2005) These conditions create delivery and access problems which are generally similar to circumstances encountered in planning future planetary bases.

Logistics and transportation to some disaster areas on Earth can also pose challenging access difficulties. Responses to major disasters require that complex management, training and logistics plans be developed and implemented to deliver services to effected locations quickly and effectively. Special arrangements must be planned to address a broad variety of critical needs. Included are requirements for search and rescue operations; emergency medical accommodations; evacuation and shelters for impacted populations; food and water replenishment; waste cleanup and pollution control; and restoration of power, communications, transportation and other vital support systems.

Terrestrial analog experiences can be useful references to assess and confirm important requirements for space mission planning. It must be recognized, however, that the space environment is very different in many respects from human terrestrial environments:

- There is total dependence on artificial systems.
- Altered gravity conditions influence most activities.

TABLE 28.2

Comparison Between Human Missions on and Close to Earth and Future Space Missions.

Missions Factors	Orbital Missions	Winter-Over In Polar Regions	Lunar Missions	Mars Missions
DURATION (months)	4–6	9–12	6	16–36
DISTANCE TO EARTH (km)	300–400	NA	350-400 thousand	60–400 million
CREW SIZE	3–6	4–100	4	6–8
DEGREE OF ISOLATION AND SOCIAL MONOTONY	Low to high	Medium	High	Very high
CREW AUTONOMY	Low	High	Medium	Very high
EVACUATION IN CASE OF EMERGENCY	Yes	No	Yes	No
Availability of In-Site Support Measures				
Outside monitoring	Yes	Yes	Yes	Very restricted
2-way communication	Yes	Yes	Yes	Very restricted
E-mail up/down link	Yes	Yes	Yes	Yes
Internet access	Yes	Yes	Yes	No
Entertainment	Yes	Yes	Yes	Yes
Re-supply	Yes	No	Restricted	No
Visitors	Yes	No	No	No
VISIBILITY OF EARTH	Yes	Yes	Yes	No

- Extreme radiation, temperature and operational conditions present hazards for people and equipment.
- Stresses related to isolation in close confinement impact crew health and morale.

To accomplish proper planning planners must understand special characteristics of space environments:

- Reduced gravity levels and their implications.
- Radiation hazards and health risks.
- Micrometeoroid/space protection requirements.
- Special lunar/Mars surface features and environmental conditions.

FIGURE 28.3
SICSA's concept of underwater experimental labs.

Barabasz, A.F., (1991) Kanas N. and Manzey, D., (2003) Ocean Analogs

Oceans, like space, offer vast and exciting frontiers for science and exploration. The lack of natural life support systems in space and offshore underwater settings force mission planners to provide artificial alternatives with control systems that conserve and protect non-renewal resources. As in space, discharges of toxic wastes can produce harmful consequences. The oceans and other water bodies that constitute a primary part of our natural life support system have a limited capacity to sustain toxic abuse.

Deep ocean diving operations present requirements and constraints that are also similar in many aspects to extra-vehicular activities (EVA) in space. Divers and astronauts depending upon artificial life support systems must perform exhausting, often hazardous work, encumbered by rigid pressure suits that limit body mobility and dexterity. Poor or harsh lighting conditions hamper visibility in performing demanding and potentially dangerous work functions.

Construction Methods and Support Infrastructures

Construction methods in extreme environments must address vital structural safety and reliability requirements and take special environmental influences into account. Included are:

- Lack of onsite equipment and limited labor personnel
- Short construction windows
- Equipment breakdowns with limited tools/spares;
- Hazardous working conditions;
- Extreme temperatures impacting thermal control and structural fatigue.

A common construction priority for extreme environments is to design structures that can be rapidly assembled and deployed under harsh conditions. Modular approaches facilitate deployment and afford immediate occupancy but usually impose internal volume constraints driven by transportability requirements. Erectable structures can overcome volume constraints but add to on-site time and labor required for readiness. Advanced technologies including inflatable and other tensile systems applied to polar and desert environments can have transferable benefits.

Conclusions

Extreme environments offer good opportunities to demonstrate and assess the practical attributes and performance of equipment and operations under rigorous and demanding circumstances. High logistics costs and transportation constraints on allowable volume and weight force designers to create systems that are small and highly efficient. Harsh climates and isolated working conditions impose requirements for ruggedness and dependability. Limited labor resources and available tools place a priority upon ease of equipment deployment and repairs. Planning and design to optimize human safety under normal and emergency circumstances takes on a special urgency.

Operations in extreme environments often place people in small isolated groups where they must learn to depend upon themselves and their team members for social companionship and support ordinarily provided by large and diverse communities. They often experience dangers and stresses that test their ability to adapt, cope and perform. They are forced to work together and be resourceful in dealing with problems and emergencies. By observing experiences in extreme environments, we can learn about fundamental human capabilities and needs that are frequently overlooked or forgotten in modern society.

Different extreme environments on Earth provide venues for testing facilities, diverse issues and influences that apply to space missions. Table 28.3 presents some correlative examples.

While underwater facilities might be considered most applicable for many space factors, other issues such as transportation and logistics may more closely relate to polar and desert environments. In return, space technology, including easily transportable and deployable habitats using new materials,

TABLE 28.3

Compatibility and Testing Abilities of Terrestrial Analog Settings for Space Applications

Factors \\ Settings	Polar Regions	Under Water	Deserts	Disaster Areas
TRANSPORTATION	Maximum	Medium	Medium	Less
ENVIRONMENT	Less	Maximum	Medium	Medium
CREW: SIZE/ ACTIVITIES/ DURATIONS	Maximum	Maximum	Medium	Medium
CONSTRUCTION METHODS	Less	Maximum	Less	Less
SAFETY AND EMERGENCY REQUIREMENTS	Less	Maximum	Less	Maximum

Maximum | Medium | Less

advanced power and power storage devices, and novel approaches to reduce and reuse waste materials can benefit all settings.

Existing terrestrial facilities such as NASA human-rated test facilities, subsea laboratories and polar camps can be used at low-cost as analogs at early stages of mission planning. To increase analog fidelity new terrestrial facilities that are specifically designed for space exploration will be necessary for future mission development. Low Earth Orbit facilities, such as the ISS, can provide a variety of space flight parameters and lunar outposts can provide analogs for future Mars missions.

In every analog, an appropriate mix of systems testing, human research, and mission operations simulation is necessary to achieve early space exploration milestones, both technical and strategic. Earth-based preflight crew training in high-fidelity simulators, geology training at appropriate locations on Earth, new ground facilities including a life support test facility, and life sciences research into human factors including psychosocial issues and habitat design can contribute to planning successful space exploration missions.

Transferable benefits from and between extreme environments can take many forms. Included are advanced technological innovations, significant scientific developments, and probably of greatest importance, enlightenment about ways humans can live and work in harmony with all environments. The ultimate benefit may be to help prevent our entire, fragile, planet Earth from eventually becoming an extreme environment.

References

IDEEA One, The First International Design for Extreme Environments Assembly, Final Conference Report, November 12–15, 1991.

Kanas N. and Manzey, D., *Space Psychology and Psychiatry*, Kluwer Academic Publishers, London, 2003.

Barabasz, A.F., Effects of Isolation on States of Consciousness, in: *From Antarctica to Outer Space: Life in Isolation and Confinement*, ed. Harrison A.A., Clearwater, Y.A., and McKay, C.P. Springer Verlag: New York, 1991.

Bannova, O., Smith, I. F. C., Autonomous Architecture: Summit Station in Greenland Design Proposal as a Test-Bed for Future Planetary Exploration. SAE 2005 Conference Proceedings, July, 2005.

SICSA space architecture seminar lecture series. Part VIII: Shelter Design and Construction. Section A: The Nature of Shelters, www.sicsa.uh.edu, 2006.

Adam, B. and Smith, I.F.C., (2007), Self-Diagnosis and Self-Repair of an Active Tensegrity Structure, *Journal of Structural Engineering*, Vol. 133, 1752–1761.

29

Surface Infrastructure Planning and Design Considerations for Future Lunar and Mars Habitation

Larry Bell

Professor/Director of SICSA

Larry Bell is a professor of architecture and endowed professor of space architecture in the University of Houston's College of Architecture, where he is also the director of the Sasakawa International Center of Space Architecture (SICSA). Larry also heads the college's MS-Space Architecture program, serving students from NASA and major aerospace companies along with professionals from many countries. In addition to achievement awards from leading aerospace engineering societies and NASA headquarters, he has received important international honors, including the Space Pioneer Award by Kyushu Sanyo University in Japan and two of the highest honors awarded by the Federation of Astronautics and Cosmonautics of the former Soviet Union for his contributions to international space development. His name was placed in large letters on the Russian Proton rocket that launched the first crew to the International Space Station.

ABSTRACT This paper discusses and compares advantages and disadvantages of different types of modular habitats that offer prospective uses to support human surface operations on the Moon and Mars. It also summarizes important considerations associated with surface placement and configuration options, including evolutionary growth capabilities, emergency egress features, launch and landing economies and deployment factors. Basic types of habitats addressed include "hard" fixed-volume modules with pre-integrated utility systems and expandable-volume "inflatable" variations. Scenarios include combinations of both types that can optimize overall site development and operational benefits.

Introduction

Recognizing the enormously high costs of delivering habitats and support equipment to any lunar or planetary surface destination it is essential to provide means to deliver the greatest amount of useful real assets in the most practical and efficient manner possible. This goal presents a number of major planning and design challenges. All elements must comply within stringent payload mass and volume limitations imposed by available launch, orbital transfer and landing vehicles. If multiple habitats are to be provided, means must be afforded to transport them from the landing area to the operational destination over rough terrain, then position them in place with secure pressure-tight interfaces. Utility systems, including power and life support, must be put in a proven state of readiness prior to occupation, potentially accomplished by automated devices. Means for safe egress should be made available in the event of a pressure failure or other emergency in any element, and opportunities for evolutionary configuration growth should extend this capability.

Representative Design Concepts

A large variety of design concepts developed by numerous international government, corporate and academic entities have proposed ways to address these and other challenges. Specific solutions are guided by different assumptions regarding transportation system capabilities, lander and surface mobility systems, space radiation and meteorite hazard mitigation strategies and other critical engineering accommodations. Given that no single definitive habitat design solution can universally satisfy all variant assumptions, this presentation will, by necessity, address three general habitat types and associated application considerations. The first is a large diameter rigid module oriented in a vertical surface placement. A second type is a vertically-oriented module that incorporates a pliable, "inflatable" pressure shell deployed at the destination site to afford greatly expanded interior volume benefits. The third type is a smaller diameter rigid module oriented in a horizontal surface placement similar in design to those used for the International Space Station (ISS) (Figure 29.1).

Although each type of module presents special surface landing requirements and challenges, these issues are beyond the practical scope of this paper. Long cylindrical modules can be expected to be particularly problematic in this regard due to difficulties in balancing the center of gravity throughout the surface decent stages (Figure 29.2).

Module Volumetric Considerations

Habitat module scaling decisions must correlate module types with launch/landing constraints and functional accommodation requirements driven by

Vertical Cylindrical Modules

Horizontal Cylindrical Module

Vertical Inflatable Module

FIGURE 29.1
SICSA module concepts.

FIGURE 29.2
SICSA launch and lander concepts.

mission applications. Each option type must be optimized for volumetric and overall design efficiency to:

- Maximize the total amount of internal space available for essential equipment systems, human operations and crew comfort/morale.
- Plan interior circulation within and between modules for efficient use, convenience and safety.
- Accommodate manifesting and delivery of support equipment and expendable logistic supplies to the extent possible.
- Enable rapid relocation, integration and change-out of utility-dependent systems during and following operational deployment at the site.

Large diameter modules and longer smaller diameter modules present different layout optimization characteristics. Vertically-oriented large diameter modules naturally lead to a "bologna-slice" internal configuration with circular floor areas. The amount of usable floor increases greatly as a function of diameter and is only practical for modules of significant cross-section (perhaps 30-ft diameter or more). Such modules may also facilitate landing by combining a wide footprint for stability and a relatively compact center of gravity during decent.

Long, smaller diameter modules naturally provide a "banana-split", rectangular floor oriented orientation that uses space efficiently and affords a longer "line-of-sight" than the bologna-slice approach. This configuration also offers considerable versatility to accommodate wall-mounted equipment systems and conventional circulation layouts (Figure 29.3).

Larger diameter and/or longer module interiors can be produced using soft, pliable layered inflatable pressure enclosures that are compactible to fit into constrained-volume payload shrouds and deployed at the destinations. Here, the chief volumetric advantage is for modules that increase in diameter and expand as a function of radius squared. Modules that extend only in length expand only by linear movements.

SICSA has undertaken numerous design studies that exploit potential capabilities of inflatable systems to greatly expand internal volumes. These schemes typically incorporate the inflatable section into a hard lower section that provides surface support structures, life support equipment and utility connections, and accommodation space for the undeployed soft upper structure during launch, orbital transfer and landing (Figures 29.4 and 29.5).

SICSA has also proposed a "pop-out" floor structure that can be incorporated into an inflatable module to eliminate the need for internal construction build-out following deployment (Figures 29.6 and 29.7).

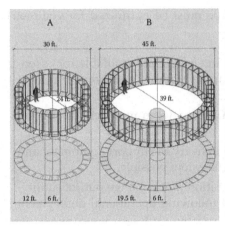

Bologna Slice Scheme
Usable floor area increases greatly as a function of module diameter.

- Total area per floor: (A) 705 sq.ft. (B) 1585 sq.ft.
- Total open floor area: (A) 450 sq.ft. (B) 1195 sq.ft.
- Usable open floor area: (A) 420 sq.ft. (B) 1165 sq.ft.
- Usable/total floor area ratio: (A) 0.59 (B) 0.73
- Maximum sight distance vista: (A) 24 ft. (B) 39 ft.

Banana Split Scheme
Usable floor area efficient for modules 15 ft. diameter and larger.

- Average area per floor: (A) 545 sq.ft. (B) 1730 sq.ft.
- Average open area: (A) 286 sq.ft. (B) 1395 sq.ft.
- Average usable area: (A) 286 sq.ft. (B) 1275 sq.ft.
- Usable/total floor area ratio: (A) 0.5 (B) 0.74
- Maximum sight distance vista: (A) 45 ft. (B) 45 ft.

FIGURE 29.3
Module layout/configuration options.

FIGURE 29.4
SICSA's pop-out floor system.

FIGURE 29.5
Inflatable module interior.

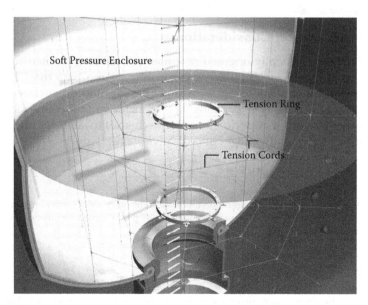

FIGURE 29.6
SICSA's pop-out floor system.

FIGURE 29.7
Inflatable module interior.

Site Configuration Considerations

Large diameter vertically-oriented modules are naturally suited for either linear, triangular or rectangular surface configuration patterns. The reference patterns illustrated below provide separate module surface access/ egress locations at central locations, and transfer tunnel connections that connect them together at perimeters. Both schemes (triangular and rectangular) afford special advantages and disadvantages (Figure 29.8).

Triangular Configuration

- Pros: A relatively compact configuration footprint at the entry airlock level can minimize the area for site surface preparation (if required). A loop dual-egress for emergencies is achieved with only three modules.
- Cons: May be more difficult to position/assemble this geometry.

Rectangular Configuration

- Pros: More conventional geometry may facilitate module site positioning
- Cons: Requires larger footprint for good site geometry and/or surface selection. Four modules are required for dual-egress from all modules.

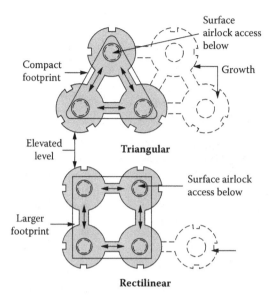

FIGURE 29.8
Large diameter module surface configurations.

Smaller diameter cylindrical modules can be configured in a variety of surface patterns most logically organized in linear, cruciform and rectilinear geometries. Connecting component elements can include separate airlock nodes for access/egress at module intersection points, and vertically-oriented hard shell or inflatable modules to form hybrid combinations. Each of these schemes presents special advantages and disadvantages associated with space launch efficiency, emergency egress opportunities, commonality of module types, evolutionary growth options and relative ease of surface positioning.

For comparison purposes, four basic schemes are presented as illustrations (Figures 29.9–29.15).

- Scheme A incorporates a combination of horizontal conventional and vertical inflatable modules to realize special advantages of each type:
  - EVA access/egress would be provided by suitlocks in each horizontal module.
  - The cruciform plan could later be expanded into a closed-loop racetrack.
- Scheme B utilizes only horizontal modules in a racetrack pattern:
  - Each module is assumed to contain an airlock which also serves as a berthing/ interface passageway.
- Scheme C utilizes a combination of horizontal conventional modules and corner berthing/ airlock nodes:
  - Suitlocks could be used, but are not presented to conserve functional module space.
- Scheme D presents a raft pattern with 2 types of horizontal modules plus separate berthing/airlock nodes:

The configuration assumes that 2 EVA access/egress airlocks will be provided.

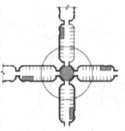
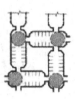

Scheme A: Cruciform with inflatable + conventional modules

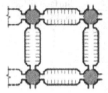

Scheme C: Conventional modules with corner airlock nodes

Scheme D: Raft with conventional modules + airlock nodes

FIGURE 29.9
Module configuration examples.

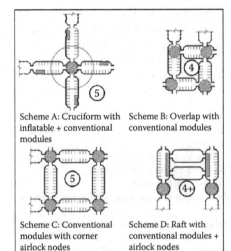

Non-functional space

Number of launches required to achieve configuration

Scheme A: Cruciform with inflatable + conventional modules

Scheme B: Overlap with conventional modules

Scheme C: Conventional modules with corner airlock nodes

Scheme D: Raft with conventional modules + airlock nodes

Scheme A:
- Suitlocks minimize non-functional space associated with conventional airlocks.
- Inflatable module greatly increases crew living/work volume over all other schemes.

Scheme B:
- Internal airlocks in all modules produce a high non-functional/useful volume ratio.

Scheme C:
- External airlocks enable full utilization of modules but impose additional launch requirements.

Scheme D:
- Special circulation modules plus external airlocks impose substantial launch requirements.

FIGURE 29.10
Space/launch efficiency.

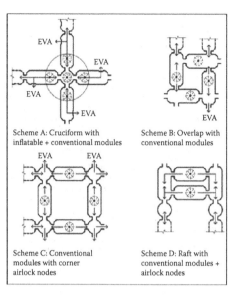

Emergency evacuation

To connecting module

EVA ← To EVA

Scheme A:
Cruciform with inflatable + conventional modules

Scheme B:
Overlap with conventional modules

Scheme C:
Conventional modules with corner airlock nodes

Scheme D:
Raft with conventional modules + airlock nodes

Scheme A:
- Direct connections, all modules.
- EVA-suitlocks in conventional modules.
- Worst case-central atrium emergency.

Scheme B:
- Connections/EVA egress through internal airlocks.
- Worst case-airlock failure prior to complete racetrack, isolating modules.

Scheme C:
- Connections/EVA egress through external airlocks.
- Worst case-airlock failure prior to complete racetrack, isolating modules.

Scheme D:
- Connections through special modules.
- EVA egress through separate nodes.
- Worst case-airlock failure prior to complete racetrack, isolating modules.

FIGURE 29.11
Emergency egress.

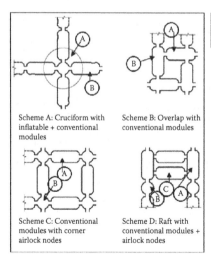

Separate Module Types

Scheme A:
Cruciform with inflatable + conventional modules

Scheme B:
Overlap with conventional modules

Scheme C:
Conventional modules with corner airlock nodes

Scheme D:
Raft with conventional modules + airlock nodes

Scheme A:
- Applies 2 module types, each with important functional support benefits (inflatable volume & conventional module pre-integration).

Scheme B:
- Uses a single standard module but with constricted volume capacity.
- For double connection interfaces the module must be modified for a 2nd berthing port.

Scheme C:
- Uses a single standard module + separate airlock element.

Scheme D:
- Uses 2 types of modules + a separate airlock element.

FIGURE 29.12
Module commonality.

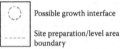

○ Possible growth interface

---- Site preparation/level area
boundary

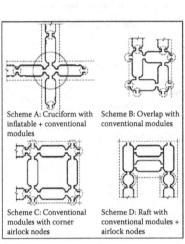

Scheme A: Cruciform with inflatable + conventional modules

Scheme B: Overlap with conventional modules

Scheme C: Conventional modules with corner airlock nodes

Scheme D: Raft with conventional modules + airlock nodes

Scheme A:
- Configuration can extend lineally & possibly replicate.
- Smallest boundary for level site requirement.
- Does not impose a requirement for more than 2 modules/launches prior to operational configuration.

Scheme B:
- Configuration can grow along 2 axes & can replicate a 2nd racetrack group.
- More compact for site preparation than Scheme C.
- Requires 4 modules/launches to achieve racetrack advantage.

Scheme C:
- Configuration can grow along 2 axes & can replicate a 2nd racetrack group.
- Imposes the largest level site requirement of all schemes.
- Requires 4 modules/5 launches to achieve racetrack advantage.

Scheme D:
- Configuration can grow along one side (unless additional airlocks are added) requiring 4+ launches, and can replicate.
- More compact for site preparation than schemes B&C.
- Requires 4 modules +2 airlocks to achieve racetrack advantage.

FIGURE 29.13
Evolutionary growth.

Establishes initial reference

Scheme A: Cruciform with inflatable + conventional modules

Scheme B: Overlap with conventional modules

Scheme C: Conventional modules with corner airlock nodes

Scheme D: Raft with conventional modules + airlock nodes

Scheme A:
- Central inflatable module establishes the site center & is not repositioned.
- Conventional modules with wheels are aligned to interface at a single point.

Scheme B:
- Conventional modules with wheels must be forward & rotationally aligned for mating at 2 berthing points.
- Placement positioning may be difficult by towing due to interference by obstructing modules.

Scheme C:
- Accurate positioning of conventional modules and nodal airlock elements may be difficult, particularly on rough, uneven sites.
- While conventional modules can have wheels, means for transferring/aligning nodal airlocks are unknown.

Scheme D:
- Accurate positioning of all 4 conventional modules to accommodate berthing interfaces may be difficult, particularly for rotational alignments of end circulation modules.
- Transport & positioning problems for nodal airlock elements are similar to Scheme C.

Forward, rotational & elevation alignments with 2 or more interfaces

Forward, rotational & leveling alignments with 2 or more interfaces

Forward, rotational & leveling alignments with 1–2 interfaces

FIGURE 29.14
Surface positioning.

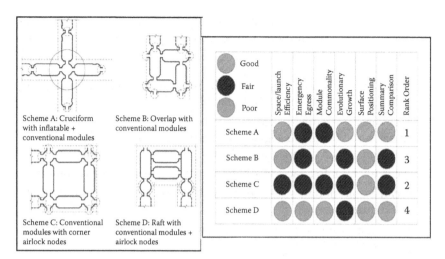

FIGURE 29.15
Summary observations.

Summary Conclusions

Guided by the configuration option comparisons, SICSA selected a reference design that combines use of a hard and inflatable hybrid combination of modules for special consideration (Figure 29.16).

FIGURE 29.16
Module hybrid combination approach.

This approach combines advantages of large interior volumes of inflatables with means to integrate utilities and equipment systems afforded by conventional modules. In addition:

- It allows conventional modules to be used to transport cargo/ equipment that can't be carried in inflatables.
- It enables conventional modules to be standardized for use as laboratories and for use as logistics carriers that can be used for lab/hab functions when emptied (excellent commonality functions).
- It can evolve into a racetrack pattern, offering dual egress capabilities.
- It can accommodate separate attachable airlocks, but potentially will not require them.
- It presents a small footprint to minimize site selection and preparation problems.

30

Settlement Site Selection and Exploration through Hierarchical Roving

Gregory Konesky

SGK Nanostructures, Inc.
Hampton Bays, New York

Gregory Konesky is co-founder and director of R&D of SGK Nanostructures, Inc., established to develop advanced materials based on cross-linked carbon nanotubes for high strength-to-weight, isotropic heat spreader, and hydrogen storage applications. He is also a member of the Board of Scientific Advisors for Bovie Medical Corp., where he is developing an atmospheric-discharge plasma beam for biomedical and defense/homeland security applications. He has authored over 49 articles on a wide range of subjects including teleoperation and telepresence, optical communications, astrobiology, and plasma decontamination. He is a member of the AIAA, IEEE, SPIE, OSA, MRS, ACS, AAAS, and NYAS, and received a BSEE from Polytechnic Institute of Brooklyn in 1977.

ABSTRACT   While orbital reconnaissance is useful for initial lunar settlement site selection, it is no substitute for in-situ ground truth, which could be easily accomplished with teleoperated roving. The Mars Exploration Rovers (MER) experience with Sojourner in 1997, and Spirit and Opportunity in 2004 show how the selection of rover size influences the capabilities and nature of their respective missions. Smaller rovers can more closely explore a complex environment and tend to be more nimble, but at a cost of reduced payload capability. Larger rovers enjoy enhanced payload capabilities, but at a cost of being somewhat ponderous and difficult to maneuver in complex environments. The best of these extremes can be optimized by a hierarchical roving approach wherein a large rover carries a hierarchy of smaller specialized rovers. The large rover, in addition to serving as a transport for the collection of smaller rovers to a remote deployment site, also acts as a communications relay link and power recharge source for the smaller rovers. In a typical operational scenario, the smaller specialized rovers are deployed at a given site and execute their collective mission. They are then recovered by the large carrier rover and transported to the next site. Some of the benefits of hierarchical roving include greater situational awareness, redundancy, spatially distributed capability, and the opportunity for self-rescue. The large rover can also serve as an anchor point for tethered roving, permitting smaller rovers to negotiate steep slopes or down-hole exploration that would otherwise be inaccessible to a wheeled vehicle.

Design and operating experience with a test model carrier vehicle containing three smaller specialized rovers is discussed, as are the design tradeoffs. In an alternate realization, a small fleet of rovers, teleoperated by Earth-bound users, is considered as a public outreach/commercial enterprise mission. The economics of an access fee structure are discussed for a baseline study of 10 rovers, each carrying 50 independently controllable camera heads, which permits a hierarchy of user interaction.

Introduction

Increasingly capable lunar orbital imaging will play a key role in anticipated settlement site selection when Man returns. However, it is no substitute for on-site ground truth reconnaissance.

The Moon enjoys a particular advantage in its being only approximately 1.3 light seconds away, permitting near real-time teleoperation on the lunar surface by Earth-bound users. In addition, there is a wealth of engineering data, and operational experience, using U.S. Apollo and Russian Luna rovers on the Moon.

For lunar settlement site selection and exploration, the physical size of the rover used has consequences in terms of tradeoffs of variables such as mission capabilities, mission duration and ability to deal with obstacles, instrument suite, and so on. The early rover exploration of Mars [1] perhaps best illustrates these tradeoffs. The first of the Mars Exploration Rovers (MER), *Sojourner*, in 1997, was necessarily limited both in size and scope of mission. It carried an Alpha Proton X-Ray Spectrometer (APXS), and associated deployment mechanism, and an imaging system. *Sojourner* lasted a few months in the Martian environment and traveled a few hundred meters until contact was lost.

By contrast, the subsequent *Spirit* and *Opportunity* MER missions, in 2004, carried a much larger array of instrumentation, including an APXS, Rock Abrasion Tool, and Microscopic Imager, which were all mounted on a common deployment mechanism, Stereoscopic Panoramic Cameras, Navigation and Hazard Avoidance Cameras, a Miniature Thermal Emission Spectrometer, a Mossbauer Spectrometer, and an array of magnets for magnetic particle detection. They each traveled several kilometers, collectively transmitted over 150,000 images back to Earth, and continue to operate today. Both *Sojourner* and *Opportunity* are shown in Figure 30.1 for a comparison of their relative sizes.

The greater size of *Spirit* and *Opportunity* provided for a larger array of instrumentation and continued endurance, when compared to the smaller *Sojourner*. However, *Sojourner*, by chance landed in a strewn rock field, where its small size was an enabling asset, allowing it to maneuver around and between this

FIGURE 30.1
Sojourner and Opportunity Mars Exploration Rovers (NASA/JPL photo).

rock field easily. If either *Opportunity* or *Spirit* had landed there, their large size would cause considerable difficulty in negotiating these obstacles. After the *Sojourner* experience, additional effort was given to reconsidering the proposed landing sites for *Opportunity* and *Spirit* [2, 3] to avoid a similar situation.

When heading into unknown terrain, the ideal situation would provide both the enhanced instrument payload capability and endurance of a large rover, with the nimbleness and agility of a small rover, especially in a complex environment. Hierarchical roving attempts to balance the tradeoffs between large and small rovers into a single combined system.

Hierarchical Roving

The operational experiences with the *Sojourner* and *Spirit/Opportunity* MERs serve to highlight the tradeoffs between small and large rovers respectively. Hierarchical roving is somewhat of a paradigm shift in the allocation and redistribution of instrumentation capability.

Rather than concentrating those capabilities in one large and ponderous rover, they are instead redistributed into a small fleet of specialized rovers which interact, both simultaneously and sequentially, to accomplish a specific mission. The limited endurance of the small specialized rover fleet is compensated for by using a large rover as a group carrier of the small rovers. They are collectively transported to a deployment site by the large rover. The large rover

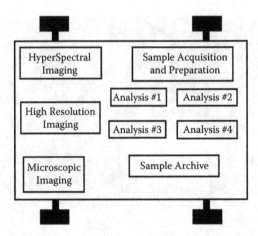

FIGURE 30.2A
Single large rover implementation.

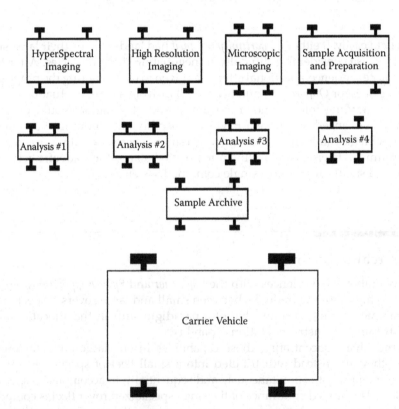

FIGURE 30.2B
Redistribution of capabilities among several small specialized rovers.

also serves as a power recharge source for the small rovers, and as a communications relay link. This redistribution in capability is illustrated in Figure 30.2.

Hierarchical roving presents several advantages [4] over the single large rover approach. They include a fail-soft redundancy where the loss of any one small rover only somewhat diminishes the overall mission capability, rather than ends it. Several small rovers can simultaneously and/or sequentially operate upon a larger area, increasing the rate of coverage of exploration as well as provide a greater situational awareness. This approach also allows for a potential self-rescue capability which is otherwise unavailable with a single large rover.

The fleet of small specialized rovers may operate collectively in an autonomous mode [5–7] or in a teleoperation mode [8], which is made feasible by the Moon's proximity. Some potential areas of specialization for the small rovers include imaging, sample acquisition and preparation, sample analysis, sample archiving, and manipulation. Imaging will always include sensors for navigation and obstacle avoidance, and will be common to all small rovers. However, additional imaging specialization would include long-distance high resolution telephoto imaging from an extended mast, with scout vehicles for path planning, hazard avoidance and so on. Hyperspectral imaging/microscopic imaging is another form of specialization to be used for identifying regions of interest to be subsequently visited by other specialized small rovers.

Once a region of interest is identified, a small rover specialized in sample acquisition and preparation provides a sample to one or more analysis-specialized small rovers, and then finally provides a sample to a sample

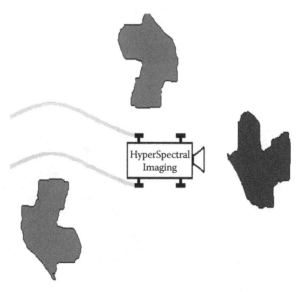

FIGURE 30.3A
Region of interest identification.

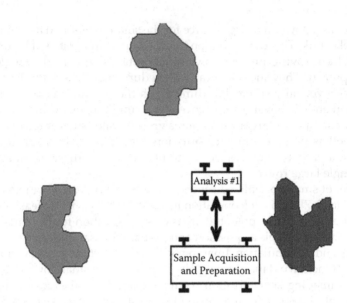

FIGURE 30.3B
Sample acquisition and preparation, handed to an analysis specialized rover.

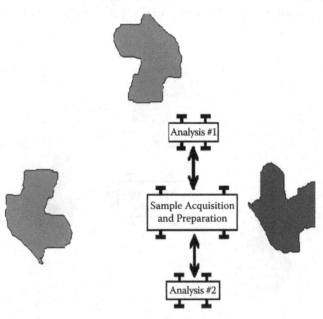

FIGURE 30.3C
Acquired and prepared sample shared by two analysis rovers.

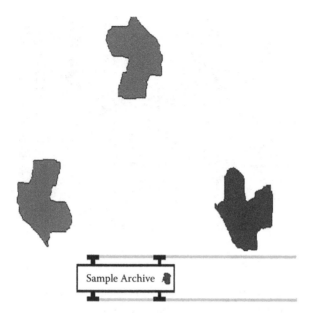

FIGURE 30.3D
Sample archived.

archive small rover. While this process is on-going, the imaging rover(s) are off identifying a new region of interest sites. The overall process is similar to an assembly line where sample site identification, acquisition and processing, analysis, and archiving all proceed simultaneously at different stages, in various sites. An example of this overall process is illustrated in Figure 30.3.

Once the collective mission of the fleet of specialized small rovers is complete, they reenter the large carrier rover and are transported as a group to the next deployment site. During this transport process, the batteries of the small rover fleet are also being recharged by the large rover, so they are ready to go upon arrival at the next mission site.

Test-Bed Prototype

In order to test some of these concepts, a test-bed prototype was constructed. Since the goal is to gain some experience with the basic concept, rather than any specific technological implementation, commercial off-the-shelf technology was used wherever possible. To that end, simple remote controlled hobbyist-scale vehicles provide readily available and capable vehicles and technologies, and at low cost.

FIGURE 30.4A
Carrier vehicle with remote steering and hazard avoidance cameras.

In the test bed prototype, carrying and deploying the small rovers was addressed by constructing a carrier bay on the large vehicle. The framework of the carrier bay also acts as a support structure for solar panels, totaling 66 Watts. Overall dimensions for this proof-of-principle prototype are 34" wide, 52" long, and 37" high. The maximum weight is 152 pounds, which includes up to 48 pounds of small rover payload. The interior dimensions of the carrier bay are 24" wide, 31" long, and 18" high. This basic construction sequence is shown in Figure 30.4.

The small specialized rovers exit and re-enter the large vehicle carrier bay through a folding ramp mechanism. The deployment sequence of this ramp is shown in Figure 30.5.

FIGURE 30.4B
Construction of the carrier bay.

FIGURE 30.4C
Addition of the solar panels and the remote control electronics.

Large-Scale Teleoperation as a Commercial Venture

The notion of large-scale hierarchical roving can be extended to a commercial venture which is potentially economically self supporting [9]. The functional hierarchy here is somewhat more simplified with a lunar lander acting as a deployment mechanism as well as a communications relay link between the rover fleet and Earth-bound users, as shown in Figure 30.6. While lunar settlement site selection and exploration may be the initial mission goal, the

FIGURE 30.5A
Ramp retracted.

FIGURE 30.5B
Ramp is being deployed.

addition of an array of independently controllable remote camera heads to each rover permits a potential public outreach add-on mission. An access fee structure is proposed that is also hierarchical, based on the level of Earth-bound user participation.

The highest level of user participation would be a remote vehicle driver, whose number is limited only by the size of the vehicle fleet. The next lower level would be the ability to remotely steer a given camera head on a given rover vehicle, which is termed an "active viewer." Since there are a large number of camera heads per vehicle, as illustrated in Figure 30.7, the total number of simultaneous remote users is much larger than the number of

FIGURE 30.5C
Ramp is fully deployed.

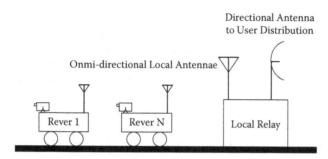

FIGURE 30.6
A fleet of small rovers linked to Earth-bound users through a local relay.

potential drivers. Finally, the lowest level of participation is a "tag along" viewer (or "passive viewer") which simply monitors any given camera head on any given vehicle, without the ability to affect the pointing direction of that camera head. Significantly, there are an unlimited number of potential passive viewer users.

Suppose access to these levels of interaction was available over a high speed internet connection, and access was allocated into 15 minute time slices. We consider a base-line study of a 10 rover vehicle fleet on the Moon, with each vehicle carrying 50 independently controllable camera heads, for a total of 500 simultaneous video feeds back to Earth. Additionally, a fee of $100 per 15 minute time slice is charged for remote vehicle drivers, $10 for active viewers, and $1 for passive viewers.

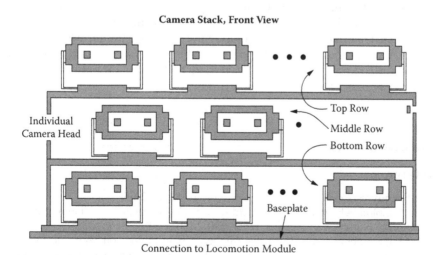

FIGURE 30.7
Illustration of the camera stack array carried by each rover vehicle per active channel from the Moon.

Assuming the entire venture is solar powered, operational availability occurs for at most about 14 Earth days per synodic month, but low Sun angle may limit this to 12 days. Since there are 13 synodic months per year, there will be 156 working Earth days per year. Given that a lunar day is in continuous sunlight 24 hours per Earth day, there are 3744 working hours per year, or 14,976 15-minute time slices

Using the proposed fee structure described previously, there will be approximately $15 million from rover vehicle driver revenue and almost $75 million from active viewers per year. Perhaps surprisingly, the largest potential revenue stream is from passive viewers, since there are an unlimited number of possible passive viewers per active video channel from the Moon. If, for example, there are 10 passive viewers per active viewer, about another $75 million will be added to the total revenue stream. If there are 100 passive viewers per active viewer, that number jumps to $750 million.

Conclusions

Hierarchical roving provides the possibility of combining the best attributes of large and small rovers into a single integrated system. It provides the ability to sense and sample the local environment from multiple mobile points simultaneously, while providing redundancy and fault tolerance so that the loss of any given small rover only diminishes the overall mission capability, rather than ends it.

A different form of hierarchical roving may be used to produce a potentially economically self-supporting exploration venture on the Moon through large-scale teleoperation and personal remote access to the surface of another world. An additional and perhaps greater benefit from this is the desire to significantly expand human presence there by the general public, the attendant consequences for space-related funding.

References

1. S. Squyres, *Roving Mars*, Hyperion, New York, NY, 2005.
2. M. Golombek, et al., "Assessment of Mars Exploration Rover landing predictions," Nature, Vol. 436, pgs. 1588–1590, 16 June 2006.
3. M. Golombek, et al., "Rock size-frequency distributions on Mars and implications for MER landing safety and operations," Journal of Geophysical Research, Vol. 108, (E12), pg 8086–8089, 2003.

4. G. Konesky, "Rovers within Rovers—a Hierarchical Approach," Proc. of *Instruments, Methods, and Missions for Astrobiology IX*, R. Hoover, G. Levin, and A. Rozanov, editors, SPIE Proc. Vol. 6309, pgs 63090F0-8, San Diego, CA, 2006.
5. B. Siciliano and K. Valavanis, "Multirobots and Cooperative Systems," *Control Problems in Robotics and Automation*, Springer, New York, NY, pgs 19–33, 1998.
6. A. Bicchi, A. Danesi, G. Dini, S. La Porta, L. Pallottino, I. Savino, and R. Schiavi, "Heterogeneous Wireless Multirobot System," IEEE Robotics and Automation, Vol. 15, No. 1, pgs 62–71, March 2008.
7. N. Michael, J. Fink, V. Kumar, "Experimental Testbed for Large Multirobot Teams," *ibid.*, pgs 53–62.
8. G. Konesky, "Group Telepresence," Proc. of *Intelligent Manufacturing*, B. Gopalakrishnan, A. Gunasekaran, and P. Orban, editors, SPIE Proc. Vol. 179–187, Providence, RI, 2003.
9. G. Konesky, "Large Scale Teleoperation on the Lunar Surface," Proc. of *Materials in Space–Science, Technology, and Exploration*, A. Hepp, J. Prahl, T. Keith, S. Bailey, and J. Fowler, editors, MRS Proc. Vol. 551, pgs 33–38, Boston, MA, 1998.

31

Integrated Lunar Transportation System

Jerome Pearson
Star Technology and Research, Inc.
Mount Pleasant, South Carolina

John C. Oldson
Star Technology and Research, Inc.
Mount Pleasant, South Carolina

Eugene M. Levin
Star Technology and Research, Inc.
Mount Pleasant, South Carolina

Harry Wykes
Star Technology and Research, Inc.
Mount Pleasant, South Carolina

Jerome Pearson is president of Star Technology and Research, Inc., a small business in Mount Pleasant, South Carolina. He was an engineer for NASA during Apollo and a branch chief for the Air Force during SDI, and invented the space elevator and the lunar space elevator. He has degrees in engineering and geology and is author of nearly 100 technical publications, including invited articles for *Encyclopaedia Britannica* and *New Scientist*. He has been interviewed many times on television about space elevators and global warming, and was featured in the Discovery Channel series, *Science of the Impossible*. He is a Fellow of the BIS, an Associate Fellow of AIAA and a member of the International Academy of Astronautics.

John C. Oldson is a senior engineer at Star Technology and Research, Inc. and resides in San Diego, California. He has an MS in engineering geosciences from UC-Berkeley, and worked for several years as a senior engineer for Energy Science Laboratories in San Diego. He has published numerous papers on non-eroding nozzles for chemical rockets, cooled composite nozzle design for VentureStar linear aerospike engine, and Stirling engine regenerator. He has done engineering work on advanced space propulsion concepts, including tethers, lunar space elevators, low-cost small launch vehicle systems, and climate control.

Eugene M. Levin, PhD, is senior scientist at Star Technology and Research, Inc. and a technical consultant on space tether dynamics in Minneapolis, Minnesota. He received his PhD from Moscow University, and worked for several years at the Institute for Mechanical Engineering Research at the Russian Academy of Sciences in Moscow. Since coming to the United States in 1993 and later becoming a U.S. citizen, he has written two definitive books on the dynamics of space tethers and Earth and lunar space elevators, and holds several patents on space tether concepts. He has consulted with NASA, the Air Force and Navy, and has presented seminars at numerous universities on space dynamics.

Harry Wykes is a technical consultant for Walt Disney Engineering and for Star Technology and Research, Inc. He has more than 30 years of experience in mechanical design, CAD drafting, and stereo lithography. He has designed electronic special effects, optics and robotics for Disney, and is an accomplished illustrator, painter, and photographer. He managed the studio and staff for The Brubaker Group in Los Angeles, and was in charge of conceptual and detailed design of products and vehicles for clients such as Learjet and North American Rockwell. He has published articles on the design of space habitats and lunar habitation modules.

ABSTRACT An integrated transportation system is proposed from the lunar poles to Earth orbit, using solar-powered electric vehicles on lunar tramways, highways, and a lunar space elevator. The system could transport large amounts of lunar resources to Earth orbit for construction, radiation shielding, and propellant depots, and could supply lunar equatorial, polar, and mining bases with manufactured items. We present a system for lunar surface transport using "cars, trucks, and trains," and the infrastructure of "roads, highways, and tramways," connecting with the lunar space elevator for transport to Earth orbit. The Apollo Lunar Rovers demonstrated a battery-powered range of nearly 50 kilometers, but they also uncovered the problems of lunar dust. For building dustless highways, it appears particularly attractive to create paved roads by using microwaves to sinter lunar dust into a hard surface. For tramways, tall towers can support high-strength ribbons that carry cable cars over the lunar craters; the ribbon might even be fabricated from lunar materials. We address the power and energy storage requirements for lunar transportation vehicles, the design and effectiveness of lunar tramways, and the materials requirements for the support ribbons of lunar tramways and lunar space elevators.

Introduction

NASA is implementing a plan for a return to the Moon, which will build on and expand the capabilities demonstrated during the Apollo landings. The plan includes long-duration lunar stays, lunar outposts and bases, and exploitation of lunar resources on the Moon and in Earth orbit.[1] Because there are apparently deposits of water ice in shadowed craters near the lunar poles, and extensive areas of lunar regolith deposits of useful minerals in the lunar maria nearer the lunar equator, it will be necessary to create an integrated lunar transportation system to connect these locations with each other and with locations in Earth orbit. Because of the large delta-V requirements for carrying rocket fuel from the Earth's surface all the way to the Moon, we examined alternative transportation systems that do not use rockets, but do use indigenous lunar materials. The system we propose here is based on presentations at the Rutgers Lunar Settlements Symposium in 2007,[2] the Moonbase conference in Venice, Italy in 2005[3] and the final report of a study for NIAC in 2005.[4]

The integrated lunar transportation system consists of a lunar space elevator (LSE) balanced about the L1 Lagrangian point and extending directly down to the lunar equator; an elevated tramway, using the same composite ribbon as the space elevator, extending to the lunar south pole; and robotic vehicles that move along this transportation system by solar power and efficient energy storage to operate through the lunar night. As part of the process of building the tramway suspended on towers located on mountain tops and crater rims, highways can also be created for robotic vehicles (and perhaps even manned vehicles) to move the 2700 km between the lunar equator and the poles. The integrated transportation system is shown schematically in Figure 31.1.

This system is designed to transport lunar polar ice over the tramway, up the space elevator, and from there into high Earth orbit, where it can be used for refueling hydrogen/oxygen rocket engines for launches to all over the solar system. The flow in the opposite direction will be supplies and manufactured goods from Earth orbit to the lunar bases and the polar mining stations. The components of the system are described in the following sections.

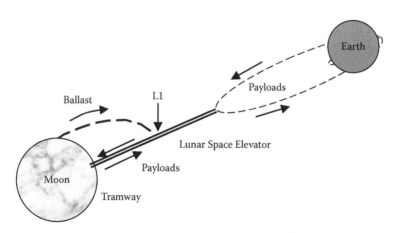

FIGURE 31.1
Integrated lunar transportation system.

System Description

The key component of the system is the lunar space elevator, shown in Figure 31.2, first published by Pearson[5,6] in 1977 and independently by Artsutanov[7] in 1979. (Tsander, the Russian visionary, looked at a lunar space elevator even earlier.[8]) The LSE is an extension of the concept of the Earth space elevator, invented by Artsutanov[9] (1960) and independently by Pearson (1975).[10]

The classical Earth space elevator is essentially a geostationary satellite that is elongated until the lower end touches the Earth at the equator, and the upper end extends to an arbitrary distance and ends in a counterweight that keeps the entire structure in balance about the geostationary orbit altitude. Unfortunately, the Earth space elevator requires materials as strong as carbon nanotubes, because of the Earth's high gravitational field. The lunar space elevator can be constructed from current composite materials, but it is more complicated to analyze, because it can only be balanced about the L1 or L2 unstable Lagrangian points of the Earth-Moon-spacecraft three-body system. These points are about one-sixth of the Earth-Moon distance from the lunar surface. The lunar transportation system uses the L1 lunar space elevator with ribbons of available high-strength composites. As shown in the figure, the lunar space elevator could be curved to touch down at points other than the lunar equator.

Figure 31.3 shows an artist's concept of the transportation system for carrying payloads of water from the lunar poles to the lunar space elevator, and from there to Earth orbit. Robotic vehicles with electric motors powered by large solar arrays and energy storage, like the one shown here, would move from the pole along the tramway, climb up the LSE, be released at the top,

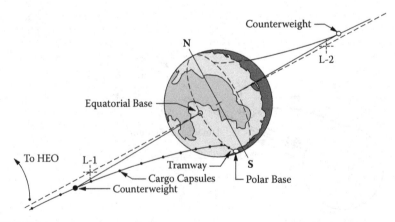

FIGURE 31.2
Lunar space elevators.

FIGURE 31.3
The lunar space elevator and tramway system.

and continue to Earth orbit using electric propulsion. The vehicles could drop lunar resources in Earth orbit, pick up supplies for the lunar polar station, and return by rendezvous with the top of the elevator and continue down to the tramway. The second major element of the system is the lunar tramway connecting the base of the LSE on the equator with the ice mines at the poles. The lunar tramway is envisioned to run over 2700 km from the equator to the poles, connecting various maria mineral deposits and regolith mining sites as well as the water ice mines at the south pole. There may be ice deposits at the north pole as well, and a second tramway could be constructed in that direction.

The tramway needs to be constructed of the same high-strength composite ribbons as the space elevator, suspended from towers located on lunar mountains and crater rims. The high strength allows long spans of scores to hundreds of kilometers, minimizing the number of support towers required. For maximum span, the support towers could be located on the rims of craters and on the tops of mountains. The tramway system is shown schematically in Figure 31.4, with spans extending up to scores of kilometers between towers in the low lunar gravity field.

The tramway terminates at the lunar polar mining camp, where the same kind of high-strength ribbons can support mining rigs suspended over shadowed polar craters, as sketched in Figure 31.5. The water is mined in the

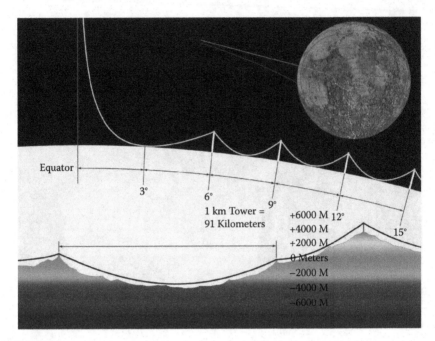

FIGURE 31.4
The lunar tramway extends from equator to pole.

crater at a temperature of less than 100K (~10^{-9} Torr), and is transported at a temperature of about 270 K (~2 Torr). Such a facility, powered by 110 kW of continuous power from a nearby sunlit area covered with solar cells, could collect about 200,000 kg of water per year. This would compose the bulk of the cargo carried by the tramway capsules to the lunar space elevator and on to Earth orbit.

The next sections discuss these components of the integrated transportation system in more detail, and provide some numbers on the system design parameters and operation.

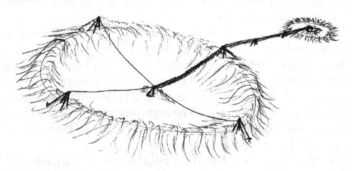

FIGURE 31.5
In-Situ lunar resource mining at a polar crater.

The Lunar Space Elevator Component

Lunar and Mars space elevators are much easier than Earth space elevators, as shown in Figure 31.6. The basic parameter is the specific strength of the material, $h = \sigma \cdot \rho / g_0$, the limiting stress times the density divided by Earth's gravity. The term h has the dimensions of length (the longest cable that can be suspended in 1 g), and has been called the "characteristic height," or "breaking height." Typical metals like aluminum and steel have h = 10–50 km, high-strength composites like M5 are about 500 km, and carbon nanotubes are about 2200 km. The design parameter for a space elevator is the area taper ratio in cross-sectional area from the maximum at the balance point to the minimum at the surface. The minimum cross-sectional area is determined by the required lifting capacity at the base, which is the stress limit divided by the area. The lunar space elevator has about 1% of the specific stress requirements of the Earth space elevator.

The space elevator must be constructed of extremely strong, lightweight materials, to support its weight over the tens of thousands of kilometers of length; even then, for minimum mass it must be tapered exponentially as a function of the planet's gravity field and the strength/density of the building material. The table below shows some candidate materials for lunar space elevators, with density, stress limit, and the breaking height. Lunar space elevators require much lower material strengths than the Earth space elevator, which will require carbon nanotubes (shown in Table 31.1 for comparison). All these materials, save the carbon nanotubes, are available now.

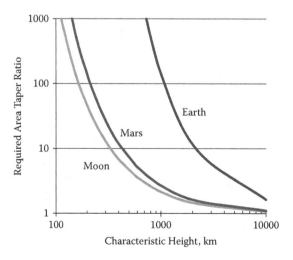

FIGURE 31.6
Space elevator taper ratios.

TABLE 31.1

Candidate Materials for LSE Compared with Carbon Nanotubes

Material	Density ρ, kg/m³	Stress Limit σ, GPa	Breaking height $h = \sigma/\rho g$, km
SWCN*	2266	50	2200
T1000G†	1810	6.4	361
Zylon PBO‡	1560	5.8	379
Spectra 2000¶	970	3.0	316
M5**	1700	5.7 (9.5)	342 (570)
Kevlar 49††	1440	3.6	255

* Single-wall carbon nanotubes (laboratory measurements)
† Toray Carbon fiber
‡ Aramid, Ltd. Polybenzoxazole fiber
¶ Honeywell extended chain polyethylene fiber
** Magellan honeycomb polymer (with planned values)
†† DuPont Aramid fiber

The design of the lunar space elevator ribbon can be made more robust and fail-safe by using multiple ribbons with alternate load paths, after Forward and Hoyt.[11] The concept is shown in Figure 31.7, with a table of the required safety factor vs. the number of ribbons. The lifetime of the LSE can be estimated from the mean time between meteor cuts: T, yrs = 6 $h^{2.6}$/L, where h is the ribbon width in mm and L is the length in km.

Once the minimum base area is set and the taper ratio is known, the total mass of the space elevator can be calculated. For a modern high-strength composite like the Magellan M5, the total system mass is shown in Figure 31.8. The mass is plotted vs. the length of the space elevator, from the surface to a point beyond L1, where it is terminated by a counterweight that keeps it in balance while it lifts loads at the surface. The CW must equal in weight the entire length of the ribbon below L1; its weight is zero at L1, where it is balanced in orbit, and rises linearly with distance above L1. If the ribbon extends to infinity, no counterweight is required. This gives the very interesting paradox that the longer the space elevator, the less the total weight,

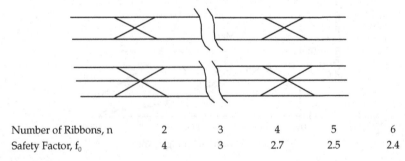

Number of Ribbons, n	2	3	4	5	6
Safety Factor, f_0	4	3	2.7	2.5	2.4

FIGURE 31.7
Fail-safe ribbon design.

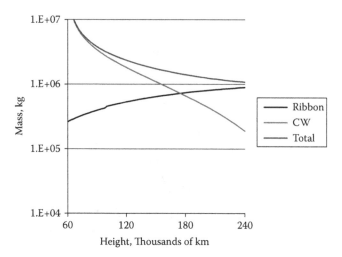

FIGURE 31.8
Mass of the LSE ribbon and counterweight for different lengths.

although the amount of high-strength ribbon goes up. The counterweight can be inert lunar regolith material, or even a space station.

The analysis of lunar space elevators by Levin[12] is the basis for these results. The L1 point is only about 58,000 km from the center of the Moon, but there are two advantages to making the lunar space elevator extend far beyond this point. The first is the drastic reduction in the mass of the CW, and therefore the total mass that must be lifted from the Moon to L1 for its construction. The second is the fact that payloads released from the top of the LSE drop into elliptical Earth orbits whose perigees will be lower the longer the LSE is. Figure 31.9 shows this effect. If the LSE reaches 130,000 km, payloads will have perigees at GEO; 180,000 km puts them in MEO, and releasing at 240,000 km would cause them to enter the atmosphere. Because the tram vehicles have plenty of electric power, they could use electrodynamic propulsion to circularize their orbits if the perigee is below about 8000 km radius. Conversely, electrodynamic propulsion vehicles leaving LEO to carry payloads to the Moon could propel themselves to the top of the LSE without rockets if the top were about 210,000 km from the Moon. For these important reasons, the lunar space elevator should be perhaps up to 200,000 km long; this will require 400,000 kg of ballast material, typically lunar regolith, in addition to the 800,000 kg of high-strength composite. The conceptually simplest way to send this material to the L1 position is with a mass driver.

Arthur Clarke[13] suggested electromagnetic launching in 1950, and the concept of the electromagnetic mass driver was revived[14] in the 1970s. However, this design requires high power, high precision, and lots of material to be shipped from Earth for its construction. A simpler system would be a rotating sling launcher after Heppenheimer,[15] Levin,[12] and Landis,[16] using the same high-strength material as the lunar space elevator. Emplacing such

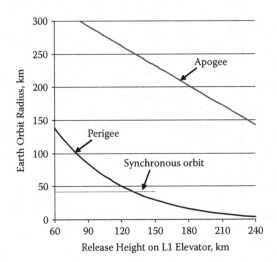

FIGURE 31.9
Earth orbit after release.

a sling near the lunar North Pole on a high point in nearly constant sunlight would allow the use of solar power to rotate the system, with dual payloads at the ends. The concept is shown in Figure 31.10, based on the Levin analysis. The power station and sling tower are on a mountain top, and the cables are extended as the rotational velocity rises to the launch velocity.

For 100 kW of total power, the tip velocity could reach 2.38 km/sec, enough to reach escape (or L1), at a length of 236 km, and the system could launch 3 tons per day to L1. The lunar sling could launch both regolith counterweight and high-strength ribbon material into L1, from which the lunar space elevator could be extended until the bottom touched the ground. The LSE could then support climbers to lift additional materials.

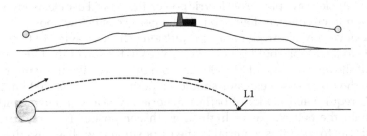

Type	h, km	r, km	V_{tip}, km/s	a_{tip}, g's	P, kW	Tons/day
Low Orbit	4	118	1.68	2.4	100	3
Escape	4	236	2.38	2.4	100	3

FIGURE 31.10
Lunar sling launcher.

Lunar Tramway and Cable Cars

The lunar tramway component must extend from the equator to the pole, about 2700 km. There is a trade-off between the height of the towers and the span, as shown in Figure 31.11. The ground clearance height is shown vs. the latitude span for support towers of 1, 2, and 3 km height. The 1-km towers allow spans over level ground of at least 3 degrees, or about 90 km. This would require only 30 towers between the equator and the pole. If the tramway takes advantage of topography, it would require even fewer towers. Towers of 2 or 3 km would reduce the number required to 20 or even 16, but their greater height might cause problems in construction and stability. Building 1-km towers on the Moon is like building 160-meter towers on the Earth, and they can be very light if made of modern composite designs, such as that shown in Figure 31.12.

One method for reducing the overall cost of the lunar space elevator is to use *in situ* lunar materials to make fibers that are strong enough to reinforce the initial ribbon. This could greatly increase the carrying capacity of the LSE, and also greatly reduce the amount of material that must be lifted out of the Earth's gravity well.

Lunar aluminum, silicon, iron and titanium are abundant. Aluminum has a relatively low density, and can be used to create high strength fibers. Its strongest form seems to be sapphire, which can be grown as long single crystals or whiskers. The processes involved might even benefit from the microgravity environment at L1. Perhaps we could grow continuous crystal strands that could go directly into the ribbon assembler. Sapphire whiskers are almost as strong as graphite whiskers, although they are more than twice as heavy.

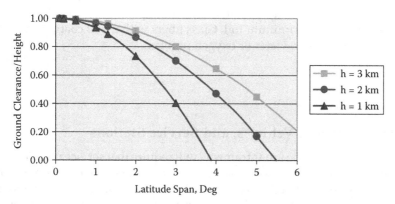

FIGURE 31.11
Tower span vs. height.

FIGURE 31.12
Lightweight composite towers.

Another material which compares favorably is quartz whisker. Silicon is plentiful and if we can generate whiskers in space they would be many times stronger than glass fibers made from the same element. Fibers in a metal matrix are also currently popular, and an application might be sapphire whiskers in glassy aluminum foil. Glass fibers with metal coatings might be used, since there is no water or oxygen problem.

Lunar Highways, Vehicles, and Service Stations

The construction of the lunar tramway will require that service vehicles travel over the length of the system, and they will need a good surface on which to travel. It is possible that the vehicles that erect the support towers could also smooth or pave a road on which ground vehicles could travel. Taylor and Meek of UT Knoxville developed a method[17] for using microwaves to sinter

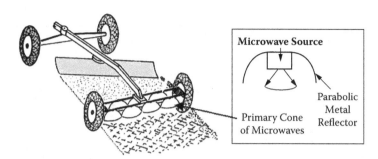

FIGURE 31.13
Lunar microwave paving machine.

lunar regolith into a hard, smooth surface like a paved road. Figure 31.13 shows a sketch of their microwave paving machine. It would have two sets of magnetrons that can be set to various microwave frequencies and power in order to effectively sinter/melt the lunar soil.

The first set would sinter the regolith to a depth of about half a meter, and the second set would melt the top 3–5 cm to create a hard, smooth road surface. The microwave process would release most of the solar wind particles imbedded in the regolith, notably hydrogen, helium, carbon, and nitrogen, which might be captured for other uses.

This could be very valuable in solving the problem of lunar dust, which has proven to be very difficult to deal with. In building the lunar tramway, we could also end up with a lunar highway for surface vehicles.

This leads into the problem of powering vehicles, whether tramway capsules or robotic ground vehicles, or eventually manned vehicles, over long distances during the lunar night. The Apollo rovers used silver-zinc batteries with storage efficiency of about 130 W·hr/kg, and achieved ranges over "dirt roads" of about 50 km.[18] More efficient storage systems, whether advanced rechargeable batteries such as lithium ion at 350 W·hr/kg, or hydrogen/oxygen fuel cells at 650 W·hr/kg, could raise the range considerably, as shown in Figure 31.14.

More importantly, the rolling resistance of wheels on the lunar regolith can be greatly reduced by preparing the surface. The hatched line on Figure 31.14 represents the Apollo lunar roving vehicles—flexible aluminum tires on unimproved regolith. The dashed line represents the rolling resistance of typical tires on concrete, which might be achieved by treating the regolith with microwaves. Finally, the solid line represents the ranges that could be achieved with metal wheels and rails, like typical railroads on Earth. These three cases represent coefficients of rolling resistance of 0.12, 0.015, and 0.005, respectively.

The best we can do with unimproved regolith is perhaps 300 km of range. That would require about 9 service centers spread over the 2700 km from the pole to the equator. By microwave treatment of the regolith, we might achieve up to 2000 km of range, requiring only one service station, in the middle of the trip. If we could create a lunar railroad, however, with solid metal-wheeled

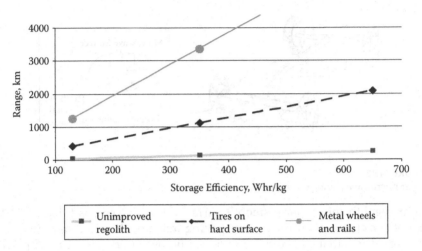

FIGURE 31.14
Surface vehicle range vs. storage efficiency.

vehicles and metal rails, then we could achieve pole-to-equator range during the lunar night, and not require service stops along the way at all, except perhaps as a backup for vehicle failures. The capsules traveling on the lunar tramway should be roughly equivalent to metal wheels on metal rail, and would not need to stop during the long lunar night. At a speed of 30 m/s, the tramway capsules could cover the entire distance in just 25 hours.

Even if service stations are required for the ground vehicles, they could be automated to charge enough batteries by solar power during the lunar day to provide charged batteries for many vehicles during each night. If required, an occasional vehicle could drop off extra batteries, leaving sufficient numbers of charged batteries each lunar night. With parallel tracks, or even occasional "passing tracks," the system could provide continuous two-way traffic.

Advanced System Operations

The system of robotic vehicles on tramway, roadway, and space elevator might be able to function nearly autonomously, carrying multiple vehicles both ways. Robotic attendants at the equator, pole, and central service station might be all that is required to keep operations smooth.

There are some possible advanced methods to improve the operation of the lunar space elevator. Building a tall tower at the base of the space elevator, creating a space elevator partly in tension and partly in compression[19] would reduce the taper ratio required, resulting in much lower mass and less counterweight. Curving the lower end of the elevator ribbon to touch down away from the equator is not an improvement; although it lowers the number of towers required to reach the pole, it drastically reduces the payload capacity of the elevator ribbon. Reaching just 15 degrees of latitude reduces the carrying capacity by 25%.

Finally, it may be possible eventually to use the tramway ribbon itself to carry electrical power along its length, allowing capsules to draw power continuously through the night. Other, less likely, methods are to beam power to spots on the tramway from solar power stations at the equator, the poles, and perhaps even in low lunar orbit.

Conclusions

The integrated lunar transportation system is a complete non-rocket transportation system for carrying lunar resources to Earth orbit and for carrying manufactured goods from Earth orbit to the lunar poles. It depends on high power from solar arrays, electric motors for propulsion along the tramway and space elevator, and electric rockets for the free-flight leg between the top of the lunar space elevator and Earth orbit. If the Earth space elevator is ever built, the system could provide two-way cargo transportation from the surface of the Earth to the surface of the Moon.

Acknowledgments

The authors are indebted to Klaus Heiss for support of the lunar development initiative and for information on *in situ* lunar resource recovery, and to Paul van Susante for information on tramway systems over the world.

References

1. Moore, C., "Exploration Systems Research and Technology," Presented at the NASA Institute for Advanced Concepts Fellows Meeting, 16 March 2005.
2. Pearson, J., Wykes, H., Levin, E., Oldson, J., Heiss, K., and van Susante P., "Lunar Frontier Transport System," Rutgers Symposium on Lunar Settlements, New Brunswick, NJ, 3–8 June 2007.
3. Pearson, J., Levin, E., Oldson, J., and Wykes, H., "Lunar Space Elevators for Cis-Lunar Transportation," presented at Moonbase: A Challenge for Humanity, Venice, Italy, 26–27 May 2005.

4. Pearson, J., Levin, E. M., Oldson, J., and Wykes, H., Lunar Space Elevators for Cislunar Space Development, Final Technical Report on NIAC Research Sub-Award 07605-003-034, May 2005.

5. Pearson, J., "Anchored Lunar Satellites for Cis-Lunar Communication and Transportation," 1st European L-5 Conference, London, 20 Sep 1977.

6. Pearson, Jerome, "Anchored Lunar Satellites for Cislunar Transportation and Communication," *Journal of the Astronautical Sciences*, Vol. XXVII, No. 1, pp. 39–62, Jan/Mar 1979.

7. Artsutanov, Yuri, "The Earth-to-Moon Highway," (in Russian), *Technika-Molodyegi*, No. 4, 21, 35, 1979.

8. Tsander, F., Selected Papers (in Russian), Zinatne, Riga, 1978.

9. Artsutanov, Yuri, "V Kosmos na Electrovoze (in Russian, Into Space on a Train)," *Komsomolskaya Pravda*, July 31 1960.

10. Pearson, Jerome, "The Orbital Tower: A Spacecraft Launcher Using the Earth's Rotational Energy," *Acta Astronautica*, Vol. 2, pp. 785–799, Sep/Oct 1975.

11. Forward, R., and Hoyt, R., "Failsafe Multiline Hoytether Lifetimes," AIAA Paper 95-2890, 31st AIAA/SAE/ASME/ASEE Joint Propulsion Conference, San Diego, CA, July 1995.

12. Levin, E., "Colonizing the Moon," Chapter 7 of *Dynamic Analysis of Space Tether Missions*, Advances in the Astronautical Sciences, Vol. 126, April 2007.

13. Clarke, A., "Electromagnetic Launching as a Major Contributor to Spaceflight," *JBIS*, Vol. 9, No. 6, pp. 261–7, November 1950.

14. O'Neill, G., and Kolm, H., "Mass Driver for Lunar Transport and as a Reaction Engine," *Journal of the Astronautical Sciences*, Vol. 15, No. 4 Jan/Mar 1976.

15. Hoppenheimer, T. A., "Rotary Pellet Launcher," in NASA *Space Settlements: A Design Study*, R. D. Johnson and Holbrow, Editors, NASA SP-4113, pp. 130–132, 1977.

16. Landis, G., "A Lunar Sling Launcher," *JBIS*, Vol. 58, No. 9/10, pp. 294–297, 16 March 2005.

17. Moore, C., "Exploration Systems Research and Technology," Proceedings of the International Lunar Conference 2003/ILEWG5, American Astronautical Society 108 (Sciences & Technology Series), 109–123, 2004.

18. Anon., "Lunar Roving Vehicle," NASM, http://www.nasm.si.edu/collections/imagery/apollo/lrv/lrv.htm

19. Landis, G., and Cafarelli, C., "The Tsiolkovski Tower Reexamined," IAF-95-V.4.07, 46th IAF Congress, Oslo Norway, 1995, *J. British Interplanetary Society*, Vol. 52, 175–180, 1999.

Section VI

Lunar Bases

32

Lunar Base Site Preparation

Florian Ruess
HE2 – Habitats for Extreme Environments
Bremen, Germany

Benjamin Braun
HE2 – Habitats for Extreme Environments
Stuttgart, Germany

Kris Zacny
Director, Drillling and Excavation Systems
Honeybee Robotics Spacecraft Mechanisms Corporation
New York, New York

Martina Pinni
Architect, ESA-EAC Crew Instructor
Koeln, Germany

Florian Ruess, Dipl.-Ing, University of Stuttgart and University of Calgary civil engineering (1988–2004); research on lunar bases at Rutgers University as part of the master thesis (2004); structural engineer with BDE engineering in Stuttgart with focus on bridge design and engineering (2004–2006), and since 2006 structural engineer at Airbus Bremen, Wing High Lift Design HE2–Habitats for Extreme Environments.

Kris Zacny, PhD, is a director of drilling and excavation systems at Honeybee Robotics Spacecraft Mechanisms Corporation, and focuses on autonomous terrestrial and extraterrestrial drilling, excavation, and geotechnical systems, as well as sample acquisition, transfer and processing technologies. In his previous capacity as an engineer in the South African mines, Dr. Zacny managed numerous mining projects and production sections. Dr. Zacny received his PhD in 2005 from the University of California, Berkeley. During his graduate work on drilling mechanics in planetary environments he discovered a sublimation effect that has tremendous impact on drilling performance on many extraterrestrial bodies. He received a B.Sc. in mechanical engineering from the University of Cape Town in 1997 and an ME in petroleum engineering from the University of California, Berkeley in 2001. He participated in two Arctic expeditions (Ellesmere Island, 2004 and Devon Island, 2006) where he performed extensive drilling tests in the Arctic permafrost and in ice. Dr. Zacny has over 100 publications, is a member of numerous committees, has been invited to speak at national and international meetings, reviewed scientific papers for various scientific journals, chaired a number of sessions and symposiums and recently completed an edited book, *Drilling in Extreme Environments: Penetration and Sampling on Earth and Other Planets* (J. Wiley & Sons). Dr. Zacny is executive committee member of the American Society of Civil Engineers Aerospace Division.

Martina Pinni is a registered architect with an interest in high-tech buildings, extreme environments, and spinoff applications of space technology in building construction. Her thesis in architecture focused on an interior design for a space habitation module (2000). Her professional experience includes 6 years of technical and commercial positions in façade engineering, architectural design, and research (2001–2007). Teaching experience includes three semesters as faculty assistant of building technology at the University IUAV of Venice (2006–2007), and current position in astronaut training—structures, thermal control, life support system, emergencies—at ESA-EAC Cologne for the International Space Station project (2008–present). She earned a *Laurea* (M. Arch.) from the University IUAV of Venice (2000), an MS from the University of Houston, Space Architecture (2004), and attended the ISU SSP'07 at BeiHang University in Beijing (2007).

ABSTRACT   The construction of a base on the Moon will require excavation and movement of large quantities of lunar soil, the regolith. This paper focuses on the initial site preparation tasks that need to be performed. Excavation technologies that control dust generation and mitigation are highlighted as being critical for lunar base operations.

Possible site preparation, excavation, hauling and dust control equipment designs are reviewed, starting with designs based on terrestrial "traditional" construction equipment. Then, innovative ideas that try to overcome the limitations of traditional approaches, among them sintering of regolith, vibratory and pneumatic excavation, are discussed. Efforts to evaluate different equipment designs are on the way but need to address the latest technology developments.

In a qualitative approach, the pneumatic and vibratory excavation methods are identified to be very promising technologies with great potential for lightweight, robust and reliable yet high-performance designs. Such technologies could help to significantly reduce Lunar Base construction cost and risk.

Introduction

The capability to excavate and move large quantities of lunar soil and rocks will be required to establish and operate a lunar base for several reasons (Eckart 2006, Ruess et al. 2006):

- ground preparation for the placement of habitation modules and other infrastructure elements
- clearing the immediate lunar base area from dust
- covering habitation modules with regolith for insulation, radiation and micrometeoroid protection
- construction of roads and launch pads
- mining and extraction of lunar resources

Initial Excavation Tasks to Establish a Lunar Base

Especially the first three tasks will have to be performed at the very start of establishing a base on the Moon.

In particular dust control is generally recognized as the most critical issue for lunar operations. The complexity of this matter was summed up by Eugene Cernan, Apollo 17 astronaut: "I think dust is probably one of our greatest inhibitors to a nominal operation on the Moon. I think we can overcome other physiological or physical or mechanical problems except dust. Dust adheres" . . . "to everything no matter what kind of material, whether it be skin, suit material, metal, no matter what it be and its restrictive friction-like action to everything it gets on. For instance, the simple, large tolerance, mechanical devices on the rover began to show the effects of dust as the EVAs (Extra Vehicular Activities) went on" (Gaier 2005).

Therefore it has been proposed to put the regolith for radiation & micrometeoroid protection in bags in order to be able to control dust mitigation (Ruess et al. 2008, Smithers et al. 2007, Lindsey 2003) among other advantages (see Figure 32.1).

Combined with clearing the immediate lunar base area from the fine top layers of regolith, which are the main contributor to dust generation, the dust problem could be minimized at least for the immediate lunar base site itself and all surrounding operations.

Lunar construction equipment designs should also consider the dust problem in terms of their own maintainability and robustness. Few moving and easily cleanable and replaceable parts will be critical for successful designs.

The immediate tasks for the lunar base site preparation equipment to achieve dust control, prepare the construction ground and excavate regolith include (Gies 1994):

- Digging (scooping, scraping, dozing, ripping, trenching)
- Hauling (loading, transport, unloading)
- Placing (spreading, piling, compacting, shaping, grading)

Other ideas call for sintering the regolith surface to control dust and construct roads and launch pads. This can be achieved for example by using microwaves as suggested by Taylor and Meek (2005).

Site Preparation Equipment

Terrestrial equipment that performs similar functions is often used as a starting point for the design of lunar construction equipment.

Among the terrestrial construction equipment proposed for lunar applications are:

- Backhoe
- (Front-end) loader
- Bulldozer
- Scraper
- Grader
- (Bucket) excavator
- (Dump) truck
- Belt conveyor

Figure 32.1 to Figure 32.14 show examples of some of these machines on Earth as well as a corresponding lunar application.

In June 2008, NASA tested the prototype of a mobile lunar platform, named CHARIOT, which may be equipped with a lightweight bulldozer blade (LANCE—Lunar Attachment Node for Construction Excavation) to perform site preparation and clearing of areas where a lunar outpost could be deployed (see Figure 32.13 and Figure 32.14). For use as a bulldozer alone, this vehicle looks much too massive but if surface leveling is only one function among many, the concept might prove suitable. For example other exchangeable nodes could be developed with the construction equipment functions listed above. However, the design of the CHARIOT leaves many questions unanswered: It is an extremely complex machine with many moving parts. How such a design will perform under lunar conditions, especially

FIGURE 32.1
Excavator in bucket-wheel configuration—terrestrial application.

FIGURE 32.2
Excavator in bucket-wheel configuration—lunar application prototype (van Susante 2008).

FIGURE 32.3
Excavator in bucket-ladder configuration—terrestrial application.

FIGURE 32.4
Excavator in bucket-ladder configuration—lunar application prototype (van Susante 2008).

FIGURE 32.5
Terrestrial clamshell excavator.

FIGURE 32.6
Lunar clamshell rover (courtesy: DigitalSpaces).

FIGURE 32.7
Terrestrial bulldozer (courtesy: Caterpillar).

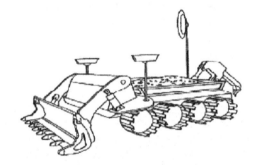

FIGURE 32.8
Lunar haul-dump vehicle with bulldozing capabilities (Podnieks and Siekmeier 1992).

FIGURE 32.9
Terrestrial front loader.

FIGURE 32.10
Lunar front loader vehicle (courtesy: DigitalSpaces).

FIGURE 32.11
Terrestrial "dump truck" hauler (courtesy: Caterpillar).

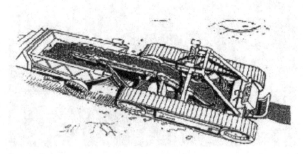

FIGURE 32.12
Lunar miner and hauler (courtesy: NASA).

FIGURE 32.13
CHARIOT—The prototype mobile lunar platform (courtesy: NASA).

in combination with the lunar dust, remains to be proven. The current design lacks even the most basic precautions like fenders for example.

As already outlined with the NASA CHARIOT design, it seems to be desirable to try and include many functions into one vehicle. Mueller and King (2008) used this guideline for their design of a small, tele-operated, multipurpose lunar excavator (see Figure 32.15).

FIGURE 32.14
CHARIOT equipped with lightweight bulldozer blade LANCE (courtesy: NASA).

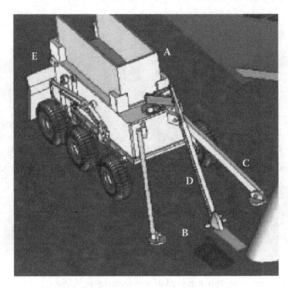

FIGURE 32.15
A multi-purpose excavator: A: Transport box, B: Bucket, C: Outrigger, D: Manipulator, E: Blade
(Mueller and King 2008).

It is obvious that although the functions to be performed on the Moon are
similar to those performed on Earth, the equipment for lunar surface operations will be different due to the constraints imposed by the lunar environment. Considerable modification of terrestrial designs will be needed to suit
the conditions on the Moon (Boles 1992, Toklu 2003).

For example equipment will most likely be electrically operated since diesel engines don't work in a hard vacuum environment.

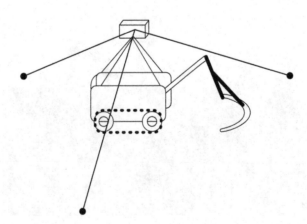

FIGURE 32.16
Lunar excavation equipment pulled down by cables (Toklu 2003).

Also low mass of the equipment is critical due to cost constraints. Bernold (1991) reported about efforts to study the unique problems related to digging and trenching on the Moon. All the common excavation technologies used on Earth depend on the effect of gravitational acceleration that turns mass into forces that are needed to cut, scoop, and move soil (Bernold 1991). Boles and Connolly (1996) reported that the mass of "terrestrial design" lunar excavators will only be half effective hinting that such designs will come with prohibitively high masses to be transported to the Moon (Boles and Connolly 1996). None of the above lunar applications in Figures 32.2 to 32.13 seems to answer the mass question. All designs look like mere scaled-down copies of their terrestrial "brothers."

Alternative approaches to really tackle this problem include:

- pulling down the machines with cables (Toklu 2003, see Figure 32.16)
- using regolith as "ballast" to increase the mass of excavation equipment in-situ
- sintering of soil using microwaves (Taylor and Meek 2007)
- pneumatic excavation systems (Zacny et al. 2008a, see Figures 32.17 to 32.18)
- vibratory excavators (Zacny et al. 2008b, see Figure 32.19 and Figure 32.20)
- loosening of regolith using explosives prior to excavation (Dick et al. 1992)

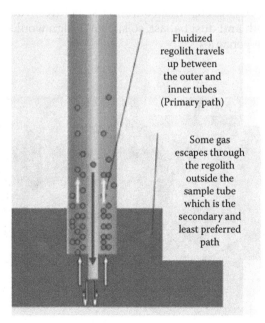

Fluidized regolith travels up between the outer and inner tubes (Primary path)

Some gas escapes through the regolith outside the sample tube which is the secondary and least preferred path

FIGURE 32.17
Pneumatic excavator: principal gas and regolith flow (Zacny et al. 2008a).

FIGURE 32.18
Pneumatic excavator "Rookie": Application for lunar base site preparation and transferring regolith into bags (Ruess et al. 2008).

The cable solution doesn't seem suitable because it massively hinders movement of the excavation vehicle to the point where it can barely move at all.

Regolith on an excavator as ballast has to get there in the first place and stay there. The ballast itself also only provides weight corresponding to 1/6 g. So large quantities would be needed in relation to the overall excavator dimensions. It can also be expected that confining structures would be needed to control the regolith and dust ballast, so such a design would still come with mass and volume penalties.

FIGURE 32.19
Honeybee Robotics percussive scoop on the Foster Miller Talon rover (Zacny et al. 2008b).

The regolith processed with the microwaves rapidly increases in temperature to as much as 2000 degrees centigrade and heating rates as fast as 140 degrees K/sec. This melts the regolith into a glassy substance with good mechanical strength. One of the interesting details of the microwave sintering process is that it can be tuned to provide surfaces of various strengths. However, possible applications are restricted to paving surfaces, e.g. for roads or landing pads.

A landing pad with a diameter of 20 m for example could be sintered within 2 lunar days consuming about 50,000 kWh of power (Taylor and Meek 2007).

Also the pneumatic excavator shows great promise. It can combine all the initially needed tasks in one machine. For example it is able to clear the immediate lunar base site from regolith dust and to fill regolith bags for habitation shielding with it in one step (see Figure 32.18).

Gas is injected (pulsed to increase efficiency) into the regolith and creates high-pressure gas cells. The easiest path for the gas to escape is straight up the nozzle and into the collection container (and not down and around the rim of the nozzle and into the lunar vacuum). Thus as the gas moves up it also carries the regolith particles with it. Impressive efficiency rates (it has

FIGURE 32.20
Honeybee Robotics percussive scoop on the iRobot PacBot (Zacny et al. 2008b).

been shown that with 1kg of gas 1000-3000 kg of regolith can be mined at 1/6g) further support this concept (Zacny et al. 2008a, Ruess et al. 2008).

The pneumatic system can also be applied into the transfer of regolith. For example other excavation methods such as backhoe, could be integrated with a pneumatic transfer system for moving lifted regolith to a mined container. This will reduce the time required to dump the regolith since now the excavation and movement of regolith can occur simultaneously.

In addition to the pneumatic mining system, another promising technology includes the percussive digger developed by Honeybee Robotics (Zacny et al. 2008b). A scoop or a blade that uses percussive system requires 90% smaller digging force. This translates directly into smaller vehicle mass and reduced traction requirements. Percussive or vibratory motion of a scoop essentially reduces sliding friction. This is because the regolith particles are in constant vibratory motion, bouncing up and down, instead of continuously pressing against the surface (see Figure 32.19 and Figure 32.20).

Dick et al. (1992) presented the result of experimental work to study another alternative or supplementary technique to traditional excavation of soil, namely the use of explosives to loosen the dense soil below 30 cm depth so it can be excavated with a limited amount of force. Although the ejection of regolith would not be acceptable on the lunar surface since the resulting dust would travel far, research showed that explosives buried deep enough would not create craters but loosen the soil very effectively.

Design Evaluation

The discussion so far clearly shows that lunar construction and especially excavation equipment design requires a highly multidisciplinary approach.

Long-term operation and maintenance, effects of radiation, temperature, vacuum, micrometeorites and dust must be considered for lunar construction equipment design. Tele-operation capabilities increase the safety and effectiveness (Podnieks 1994).

Flexibility and commonality of mining equipment should also be considered, especially for the early stages of lunar base development before specialized equipment is introduced.

For the selection of an appropriate set of lunar construction equipment, a methodology needs to be developed that includes considerations of performance in terms of productivity and life-cycle launch mass requirements such as re-supply and maintenance. Criteria for lunar construction equipment selection include robustness, modularity, servicing and maintenance requirements (Gies 1996).

Several scientific studies have tried to evaluate lunar excavation and mining technologies. While the bucket-wheel excavator is favored many times,

TABLE 32.1

Construction and Mining Task Comparisons: Earth vs. Moon (Connolly 1994)

Tasks	Earth	Moon
Bulk cargo loading/ unloading	Shovel, backhoe, track loader, excavator, scraper	Front-end wheel loader, wheel excavator
Bulk cargo hauling	Truck, belt conveyor	Rover/truck, front-end loader
Rock breaking/blasting	Explosives, pneumatics	Thermal breaking, explosives
Site leveling and grading	Scraper, bulldozer, grader	Scaled-down scraper
Digging and trenching	Shovel, draglines, excavator, loader	Front-end loader, bucket wheel excavator
Regolith excavation	Excavator, backhoe	Bucket wheel excavator
Raw materials transport	Dump truck, scraper	Rover/truck
Underground mining	Tunnel boring machine, drum-type continuous miner, roadheader, hydraulic rock splitter, explosives	Scaled-down roadheader, hydraulic rock splitter, tunnel boring machine, explosives

Connolly (1994) for example concluded that the versatility of the front-end loader would probably be best for low-volume surface mining. The same study compared construction tasks and the corresponding equipment on Earth vs. on the Moon (see Table 32.1).

Mueller and King (2008) have introduced an evaluation method for lunar excavation concepts. The resulting decision matrix is based on a weighted points system taking into account capabilities, productivity, reliability, dust generation, power efficiency, maintainability, supportability and versatility. Suitable designs should yield a high point value. It has to be noted though, that almost all concepts under consideration were based on terrestrial excavation designs. The only exception, a pneumatic concept, was not evaluated in all criteria and also not with respect to the latest developments as presented in Zacny et al. (2008a) and Ruess et al. (2008).

An evaluation including the newest developments in microwave sintering, pneumatic and vibratory excavation systems seems advisable and only consequent.

Conclusion

The settlement of new worlds such as the Moon will be difficult at best. To succeed in such a task from a technological as well as a budgetary standpoint we have to use the resourceful ingenuity of the human mind to the largest degree possible. Simple solutions based on terrestrial experience as well as completely new ideas have to be taken into account, weighed and the best solution for the task under consideration has to be chosen.

In the paper at hand, the authors detail the preparatory construction and excavation tasks for a permanent lunar base. Initially large quantities of regolith will have to be moved and used for construction. Lunar dust poses one of the highest obstacles for human and robotic operations on the Moon, so it is proposed to clear the immediate lunar base site of the upper fine layers of regolith, which are the main contributor to lunar dust.

A fine collection of excavation equipment proposed for these tasks is reviewed starting with concepts based on terrestrial applications and finally discussing innovative ideas that try to better account for the special environmental conditions on the Moon as well as the mass restrictions for all equipment that needs to be transported to Earth's companion.

Two particularly interesting technologies, namely the pneumatic and vibratory excavation methods, are identified to be very promising. They show great potential in terms of providing lightweight, flexible, robust and reliable systems with high performance for the tasks needed to prepare a lunar base construction site. Such technologies could help to significantly reduce Lunar Base construction cost and risk.

References

Bernold, L. (1991) "Experimental studies on mechanics of lunar excavation." *Journal of Aerospace Engineering,* 4 (1), pp. 9–22, ASCE, NY.

Boles W. (1992) "Performance-Based Evaluation of Lunar Base Construction Equipment and Methods." *Proceedings of SPACE 92,* Vol. I, pp. 332–340, ASCE, NY.

Boles W., Connolly, J. (1996) "Lunar Excavating Research." *Proceedings of SPACE 96,* pp. 699–705.

Connolly, J., Shoots, D. (1994) "Transferring Construction Technology to the Moon and Back" *Proceedings of SPACE 94,* pp. 1086–1096, ASCE, NY.

Dick, R., Fourney, W., Goodings, D., Lin, C.-P., and Bernold, L. (1992) "Use of explosives on the Moon." *Journal of Aerospace Engineering,* 5 (1), pp. 59–65, ASCE, NY.

Eckart, P. (2006) *The Lunar Base Handbook* 2nd Ed. McGraw-Hill.

Gaier, J. (2005) "The Effects of Lunar Dust on EVA Systems During the Apollo Missions." *NASA/TM–2005-213610,* NASA Center for Aerospace Information, Hanover, http://gltrs.grc.nasa.gov

Gies J. (1994) "Design Criteria for Lunar Construction Equipment." *Proceedings of SPACE 94,* Vol. II, pp. 1237–1255, ASCE, NY.

Gies J. (1996) "The Effect of the Lunar Surface Environment upon Machinery." *Proceedings of SPACE 96,* Vol. I, pp. 639–645, ASCE, NY.

Lindsey N. (2003) "Lunar Station Protection: Lunar Regolith Shielding." *Proceedings of the International Lunar Conference,* American Astronautical Society, Science and Technology Series, Vol. 108.

Mueller, R. and King, R. (2008) "Trade Study of Excavation Tools and Equipment for Lunar Outpost Development and ISRU", *Space Technology and Applications International Forum—STAIF 2008*, pp. 237–244, AIP.

Podnieks E., Siekmeier J. (1992) "Lunar Surface Mining Equipment Study." *Proceedings of SPACE 92*, Vol. I, pp. 1104–1115, ASCE, NY.

Podnieks E., Siekmeier J. (1994) "Role of Mining in Lunar Base Development." *Journal of the British Interplanetary Society*, Vol. 47, pp. 543–548, BIS, London.

Ruess, F., Braun, B. and Zacny, K. (2008) "Lunar in-situ resource utilization—Regolith bags automated filling technology" Paper No: AIAA-2008-7678 and Presentation, *AIAA SPACE 2008 Conference*, 9–11 Sep 2008, San Diego, California.

Ruess, F., Schaenzlin, J., Benaroya, H. (2006) "Structural Design of a Lunar Habitat." *Journal of Aerospace Engineering* 19, No. 3, pp. 133–157, ASCE, NY.

Smithers, G., Miller, J., Broughton, R. and Beale, D. (2007) "A one-piece lunar regolith-bag garage prototype." *Lunar Settlements Symposium*, Rutgers University, NJ.

Taylor, L. and Meek, T. (2005) "Microwave Sintering of Lunar Soil: Properties, Theory, and Practice" *Journal of Aerospace Engineering*, Vol. 18, No. 3, pp. 188–196, ASCE, NY.

Toklu, Y., Järvstråt, N. (2003) "Design and construction of a self sustainable lunar colony with in-situ resource utilization." *CE: The Vision for the Future Generation in Research and Applications*, pp. 623–628.

van Susante, P. (2008), copyright Paul van Susante, Colorado School of Mines. For more information contact: paulvans@mines.edu.

Zacny, K., Mungas G., Mungas C., Fisher D., and Hedlund M. (2008a) "Pneumatic Excavator and Regolith Transport System for Lunar ISRU and Construction" Paper No: AIAA-2008-7824 and Presentation, *AIAA SPACE 2008 Conference*, 9–11 Sep 2008, San Diego, California.

Zacny, K., Craft, J., Wilson, J., Chu, P., and Davis, K., (2008b) "Percussive Digging Tool for Lunar Excavation and Mining Applications" Abstract 4046, *LEAG-ICEUM-SRR*, 28–31 October 2008, Cape Canaveral, FL.

33

A Review of Technical Requirements for Lunar Structures: Present Status[1]

Alexander M. Jablonski
RAST
Defence R&D Canada Ottawa
Ottawa, Ontario, Canada

Kelly A. Ogden[2]
University of Waterloo
Waterloo, Ontario, Canada

Alexander M. Jablonski, PhD is program manager/defense scientist at the Defence R&D Canada Ottawa (DRDC Ottawa), on secondment from the Canadian Space Agency (CSA). He received his BSc and MS degrees in civil engineering (structures) at the Technical University of Kracow, Poland, MS in mechanics and materials engineering at the University of Illinois at Chicago, and PhD in structural dynamics at Carleton University in Ottawa, Canada. He has more than 28 years experience in the R&D and reconnaissance projects, strategy planning, and project management. He worked in Poland, Finland, Germany, and in the United States and now since 1982 he works in Canada. His initial career in Canada started in 1988 in the field of structural dynamics and earthquake engineering at the Dynamics Laboratory, Institute for Research in Construction, National Research Council (NRC) in Ottawa. In 1992, he joined Directorate of Space Mechanics, Space Technology Branch, Canadian Space Agency, where he worked on various space projects as dynamics research engineer and research scientist (the Lens Antenna Deployment Demonstration (LADD) test article, Observations of Electrical field Distribution in Ionospheric Plasma Unique Strategy (OEDIPUS C) and Bistatic Observations with Low Altitude Satellites (BOLAS) missions). He was Tether Technology Project Leader in 1993–2000. During the period 1998–1999 he was one of eight managers working for Space Plan Task Force (SPTF), CSA, to formulate Canada's space plan under the leadership of Dr. Karl Doetsch,

---

[1] This paper is a new version of the paper originally presented at the International Lunar Conference 2005 in Toronto, ON, Canada and includes updated results from the paper entitled "Technical Requirements for Lunar Structures," *Journal of Aerospace Engineering*, Vol. 21, No. 2, April 1, 2008. pp. 72–90.
[2] Formerly Research Manager, Canadian Space Agency, Saint-Hubert QC, Canada.

VP and chairman of SPTF. Then he worked for Space Science Program as academia and research institution liaison, being responsible for the CSA Grants and Contributions Program. Dr. Jablonski is the author and the co-author of more than 50 design projects of structures built for various industries and more than 85 publications (papers, reports, manuals and guidelines) on various aspects of space engineering and dynamics. He is recipient of a number of professional awards and is Associate Fellow of the American Institute of Aeronautics and Astronautics (AIAA) and the Canadian Aeronautics and Space Institute (CASI).

ABSTRACT The Moon has recently regained the interest of many of the world's national and international space agencies. Lunar missions are the first steps in expanding manned and unmanned exploration inside our solar system. The Moon represents various options. It can be used as a laboratory in lower gravity (1/6 of the Earth's gravity field), it is the closest and most accessible planetary object from the Earth, and it possesses many resources that humans could potentially exploit. This paper has two objectives: to review the current status of the knowledge of lunar environmental requirements for future lunar structures, and to attempt to classify them on the current knowledge of the subject. This paper divides lunar development into three phases. The first phase is building shelters for equipment only; in the second phase, small temporary habitats will be built, and finally in the third phase, habitable lunar bases will be built with observatories, laboratories, or production plants. Initially, the main aspects of the lunar environments that will cause concerns will be lunar dust and meteoroids, and later will include effects due to vacuum environment, lunar gravity, radiation, a rapid change of temperature, and the length of the lunar day. This paper presents a classification of technical requirements based on the current knowledge of these factors, and their importance in each of the phases of construction. It gives recommendations for future research in relation to the development of conceptual plans for lunar structures, and for the evolution of a lunar construction code to direct these structural designs. Some examples are presented along with the current status of the bibliography of the subject. The specific sources of lunar information are also presented.

Introduction

The Moon has recently regained the interest of many of the world's national space agencies. The Moon has developed a dual role in human thought. First, through pre-*Apollo* years (1958–1969), *Apollo* exploration (1969–1972), and the current post-*Apollo* era (1972 to date), it has been shown that the Moon is a scientifically important planet. Second, the Moon, because of its closeness to

Earth, is a natural target for the first step of human exploration beyond our planet, including future utilization of its vast resources (see Jolliff et al. 2006, p. 619) and even colonization. The knowledge of the Moon's environmental conditions and resultant technical requirements for lunar structures is key for their successful designs depending on the phase of construction. This paper proposes three phases of construction on the Moon:

- Phase 1 (2010–2020): Designs for equipment shelters during initial unmanned robotic or manned missions;
- Phase 2 (2020–2030): Development of structures for medium length of stay (up to several months);
- Phase 3 (2030 on): Long-term construction of lunar structures for different purposes (primarily for long term habitats, resource use, laboratories, and finally permanent lunar bases).

First, a brief description of each phase is presented. Then the Moon's environmental conditions and technical requirements for lunar structures are discussed based on available data. They include temperature, radiation, gravity, atmosphere/pressure, the lunar day, lunar surface conditions (lunar dust), lunar seismicity, and meteoroids. The impact on materials, shapes and location of lunar structures is also assessed. Then, a short review of recent findings of lunar missions and how they can affect future lunar structures is provided. Science objectives of selected planned missions that will be useful for lunar structure design are also described. Finally a set of recommendations for future research is presented.

Moon: The Nearest Important Destination

The Moon is the only celestial body, beyond the planet Earth, which has been systematically sampled and studied. Physical characteristics of the Moon and Earth are compared below (see Table 33.1).

Although exploration of the Moon is still incomplete, the overall effort has been extensive. The first Russian imaging missions (*Luna, Zond*) and American missions (*Ranger*) were flown between 1959 and 1965. At the same time, a systematic Earth-based mapping of the Moon by telescope began, and it led to determination of lunar stratigraphy. The *Luna 10* mission provided the first orbital gamma-ray chemical data in 1966. *Lunar Orbiter* returned images in preparation for the U.S. manned *Apollo* missions. Unmanned soft landings and surface operations took place between 1964 and 1976, with the first data on soil physics and chemistry sent by radio back from lunar service in 1966 and the first Soviet robotic collection of lunar samples returned by *Luna 16* in 1970. However, the greatest achievements in lunar sampling were

TABLE 33.1

Comparison of the Physical Characteristics of the Moon and Earth

Property	Moon	Earth
Mass	7.353×10^{22} kg	5.976×10^{24} kg
Radius (spherical)	1,738 km	6,371 km
Surface area	37.9×10^6 km^2	510.1×10^6 km^2
Flattening*	0.0005	0.0034
Mean density	3.34 g/cm^3	5.517 g/cm^3
Gravity at equator	1.62 m/s^2	9.81 m/s^2
Escape velocity at equator	2.38 km/s	11.2 km/s
Sidereal rotation time	27.322 days	23.9345 hr
Inclination of equator/orbit	6°41′	23°28′
Mean surface temperature	107°C (day); −153°C (night)	22°C
Temperature extremes (see also Table 2)	~233°C (?) to 123°C	−89°C to 58°C
Atmosphere	~10^4 molecules/cm^3 (day)	
200×10^3 molecules/cm^3 (night)	2.5×10^{19} molecules/cm^3 (STP)	
Moment of Inertia (I/MR^2)	0.395	0.3315
Heat flow (average)	~29 mW/m^2	63 mW/m^2
Seismic energy	20×10^9 (or 10^{14}?)J/yr**	$10^{17} - 10^{18}$ J/yr
Magnetic field	0 (small paleofield)	24 − 56 A/m

*(Equatorial-ideal)/ideal radii
**These estimates account for Moonquakes only and do not account for seismicity from meteoroid impacts
Heiken et al. 1991, Table on p. 28, reprinted with permission of Cambridge University Press.

the six *Apollo* manned landings between 1969 and 1972 (NASA 1969, 1970, 1971a, 1971b, 1972, 1979). Sites of six *Apollo* and three *Luna* sample-return sites are depicted in Figure 33.1 (Spudis 1999, p. 126).

Lunar research has brought to attention a rich output of information achieved from the past lunar exploration scientific data; there are several depositories of lunar samples and scientific results. The most important are listed below:

Lunar and Planetary Institute (LPI)

National Space Science Data Center (NSSDC)

National Technical Information Service

NASA Johnson Space Center History Office

NASA Johnson Space Center Lunar Sample Curatorial Facility

These data have provided the database for various scientific aspects of the Moon. Some of these data are the basis of the current knowledge on derived technical requirements for future lunar structures (Heiken et al., 1991, p. xix).

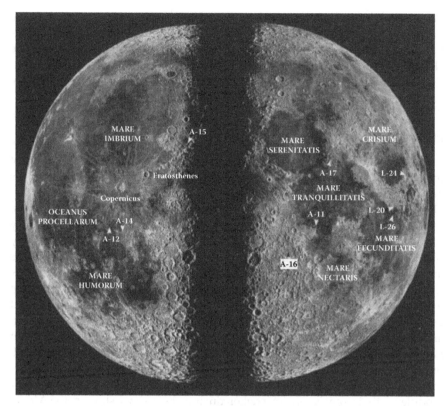

FIGURE 33.1
A pair of Lick Observatory (USA) photographs with labels showing selected lunar features and the location of the six *Apollo* (A) and three *Luna* (L) sample-return sites (photograph © UC Regents/Lick Observatory).

Future lunar structures will serve to achieve a permanent utilization of the Moon and its resources. The Moon will also serve as a place for future laboratories, astronomical observatories, testing grounds, and manufacturing plants. Finally, it will be an important stepping-stone to reach to other planets from our solar system. The comprehensive scientific context for exploration of the Moon is provided in the current report prepared by the Space Studies Board, National Research Council of the National Academies (Space Studies Board, 2007).

Three Phases of Construction

Lunar base development has previously been categorized into stages, usually by the degree of human presence on the Moon (Eckart, 1999; Toklu, 2000; Ruess et al., 2004). A similar approach is taken here, although the phases are defined by the uses of the different structures in each phase, rather than the human presence, thus focusing on the technical requirements for construction. The evolution of the structures that will be used on the Moon can be classified into three general phases: those of support and shelters for scientific equipment, temporary habitats for conducting science and exploration, and long-term settlements primarily for resource utilization.

Phase 1

The first phase involves the structures that are closest to realization. It is not included as a stage in the previously mentioned lunar base plans because it does not involve inhabitants; however in lunar structure evolution, it is an important phase. Structures that will support or contain scientific equipment are the basis of this phase. Phase 1 will begin around 2010, and extend exclusively until around 2020; after that time, it will continue, running concurrently with phases two and then three. An example of a structure in this phase is an assembly that would house a Lunar Liquid Mirror Telescope, which was recently proposed by Dr. Roger Angel and his team. A possible structure that would surround the LLMT and a rigidizable structure developed by the Jet Propulsion Laboratory (JPL) and L'Garde are shown in Figure 33.2 (Cadogan and Scarborough 2001; Angel, 2005). These structures will be built entirely on Earth and transported to the Moon, where they will be automatically deployed, or set up by robots or humans. These structures will not be inhabited by humans; if people are required to erect the structures, they will use the lunar module of their spacecraft for shelter during their short stay on the Moon, which is not considered part of the phase. Any extended time spent on the Moon for which a separate shelter is required is considered part of Phase 2.

FIGURE 33.2
Possible structure to surround the LLMT (left) (Angel, 2005; drawn by Tom Connors), and a rigidizable support structure developed by JPL and L'Garde in the 1990s (right) (Cadogan and Scarborough, 2001 with permission from ILC Dover).

The function of the structures in this phase will be to protect the equipment from dust, meteoroids, and radiation, as well as to provide structural support. The temperature fluctuations and seismic activity will also be important considerations that will affect the design of the structures.

Phase 2

Phase 2 of construction begins with the first structures that are deployed on the Moon and inhabited by humans, which will start around 2020 (Hawes, 2005). This phase is similar to the late part of Eckart's *Pioneering Phase* and the *First Lunar Outpost* (Eckart, 1999 p. 236), as well as to Toklu's *Pre-Fabricated* classification (Toklu, 2000). The purpose of the structures in this phase will be to conduct science, allowing people to work with the equipment that has already been placed there, and to investigate and prepare possible locations for a permanent lunar base. They will be intended for only a short time on the Moon, up to several months. As well, they will be designed for few people, up to approximately 10.

In this phase, the structures will be inflatable to maximize the final volume of habitable space while minimizing the initial, compacted volume and weight because these structures will also be constructed on Earth before relocating them to the Moon; as well, they should be modular to allow the lunar base to be expanded, eventually leading into Phase 3. An example of an inflatable habitat, possible in Phase 2, is shown in Figure 33.3 (Criswell and Carlson 2004). Although they will initially be constructed with resources from Earth, Phase 2 will involve some lunar resource utilization. This will mainly be the use of regolith for shielding the habitat from radiation, thermal extremes and cycling, and meteoroids.

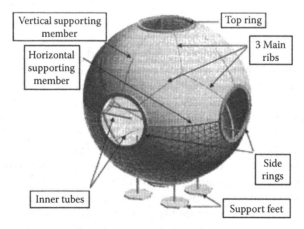

FIGURE 33.3
Example of an inflatable habitat, possible in phase two (Criswell and Carlson, 2004, ASCE).

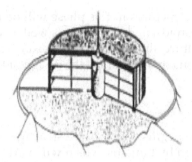

FIGURE 33.4
A Lunar concrete structure that would be constructed under a pressurized dome; Phase 3 example (Eckart, 1999, p. 297).

These requirements, unique from those of Phase 1, are impacted by additional environmental conditions, which must be considered in their design. As well as dust, meteoroids, radiation, temperature, and seismic activity, the effects of gravity, pressure, and the length of the lunar day will also become significantly more relevant because human safety is involved. As well, each of the conditions that were important in Phase 1 will affect design in Phase 2 in a different, often more significant way.

Phase 3

The final phase of lunar construction, as the farthest from realization, is also the least rigidly defined. It will begin with more permanent, habitable lunar bases, as in the *Consolidation* and *Settlement* phases of Eckart, and the *In-Situ Resource Construction* of Toklu (Eckart, 1999, p. 236; Toklu, 2000). This phase will develop through a gradual transition from Phase 2, by about 2030 (Hawes, 2005). Instead of relocating the structures to the Moon, additions to Phase 2 structures will be constructed on the Moon from in-situ resources such as lunar concrete (see Figure 33.4). They will be able to house many people comfortably, as their inhabitants will stay in them for extended periods of time; as well, it should be possible to build on and expand them to increase capacity.

The purposes of this phase will be to continue science involving the Moon and to increase lunar resource use. After phase three begins, processing and production plants will be built to develop in-situ resource utilization. Lunar resources will be used to a further extent than simply for regolith shielding; they will be used to expand the previously existing bases and reduce dependence on supplies from the Earth. At this time, the lunar habitats should have little dependence on supplies from Earth for survival.

The environmental conditions that are important in this phase will be the same as those that were considered in phase two; however, they will be dealt with differently. Specifically, radiation shielding will be more advanced and location will be selected to minimize the effects of radiation, temperature, and meteoroids. As well, all of the conditions must be addressed so that the inhabitants are comfortable, rather than simply surviving, because they will remain in the habitats for a longer time.

Environmental Conditions and Resultant Requirements

Unique technical requirements exist for construction on the Moon, and are primarily caused by the harsh lunar environment. Between the Earth and the Moon, there are several important differences to consider, and while most of these are challenges to the design, some, such as seismicity, are actually less important considerations than they would be on Earth. The most significant factors that differ lunar construction from construction on Earth are the following:

- Temperature
- Radiation
- Atmosphere and pressure
- Meteoroids
- Gravity
- The length of the lunar day
- Dust
- Seismicity

Each of these must be considered to varying degrees with respect to each phase of construction, although the most important in all phases are temperature, radiation, pressure, and meteoroids (see Appendix A for importance of and requirements due to each condition). The *Lunar Construction Code* should include information on these conditions and the structural requirements that they create.

Temperature

One of the most important environmental differences between construction on the Moon and construction on Earth is the temperature ranges of the two planets. Because of the long lunar day and thin atmosphere, the temperature on the Moon varies greatly, by up to 280 K at the equator. Further, at the equator it has an average temperature of only 255, or −18°C (Aulesa et al., 2000). This means any structure placed on the Moon must be able to sustain very cold temperatures, as well as severe thermal strain caused by the fluctuation.

Although the equator represents the largest temperature variation on the Moon, the coldest temperatures occur in the permanently shadowed parts of craters at the poles. There, the temperatures are constant at 40 K, or −233°C (Aulesa et al., 2000). The most moderate temperature conditions, considering

TABLE 33.2

Temperature Ranges and Variations on the Moon

	Permanently Shadowed Polar Craters		Other Polar Areas		Equatorial Zone		Mid-Latitudes	
Average Temperature	40 K	–233°C	220 K	–53°C	255 K	–18°C	237.5 K	–35.5°C
Thickness of Regolith Cover (m)	Monthly Variation and Range (°C)							
	Variation	Range	Variation	Range	Variation	Range	Variation	Range
0.0	0	–233	+/–10	–63 to –43	+/–140	–158 to 122	+/–50	–85.5 to 14.5
0.5	0	–233	+/–3.9	–56.9 to –49.1	+/–55.8	–73.8 to 37.8	+/–19.6	–55.1 to –15.9
1.0	0	–233	+/–1.2	–54.2 to –51.8	+/–16.6	–34.6 to –1.4	+/–5.8	–41.3 to –29.7
1.5	0	–233	+/–0.5	–53.5 to –52.5	+/–7.5	–25.5 to –10.5	+/–2.7	–38.2 to –32.8
2.0	0	–233	+/–0.3	–53.3 to –52.7	+/–4.3	–22.3 to –13.7	+/–1.5	–37.0 to –34.0
2.5	0	–233	+/–0.2	–53.2 to –52.8	+/–2.8	–20.8 to –15.2	+/–1.0	–36.5 to –34.5

Derived from Aulesa et al., 2000.

both the range and average temperature, are at the polar areas other than permanently shadowed craters, where the average is –53°C and varies by +/–10°C, or the mid-latitudes where the average is warmer at –35.5°C but varies by +/–50°C (Aulesa et al., 2000) (see Table 33.2).

Also summarized in Table 33.2 are the results of one method of temperature shielding; that is a regolith cover (Aulesa et al., 2000). Regolith provides thermal insulation and can be used to shield a habitat from temperature, as well as radiation and meteoroids. Two metres of regolith shield bring the ranges in all locations, except at the equator, to a small variation.

Using a regolith shield, however, is only a feasible solution for Phases 2 and 3; in Phase 1, when structures will be used as support for scientific equipment such as telescopes, covering them in regolith is not an option. As well, while a regolith shield could be used to keep the internal temperature of the structure more consistent, the structure itself still must be able to withstand the cold, at least during set-up. Therefore, in all phases, the materials used in construction must maintain their properties in the temperatures of the environment without shielding, such as Kevlar®, which does not become brittle until −196°C (Kennedy et al., 2001, p. 540) and could therefore be used in all areas of the Moon except the permanently shadowed craters. In Phase 1 this is required for the life of the structure, and in Phases 2 and 3, it is important while the structure is being deployed, built, or repaired.

Radiation

The hazardous radiation that reaches the Moon comes primarily from two sources; they are galactic cosmic rays (GCR) and solar energetic particle (SEP) events (Parnell et al., 1998). This radiation is considered to be a significant threat to human life in addition to having a negative impact on equipment. To protect human life in a habitat, radiation shielding is necessary in Phases 2 and 3. It is also important in Phase 1; the degree to which it is important depends on the purpose of the structure (for radiation analysis, see Appendix B).

Phase 1

Because structures in Phase 1 will not be designed to shelter humans, biological radiation effects are not a concern; however, radiation also has a negative impact on equipment. The state of electronic equipment can be altered by an ion-induced charge from radiation particles, and radiation can create extra noise for sensors; it also breaks down materials and reduces power output from solar panels (Parnell et al., 1998). For some of these problems, radiation shielding can be used, and the vulnerable parts can be sheltered by the structure or by regolith, which would reduce the effects of the radiation, as is suggested for the Lunar Liquid Mirror Telescope (Angel, 2005). Most shielding in this phase will be part of the structure, and included in the construction on Earth, rather than added during set up or deployment. Few structures are likely be buried or covered in regolith. Other approaches, where shielding is not possible, include redundant circuits for electronics and larger than required solar panels (Parnell et al., 1998). Structures in this phase must use materials that are relatively resistant to radiation, or radiation-hardened materials (Parnell et al., 1998).

Radiation Limits for Phases 2 and 3

The average radiation dose on Earth is 0.0036 Sv/yr; however, on the Moon, the average dose, caused almost entirely by galactic cosmic rays (GCR) and Solar Energetic Particle (SEP) events, is considerably larger at 0.25 Sv/yr (Lindsey, 2003) or greater (ISU 2000). Further, individual SEP events can deliver up to 1000 Sv (Lindsey, 2003). Humans on the Moon must be protected from this significant increase in radiation, although how much protection is required depends on factors such as the length of stay, and a definitive limit is still undetermined. A maximum dose of 0.5 Sv/yr is recommended by the Space Studies Board in 1996 (Aulesa, 2000), which is consistent with the National

Council on Radiation Protection's (NCRP) recommendation in 1989 (Parnell et al, 1998). However, the career maximum for blood forming organs is 1 to 4 Sv, so even that dose can only be endured for less than two to eight years, considering other sources of radiation throughout the person's life. For very long-term bases, this will not be an acceptable limit. Also, sufficient protection must be provided from solar flares because all of the maximum radiation dose should not be received over the few hours of a solar flare. Further, for nuclear power plant workers, the International Commission of Radiological Protection recommends an annual limit of 0.05 Sv (Parnell et al, 1998). Because the nuclear power plant workers will be exposed to the radiation for years, rather than just a short mission considered by the Space Studies Board, their recommended dose is lower. This indicates that structures in phase two may not need significant shielding; however toward the end of the second and into the third phase, more effective shielding will be required.

Phase 2

Large-scale radiation shielding will be required, beginning in Phase 2, to protect humans spending time in lunar structures. The most commonly suggested type of radiation shielding is a regolith cover because it includes the advantages of in-situ resource utilization, as well as provides meteoroid and thermal shielding. A regolith shield increases the amount of mass through which radiation particles must pass, and with enough shielding, can stop the particles or slow them to an acceptable energy level. However, a sufficient layer of shielding must be provided because initial collisions create high-energy particles, or *brehmsstrahlung* radiation, and with just a thin layer, the inhabitants of the structure will be exposed to these high-energy particles, increasing the radiation damage that occurs rather than reducing it (Buhler and Wichmann, 2005). The minimum suggested amount of shielding varies between sources, from two metres of regolith to over five; the thickness required depends on the radiation limit that is recommended, and the density of regolith. The Earth's atmosphere provides 1000 g/cm^2 of shielding at sea level, so equivalent protection on the Moon is ideal (Heiken, 1991, p. 53); however, 700 g/cm^2 is considered acceptable, as inhabitants will not be on the Moon for extended periods of time (Aulesa et al., 2000). If the lowest density of regolith is approximately 1.3 g/cm^3, the regolith shield will have to be 5.4 m thick to provide 700 g/cm^2 of protection (Aulesa, 2000).

Phase 3

In Phase 3, there are further methods of radiation shielding to consider. Other than only regolith, an electrostatic radiation shield used with regolith (Buhler and Wichmann, 2005) is possible, or, if the habitat is constructed in a lava tube, the ceiling of the lava tube itself may be used. In this phase, radiation shielding is extremely important because an increased amount of time will be spent on the Moon. The amount of radiation received should be as low as possible; ideally, it should be as low as that received on Earth to avoid any increase in the inhabitants' probability of cancer.

Atmosphere and Pressure

The near zero pressure of 3 nPa on the Moon increases the severity of several of the environmental conditions that are relevant to lunar construction. The thin atmosphere of at most 2×10^5 molecules/cm^3, which occurs at night, provides little thermal insulation, contributing to the significant temperature range and thermal cycling that exist on the Moon. A dense atmosphere would provide some radiation shielding, reducing the measures required to protect equipment and humans from it. Finally, a thicker atmosphere would burn up meteors, and fewer meteoroids would reach the surface.

The thin atmosphere also results in specific structural requirements in the second and third phases. In these phases, the internal pressure in the structure must be sufficient to sustain human life, so the structure becomes a pressure vessel. The internal pressure must be at least 26 (Aulesa, 2000) to 30 kPa (Langlais and Saulnier, 2000) to support human life and avoid altitude sickness, if it is composed purely of oxygen. Realistically, however, the pressure must be higher to make the living environment comfortable and avoid such effects as difficulty speaking and ineffective coughing (Eckart 1999, p. 276), and most importantly to reduce the extreme fire hazard that pure oxygen would cause.

The internal pressure of the structures in the second and third phases will create substantial tensile loads on the structure. The regolith piled on the structure for shielding in Phase 2 will somewhat counter the load on the top. However due to the reduced gravity, the pressure caused by its weight will not be greater than the internal pressure. If the regolith shield is 5.4 m high, gravity is 1.62 m/s^2, and the density of regolith is between 1.3 (Aulesa et al., 2000) and 1.75 g/cm^3 (Sadeh et al., 2000), the pressure created by the regolith will be between 11.4 and 15.9 kPa, which is much less than the minimum required internal pressure for human life, of 26 to 30 kPa (Aulesa, 2000; Langlais and Saulnier, 2000). As well, there will be horizontal loads on the structure due to the pressure, which the regolith will not counter.

Meteoroids

Meteoroid bombardment since the formation of the Moon has resulted in the present lunar topography. Meteors are a threat to structures on the Moon because there is almost no atmosphere on the Moon to burn them up or even slow them down. As a result, meteoroids impact the Moon with their full velocity, which can range from 10 to 72 km/s (Coronado et al. 1987, p.12).

For some structures in Phase 1 such as a lunar telescope, the threat may simply have to be accepted because creating a shield that would protect equipment such as a telescope but not interfere with its view may not be possible. Further, as Gorenstein says in an analysis of a proposed lunar observatory, "disturbances would be meteoroid impacts but the probability that an impact would affect the observatory is small" (Gorenstein, 2002, p.46). Particularly for larger particles, meteoroid flux is very low (see Appendix C).

However, even small particles constitute a threat in Phases 2 and 3; meteoroid impact is a more significant concern because a leak in a structure in one of these phases could be catastrophic if not repaired quickly. Therefore, meteoroid shielding is required. One solution is a regolith covering, which would absorb the impact of meteoroids, preventing them from reaching the structure.

Lindsey analyzes, using the *Fish-Summers Penetration Equation*, the thickness of the regolith layer required to protect a structure from meteoroids of diameter 7 cm or smaller, and finds it to be 45.9 cm (Lindsey, 2003). The flux of a meteoroid this large is 1.76×10^{-4} impacts/km^2/yr and decreases for larger particles, making a catastrophic impact very unlikely with this amount of shielding (Lindsey, 2003). Further, using this equation (Hayashida and Robinson, 1991), the effectiveness of a regolith shield of 5.4 m can be approximated. This is also the recommended thickness for sufficient radiation shielding, and would provide protection from meteoroids with a diameter of 52 cm or smaller, reducing the flux of a penetrating impact on the structure to between 10^{-8} and 10^{-7} impacts/km^2/yr (see Appendix C) (Lindsey, 2003; Eckart, 1999, p. 148).

Although meteoroids are an important concern in lunar structural design, the probability of impact is very low. Because of this, meteoroid shielding is primarily a concern in the later phases of construction. A regolith shield that will provide temperature and radiation protection is also sufficient to provide meteoroid protection; however more advanced solutions may be investigated for late in Phase 3. For additional meteoroid shielding, a layer of the structure should provide some protection, as suggested for the TransHab (Kennedy et al., 2001, p. 545).

Gravity

The effects of the reduced gravity on the Moon compared to the Earth significantly alter the loads that must be considered in lunar construction. Self-weight of the structures is much less of a concern, as the gravity on the Moon is only approximately 1.62 m/s^2 (ISU 2000), or 1/6 of the Earth's gravity, and varies very slightly due to mass concentrations. In Figure 33.5, the variations on the nearside (left) and far side (right) are depicted, with variations in 10^{-5} m/s^2 (Carroll et al. 2005; Konopliv et al. 1998). In Phase 1, this will primarily be a benefit in design because the structure will simply have to support less weight. In Phases 2 and 3, however, lower gravity will not significantly counter the net vertical tensile loads caused by the internal pressure required to sustain human life.

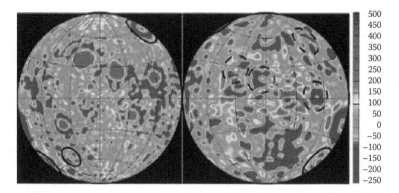

FIGURE 33.5
Gravity variations on the nearside (left) and far side (right) with variations in 10^{-5} m/s^2 (Konopliv et al. 1998., reprinted with permission from the American Association for the Advancement of Science).

Length of Lunar Day

One lunar day, the time from one new Moon to the next, is 29.53059 Earth days (ISU, 2000). This causes some substantial environmental differences between the Earth and Moon. Primarily, the long lunar day gives the Moon more time to heat up while exposed to the sun, and then more time to cool during the night, contributing, along with the lack of atmosphere, to the extreme thermal cycling. This drastic difference from Earth requires some changes in the way that lunar construction must be approached; lunar structures must be able to withstand extremely cold temperatures and their sharp variation. Specifically, the materials used must retain similar properties over the range of temperatures, and also must not fail easily due to thermal fatigue.

To achieve this, insulation such as lunar regolith will be used; however, the insulated parts of the structure must still be exposed to the environment while the structure is being deployed or constructed. Here, the length of the lunar day is beneficial; it is long enough so that missions to set up structures on the Moon can be timed such that the structure is set up in relatively mild temperatures, and thermal shielding can be in place before extreme temperatures occur.

Dust

Dust, while it does not create any specific technical requirements, is a feature of the lunar surface that complicates construction. The lunar sunrise and sunset create a photoelectric change in the conductivity of the dust particles, which causes them to float in the air and allows them to adhere to surfaces (Eckart, 1999, p. 139). This could interfere with any mechanical parts on the outside of a structure, the deployment of the structure, or the observations of scientific equipment. To counter this, abrasion resistant materials must be used, and any moving parts should not be easily deterred by dust.

Seismicity

Although lunar seismicity is a consideration in the design of lunar structures, the annual seismic energy released on the Moon is significantly less than that on Earth; unlike Earthquakes, Moonquakes will have little impact on lunar structural design.

In the *Apollo 15 Preliminary Science Report*, it is estimated that the annual seismic energy release on the Moon is 10^{11} to 10^{15} ergs, whereas on Earth it is approximately 5×10^{24} ergs (NASA 1971b); the seismic energy released on the Moon is $1/(5 \times 10^9)$ of that released on Earth. Further, the largest recorded Moonquake in the *Apollo* data was 2 to 3 on the Richter scale, and the usual magnitudes of Moonquakes were only 1 to 2. A further source, Toklu, supports 1 to 2 magnitude on the Richter scale as the average, but suggests the one larger quake may have been as much as 4 (Toklu, 2000), although this is still significantly lower than many of those on Earth.

There were five sites with special seismographs placed on the Moon during the landings of Apollo 12, 14, 15, 16, and 17 (NASA 1969, 1971a, 1971b, 1972, 1973). For many millions of years, the Moon has been a dynamically quiet planetary body with no known plate motions, no active volcanoes, and no ocean trench systems in place. It was very special to find out that each lunar seismograph detected 600–3,000 Moonquakes every year, but most of them were very small; up to about 2 on the Richter scale. The lunar seismic events could be divided into three distinct groups: deep Moonquakes (with their foci at depths of 600–900 km in the Moon), shallow Moonquakes (less frequent), and Moonquakes artificially induced by a meteoroid impact.

Much of the data collected in the *Apollo* seismic experiments was due to meteoroid impact, demonstrating that meteoroids will be a much more significant concern, [See Figure 33.6 (Latham et al. 1972)]. Also due to the small magnitude of lunar seismicity, the main structural loads to consider in lunar construction will be due to the self-weight of the structure and equipment in Phase 1, and the internal pressure of a habitat in Phases 2 and 3, rather than Moonquakes.

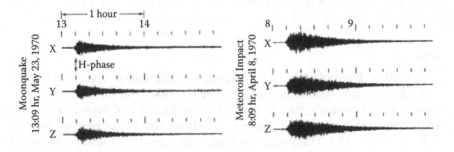

FIGURE 33.6
Moonquake and meteoroid seismic activity (Latham et al. 1972, p. 377, Fig. 2, printable license is given with permission from Springer Science and Business Media).

Materials

The materials used in lunar structures must be resistant to the environmental conditions and retain their mechanical properties for which they were chosen. However, as the purpose of the lunar structures evolves, the type of material that will best serve the purpose will also change, so a variety of acceptable materials are required. One common factor influencing the material choice in all phases is mass and volume during transportation. The mass and volume that is transported should always be minimized due to the high cost and amount of fuel required. Although limitations for individual missions will vary and are not discussed here, minimizing these properties is expected to remain important for all phases.

Phase 1

In Phase 1, the materials used must be able to function in the temperature extremes, maintain their properties while exposed to radiation, and be resistant to abrasion. The structure should also be as compact and lightweight as possible to reduce cost of transportation, and be deployed on the Moon to take on its final shape. Options in the phase include solid collapsible structures, inflatable fabric, plastic, or metal film structures, and rigidizable inflatable structures (Kennedy et al., 2001; Cadogan and Scarborough, 2001). The best option of these is rigidizable material because it can have the advantage of small volume and weight when stored, and can be deployed on the Moon to a larger size. Further, they do not depend on maintaining internal air pressure to keep their shape, as other inflatable structures do. Rigidizable materials provide the benefits of inflatable structures, without the risk of a catastrophic failure if a small puncture occurs (Cadogan and Scarborough, 2001). This is an important property in this stage because meteoroid shielding will not be provided.

Phase 2

The internal air pressure of the structure must be maintained in Phase 2 because it is necessary to sustain human life. For this reason, rigidizable inflatable structures may still be used, although not all of the advantages that they provide in Phase 1 apply. However, the structures in this phase may still be inflatable because they will be transported from the Earth and therefore must be small and lightweight for travel, but able to expand to provide adequate space. The membranes of these structures will be multi-layered, including a liner, bladder, restraint layer, insulating layer, and protective layer for meteoroids (see Figure 33.7) (Kennedy et al., 2001, pp. 535–548; Langlais and Saulnier, 2000).

Testing has been performed on materials chosen for the TransHab for low Earth orbit (Kennedy et al., 2001, pp. 548–552), although before a material can be used, it must be tested to ensure that it retains its properties in the lunar

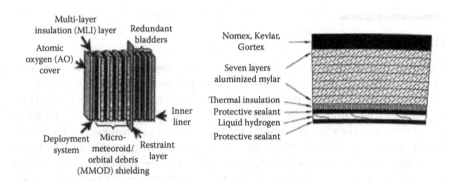

FIGURE 33.7
Layers of the inflatable habitats TransHab (left) (Kennedy et al., 2001, p. 535, with permission of American Institute of Aeronautics and Astronautics), and the proposed construction dome (right) (Langlais and Saulnier, 2000).

environment, including extreme temperatures, radiation, and abrasion. The former two conditions are most important during deployment, as regolith will later be used for radiation, temperature, and meteoroid shielding; abrasion resistance is necessary throughout the life of the structure. In this phase, lightweight, rigid, collapsible metal or alloy structures could also be used.

Phase 3

In-situ resource utilization will be the main source of the materials used in Phase 3. In addition to regolith for shielding, lunar resources will be used to construct the main structure of the habitat. To build the structure, lunar materials that may be used include lunar concrete (Lin 1987; Lin et al. 1991; Eckart 1999, p. 656), sulphur-based concrete (Casanova and Aulesa, 2000; Eckart 1999, p. 656), and cast basalt (Greene, 2004).

Lunar concrete, although it may be possible, presents several difficulties; it would require water transported from Earth or obtained from the Moon, which would also evaporate more quickly in the thin atmosphere, resulting in weaker concrete. To prevent the water from evaporating from the mortar, epoxy binders could be added to the concrete, or the concrete could be preset in a pressurized environment to prevent evaporation (Eckart, 1999, p. 657), such as the pressurized construction dome suggested by Langlais and Saulnier (2000). Lunar concrete presents difficulties, which may be overcome. An alternative to lunar concrete that would eliminate these problems is sulphur concrete, which does not require water, needs less energy to manufacture, and can be produced in cold environments (Casanova and Aulesa, 2000).

Finally, because basalt is widely available on the Moon, cast basalt is suggested as a lunar construction material (Greene, 2004). It also does not require water to cast, but uses heat, which can be obtained from concentrated sunlight (Greene, 2004).

All of these materials share one major drawback. They are stronger under compression than tension; the tensile strength of concrete is only approximately 10% of its compressive strength (Casanova and Aulesa, 2000). However because of the internal pressure required in a lunar habitat, the net loads on the structure will be tensile. This means their optimum properties are not taken advantage of, which must be considered in the structural design. When phase three is in development, these and other possible materials must be more closely examined with respect to the specific needs of the habitat.

Shape

In phase one, the shape will be determined primarily by the requirements of the scientific equipment that will be supported or protected by the structure. Before phase two, there will be little choice in possible shapes for the structure. When different shapes are options, there are two important areas to consider in deciding the shape of a lunar habitat. They are the human requirements and the structural stresses. The practicality of the space available necessarily determines what shapes are options, and the stresses that the shape would have to sustain determine the best choice of these options. Here, shapes will be considered separately for phase two.

In the second phase, inflatable structures will be used because they are lightweight and small when compacted, and can provide a large volume when deployed. Some of the most efficient in mass to final volume are sphere, cylinder, and toroid shells; these structures also eliminate sharp corners that would concentrate shell stresses (see Figure 33.8). Another possibility is an inflatable dome with an anchored base (Langlais and Saulnier, 2000). Structures that have

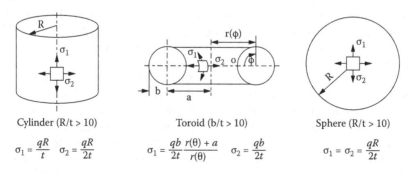

Cylinder (R/t > 10)

$$\sigma_1 = \frac{qR}{t} \quad \sigma_2 = \frac{qR}{2t}$$

Toroid (b/t > 10)

$$\sigma_1 = \frac{qb}{2t} \frac{r(\theta) + a}{r(\theta)} \quad \sigma_2 = \frac{qb}{2t}$$

Sphere (R/t > 10)

$$\sigma_1 = \sigma_2 = \frac{qR}{2t}$$

FIGURE 33.8
Directions of stresses in shells of various shapes (Kennedy et al., 2001, p. 537, reprinted with permission from the American Institute of Aeronautics and Astronautics).

been proposed or built for space or planetary habitats that use these structures are the TransHab, a combination of a toroid and cylinder (p. 528, Kennedy et al., 2001), the spherical inflatable habitat proposed by Criswell and Carlson (see Figure 33.3) (Criswell and Carlson 2004), the Astrophytum, consisting of four spheres arranged radially around a cylinder (Borin and Fiscelli, 2004), and a hemispherical inflatable construction dome (Langlais and Saulnier, 2000).

A sphere does not allow a very efficient use of space, although it gives the most habitable volume for the material used. The space created with the shape of a cylinder or toroid would be much more efficient, although they would both create more volume and mass when compacted.

Structures in the third phase will most likely be dome shaped or rounded because, while they will probably be built from concrete or a similar material, they will still include an airtight membrane, so sharp corners should be avoided.

Location

In Phase 1, the scientific goal of the mission will be the most important factor in determining the location of the lunar structure. For example, the south pole (See Figure 33.9), including the South Pole Aitken Basin, is a geologically interesting location that is the goal of a proposed NASA sample return mission (Koelle et al., 2005), and is also an important target in the National Research Council's Solar System Exploration Strategy (Smith, 2002, Chap 2); it may also become a desired location for scientific experiments requiring structures. For astronomy, a location on the far side of the Moon may be ideal because it is shielded from radio noise from Earth, or to investigate the existence of ice water, a polar location would be required.

Scientific areas of interest will still be important in Phase 2, and additionally, locations that have resources will be important to investigate the possibility of in-situ resource utilization. These resources include ^{3}He, which is rare on Earth but could be used as fuel in nuclear fusion, oxygen for rocket fuel, and materials such as concrete for lunar construction.

As the purpose of lunar structures in Phase 3 evolves toward establishing a human colony on the Moon, locations that make design and construction easiest and increase the lifetime of the structure will be used. Some of these locations include the inside of a lava tube, which would provide excellent environmental protection, or the poles, where peaks of eternal light might be located. While these locations may not be options in earlier phases due to the purpose of the mission, the needs of an inhabited, semi-permanent base, such as reliable shielding and power production, will be more important in phase three. The choice of location of lunar structures will be greatly influenced by the results of future lunar missions.

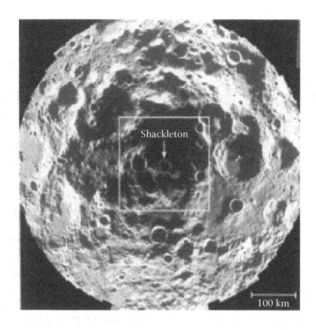

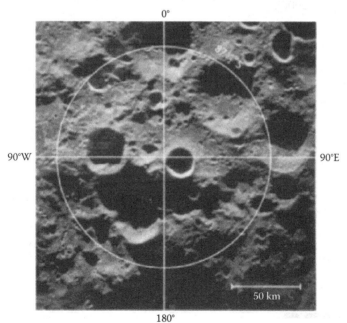

FIGURE 33.9
The Lunar South Pole, including Shackleton Crater by Clementine (left), and the Arecibo Observatory in Puerto Rico (Spudis, 1999, p. 127, reprinted with permission of Cambridge University Press).

Lava Tubes

The idea of locating a lunar base inside a lava tube is decades old. It was revisited in 1988 by Coombs and Hawke, who stated that lava tubes would be a good location for a base if their existence and exact location could be confirmed before the mission was sent (Coombs and Hawke, 1988). Still, as late as 1999, Eckart says in his lunar base handbook that lava tubes would be a good location if they were found (Eckart, 1999, p. 118). Although lava tubes offer relief from several of the severe environmental conditions present on the Moon, further research into their exact location and properties is required; however, because they would not be used until Phase 2 or 3, this investigation could be done in Phase 1 and 2.

Locating the lunar base in a lava tube would fulfill several of the previously determined technical requirements that are unique to the Moon. The ceiling of the tube would provide natural meteoroid and radiation shielding significantly more than the 5.4 m required. The inside of the lava tubes may also have less dust, minimizing the complications that dust can cause (Billings et al., 2000), and what dust there is will not be exposed to the sun, reducing the levitation due to photoelectric change. Most probably, the temperature inside a lava tube is almost constant at approximately −20°C (Billings et al., 2000).

Peaks of Eternal Light

Some points on the lunar north pole may receive almost constant low angle sunlight due to their elevation and the small tilt of the Moon's axis. Previously, locations such as Malapert Mountain near the south pole were considered, which may also receive sunlight for more than half of the lunar day (Kruijff, 2000; Sharpe and Schrunk, 2002). More recently, locations on the rim of Peary Crater near the north pole were suggested, which receive constant sunlight during the summer, although the percent of time during which sunlight is received in the winter is unknown (Bussey et al., 2005). Peaks of eternal light would be useful locations for a lunar base because temperature is relatively constant at about −50°C, +/−10°C (Bussey et al., 2005). Also, solar power would be available for most of the time, and finally, peaks of eternal light are probably located near permanently shadowed craters, which may contain ice water. Also after further investigation, peaks of eternal light might provide excellent lunar base sites.

Recent Findings

There were eight missions since 1976, one of which was not initially dedicated to the Moon. Hiten was the first Japanese orbital lunar mission with a controlled crash on the Moon's surface in 1990. There were two American missions, Clementine (1994) and Lunar Prospector (1998), ESA sponsored mission, SMART 1 (2003), second Japanese mission SELENE (2007), first Chinese mission called Chang'e 1 (2007) and first Indian mission, Chandrayaan (2008). Finally, there was also one unexpected flyby mission when the faulty communication satellite AsiaSat 3/HGS 1 was directed to do two flybys to place it in geosynchronous orbit. They all contributed at various levels to the current knowledge of the lunar environmental conditions. Hiten was more focused on a series of technology demonstrations for future lunar missions. Three other dedicated lunar missions (two American and one European) concentrated on two major aspects of lunar studies: remote sensing of the lunar environment including optical imaging of the lunar surface and lunar mapping activities. Both are of utmost importance to the future missions and also related to the derivation of specific technical requirements of lunar structures.

Clementine was launched in 1994 and achieved the mapping of the lunar surface in the same year. It brought to attention a large variation of topography of the lunar poles, especially of the Moon's south pole and its South Pole-Aitken basin, and revealed the presence of this extensive depression caused by the impact of an asteroid or comet (See Figure 33.10). There is also a permanent dark area around the pole, which is sufficiently cold to trap water of cometary origin in the form of ice. The laser altimeter on Clementine gave, for the first time, comprehensive images of the lunar topography (See Figure 33.10) (Williams, "Clementine Project," 2005; Williams, "Clementine," 2000). The near side is relatively smooth in comparison to the far side, which has extreme topographic variation. The large circular feature centred on the southern far side is the South Pole-Aitken basin, which is 2,600 km in diameter and over 12 km in depth (Williams, "Clementine Project," 2005; Williams, "Clementine," 2000). In general, there is a similar range of elevation on the Moon's surface as the range exhibited by the Earth. Gravity mapping obtained from Clementine also revealed the crustal thickness, which has an average of 70 km, and varies from a few tens of kilometres on the mare basins to over 100 km in the highland

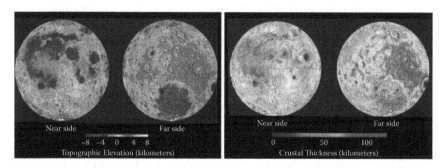

FIGURE 33.10
Lunar topographic elevation (left) and crustal thickness (right) as determined by Clementine (Spudis, 1999, p. 135, reprinted with permission from Cambridge University Press).

areas (See Figure 33.10) (Williams, "Clementine Project," 2005; Williams, "Clementine," 2000). The orbiting spacecraft also experienced a slight change in the velocity passing over mass concentrations called *mascons* and confirmed earlier findings of the local irregularities in the Moon's gravity field. All of these findings are related to the environmental conditions on the Moon, such as the distribution of the temperature due to topographic variation of the Moon's surface and some aspects of the change in gravity field, but also place high importance on the impact of the location on the structural technical requirements (Williams, "Clementine Project," 2005; Williams, "Clementine," 2000).

Lunar Prospector continued mapping of the lunar surface including low polar orbit investigation, and made studies associated with potential lunar resources: minerals, water ice and certain gases. It mapped the Moon's gravitational field anomalies (first encountered during the *Apollo* era). However, the controlled crash of the Lunar Prospector spacecraft into a crater near the south pole of the Moon on July 31, 1999 did not produce any observable signature of water based on the astronomical observations made with telescopes (Williams, "Lunar Prospector," 2005).

SMART 1 (Small Missions for Advanced Research in Technology) continues efforts of previous missions including the search for water ice at the south pole using NIR (near-infrared) spectrometry (ISU, 2003). It also made further advances in the geology, morphology, topography, mineralogy, and geochemistry of the lunar surface. It measured minerals and chemical elements using visible light, near-infrared and X-ray spectroscopy. The original mission of a 6-month lifetime was extended until August 2006 (Williams, "SMART 1," 2004).

The second Japanese mission after cancelled **Lunar-A** mission was named **SELENE**. This mission was launched on September 14, 2007 and it brought more instruments into lunar orbit and was equipped with two additional small satellites of 50 kg each. Thus, the mission consisted of a main orbiter (150 kg) called **Kaguya**, a small relay satellite (for communications), and a small VLBI (Very Long Baseline Interferometry) satellite (Williams, "Missions Under Consideration," 2005). The latter conducted careful investigation on the position and precession of the Moon's orbit. The orbiter performed a suite of experiments, some of which were similar to other scientific experiments performed by earlier American missions, as well as some that supplemented other planned efforts (Williams, "Missions Under Consideration," 2005). The overall impact of all of these studies from orbit are substantial but not with respect to technical requirements for future lunar structures, except some aspects associated with determining prime areas of future in-situ investigation of the lunar surface. This was the largest Moon mission since the *Apollo* program on an H-2A rocket.

The first Chinese mission to the Moon, **Chang'e 1**, was devoted to testing several technologies for future missions, and it could be treated as the Chinese reconnaissance mission. It was launched on October 24, 2007 from the Xichang Satellite Launch Center. It is intended to study the lunar environment through a similar suite of instruments as the Japanese missions, with one addition of a microwave radiometer to study the thickness of the lunar regolith. There are also instrumentation to study the solar wind and near-lunar region, the area just above the lunar surface (Williams, "Missions Under Consideration," 2005).

The first Indian lunar mission **Chandrayaan-1,** sponsored by ISRO, has an extensive suite of instrumentation, solicited partially through announcements of opportunities. The **Chandrayaan-1** mission was launched on October 22, 2008. Its focus is on studying 3D topography of the Moon and distribution of the minerals and elemental chemical species. However, the concentrated effort is directed to high-resolution mineralogical and chemical imaging of the lunar poles and chemical stratigraphy of the lunar crust. A unique effort will be devoted to map the height variation of the lunar surface features ("India's First Mission," 2004; Williams, "Chandrayaan-1," 2005). This will result in better choices of future landing sites. However, it will carry some value in the future mapping efforts of various lunar features, as they might affect the construction efforts, especially in phases two and three.

In summary, achievements of the above listed missions have contributed substantially to the knowledge of the Moon, although they do not bring much to the current knowledge with respect to technical requirements for future lunar structures. However, they have proven that the major impact of the individual technical requirements depends on the choice of location for the specific structure, regardless of the construction phase. Further advancements in the assessment of the technical requirements will be achieved through robotic missions with landing potential on the lunar surface and establishment of permanent stations to further quantify all parameters for the future lunar design code.

Planned Missions

The series of planned lunar missions shows some aspects of competition between various space-faring nations and poses some questions to streamline different efforts into a unified effort for future Moon utilization and eventual colonization. Some missions might contribute to extending the knowledge of the lunar environment and its impact on the future structural requirements. The leading efforts with respect to the planned lunar missions represent the United States' new exploration strategy of planned return to the Moon by 2020 with aggressive planning from the technology side (Berger, 2005), partially known Chinese efforts associated with their human flight program, and Indian and European ambitious plans concentrating on the scientific investigations of the Moon's environmental features. All of these efforts will contribute to further developing better technical requirements for all three phases of lunar construction. Some aspects are discussed below based on the available description of the current missions and their objectives.

The planned and/or approved missions include five American, one private (Trailblazer) and four by NASA (Lunar Reconnaissance Orbiter (LRO), the Gravity Recovery and Interior Laboratory mission GRAIL, the Lunar Atmosphere and Dust Environment Explorer LADEE mission, and the International Lunar Network (ILN) mission), one Chinese named Chang'e 2 (follow on to Chang'e 1 mission), one Indian named Chandrayaan-2 (follow on to the currently launched mission by ISRO) and one small university based German mission called Baden-Württemberg 1 (BW1) after the provincial program from this state of Germany and led by the Technical University of Stuttgart (Williams, "Timeline," 2005; Laufer and Roeser, 2005). The lunar exploration can be subject to substantial changes due to the current financial world crisis and in the political climate in leading space-faring nations.

Major contributions will be possible through the larger missions and those sponsored or funded by the national agencies, although the new interest from private or university sectors might be also important based on initial success from the Trailblazer or BW1 missions. It is also expected that the eventual series of landing lunar missions in the next decade will start detailed in-situ investigations of the lunar environment.

The NASA **Lunar Reconnaissance Orbiter (LRO)** mission is mainly devoted to identify landing sites for future robotic and human explorers. LRO is planned to be launched on April 24, 2009. It will also bring a suite of important instruments (some for the first time) to relatively low 30–50 km polar circular orbit. A large effort will be placed on studies of the Moon's radiation environment, lunar topography and scanning of the resources in the polar regions. Mapping of the composition of the lunar surface will continue (Lunar Reconnaissance Orbiter (LRO), Williams, 2005). There are three other planned American missions: **the Gravity Recovery and Interior Laboratory** mission **GRAIL** (2011), **the Lunar Atmosphere and Dust Environment Explorer LADEE** mission (2011) and the **International Lunar Network ILN** mission (2013). First two are lunar orbital missions under study by NASA (Williams, "Chronology of Lunar and Planetary Exploration (Future Missions)," 2008) and the third one is associated with the robotic portion of the lunar exploration architecture. This concept called the International Lunar Network (ILN) aims to provide an organizing scheme for all landed science missions in the next decade by involving each landed station as a node in a future lunar geophysical network. Ultimately there will be 8–10 nodes placed on the far side of the Moon. However, this concept requires a lunar communication relay satellite (NASA International Lunar Network Concept, 2008). The impact on the technical requirements for lunar structures from these studies can be foreseen, especially with respect to future radiation shielding or other arrangements for both human and structural protection during all construction phases.

Conclusion and Recommendations

This paper introduces a new classification of lunar structures into three phases. It includes most of the known important lunar environmental conditions and shows their quantitative representations as well as their impact on different technical requirements associated with each construction phase. These phases are mainly associated with the lunar structure use, and somewhat with their designed service duration. This review paper is not a fully comprehensive one as the current new wave of lunar research, in the direction of the utilization of the Moon or future colonization of the Moon, has just started. The major portion of the overall knowledge on lunar environmental conditions used for derivation of the technical requirements for lunar structures is based equally on in-situ investigations from the *Apollo* era and from the *post-Apollo* era of recent orbital missions.

The overall recommendations can be divided into three distinct groups. The first group is strategic recommendations; it outlines how to develop the construction code, which encompasses all proposed construction phases, and lists the data important to lunar construction that should be obtained through future lunar missions. The second group includes specific recommendations for baseline requirements for individual phases; however, it must be noted that all technical requirements from the previous phases should also be included (see the table in Appendix A with importance factors included for each phase). Finally, additional recommendations are made for areas that require further research in relation to lunar structures.

Strategic Recommendations

A concentrated international and multidisciplinary effort is proposed to evolve, based on the current and newly acquired knowledge, the *Lunar Construction Code* [as some efforts in this direction have already been attempted in the literature, *The Lunar Sourcebook* (Heiken et al. 1991) and *The Lunar Base Handbook* (Eckart 1999)]. The *Lunar Construction Code* will include a description of the lunar environmental conditions from an engineering point of view, with their quantification and available mapping. It will also include guidelines for the development of lunar structures as proposed in the three construction phases for future Moon utilization and colonization. This code will be upgraded on an annual basis, taking into account all available findings from the current missions.

Data in the following areas should be gathered in future lunar missions to include in the *Lunar Construction Code*:

- Local lunar soil conditions,
- Additional information about temperatures,

- More accurate and precise meteoroid flux data,
- Radiation fluctuation, depending on location and time,
- Gravity field irregularities, and
- Mapping development for seismic activity of meteoroids and Moonquakes, topography, and stratigraphy.

These mission objectives should be internationally coordinated for proper accumulation of all data.

Specific Recommendations for Technical Requirements

The specific recommendations include the technical requirements derived from the lunar environmental conditions. These requirements, which should be included in the *Lunar Construction Code*, are listed in more detail in Appendix A and summarized below (see Appendix A).

Phase 1

The materials used in phase one must be able to sustain very low temperatures, in some cases as low as $-233°C$. As well, the materials must be abrasion resistant against the lunar dust. Redundancy is required in electronics and solar panels to minimize the damage caused by radiation.

Phase 2

In addition to the requirements of phase one, further shielding is needed to protect the inhabitants of the structures; at least 700 g/cm^2 of regolith is required for radiation shielding. This regolith is also enough to lower the internal temperature range in most areas to 2°C or less, and it will provide sufficient meteoroid shielding to lower the risk of penetration to 9×10^{-8} impacts/ km^2/yr. The requirements due to the thin lunar atmosphere are also important in this phase. A minimum of 26 kPa of pure oxygen is required to sustain human life, although the pressure inside the structures will be greater with a mixed gas atmosphere to reduce the risk of fire. Because of this, the structures in this and the next phase will be pressure vessels; combined with the low gravity, these conditions result in large tensile stresses on the structure. Therefore the shapes of the structures in this phase will be spherical, cylindrical, or toroidal shells to eliminate corners and avoid stress concentrations.

The inflatable structures in this phase will be multi-layered, distributing some of the requirements such as temperature control, additional meteoroid shielding, and pressure containment over several different layers. These layers must include a bladder, restraint layer, and inner liner, and may also include a deploying mechanism and thermal insulation.

Phase 3

Most of the conditions from Phase 2 carry over to Phase 3, with a few more options. More shapes are possible because the structure will be constructed from in-situ resources. Materials produced in-situ, such as lunar concrete, must be tested before use to determine their properties; lunar concrete should also be manufactured in a pressure vessel (such as the construction dome) to prevent the water from evaporating and weakening the concrete. Radiation shielding in this phase should provide a minimum of 1,000 g/cm², as the Earth's atmosphere does, due to the longer duration stays on the Moon in this phase. This may be provided through electromagnetic shielding; if it is, alternate meteoroid shielding must then be used.

Additional Recommendations

In this category of recommendations, areas of suggested research associated with future lunar structures are indicated. Advances have already been made in some of these areas.

- Develop rigidizable pneumatic structures that are highly resistant to radiation, temperature and abrasion.
- Research vacuum multi-layered pressure vessels (with external vacuum conditions).
- Develop new materials that are highly temperature and radiation resistant.
- Develop a suite of initial in-situ structural experiments with regolith shielding, to be performed during the first robotic lunar missions.
- Develop a suite of initial in-situ investigations of lunar soil for lunar civil engineering applications, to be performed during the first robotic and later manned lunar missions.
- Develop radiation shielding for lunar structures.
- Develop a suite of ground-based experiments including simulations of lunar conditions.

Acknowledgment

The authors are thankful for comments from Dr. Victoria Hipkin and Dr. Bjarni Tryggvason of the Canadian Space Agency, and Dr. Harold Ogden of Saint Mary's University.

References

Angel, R. (2005). "A Deep Field Infrared Observatory Near the Lunar Pole." Phase 1 report. [On-line] Available: http://www.niac.usra.edu/files/studies/final_report/1006Angel.pdf, from NASA Institute for Advanced Concepts, Atlanta, Georgia. 2005.

Aulesa, V. (2000). "Architecture of Lunar Habitats." in *Proceedings of the Fourth International Conference on Exploration and Utilization of the Moon*, edited by B.H. Foing and M. Perry, ESA, ILEWG, The Netherlands, 2000, pp. 289–292.

Aulesa, V., Ruiz, F., Casanova, I. (2000). "Structural Requirements for the Construction of Shelters on Planetary Surfaces," in *Space 2000*, edited by K.M. Chua et al., American Society of Civil Engineers, Albuquerque, New Mexico, 2000, pp. 403–409.

Bell, E.V. (2005). "AsiaSat3." [On-line] Available: http://nssdc.gsfc.nasa.gov/database/MasterCatalog?sc=1997-086A. NASA Goddard Space Flight Center. Greenbelt, MD 20771. 2005.

Berger, B. (2005). "NASA Rolling Out Lunar Exploration Study Results," in *Space News*, Vol. 16 Issue 26, July 4, 2005, pp. 1,4.

Billings, T.L., Walden, B., York, C.L. "Lunar Lavatube Base Construction," in *Space 2000*, edited by K.M. Chua et al., American Society of Civil Engineers, Albuquerque, New Mexico, 2000, pp. 631–637.

Borin, A., Fiscelli, M. (2004). "An Inflatable Living Concept." in *Engineering, Construction, and Operations in Challenging Environments*, edited by Maji, A., Malla, R. B., ASCE, Houston, TX, 2004, pp. 797–804.

Buhler, C.R., Wichmann, L. (2005). "Analysis of a Lunar Base Electrostatic Radiation Shield Concept." Phase 1 report. [On-line] Available: http://www.niac.usra.edu/files/studies/final_report/921Buhler.pdf, from NASA Institute for Advanced Concepts, Atlanta, GA. 2005.

Bussey, D.B.J., Fristad, K.E., Schenk, P.M., Robinson, M.S., Spudis, P.D. (2005). "Constant illumination at the lunar north pole," in *Nature*, edited by P. Campbell, Vol. 434, 14 April, 2005, p. 842.

Cadogan, D.P., Scarborough, S.E. (2001). "Rigidizable Materials for use in Gossamer Space Inflatable Structures." [On-line] Available: http://www.ilcdover.com/products/aerospace_defense/supportfiles/AIAA2001-1417.pdf, 42nd AIAA/ASME/ASCE/AHS/ASC, Structures, Structural Dynamics, and Materials, Conference & Exhibit, AIAA Gossamer Spacecraft Forum, April 16–19, 2001 / Seattle, WA. 2001.

Carroll, K.A., Spencer, H., Arkani-Hamad, J., Zee, R.E. (2005). "Lunette: An Affordable Canadian Lunar Farside Gravity Mapping Mission," Canadian Space Exploration Workshop 5, CSA HQ, St-Hubert. 2005.

Casanova, I., Aulesa, V. (2000). "Construction Materials from In-Situ Resources on the Moon and Mars," in *Space 2000*, edited by K.M. Chua et al., American Society of Civil Engineers, Albuquerque, NM, 2000, pp. 638–644.

Coombs, C.R., Hawke, B.R. (1988). "A Search for Intact Lava Tubes on the Moon: Possible Lunar Base Habitats." *The Second Conference on Lunar Bases and Space Activities of the 21st Century*. NASA Conference Publication 3166, Vol. 1, Houston, TX. 1988.

Coronado, A. R., Gibbins, M. N., Wright, M. A., and Stern, P. H. (1987). "Space Station Integrated Wall Design and Penetration Damage Control," *Rep. No. D180-30550-4, Final Report, Contract NAS8-36426*, Boeing Aerospace Company, Seattle.

Criswell, M.E., Carlson, J.S. (2004). "Concepts for the Design and Construction of a Modular Inflatable Habitat," in *Engineering, Construction, and Operations in Challenging Environments,* edited by Maji, A., Malla, R. B., ASCE, Houston, TX, 2004, pp. 9–16.

Eckart, P. (1999). *The Lunar Base Handbook,* edited by W.J. Larson, McGraw Hill, Montreal, 1999.

Eckart, P. (2000). "Lunar Base Development Issues, Technological Requirements, and Research Needs," in *Proceedings of the Fourth International Conference on Exploration and Utilization of the Moon,* edited by B.H. Foing and M. Perry, ESA, ILEWG, The Netherlands, 2000, pp. 167–171.

Gorenstein, P. (2002). "An Ultra High Throughput X-Ray Astronomy Observatory With a New Mission Architecture." [On-line] Available: http://www.niac.usra.edu/files/studies/final_report/380Gorenstein.pdf, Prepared for NIAC, Smithsonian Institution Astrophysical Observatory, Cambridge, MA. 2002.

Greene, K. A. (2004). "Design Note on Post-Tensioned Cast Basalt," in *Engineering, Construction, and Operations in Challenging Environments,* edited by Maji, A., Malla, R. B., ASCE, Houston, TX, 2004, pp. 45–50.

Hawes, M. (2005). *Lunar Strategic Roadmap Status, Briefing to the ISS Strategic Roadmap Committee.* NASA Space Operations, 2005.

Hayashida, K.B., Robinson, J.H. (1991). "Single Wall Penetration Equations." [On-line] Available: http://ntrs.nasa.gov/archive/nasa/casi.ntrs.nasa.gov/19920007464_1992007464.pdf. NASA Technical Memorandum 103565. George C. Marshall Space Flight Center, NASA. 1991.

Heiken, G.H., Vaniman, D.T., French, B.M. (1991). "Lunar Databases and Archives," in *Lunar Sourcebook.* Cambridge University Press, New York. 1991, p. xix.

ISU. *Autonomous Lunar Transport Vehicle.* International Space University, Design Project Report. 2000.

Jolliff, B. L., Wieczorek, M. A., Shearer, C. K., and Neal, C. R., eds. (2006), *New Views of the Moon. Reviews in Mineralogy & Geochemistry,* Vol. 60, Mineralogical Society of America, Chantilly, VA.

Kennedy, K.J., Raboin, J., Spexarth, G., Valle, G., "Inflatable Habitats," in *Gossamer Spacecraft: Membrane and Inflatable Technology for Space Applications,* edited by Jenkins, C.H.M., American Institute of Aeronautics and Astronautics, Inc., VA. 2001, pp. 527–552

Koelle, H., Benaroya, H., Laufer, R. "Lunar Development Forum," *Lunar Base Quarterly.* Vol.13, No.2/April 2005 [On-line] Available: http://vulcain.fb12.tu-berlin.de/koelle/LBQ/LBQ_index.html. 2005.

Konopliv, A. S., Binder, A. B., Hood, L. L., Kucinskas, A. B., Sjogren, W. L., and Williams, J. G. (1998). "Improved Gravity Field of the Moon from Lunar Prospector," *Science,* 281, pp. 1476–1480.

Kruijff, M. (2000) "The Peaks of Eternal Light on the Lunar South Pole," in *Proceedings of the Fourth International Conference on Exploration and Utilization of the Moon,* edited by B.H. Foing and M. Perry, ESA, ILEWG, The Netherlands, 2000, pp. 333–336.

Langlais, D.M., Saulnier, D.P. "Reusable, Pressurized Dome for Lunar Construction," in *Space 2000,* edited by K.M. Chua et al., American Society of Civil Engineers, Albuquerque, NM, 2000, pp. 791–797.

Latham, G. V., et al. (1972), "Moonquakes and Lunar Tectonism," *Earth, Moon, Planets,* 4 (304), 1972, pp. 372–382.

Laufer, R., Roeser, H-P, "An Academic Small Satellite Mission Beyond Low Earth Orbit—The Lunar Mission BW1 of the University of Stuttgart." [On-line] Available: http://www.dlr.de/iaa.symp/archive_5/pdf/0508P_Laufer.pdf. University of Stuttgart, Institute of Space Systems, Stuttgart, Germany. 2005.

Lin, T. D. (1987). "Concrete for Lunar Base Conctruction," *Concr. Int.: Des. Constr.,* 9(7), 48–53.

Lin, T. D., et al. (1991), "Stresses in Concrete Panels Exposed to the Sun on the Moon," *Lunar Concrete,* edited by R. A. Kaden, American Concrete Institute, Detroit, MI pp. 141–154.

Lindsey, N.J. (2003). "Lunar Station Protection: Lunar Regolith Shielding," in *Proceedings of the International Lunar Conference 2003/International Lunar Working Group 5 – ILC2003/ILEWG 5,* edited by Durst, S.M., Bohannan, C.T., Thomason, C.G., Cerney, M.R., Yuen, L., AAS, San Diego, CA, 2003, pp. 143–148.

National Aeronautics and Space Administration (NASA). (1969). Apollo 11 Preliminary Science Report. Reviewed by West, J.M., Bell, P.R., Calio, A.J., Harris, J.W., Schmitt, H.H., Simpkinson, S.H., Stephenson, W.K., Wiseman, D.G. http://www.history.nasa.gov/alsj/a11/as11psr.pdf (July, 2005), NASA, Washington, DC.

National Aeronautics and Space Administration (NASA). (1970). Apollo 12 Preliminary Science Report. Reviewed by Calio, A.J., Harris, J.W., Langford, J.H., Mercer, R., Moon, J.L., Simpkinson, S.H., Stephenson, W.K., Warner, J.L., West, J.M. June 1, 1970. http://www.history.nasa.gov/alsj/a12/as12psr.pdf (July, 2005), NASA, Washington, DC.

National Aeronautics and Space Administration (NASA). (1971a). Apollo 14 Preliminary Science Report. Reviewed by Chapman, P.K., Duke, M.B., Foley, H.N., Harris, J., Herbert, F.J., Mercer, B., Simpkinson, S.H., Stull, P.L., Ward, M. June 1, 1971. http://www.history.nasa.gov/alsj/a14/as14psr.pdf (July, 2005), NASA, Washington, DC.

National Aeronautics and Space Administration (NASA). (1971b). Apollo 15 Preliminary Science Report. Reviewed by Allen, J.P., Anderson, K.F., Baldwin, R.R., Cox, R.L., Foley, H.N., Giesecke, R.L., Koos R.H., Mercer, R., Phinney, W.C., Robertson, F.I., Simpkinson, S.H. Dec. 8, 1971. http://www.history.nasa.gov/alsj/a15/as15psr.pdf (July, 2005), NASA, Washington, DC.

National Aeronautics and Space Administration (NASA). (1972). Apollo 16 Preliminary Science Report. Reviewed by Brett, A., England, A.W., Calkins, J.E., Giesecke, R.L., Holman, D.N., Mercer, R.M., Murphy, M.J., Simpkinson, S.H. Nov. 10, 1972. http://www.history.nasa.gov/alsj/a16/as16psr.pdf (July, 2005), NASA, Washington, DC.

National Aeronautics and Space Administration (NASA). (1973). Apollo 17 Preliminary Science Report. Reviewed by Parker, R.A., Baldwin, R.R., Brett, R., Fuller, J.R., Giesecke, R.L., Hanley, J.B., Holman, D.N., Mercer, R.M., Montgomery, S.N., Murphy, M.J., Simpkinson, S.H. 1973. http://www.history.nasa.gov/alsj/a17/as17psr.pdf (July , 2005), NASA, Washington, DC.

National Aeronautics and Space Administration (NASA). (2008) International Lunar Network (ILN), NASA Science Missions. http://nasascience.nasa.gov/missions/iln. (2008), NASA, Washington, DC.

Parnell, T.A., Watts, Jr., J.W., Armstrong, T.W. (1998) "Radiation Effects and Protection for Moon and Mars Missions," in *Space 98* Conference Proceedings, pp. 232–244, [On-line] Available: http://science.nasa.gov/newhome/headlines/space98pdf/cosmic.pdf. 1998.

Ruess, F., Kuhlmann, U., Benaroya, H. (2004) "Structural Design of a Lunar Base," in *Engineering, Construction, and Operations in Challenging Environments*, edited by Maji, A., Malla, R. B., ASCE, Houston, TX, 2004, pp. 17–23.

Sadeh, E., Sadeh, W., Criswell, M., Rice, E.E., Abarbanel, J. (2000) "Inflatable Habitats for Lunar Development," in *Proceedings of the Fourth International Conference on Exploration and Utilization of the Moon*, edited by B.H. Foing and M. Perry, ESA, ILEWG, The Netherlands, 2000, pp. 301–304.

Sharpe, B.L., Schrunk, D.G. (2002) "Malapert Mountain Revisited," in *Space 2002 and Robotics 2002*, edited by Laubscher, B.E., Johnson, S.W., Moskowitz, S.E., Richter, P., and Klingler, D. American Society of Civil Engineers, Albuquerque, NM, 2002, pp. 129–135.

Smith, D.H. (2002) (study director) "New Frontiers in the Solar System: An Integrated Exploration Strategy." [On-line] Available: http://www.aas.org/dps/decadal/. From the Solar Systems Exploration Survey, Space Studies Board, National Research Council. National Academy of Sciences. Washington, DC. 2002.

Space Studies Board (2007), "The Scientific Context of the Moon," Committee on the Scientific Contect for Exploration of the Moon, Space Studies Board, National Research Council of the National Academies, The National Academies Press, Washington, DC, 2007. http://books.nap.edu/openbook. php?record_id=11954&page=R1

Spudis, P.D. (1999). "The Moon," in *The New Solar System*, edited by Beatty, J.K., Petersen, C.C., Chaikin, A., Cambridge University Press. 1999, pp.125–140.

Toklu, Y.C. (2000). "Civil Engineering in the Design and Construction of a Lunar Base," in *Space 2000*, edited by K.M. Chua et al., American Society of Civil Engineers, Albuquerque, NM, 2000, pp. 822–834.

Williams, D.R. (2004, 2008) "Chandrayaan-1 Lunar Orbiter." http://nssdc.gsfc.nasa. gov/database/MasterCatalog?sc=CHANDRYN1

"Change'e 1" http://nssdc.gsfc.nasa.gov/planetary/prop_missions.html#change1

"Clementine Project Information." http://nssdc.gsfc.nasa.gov/planetary/clementine.html

"Clementine." http://nssdc.gsfc.nasa.gov/database/MasterCatalog?sc=1994-004A

"Chronology of Lunar and Planetary Exploration (Future Missions)." http://nssdc. gsfc.nasa.gov/planetary/chrono_future.html

"Lunar Exploration Timeline." http://nssdc.gsfc.nasa.gov/planetary/lunar/lunar-timeline.html

"Lunar Prospector." http://nssdc.gsfc.nasa.gov/planetary/lunarprosp.html

"Lunar Reconnaissance Orbiter (LRO)." http://nssdc.gsfc.nasa.gov/database/MasterCatalog?sc=LUNARRO

"Lunar-A." http://nssdc.gsfc.nasa.gov/database/MasterCatalog?sc=LUNAR-A

"Planetary and Lunar Missions Under Consideration." http://nssdc.gsfc.nasa.gov/planetary/prop_missions.html

"SELenological and ENgineering Explorer (SELENE)." http://nssdc.gsfc.nasa.gov/database/MasterCatalog?sc=SELENE

"SMART 1." http://nssdc.gsfc.nasa.gov/database/MasterCatalog?sc=2003-043C

"The Moon." http://nssdc.gsfc.nasa.gov/planetary/planets/Moonpage.html

NASA Goddard Space Flight Center. Greenbelt, MD, 2004, 2008, Mission Homepages and Main Websites:

NASA Moon Homepage http://nssdc.gsfc.nasa.gov/planetary/planets/Moonpage. html;

Hiten http://www.isas.ac.jp/e/enterp/missions/complate/hiten.shtml, http://spaceinfo.jaxa.jp/db/kaihatu/wakusei/wakusei_e/hiten_e.html, http://spaceinfo.jaxa.jp/db/kaihatu/wakusei/wakusei_e/hagoromo_e.html;

Clementine http://nssdc.gsfc.nasa.gov/planetary/clementine.html, http://nssdc.gsfc.nasa.gov/database/MasterCatalog?sc=1994-004A;

AsiaSat3/HGS-1 http://nssdc.gsfc.nasa.gov/database/MasterCatalog?sc= 1997-086A;

Lunar Prospector http://nssdc.gsfc.nasa.gov/planetary/lunarprosp.html;

SMART 1 http://smart.esa.int/science-e/www/area/index.cfm?fareaid=10;

Trailblazer http://www.transorbital.net/TB_mission.html;

Lunar-A http://www.jaxa.jp/missions/projects/sat/exploration/lunar_a/index_e.html (now cancelled);

SELENE http://nssdc.gsfc.nasa.gov/planetary/prop_missions.html, http://nssdc.gsfc.nasa.gov/database/MasterCatalog?sc=SELENE;

Chang'e 1 http://nssdc.gsfc.nasa.gov/planetary/prop_missions.html#change1;

Chandrayaan-1 http://www.isro.org/chandrayaan-1/announcement.htm;

Lunar Reconnaissance Orbiter (LRO) http://nssdc.gsfc.nasa.gov/database/MasterCatalog?sc=LUNARRO;

BW1 http://www.dlr.de/iaa.symp/archive_5/pdf/0508P_Laufer.pdf

Appendix A: Technical Requirements

Condition	Quantification	Phase One Importance	Phase One Technical Requirements	Phase Two Importance	Phase Two Technical Requirements	Phase Three Importance	Phase Three Technical Requirements
Temperature	Temperature range = 280 K	High	Material must not be brittle above −233°C in permanently shadowed craters, −150°C at the equator, −85.5°C at mid-latitudes, and −63°C around the poles.	High	Conditions from phase one Additionally, insulation or shielding required, 2.5 m or more of regolith	High	Conditions from phase one and two
Radiation	Average dose = 0.25 Sv/yr	Medium	Electronics and solar panels should be redundant in case of radiation damage	High	Conditions from phase one Additionally, minimum 700 g/cm² shielding must be provided (Approximately 5 m of regolith)	High	Conditions from phase one 1000 g/cm² shielding should be provided, or equivalent electromagnetic shielding

Continued

(Continued.)

		Phase One		Phase Two		Phase Three	
Condition	Quantification	Importance	Technical Requirements	Importance	Technical Requirements	Importance	Technical Requirements
Atmosphere/ Pressure	3 nPa (p.15, ISU, 2000)	Medium	Instrumentation must be vacuum tested	High	Conditions from phase one; Structure must be pressure vessel with minimum 26 kPa of internal pressure, causing high tensile stresses on the structure	High	Conditions from phase one; Concrete must be manufactured in pressure vessel to prevent water evaporation and resultant weakening
Meteoroids	Micrometeoroids $v = 13$ to 18 km/s	Low	Little if any defence is possible	High	Meteoroid shielding must be provided, minimum 0.5 m regolith	High	Meteoroid shielding must be provided separately from radiation shielding if electromagnetic radiation shielding is used
Dust	More than 50% of particles between 20 and 100 μm, (Toklu, 2000)	Medium	Material must be abrasion resistant, mechanical parts must be sturdy, able to operate in dusty conditions	Medium	Conditions from phase one	Medium	Conditions from phase one

Length of Lunar Day	Lunar day = 29.53059 Earth days	Low	Effect depends on mission objectives	Medium	Missions should arrive at beginning of lunar day, make use of light and warm temperatures	Low	Conditions from phase two
Gravity	g =1.62 m/s²	—	Self-weight of structure is less	Low	Tensile stresses are not countered by gravity	Low	Conditions from phase one
Seismicity	Maximum quake recorded 4 on Richter Scale, average 1–2	—	Lower risk than on Earth	—	Conditions from phase one	—	Conditions from phase one

Notes: The location of the lunar base, specific to the mission objectives and phase, will alter the importance of the environmental conditions, changing the technical requirements of the structure. Particularly in phase two, it must be considered in conjunction with the environmental conditions to determine the technical requirements of the structure. Some locations that will have a significant impact include the following:

Lavatubes *provide natural temperature, radiation, and meteoroid shielding, shifting the focus of environmental conditions to the near vacuum conditions, dust, and gravity.*

Peaks of Eternal Light *provide relatively constant temperatures, as well as other benefits. For this location, radiation, pressure and meteoroids become more significant considerations.*

Appendix B: Radiation Analysis

The significant biological risk caused by radiation on the Moon demands substantial structural shielding beginning in Phase 2, when humans begin to spend time in lunar structures. Currently, there are no specific radiation limits, and suggested doses depend on factors such as length of exposure, area exposed, and age at the start of exposure. Table A-1 gives suggested limits from the National Council on Radiation Protection for blood forming organs, eyes, and skin, over 30 days, a year, and a career (See Table A-1).

For missions early in the second phase, these limits can be used; however, as individuals spend more time on the Moon, more specific annual limits are required.

The NCRP also gives an equation for the maximum radiation that should be received, depending on the age at which the dose starts (see equation B-1, (Parnell et al., 1998)). From this equation, maximum career doses are calculated (see Table A-2).

TABLE A-1

NCRP Recommended Ionizing Radiation Exposure Limits for Flight Crews[a] (Parnell et al., 1998)

	Blood Forming Organs	Eye	Skin
Depth (cm)	5.0	0.30	0.01
30 Days (Sv)[b]	0.25	1.0	1.5
Annual (Sv)	0.50	2.0	3.0
Career (Sv)	1.0 to 4.0	4.0	6.0

[a] The career depth dose-equivalent is based upon a maximum 3% lifetime risk of cancer mortality. The dose equivalent yielding this risk depends on sex and on age at the start of exposure. The career dose equivalent is nearly equal to 2.0+0.075 (AGE-30) Sv for males and 2.0+0.075 (AGE-38) Sv for females, up to 4.0 Sv. Limits for 10 years exposure duration: "No specific limits are recommended for personnel involved in exploratory space missions, for example, to Mars" (NCRP No. 98, 1989, p.163).
[b] Sievert-Equivalent dose determined by multiplying the absorbed dose at each energy deposition value (Linear Energy Transfer (LET)) by the corresponding quality factor for each ion and energy.

TABLE A-2

Allowable Career Dose (in Sieverts) by Age at the Start of Radiation, and Gender

Age at start of Career Dose	20	30	40	50	60	70	80	90
Male	1.25	2.00	2.75	3.50	4.25	5.00	5.75	6.50
Female	0.65	1.40	2.15	2.90	3.65	4.40	5.15	5.90

Table derived from equation (1), given by NCRP (Parnell et al., 1998)

$$\beta = 2.0 + 0.075 \times (\alpha - T_0), \qquad \text{(B-1)}$$

$T_0 = 30$ for males, 38 for females, $\alpha = $ Age at start of dose, $\beta = $ Career dose maximum

Beginning in phase two, regolith will be the primary method of radiation shielding, and in phase three, it may be complemented with other methods such as electromagnetic shielding. By increasing the mass through which the radiation particles must pass, regolith decreases the energy of the particles, eventually stopping them and reducing the dose (Parnell et al., 1998). However, a thin layer of shielding breaks down some particles and creates high-energy particles, or brehmsstrahlung radiation, which is more harmful than the initial radiation (Buhler and Wichmann, 2005; Aulesa, 2000). For this reason, if any shielding is used, it must be sufficiently thick. Figure A-1 shows the changes in dose of GCR and SEP events caused by different thickness of shield. As well, the change in flux of different radiation particles is seen in Figure A-2. While the dose of GCR with no shielding is slightly lower than that with 500 g/cm^2, shielding is required to reduce the dose caused by SEP events; therefore, more than 500 g/cm^2 of shielding should be used. Further, the atmosphere on Earth provides 1000 g/cm^2 of shielding, so this level of protection is desired on the Moon as well to reduce the radiation to Earth levels. For shorter stays on the Moon, particularly in phase two, 700 g/cm^2 is considered adequate (Aulesa et al., 2000). If the density of the regolith is between 1.3 and 1.5 g/cm^3, the regolith shield will have to be between 4.7 and 5.4 m thick.

Presently, there are no set radiation limits for lunar missions; the NCRP suggestions should be adhered to, although radiation doses should also be kept as low as possible. To achieve this, a minimum of 5.4 m of regolith is

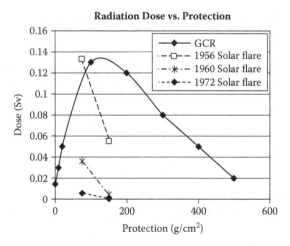

Radiation Dose vs. Protection

FIGURE A-1
Radiation dose as thickness of regolith increases, derived from Aulesa, 2000; Lindsey, 2003.

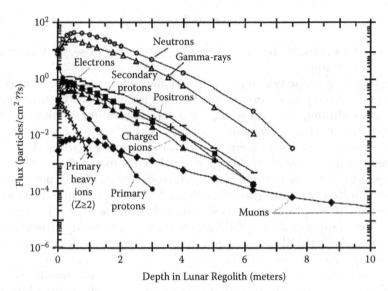

FIGURE A-2
Flux of radiation particles through depths of regolith (Parnell et al., 1998).

suggested for phase two, and in phase three, the shielding should provide protection equal to that of the Earth's atmosphere, at 1000 g/cm².

Appendix C: Meteoroid Analysis

Although the effectiveness of a regolith shield in preventing meteoroid damage is unknown and experimental testing must be done before such a shield can be assumed to be effective, the Fish-Summers Penetration Equation is used here to determine the approximate protection that will be provided with a layer of regolith 5.4 m thick. This thickness is used because it is the recommended amount for radiation protection, beginning in phase two. The Fish-Summers Equation is given below (see Equation C-1) (Hayashida and Robinson, 1991).

$$t_t = k_t \times m_m^{0.352} \times v_m^{0.875} \times \rho_m^{1/6} \qquad \text{(C-1)}$$

t = thickness of target (cm)
k = constant for target material
m = mass (g)
v = velocity (km/s)
ρ = density (g/cm³)

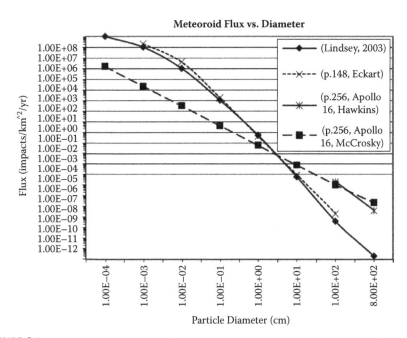

FIGURE C-1

Meteoroid Flux versus Particle Diameter, derived from Lindsey, 2003; Eckart, 1999; NASA 1972.

The subscript t stands for target, m for meteoroid, Al for aluminum, and re for regolith

$k_{Al} = 0.57$

$v_m = 18$ km/s (maximum velocity of probable range (p. 148, Eckart, 1999; p.15, ISU, 2000) used to increase safety)

$\rho_m = 0.5$ g/cm³

$\rho_{Al} = 2.7$ g/cm³

$\rho_{re} = 1.3$ g/cm³ (minimum value (Aulesa, 2000) used to increase safety)

$k_{re} = 1.18$ (using $t_{Al} \times \rho_{Al}/\rho_{re} = t_{re}$ (Lindsey, 2003))

Assuming a regolith thickness of 5.4 m, the equation shows that a meteoroid of maximum mass of 37 kg, or diameter of 52 cm, could impact the shield and not penetrate through spallation (Hayashida and Robinson, 1991). This lowers the flux of penetrating meteoroids to between 10^{-8} and 10^{-7} impacts/km²/yr (see Figure C-1) (Lindsey, 2003; p.148, Eckart, 1999). Therefore, 5.4 m of regolith should be more than sufficient shielding. However, this calculation can only be used as guide, as the equation is not intended for evaluating regolith as a shield.

34

Bidu Guiday: Design Concept for the First Manned Lunar Base

BIDU GUIDAY – "BEAUTIFUL MOON" IN CHARRUA, NATIVE LANGUAGE FOR URUGUAY'S INDIANS

Giorgio Gaviraghi

Executive Vice President, eDL

Giorgio Gaviraghi, architect, graduated from Milan Polytechnic in 1968. He is responsible for several advanced designs for aerospace and innovative products as an architect and designer. Founder with Maria Adela Gimenez of eDL, an advanced design company in architecture, city planning, and system design. He has presented over 20 papers to major space conventions, including proposals for space construction systems, asteroid deflection, Martian city and space settlements. As an instructor for high school student teams, he was the winner of over ten international awards in space settlement and advanced aerospace design. He is author of the science fiction book *First Contact*.

Introduction

The first manned lunar base as planned by the latest NASA's Vision for Space Exploration requires an all new approach and analysis of all design parameters with an advanced and different point of view than in the past. A lunar base is not another base similar to those existing in Earth's Poles or in other extreme conditions. A lunar base must be a terrestrial micro-ecosystem transplanted to the Moon and it must assure comfortable living conditions to its crew as well as many other necessary functions. In this paper we will analyze most requirements that will be illustrated by several design proposals that represent state of the art condition.

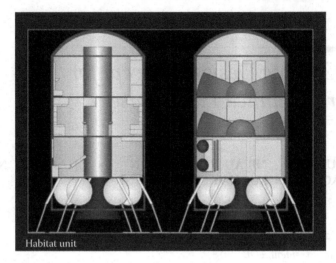

Habitat unit

FIGURE 34.1
Bidu Guiday, the first lunar base.

Goals of the Proposal

The first manned base has the following requirements:

- safety of operations
- minimum number of missions
- maximum flexibility and expansion capabilities
- ease of assembly
- maximum utilization of local resources.

Bidu Guiday is composed of two basic units entirely built on Earth, the habitat and the cargo module that will be delivered unmanned to the selected lunar site.

The habitat unit contains a self sufficient three story facility. The lower level is equipped with an airlock, dust cleaning and lockers areas for EVA activities, plus a medical facility with lab and facilities for first aid emergencies and a chemical lab.

The intermediate level contains the common facilities and equipment such as for food preparation, dining, living, meeting, including command control and communication spaces and for fitness.

Furthermore a special area at a lower level is radiation proofed for emergencies due to high solar flares events. The upper level contains all personal spaces including maintenance storage and support areas. The cargo unit contains all outdoors systems and equipment, including the connectors, the airlocks, life support systems, power generator, communications, material transformation requirements and four domes for different support activities to be connected in situ with the habitat module. Once unloaded the cargo module can be reutilized as an additional base or for greenhouse functions.

Design Parameters

In order to optimize the design the following parameters were considered:

- dimensions to fit the latest Ares V version (5.5 m diameter, approx 7 m height)
- minimum number of missions
- manned or robotic simple installation
- transformable space to allow most functions
- expansion possibilities
- maximize transportable space
- minimize in situ assembly activities
- non inflatable components
- minimum risk mission architecture
- total weight 10 tons
- maximize dust control.

Bidu Guiday satisfies all design parameters and consists of a habitat module for a total of 60 sq m on three levels, 20 m long connectors, and four domes of 5.5 m diameter to house the rover maintenance facilities, the greenhouse and animal breeding facilities, and a lunar lab plus EVA support.

The cargo module will be successively overhauled and utilized as an additional operational space for different activities, including food production. Total base components are carried by two unmanned missions, one for the habitat and the second one for the cargo. Manned or robotic assembly and unloading the cargo module are optional and final decisions will be taken in accordance with the robotic state of the art at the assembly time.

Total assembly time is estimated to be 30 hours while the entire two missions can be undertaken in a single month schedule.

Recommendations

This is a preliminary architectural design, not based on latest NASA lunar base specifications and is used as a reference model for further improvements.

Our goal is to build a simulated lunar base, following a redesign to include all eventual NASA or other parties' comments, feedback and specifications, if possible. The simulated base, while movable anywhere if necessary, will be built in Uruguay and can be utilized as a working mock up for a working design lab to define and prove each system and subsystem or for scientific tests and simulated lunar missions deemed necessary. At the end of its useful mission, it will be sent as a permanent addition to the Uruguayan Antarctic Base.

The Mission and Assembly Sequence

The entire lunar base system is contained in two main components, the habitat and cargo modules, both to be sent unmanned to the selected lunar site before any manned mission. The assembly sequence will be the following:

1. Habitat module landing in the selected site
2. Cargo module landing at approximately 50 m from habitat module. At this stage optional robotic or manned installation can be forecasted
3. Rover unloads lower connector
4. Rover installs lower connector
5. Vertical telescoping component exits from lower component
6. Rover installs upper module over lower one
7. Bridge connection to habitat module airlock with telescoping parts of upper module
8. Rover unloads and installs dome units
9. Telescoping connector preparation
10. Dome floor installation
11. Dome opened
12. Dome connected to final connector module
13. Same procedure for remaining 3 domes.

At this stage the base is all installed. If not installed by manned activities, then the first manned mission can land near the base. Once the manned crew is present the cargo module tanks and engines can be disassembled and carried away. The tanks near the landing pad are to contain lox fuel that is to be manufactured in situ while the engine can be stored for future requirements. The cargo module lower part can be closed and the entire module connected to the base and reutilized as a greenhouse and storage facility.

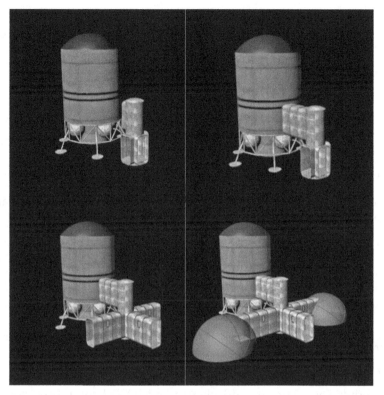

FIGURE 34.2
The assembly.

Habitat Unit

This unit is divided in three circular levels of 5.5 m diameter and 2.1 m height. Its design is based on a circular module of 6 units per level in order to allow for standardization of components and dimension flexibility. The entire system is composed of single, half, and one and a half modules, as per the chart.

Base Components

Habitat Module

The habitat unit is a three story facility to house the crew. The unit is divided into the following modules:

- Lower level
  - airlock, dust cleaning and locker facilities
  - medical lab and first aid facility

- general lab
- personal care
- maintenance
- Middle level
  - food preparation system
  - living/meeting area
  - command, control and communications
  - fitness center
- emergency toilet

This level is equipped with extra shielding to avoid radiation and can be utilized during solar flares.

- Upper level
  - four personal space modules
  - personal care module
  - maintenance module

Lower Level

Lower Floor/Dust Control and Interchange System

This module consists of the airlock area and related equipment. For extra safety reasons a dust collecting area, equipped with laminar flow overhead and a handheld air shower for boots and other critical parts, and a liquid trap for dust protected by a metal grille is included.

The overall areas of the base are divided in three parts for contamination purposes:

- red area, the risk area consisting of the personnel and rover airlocks
- the yellow area, safe but with some contamination risks due to its functions, all lockers, greenhouses, rover maintenance and laboratories located outside the habitat unit
- green area, the safest in all levels of the habitat unit excluding the interchange area.

Procedure for incoming EVAs:

- enter airlock
- enter dust control area where crewmembers will blow away all dust in the suit and helmet, will deposit in lockers all outside suits, boots and helmet

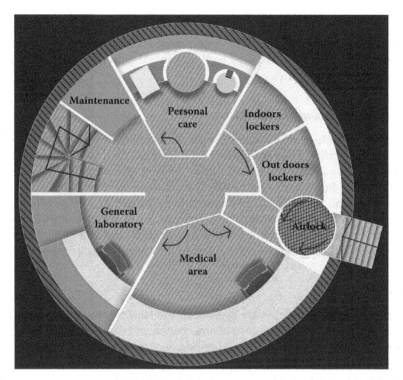

FIGURE 34.3
Lower level plan.

- enter lockers area where they will store inner part of suits and change to normal suits
- additional handheld air shower before leaving the area

All levels' baseboards will be rounded for easy cleaning and to avoid dust formation.

Medical Module

This module consists of all requirements for medical assistance. Emergency, first aid in case of injury or other common diseases can be treated with this equipment.

The module consists of a bendable bed for patient care, with wall panel containing monitors and displays showing all the vital information readings. Included in the wall panel is equipment such as oxygen or phleboclysis, and in the closet most common medical equipment. The medicine cabinet contains all needed pharmaceuticals and assistance equipment for daily check ups.

Lower Floor General Lab

This module consists of all utilities, equipment and containers to perform lab tests on geological specimens as well as general activities as photography, medical or other activities.

Circulation Stairs and Maintenance

Half of this module is dedicated to the stairway connecting the different levels of the facility. The remaining half module is dedicated, in accordance to maintenance in the lower and upper levels and emergency toilet in the middle level. Maintenance in the upper level consists of a washing machine, a press and ironing system and containers for bleach as well as other products for washing and ironing. The lower part of this maintenance module contains a roomba type robot to perform vacuum cleaning and washing of all floors on a daily basis to assure maximum cleanliness to the facility.

Personal Care

Two personal care modules are present in the facility, one at the lower level mostly for EVA assistance in case of emergency and the impossibility to raise an inert body for two levels. The other personal care module is located in the upper floor as support to the personal space modules.

The main systems of the personal care modules are:

- Shower unit – This unit is designed for maximum safety, without any edges and all subsystems are recessed. Water flow is controlled by an electronic wireless device that allows its positioning in an optimal location from outside and inside the shower pod; the same device controls the temperature and the nozzle types. Liquid soap or shampoo is delivered premixed with water to the interested body parts and is controlled by the same device. For drying purposes warm air is blown through the walls to the various body parts interested to the process avoiding the use of towels or other physical equipment. Used water is delivered to a treatment system in order to be cleaned and reutilized in the toilet or irrigation system.

- Toilet unit – This unit is entirely paperless and avoids any contact between the interested body parts for cleaning purposes. A jet of warm water is delivered to the interested parts controlled by the user and recessed in the toilet seat. Such water jet is followed by an air jet to the same body parts for drying purposes.

- Wash basin unit – This unit has been designed to allow washing most body parts without the need of the shower and its higher water consumption. For this purpose the wash basin and its controls are movable, up and down for about 60 cm, allowing the washing of the

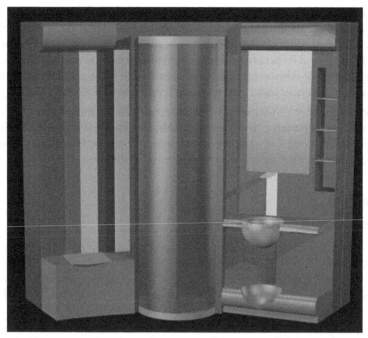

FIGURE 34.4
Shower/basin unit.

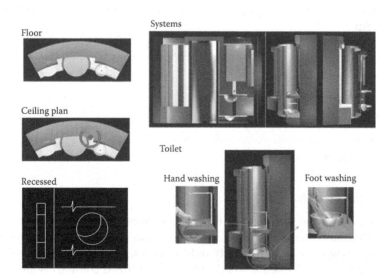

FIGURE 34.5
Layout.

head as well as the feet with the same basin. Controls are electronic wireless controlling the water flow, temperature and liquid soap mix, while an air flow allows the drying of the body parts without the need of towels and its potential danger for contamination and maintenance requirement. The basin unit is equipped with most needed items such as shaving equipment and personal color coded compartments.

- Storage system – Overhead storage container for most used supplies and two personal color coded containers, one wall mounted behind the basin unit for small personal items and another bigger one behind the toilet unit for crewmembers' utilization.

Middle Level

Middle Level Food Preparation System

Food preparation requires a complex system composed of many specialized subsystems, each one integrated by independent microsystems. Since food production will not start immediately but will progress in time, maximum flexibility is needed to add what may be required in the future with later missions, which must be easily integrated in the main system. For that reason each microsystem can be added or changed in time since it is based on a standard 20 × 20 × 20 cm micromodule. Another important factor to be considered is that most food, at least in the beginning, will be lyophilized or frozen, coming from Earth, while the food produced in situ will require processing not usually required in Earth kitchens, since the entire cycle, from seeding, harvesting, cleaning and processing, will be involved.

The one and a half food preparation module is composed by the following systems:

- storage
- wet preparation
- dry preparation
- cooking
- eating
- maintenance

Systems and Subsystems

The chart shows the different systems and subsystems. Every subsystem is independent and, before in situ food production could be considered on a permanent basis and with a variety of food types, most subsystems will not be installed. They will be installed in their specialized compartments, to be used and delivered as storage or other functions before fully operational. As an example of subsystems, we are showing the utensil module and the meal module. The first one is part of the dry preparation module while the meal module is stored in the dining module for ease of use.

The utensils module is composed of:

- cutter (no knife for safety and ease of use)
- mixer general server
- divider
- pasta server

All utensils, with the exception of the cutter rotating blade, but without personal contact, will be in lightweight plastic, color coded if necessary. Also a slicer and a grater machine are included in the other half of the micromodule.

Meal Subsystem

This subsystem is composed of a personal color coded meal and snack tray in lightweight plastic. The bigger meal tray contains a main plate and two service plates, a glass for beverages and a spork, a combination of fork and spoon for personal use. The snack tray contains a snack compartment and a warm beverage cup. All trays are color coded by crewmember and are stored in the dining module countertop for ease of use. Warm and cold beverage dispensers are located in the supporting walls of the dining module for ease of use while the upper part contains the water tanks to feed such subsystems.

Living and Meeting Area

This area is a transformable one for flexibility of functions. It can be used as a meeting room with chairs and table or as a living area with soft chairs equipped with PC screens. Four folded chairs and a table are stored in the wall module while not in use and the soft chairs are stored under the meeting table.

Fitness Center

This module is composed of a wall recessed cyclette and running belt, plus shoulder and arm exercise equipment. Due to the limited space, each unit can be used alone.

Command Control Communications

This module controls and manages the entire lunar base. Two complete work stations are designed with all controls and screens recessed in the counter-top or wall.

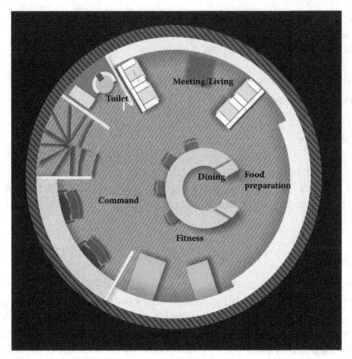

FIGURE 34.6
Middle level plan.

Upper Level

Personal Space

This personal space allows maximum privacy for crewmembers. Three alternative types are included for test purposes with the same features but a different layout and organization:

Type A	Overhead fixed bed, closet and work station at lower level
Type B	Overhead and lateral closet, folded bed under work station
Type C	Overhead storage, folded bed under work station

Exterior Components

Connectors

The connectors are outside telescoping structures that allow all connections between modules. In accordance with their location there are three different types of connectors:

Type A	One vertical modular extension and one horizontal 3-modules extension

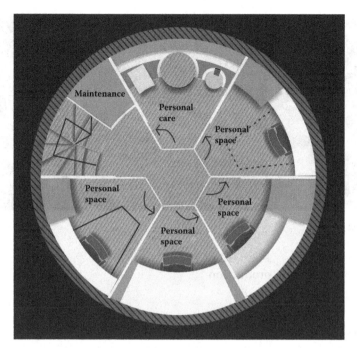

FIGURE 34.7
Upper level plan.

Type B   Four horizontal directions with two and three modular extensions

Type C   Three horizontal directions with two and three modular extensions

A rover connector with three modular horizontal extensions is also included.

Domes

This dome system is designated the AstroHab 3; since the first two are similar and designed by the same group, this represents an evolution compared to the previous ones since instead of separate modules each unit consists in a telescoping hemisphere that allows via rotation the closure of the dome.

Interior modules, attached to the walls and successively filled with regolith, for antiradiation purposes, will guarantee their safety and functionality. Each dome has a different function and in particular:

Type 1   Agricultural greenhouse

Type 2   Rover maintenance and workshop

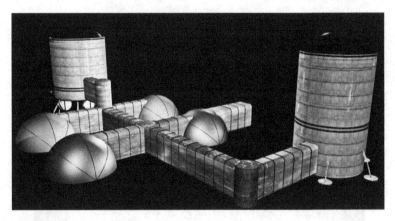

FIGURE 34.8
Finished base.

Type 3   Food production
Type 4   Laboratories

Dome Installation Sequence

The sequence of the dome installation is as follows:

1. half floor extension
2. full floor completion
3. dome extension
4. dome completion
5. interior panel installation
6. regolith wall filling

Cargo Unit

This unit, to follow the habitat unit, will deliver all necessary equipment, hardware and components to complete the lunar base facility. The cargo unit is divided in three levels:

- Level 1 contains the rover vehicle with mechanical arm for heavy loads transportation and the personal connectors.
- Level 2 contain two AstroHab 3 domes and the rover connector
- Level 3 contains the additional two AstroHab domes plus all interior panels and related equipment

Cargo Module Reutilization

Once unloaded, in a following manned mission, the cargo unit could be reutilized, after proper refurbishment for other functions. While the tanks will be carried to the spaceport fuel facilities for utilization by in situ manufactured lox, the engine will be stored for potential future utilization. The cargo area will be reutilized for food production in several levels of the aeroponics facilities. The upper part could be equipped with a solar mirror for agricultural purposes transmitted to the various levels by fiber optics that reach the light diffusers.

Finished Base

After the first phase, the finished base will consist of the two connected habitats and cargo modules, their connectors and four domes for multiple purposes.

Expansion Possibilities

Any type of expansion can be planned since the connectors allow any type of directions for future components.

Future Bases and Expansion

AstroHab

After the first base is established completely different approaches and strategies must be followed. While the first base, mostly for safety reasons can be mostly manufactured on Earth, future expansions can rely on construction and assembly performed on the Moon utilizing technologies and materials that maximize local participation.

Our proposal for future expansions is based on the AstroHab system, a modular construction system utilizing mostly in situ materials. Such systems, in their variations AstroHab1, 2 and 3 (already used in the first base) allow for modular construction, in situ assembly and utilization of local regolith to fill the modular forms to assure radiation protection, stability and insulation to the system.

Schedule of Activities

The schedule of activities can be so summarized:

1. Concept design	3 months
2. NASA's comments and feedback	2 months

3. Simulated base construction   8 months
4. Preliminary design of final configuration 1 year
5. Prototype manufacturing   3 years

This plan will allow the building of the base through a self financing scheme based on sponsorships and royalties from new products that can be derived from the concept design of the lunar base. This is because it contains over fifty patents for new and innovative products that could be sold on Earth and represents the results of the advanced designs necessary to satisfy space requirements.

While every supplier will deliver its part as a participating sponsor, design royalties will pay all expenses and volunteer work will support most personnel costs since this project is considered a grassroots effort by an entire country.

The design of Bidu Guiday has been prepared in cooperation with the architectural faculty of the ORT University and the UTU educational organization in Montevideo, including several volunteers, technicians and graphic designers coordinated by the architect Maria Adela Gimenez supported by eDL, a design company with offices in Milan and Montevideo headed by the author.

35

Design and Construction of a Modular Lunar Base

Werner Grandl

Werner Grandl was born in Vienna, Austria. He attended the Technical University Vienna, where he earned a degree in architecture in 1984; in 1985 he was in the military service of the Austrian Air Force; from 1986 to 1993 he had experience in some engineering offices; from 1994 to 2008 he was a freelance architect and civil engineer, and since 1986 he has conducted and published studies on space stations and space colonies.

ABSTRACT   Which is the favorable structural design for a Lunar Base? Is it an inflatable pneumatic structure, covered with lunar material (regolith), or a simple metal-frame structure with layers for thermal insulation? A Lunar Base, however, has to provide shelter against cosmic rays, solar flares, micrometeorites and huge temperature amplitudes. It should be easy to build and the risks for humans have to be minimized. Human safety and possible rescue operations imply a stiff construction and structural redundancy.

This paper compares a proposed double-shell structure for Lunar Base modules with single-shell designs and inflated structures, focusing on shielding and thermal protection. As an in situ resource lunar material is used for shielding. Due to its modular design and to semi-robotic assembling the proposed Lunar Base can be enlarged by stages easily.

The paper is mainly based on the Lunar Base 2015 Stage 1 Preliminary Design Study, W. Grandl/Acta Astronautica 60(2007) 554–560. [1]

Introduction

Since the beginning of the space age many proposals to construct space stations have been made, either on planetary surfaces or in the orbits of celestial bodies. In the 1950s Wernher von Braun suggested a pneumatic torus, similar to a tire, which should have been inflated in Earth orbit and finally be covered with an aluminum shell. In those days von Braun's station was a very advanced design, based on a high level of space technology. But scientists then didn't know very much about the dangerous environment in space, with meteorite impacts, solar flares and cosmic rays.

In the 1960s NASA preferred modular designs with cylindrical modules, using especially the payload capacity of the Saturn V launcher.

Later the Russian MIR-orbital station and the present ISS were built mainly of cylinders and nodes.

Nevertheless some proposals for inflatable structures were made, e.g. by Vanderbilt et al. (1988), Novak et al. (1990) and others. Chow and Lin (1988, 1989) proposed a lunar base built of double-skin membranes, filled with structural foam [2].

In 2006 Petra Gruber and Barbara Imhof presented a study on bionic (biomimetic) inflatable structures [3].

Metal Frame Cylinders versus Inflatable Structures

Any space station in orbit or on lunar and Martian surfaces is penetrated constantly by high-velocity micrometeorites and hit by cosmic rays. The 11-year cycle of solar flare eruptions can cause lethal danger for astronauts.

To construct a Lunar Base does not only mean to build a habitat but also to provide a shelter for humans. The first stage of a Lunar Base should be easy to build using just small robotic machinery. Any risks for humans should be minimized. This implies structural redundancy, shelter and easy rescue in case of damage.

The building of inflatable (pneumatic) structures will cause many risks. During inflation the construction may easily be damaged by small meteorites and cosmic rays. Because of extremely high or low temperatures brittle fractures may occur in the pneumatic skin before it is covered with regolith. To cover inflatables with a 1m thick layer of regolith on top of the structure much lunar material and a big crane is necessary. In case of meteorite impact the whole bubble will deflate immediately. Humans inside the habitat would have no realistic chance to survive and a repair of the damage is nearly impossible. Last but not least equipment and furniture can just be put into the habitat through the airlocks.

In contrast to this metal-frame modules have many advantages:

The semi-robotic assembling of the modules on the lunar surface will take a short time and no big machinery, mining or industrial facilities are necessary.

The first two or three modules can be used immediately as a shelter for astronauts during construction. After assembling, every module is divided from each other by airlocks or fire doors, which is essential for safety and rescue operations.

Double-Shell Structures

As a conclusion of the above considerations we propose to build the first Lunar Base by the use of cylindrical modules, each one approximately 17m long and 6m in diameter, according to the current payloads of the European launcher Ariane 5 ESC-B or similar US, Russian or Chinese launchers. Each module is made of thin aluminum-sheets and trapezoidal aluminum sheeting, and has a weight of approximately 10.2 tons, including the interior equipment and furnishing. (Figure 35.1)

The outer wall of the cylinders is built as a double shell system, stiffened by radial bulkheads. This construction ensures stability and structural redundancy during transportation, landing and assembling on the lunar surface. (Figures 35.2, 35.3)

To protect the astronauts from micrometeorites, radiation and for thermal protection, the space between the two shells is filled with a 0.65m thick layer of regolith in situ by a small teleoperated digger vehicle (Figure 35.4). Thus the amount of regolith can be minimized. To cover a single-shell cylinder of, e.g., 15m length and 4m in diameter with regolith we need approximately 310m3 of material and a big crane.

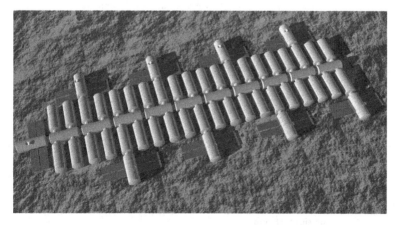

FIGURE 35.1
Lunar base for 60 inhabitants.

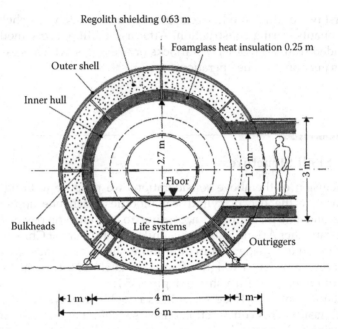

FIGURE 35.2

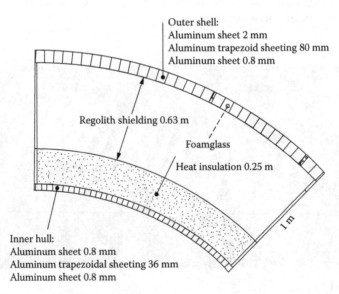

FIGURE 35.3

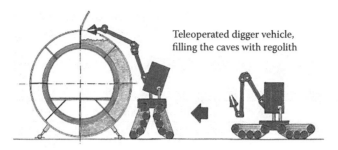

Teleoperated digger vehicle, filling the caves with regolith

FIGURE 35.4

Using the proposed double-shell structure the amount of regolith can be reduced to 130m3 for each cylinder.

Thermal Protection

We assume an average lunar surface temperature of –170°C during the lunar night, which lasts 14 days, and 21°C and 60% humidity of air inside the habitat.

Referring to K. Gösele and W. Schüle [4] we assume:

a_s	heat radiation dose of outer surface	4.0 W/m²K (Aluminum)
		5.0 W/m²K (Regolith)
a_i	heat transmission coefficient	8.13 W/m²K
T	temperature	(°C)
T_i	internal temperature	+21 °C
T_e	external temperature	–170 °C
dT		191 °K
Ts_i	internal surface temperature	(°C)
l_i	thermal conductivity of proposed layers of material	(W/m²K)
d_i	thickness of proposed layers	(m)
C	thermal conductivity of the entire construction	(W/m²K)
	temperature difference of layers:	$dT_i = 191 \times C \times (d_i/l_i)$

The condensation of water on the inner shell surface depends on temperature and humidity inside the habitat. At an inner surface temperature of +13°C or less and 60% humidity of air condensed water will occur. (Figures 35.5, 35.6)

Double-shell structure

Material	d_i(m)	a,l_i(W/m²K)	$1/a,d/l$	dT_i(°K)	T(°C)
					−170
Space as		4.0	0.25	4.50	
					−165.5
Aluminum	0.002	-	-	-	
Foamglass	0.08	0.035	2.29	41.19	
					−124.3
Aluminum	0.0008	-	-	-	
Regolith	0.65	2.0	0.315	5.67	
					−118.6
Foamglass	0.25	0.035	7.143	128.47	
					+9.83
Aluminum	0.0008	-	-	-	
Air	0.036	0.07	0.514	9.24	
Aluminum	0.0008	-	-	-	
					Ts_i + 19.2
a_i		8.13	0.123	2.21	
					+21.3

$C = 0.0942$ W/m²K $1/C = 10.62$ m²K/W

Single-shell structure with 0.7 m regolith covering

Material	d_i(m)	a,l_i(W/m²K)	$1/a,d/l$	dT_i(°K)	T(°C)
					−170
Space as		5.0	0.20	12.9	
					−157.1
Regolith	0.7	2.0	0.35	22.56	
					−134.54
Aluminum	0.002	-	-	-	
Foamglass	0.08	0.035	2.29	147.61	
					+13.0
Aluminum	0.004	-	-	-	
					Ts_i + 13.0
a_i		8.13	0.123	7.93	
					+20.93

$C = 0.3375$ W/m²K $1/C = 2.963$ m²K/W

Inflated structure with 1m regolith covering

Material	d_i(m)	a, l_i(W/m²K)	$1/a, 1/l$	dT(°K)	T(°C)
					−170
Space as		5.0	0.20	45.05	
					−124.95
Regolith	1.0	2.0	0.50	112.62	
					−12.33
Pneumatic skin	0.005	0.20	0.025	5.63	
					Ts_i − 6.70
A_i		8.13	0.123	27.70	
					+21.0

$C = 1.18$ W/m²K $1/C = 0.848$ m²K/W

On the inner surface of the pneumatic skin condensed water will occur and freeze.

The calculations demonstrate the conclusive advantage of the proposed double-shell structure, which provides a comfortable climate inside the habitat during the lunar night. The temperature difference between the inner surface and the internal air is about 2°K – similar to a terrestrial building.

A single-shell hull with 0.08m foamglass insulation, covered with 0.7 m regolith will cause condensed water on the inner surface at 60% humidity of air. In this case the temperature difference between the inner surface and the internal air will be approx. 8°K – uncomfortable for humans.

On the inner surface of an inflated structure covered by a 1m layer of regolith, condensed water will freeze. The temperature difference between inner surface and internal air is approx. 27°K.

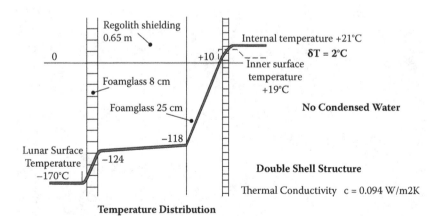

Temperature Distribution

FIGURE 35.5
No condensed water at 60% humidity of air.

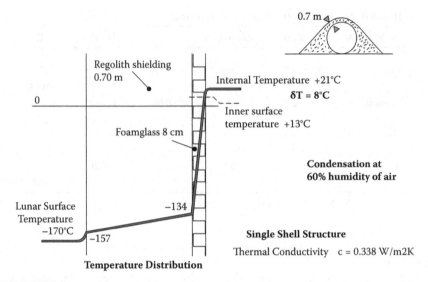

FIGURE 35.6
Condensed water on the inner surface at 60% humidity of air.

Conclusions

To minimize risks for humans and to reduce the time of construction on the lunar surface we propose the use of double-shell cylindrical modules to build an initial lunar outpost. Thus the building site on the moon just needs to be roughly prepared and the use of regolith can be limited.

FIGURE 35.7

Due to its modular design, the station can be enlarged by stages, finally becoming an "urban structure" for dozens of astronauts, scientists and space tourists (Figure 35.7).

The proposed modular design can also be used to build bases on other celestial bodies of similar size – like the moons of Jupiter. In a "second step" of lunar settlement after some years of accommodation and the establishment of mining and industrial facilities, advanced structural concepts like bionic inflatables, concrete structures or the use of lava tubes will be reasonable.

References

1. W. Grandl, Lunar Base 2015 Preliminary Design Study, Acta Astronautica 60, 2007, 554–560.
2. H. Benaroya, L. Bernold and K.M. Chua, Engineering, Design and Construction of Lunar Bases, Journal of Aerospace Engineering 2002, 33–45.
3. P. Gruber and B. Imhof, Transformation: Structure/space studies in bionics and space design, Acta Astronautica 60, 2007, 561–570.
4. K. Gösele, W. Schüle, Schall.Wärme.Feuchtigkeit, Bauverlag Wiesbaden-Berlin 1973.

Further Reading

5. AIAA-Space Manufacturing 9, The High Frontier: Accession, Development and Utilization, 11th Princeton Conference, May 12–15, 1993; A. Germano and W. Grandl, Astropolis-Space Colonization in the 21st Century.

36

Advanced Systems Concept for Autonomous Construction and Self-Repair of Lunar Surface ISRU Structures

Haym Benaroya

Department of Mechanical and Aerospace Engineering
Rutgers University

Haym Benaroya is Professor of Mechanical and Aerospace Engineering at Rutgers University. He is also founder and director of the Center for Structures in Extreme Environments, a center focusing on the conceptualization and analysis of structures placed in challenging environments. We have considered offshore drilling structures, aircraft structures and luggage containers subjected to explosions, nanostructures and lunar surface structures for manned habitation. Professor Benaroya earned his BE degree from The Cooper Union in New York, and his MS and PhD (1981) from the University of Pennsylvania.

ABSTRACT Manned exploration calls for a return to the Moon, requiring the construction of structures for habitation, challenging the engineer and the astronauts that must erect the structure.

Most concepts depict second and third generation facilities. It is more important to be able to erect the FIRST lunar structure. Given the costs associated with bringing material to the Moon, what is the best way to construct habitable structures on the lunar surface? The "grand vision" presented here is the design of a layered manufacturing machine that operates under solar power and will construct the structure in advance of astronaut arrival.

This paper is a reduced version of a proposed study submitted in 2006 to NIAC (NASA Institute for Advanced Concepts), which no longer exists. We are working on these ideas presently. The parts of this paper that discuss the specific "igloo" structure is almost completed based on the work reported in Ruess, Schänzlin and Benaroya [30] and in full detail in Ruess [29].

Advanced Concept Description

The next step for manned exploration and settlement is a return to the Moon. Such a return requires the construction of structures for habitation as well as for manufacturing, farming, maintenance and science. The most challenging of these is the construction of structures that can be used for habitation, although other challenges exist for the design engineer and, until now, the astronaut construction team that must erect the structures.

The proposed study focuses on an engineering evaluation of freeform fabrication technologies as the basis for building autonomous, solar-powered, mini-robots that can, as a team (or eventually swarm), build a first lunar structure for habitation, using primarily in-situ resources over a six to twelve month period prior to the arrival of astronauts.

Structures for manned habitation must be designed to protect against the extreme lunar surface environment. Some argue that manned structures be buried – and they may be one day as the infrastructure is created on the Moon – but initial forays to the Moon will be as modest as our capabilities to build on such a hostile environment.

Many grand and exciting visions have been created for how cities on the Moon will look like, but they invariably depict second and third generation facilities. As wonderful as it is to behold these creations, it is more important to be able to formulate a framework for the erection of the FIRST lunar structure for habitation.

The question is the following: Given the costs associated with bringing material to the Moon (assuming no Space Elevator exists before man returns to the Moon), how do we design and construct habitable structures on the lunar surface in a way that is feasible from mass and energy constraints and minimizes astronaut construction time? What is needed is a small "machine" that can be brought to the Moon that can build structures utilizing primarily in-situ resources and can operate via solar power.

The "grand vision" is the design of freeform fabrication machines that operate almost completely under solar power. Of course, this creates limitations due to the reliance on in-situ resources and the relatively low power availability from solar power.

The proposed study will be comprised of the following aspects:

1. Select a benchmark lunar structure for analysis and which is to be built using the autonomous technologies proposed.

2. Establish structural strength of "blocks" built from in-situ/regolith material (what can be expected from such a process ideally and realistically, knowing the kind of efficiencies one can obtain on Earth).

3. Determine energy/power requirements and feasibility (can current and anticipated solar energy conversion technology provide the needed power to drive such a machine, and if not, what are our other options).

4. Rate of construction as function of above (how fast can the most likely such robot build a facility that can then be organized for human habitation).

5. Examine limitations on the possible complexity of such a structure (if then it is possible to build a habitable surface lunar structure, how complicated can it be made—can the fabrication process also include holes for pipes and power lines, for example).

6. Examine the transferability of the technology to the Martian surface.

While others have suggested sintering regolith, utilizing microwaves, and some form of automated habitation construction, technical issues have been left vague if not completely omitted. We are proposing a framework based on verifiable processes and procedures. It is also anticipated that the rate at which such an autonomous system can operate is low. It may take many months for a team of autonomous systems to erect a simple igloo-like structure.

In order to prove our concept, we need a realistic lunar structure upon which to apply our ideas. Thus, what we propose has two distinct components:

1. Conceptual and actual design of a lunar structure—a benchmark lunar surface structure.

2. Assessing the issues that need to be resolved for the design of autonomous construction mini-robots, as described herein.

A Benchmark Lunar Surface Structure for Autonomous Construction

Key environmental factors affecting lunar structural design and construction are: 1/6 g, the need for internal air pressurization of habitation-rated structures, the requirement for shielding against radiation and micrometeorites, the hard vacuum and its effects on some exotic materials, a significant dust mitigation problem for machines and airlocks, severe temperatures and temperature gradients, and numerous loading conditions—anticipated and accidental. The structure on the Moon must be maintainable, functional, compatible, easily constructed, and made of as much local materials as possible.

Cast regolith has been suggested as a building material for the Moon. The use of cast regolith (basalt) is very similar to terrestrial cast basalt. The terms have been used interchangeably in the literature to refer to the same material. It has been suggested that cast regolith can be readily manufactured on the Moon by melting regolith and cooling it slowly so that the material crystallizes instead of turning into glass. Virtually no material preparation is needed. The casting operation is simple requiring only a furnace, ladle and molds. Vacuum melting and casting should enhance the quality of the end product. More importantly, there is terrestrial experience producing the material; but it has not been used for construction purposes yet.

Cast basalt [30] has extremely high compressive and moderate tensile strength. It can easily be cast into structural elements for ready use in prefabricated construction. Feasible shapes include most of the basic structural elements like beams, columns, slabs, shells, arch segments, blocks and cylinders. Note that the ultimate compressive and tensile strengths are each about ten times greater than those of concrete.

Cast basalt also has the disadvantage that it is a brittle material. Tensile loads that are a significant fraction of the ultimate tensile strength need to be avoided. The fracture and fatigue properties need further research. One possibility is to reinforce it with high tensile strength elements such as carbon nanotubes. Another is to use it in prestressed applications.

Since it is extremely hard, cast regolith has high abrasion resistance. This is an advantage for use in the dusty lunar environment. It may be the ideal material for paving lunar rocket launch sites and constructing debris shields surrounding landing pads. The hardness of cast basalt combined with its brittle nature makes it a difficult material to cut, drill or machine. Such operations should be avoided on the Moon.

Production of cast regolith is energy intensive because of its high melting point. The estimated energy consumption is 360 kWh/MT. Regarding its use in automated ISR construction of the type suggested here, it may be possible to slow the construction process sufficiently so that energy constraints can be met.

The structure in Figure 36.1 below is an update of an earlier concept in a set of papers by Benaroya [2, 3], and Ettouney and Benaroya [14, 15]. Such a structure has been recently analyzed and designed by Ruess, Schänzlin and Benaroya [30], but using more standard structural materials. All the figures in this paper are by Ruess from the reference Ruess et al. [29, 30].

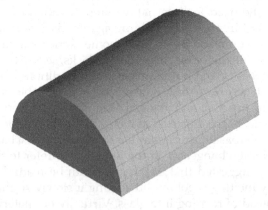

FIGURE 36.1
Rendering of a lunar habitat module.

Our goal is to do a preliminary analysis and design of an autonomous system that is capable of constructing such a structure on the surface of the Moon. This structure will be ready for the first astronaut team that arrives on the Moon to fit it with systems necessary for human habitation. In essence, this structure is envisioned to be a "shell" into which pipes, wiring, windows (if any), and equipment can be installed by the astronaut team. The appropriate volumes will be already in the structure. A preliminary analysis of this structure is provided and this will form the basis for evaluating the autonomous construction method proposed.

The structure will be human-rated, meaning that it will be shielded and can be pressurized upon a human presence. The presence of a structural shell on the Moon, awaiting human arrival, has enormous implications on the logistical planning of man's return to the Moon. All the volume that would normally be allocated for bringing structural materials to the Moon can now be replaced with other items. This leads to an enormous saving in time and money.

Determining the dimensions of a lunar base habitat is a very complex task. Numerous factors like crew size, mission duration and function of the base as an industrial or scientific outpost influence the necessary habitat size. Hence, a global approach considering the necessary habitable volume per person will be pursued. Habitable volume is interpreted as free volume, excluding volume occupied by equipment or stowage.

As per [29], the Gemini missions, which were of relatively short duration missions of up to two weeks, were endured by a person restrained to a chair most of the time. The habitable volume per crewmember in Gemini was 0.57 m³. Currently, the NASA Man Systems Integration Standards (NASA STD 3000) recommend a minimum habitable volume at which performance can be maintained for mission durations of four months or longer of about 20 m³. Despite this recommendation, a design volume (living and working areas) of 120 m³ per person for a lunar habitat has been recommended, based on research of long-term habitation and confined spaces. This value is about equivalent to the volume per crewmember onboard the International Space Station.

The next question is to find an optimum floor height. Proposed floor heights for lunar habitats range from 2.44 m to 4.0 m. People moving in low gravity will certainly require more vertical space than on Earth. They will lift off the floor higher while walking and especially when trying to run.

TABLE 36.1

Typical Properties for Cast Regolith

Property	Units	Value
Tensional Strength	N/mm²	34.5
Compressive Strength	N/mm²	538
Young's Modulus	kN/mm²	100
Density	g/cm³	3
Temperature Coefficient	10^{-6}/K	7.5–8.5

TABLE 36.2

Total Needed Floor Area with Respect to Crew Size

Crew size	6	8	10	12
Habitable area [m²]	206	275	343	412
+20% for equipment and stowage [m²]	41	55	69	82
Total area (rounded up) [m²]	250	320	415	500

Therefore, a floor height of 4.0 m seems most suitable and will therefore be used henceforth.

However, floor height is not equal to clear height. Support systems like lighting and ventilation will use 0.5 m up to 1.0 m of this space. This leaves in most cases about 3.5 m for the actual habitable volume. With these numbers fixed, one ends up with 34.4 m2 floor area per person. The total floor area depends not only on crew size but also on the amount of equipment and stowage space that is needed. A summary for different crew sizes is given in Table 36.2.

Now, having determined the total floor area, one can begin to size the structure. Depending on the chosen structural system, one has to find the most efficient span of the main structure and, depending on the structural system chosen, the spacing between primary structural elements. The necessary clear floor height can for some concepts, e.g., arches, govern the span. The layout of the habitat is also very important at this point.

Regolith, upon which the lunar structure is built, needs to be understood from a foundation engineering view. The bulk density of regolith ranges from 0.9 to 1.1 g/cm^3 near the surface and reaches a maximum of 1.9 g/cm^3 below 20 cm. The average is at 1.7 g/cm^3.

- The porosity of the regolith surface is about 45.
- Cohesion of undisturbed regolith is c = 0.1 to 1.0 kN/m^2.
- The friction angle is about 30° to 50°.
- The regolith's modulus of subgrade reaction is typically 1000 $kN/m^2/m$.
- The compressibility ranges from C_c = 0.3 (loose regolith) to 0.05 (dense regolith).
- Interparticle adhesion in the regolith is high. It clumps together like damp beach sand.

This work idealizes the lunar soil using the modulus of subgrade reaction. All structural analysis calculations are thus done with the soil simulated by springs of stiffness C_c = 1000 $kN/m^2/m$. It is a simplified method and a more detailed study of the regolith mechanics might be needed in the future.

The arch structure in Ruess et al. [30] is chosen for this preliminary study on automated ISR construction. First, the shape and rise of the arch were determined. From that study, a single floor layout is preferred to avoid

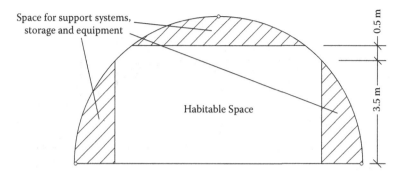

FIGURE 36.2
Space use within the proposed structure.

additional structural mass for internal flooring and reduce the size of the main structural members at the same time. Therefore, a rise of 5 m was chosen for the arch. Figure 36.2 shows how the space within the arch will be divided into the different functional areas.

On the Moon, since the governing load is not gravitational, but rather internal pressurization, the circular arch is the more suitable structure because no bending moments are introduced in the arch. An in-plane two-dimensional analysis is found to be sufficient; no major three-dimensional effects are expected since the structure runs continuously in the third direction only. Internal forces, member stresses and deflections are calculated using finite element software.

The bending moment in the tie is a result of soil structure interaction, as shown in Figure 36.3. It depends on the ratio of foundation to soil stiffness. The final bending moment in the tie can only be determined iteratively because every change in tie stiffness results in a change in bending moment that in turn may require a different tie cross-section. Thus, the final bending moment distribution will only be available after the structural design is finished.

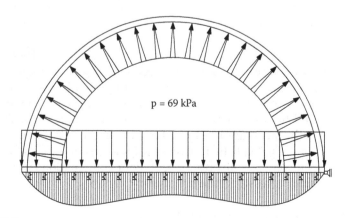

FIGURE 36.3
Bending moments for the circular arch.

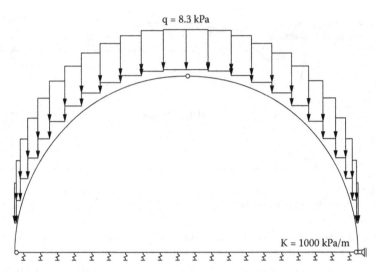

FIGURE 36.4
Regolith cover load.

On the Moon the designer has to be careful when applying gravitational loads calculated using the lunar gravitational acceleration. The resulting loads will be only about 1/6 of similar ones on Earth. For the static analysis of the structure, the main load cases identified in addition to the structure's self weight are:

1. Internal pressure $p = 69$ kPa
2. Regolith covering the whole structure $q = 8.3$ kPa.

The loads for the regolith cover, as shown in Figure 36.4, assume the regolith can be placed uniformly on the structure. If instead loose soil is simply heaped upon the top of the structure, the resulting load will be trapezoidal, not uniform.

Most of the loads described above may act at the same time. There are also a number of different scenarios that the designer needs to account for: starting with construction stages, the structure being initially pressurized with the regolith not yet on top of it, next the regular operational mode with all loads acting, and finally a planned or accidental decompression. The maximum effect on the structure has to be found using load combinations. For each scenario only the loads that increase the stresses in the structure are to be included. Self-weight is always present. Four combinations were used to find the maximum stresses in the members:

1. Internal pressure plus floor loads
2. Regolith cover plus installation loads

FIGURE 36.5
Front view of the general tie shape

3. All loads

4. Half the regolith cover (during construction)

Two main conclusions result from the preliminary structural analysis [29, 30]. First, the arch segments can have a uniform cross-section. It is possible but not necessary to adjust the arch cross-section to the distribution of internal forces since these are almost uniform. Second, in order to get an efficient cross-section for the tie it has to be adjusted to the distribution of internal forces. The bending moment has the shape of a parabola, so it was decided to give the tie a similar shape. Figure 5 shows the principal shape of the tie/floor/foundation.

The regolith foundation will need to be sintered before the structure is fabricated and will result in a higher modulus of subgrade reaction and therefore lower deflections of the tie. It does not affect the arch deflections. Calculations show that the modulus of subgrade reaction would have to be increased about tenfold to get in the range of desired deflections. This is very likely not possible to be achieved by sintering the regolith. Some additional reinforcing is needed. More research data is needed for this topic.

Advanced Concept Development Work Plan

The study under way focuses on an engineering evaluation of the use of freeform fabrication technologies as the basis for building autonomous mini-robots that can, as a team, build a first lunar structure for habitation, using in-situ resources and primarily solar power, over a six to twelve month period prior to the arrival of astronauts. The proposed study will be comprised of the following aspects:

1. An examination of the structural strength of "blocks" built from in-situ/regolith material (what can be expected from such a process ideally and realistically, knowing the kind of efficiencies one can obtain on Earth)

2. An examination of the energy/power requirements and feasibility (can current and anticipated solar energy conversion technology provide the needed power to drive such machines, and if not, what are other options for power)

3. Establish the rate of construction as function of above (how fast can the most likely ensemble of such robots build a facility that can then be organized for human habitation)

4. What is the appropriate mix of mini-robots, that is, how many different functions are needed and what is an optimal ensemble

5. How complicated a lunar structure can be erected (how difficult is it for the fabrication process to include holes for pipes and power lines, for example)

6. Address briefly the transferability of this technology to the Martian surface.

Imagine this: Several mini-robots (size of a lawn mower) are landed on the Moon, perhaps six to twelve months before the arrival of humans. Their purpose is to prepare the site and erect a simple structure that the arriving astronauts can then fit for habitation. This means installing (in a modular and easy way) life support and maintenance equipment.

Once the mini-robots land, they begin to prepare the regolith, first smoothing out the site, some sintering with attention to reinforcement (perhaps using carbon nanotubes). Once the site is prepared, the mini-robots begin to build the lunar structure in layers, a structure that has been designed *a priori*, leaving open volumes for mechanical and electronic equipment that is to be placed subsequently by astronauts. Of course, there would be a large volume, as shown in the previous figures, for habitation.

Studies will determine what kind of reinforcing the ISRU-based structure requires, possibilities being a glass fiber-reinforced or a nano-composite matrix. Some studies have shown that such a layered manufacturing process can be achieved in a number of ways. One promising approach is based on microwave sintering. Much of the reinforcing material can be found in the regolith.

Some suggestions have been made for autonomous construction on the Moon, but no study exists for autonomous construction using freeform manufacturing technologies utilizing in-situ resources, or for the erection of the first habitable structure on the Moon via such technologies.

Our vision is a suite of freeform/rapid prototyping manufacturing robots, each with its advantages, since the lunar habitat is not going to be a homogeneous structure. This has been discussed above in our preliminary design.

Rapid prototyping processes are a relatively recent development. The first machine was released onto the market in late 1987. While rapid prototyping is the term commonly applied to these technologies the terminology is now a little dated, reflecting the purpose to which the early machines were applied. A more accurate description would be layered manufacturing processes. An alternative term is freeform fabrication processes. These processes work by building up a component layer by layer, with one thin layer of material bonded to the previous thin layer. There are several different processes. The main ones are:

- stereolithography;
- laser or microwave sintering;
- fused deposition modeling;
- solid ground curing; and
- laminated object manufacturing.

In addition there are a number of newer processes, such as ballistic particle manufacturing and three-dimensional printing, which have appeared on the market. Some ballistic particle manufacturing technology use piezo-electric pumps that operate when an electric charge is applied, generating a shock wave that propels particles. All these processes essentially start with nothing and end with a completed part. Rapid prototyping processes are driven by instructions that are derived from three-dimensional computer-aided design (CAD) models. CAD technologies are therefore an essential enabling system for rapid prototyping. (Other enabling technologies are for mini-rovers, energy beams, solar power systems, and structural reliability/durability.)

The processes use different physical principles, but essentially they work either by using lasers or microwaves to cut, cure or sinter material into a layer, or involve ejecting material from a nozzle to create a layer. Many different materials are used, depending upon the particular process. Materials include thermopolymers, photopolymers, other plastics, paper, wax, or metallic powder, for example. The processes can be used to create models, tooling, prototypes, and even in some cases to directly produce metal components. As such, one may view these capabilities and laying the groundwork for self-repairing systems.

Our vision is that an optimally grouped set of mini-robots will work as a team to autonomously erect a habitable volume for finishing by the first astronaut team on the Moon. Issues that will be resolved in a preliminary study are the appropriate mix of robots by function, energy/power demands as a function of rate of construction, and redundancy needs for reliability. A point of study will be to examine whether solar and other radiation that hits the lunar surface can be factored into the "curing" process. Also, can the hard vacuum on the Moon be used to some advantage?

The appropriate mix of robots refers to the fact that different free-form manufacturing processes will be needed. We know that different parts of the structure require different construction approaches. For example, ballistic particle manufacturing may be appropriate for creating the foundation layer. For such ejection-type systems, a filtering process is needed so that relatively uniform sequences of regolith particles are fed. After a preliminary layer is prepared, some additional sintering may be required, perhaps via microwave as has been suggested in the literature.

In the end, we envision a team of such freeform manufacturing mini-robots that will work as a team, with the group comprising the skill set of capabilities needed to construct all the parts of the structure. As an example, the ballistic particle manufacturing robot can be placed on a ridge or be able to elevate itself to project particles to the higher locations of the lunar structure.

Concluding Summary

This paper proposes an outline of an approach to erecting a structure on the lunar surface autonomously using in-situ resources and layered manufacturing technologies. A simple structure is proposed as a prototype. Only the essential structure would be manufactured in this way, leaving open spaces and volumes where the astronauts would complete the construction by inserting HVAC, doors and internal equipment.

References

Benaroya, H. and Ettouney, M., (1992a) "Framework for Evaluation of Lunar Base Concepts," Journal of Aerospace Engineering, Vol. 5, No. 2, pp. 187–198.
Benaroya, H. and Ettouney, M., (1992b) "Design and Construction Considerations for a Lunar Outpost," Journal of Aerospace Engineering, Vol. 5, No. 3, pp. 261–273.
Ettouney, M. and Benaroya, H., (1992a) "Regolith mechanics, dynamics and foundations," Journal of Aerospace Engineering, Vol. 5, No. 2, pp. 214–229.
Ettouney, M., Benaroya, H., and Agassi, N., (1992b) "Cable structures and lunar environment," Journal of Aerospace Engineering, Vol. 5, No. 3, pp. 297–310.
Ruess, F., (2004) Structural Analysis of a Lunar Base, Master's thesis, Universität Stuttgart/Rutgers University, May.
Ruess, F., Schänzlin, J. and Benaroya, H., "Structural Design of a Lunar Habitat," J Aerospace Engineering, Vol. 19, No. 3, July 2006, 133–157.

Bibliography

Alexander, P., and Dutta, D., (2000) "Layered manufacturing of surfaces with open contours using localized wall thicknesses," Computer-Aided Design 32, pp. 175–189.
Benaroya, H., (1994) "Reliability of structures for the Moon," Structural Safety, Vol. 15, pp. 67–84.
Benaroya, H., (1998) "Economic and technical issues for lunar development," Journal of Aerospace Engineering, Vol. 11, No. 4, pp. 111–118.

Benaroya, H., Bernold, L., and Chua, K. M., (2002) "Engineering, design and construction of Lunar Bases," Journal of Aerospace Engineering, Vol. 15, No. 2, pp. 33–45.

Benaroya, H., (2002) "An overview of Lunar Base structures: past and future," AIAA Space Architecture Symposium, AIAA, Houston, Texas, pp. 1–12.

Braun, B., (2003) "Faserverbundkunststoffe (FVK) als tragende Struktur," Tech. rep., University of Stuttgart, Institute of Structural Design (KE).

Cohen, M. M., (2002) "Selected precepts in lunar architecture," Tech. rep., 53rd International Astronautical Congress, The World Space Congress.

Criswell, M. E., Sadeh, W. Z., and Abarbanel, J. E., (1996) "Design and performance criteria for inflatable structures in space," SPACE 96, ASCE, pp. 1045–1051.

Duke, M. B. and Benaroya, H., (1993) "Applied mechanics of lunar exploration and development," Applied Mechanics Review, Vol. 46, No. 6, pp. 272–277.

Eckhart, P., (1999) The Lunar Base Handbook, McGraw-Hill, New York.

Eichold, A., (2000) "Conceptual design of a crater Lunar Base," Proceedings of Return to the Moon II, AIAA, pp. 126–136.

Graf, J. C., (1988) "Construction operations for an early Lunar Base," SPACE 88, ASCE, pp. 190–201.

Happel, J. A., (1992a) The design of lunar structures using indigenous construction materials, Master of Science in Civil Engineering, University of Colorado.

Happel, J. A., (1992b) "Prototype Lunar Base construction using indigenous materials," SPACE 92, ASCE, pp. 112–122.

Happel, J. A., (1993) "Indigenous materials for lunar construction," Applied Mechanics Reviews, Vol. 46, No. 6, pp. 313–325.

Huntsman, "Huntsman Advanced Materials, Verslaan 45, B3078 Everberg, Belgium, http://www.huntsman.com," 2004.

Johnson, S. W., Chua, K. M., and Carrier III, W. D., (1995) "Lunar soil mechanics," Journal of the British Interplanetary Society, Vol. 48, No. 1, pp. 43–48.

Karalekas, D., and Antoniou, K., (2004) "Composite rapid prototyping: overcoming the drawback of poor mechanical properties," J. Mat. Proc. Tech. pp. 526–530.

Kennedy, K. J., (1992) "A horizontal inflatable habitat for SEI," SPACE 92, ASCE, pp. 135–146.

Kennedy, K. J. and Adams, C. M., (2000) "ISS TransHab: An Inflatable Habitat," SPACE 2000, ASCE.

Kennedy, K. J., (2002) "The vernacular of space architecture," Tech. rep., Space architecture symposium, October.

Kirihara, S. et al. (2005), "Strong localization of microwave in photonic fractals with Menger-sponge structure," Journal of the European Ceramics Society, in press.

Reynolds, K. H., (1988) "Preliminary design study of lunar housing configurations," NASA conference publication 3166, NASA, pp. 255–259.

Roberts, M., (1988) "Inflatable Habitation for the Lunar Base," NASA Conference Publication 3166, NASA, April.

Spudis, P. D., (2003) "Harvest the Moon," Astronomy, Vol. 6, pp. 42–47.

Taylor, L. A. and Meek, T. T., (2005) "Microwave Sintering of Lunar Soil: Properties, Theory and Practice," Journal of Aerospace Engineering, ASCE, Vol. 18, No. 3, pp. 188–196.

Zippert, H., (2004) "Zimmer mit Erdblick," GEO special Der Mond, Vol. 6, pp. 90–99.

37

A Reliability-Based Design Concept for Lunar Habitats

Benjamin Braun
HE2 - Habitats for Extreme Environments
Stuttgart, Germany

Florian Ruess
HE2 - Habitats for Extreme Environments
Bremen, Germany

Florian Ruess, Dipl.-Ing, University of Stuttgart and University of Calgary civil engineering (1988–2004); research on lunar bases at Rutgers University as part of the master thesis (2004); structural engineer with BDE engineering in Stuttgart with focus on bridge design and engineering (2004–2006), and since 2006 structural engineer at Airbus Bremen, Wing High Lift Design HE2–Habitats for Extreme Environments.

ABSTRACT   The key objective of every structural design on the Moon is to develop a structural system which is able to fulfill its functional requirements in a highly economical way due to initial transportation issues. Many known conceptual lunar designs use global or partial safety factors to ensure the reliability of the structural members. However, a suitable approach for traditional design on Earth is not necessarily considered appropriate for novel and unique structures such as lunar habitats. Uncertainties exist in lunar base design that can be addressed only on the basis of a structural reliability analysis in order to assess safety in a quantitative manner. As a result, an optimized structure can be designed from a holistic point of view, which is beyond that comparable to other designs also in terms of safety. Based on an existing structural design of a lunar habitat, this paper describes not only the fundamental requirements and benefits if a reliability-based design approach is chosen, but it also highlights the difficulties that are inevitably encountered.

Introduction

The erection of an outpost on the Moon is not only a question of how to choose the most suitable structure for a lunar habitat as summarized and presented in Ruess et al. (2006). The risk and safety of such structures need also to be assessed in an appropriate manner. For the design on Earth, standards and codes e.g., the family of American Standards or Eurocodes, exist that help the designer to analyze and design a safe structure. However, special uncertainties exist in lunar base design that can be assessed as a first step only on the basis of a decision analysis, namely a structural reliability analysis. It offers a way to allocate the available resources both most efficiently and in a sufficiently safe manner. Fundamentally, uncertainties can arise from an inherent randomness in loads and resistances, or inadequate knowledge, e.g., influence of radiation on material properties and statistical uncertainty due to sparse information. Such uncertainties imply the need to use statistical and probabilistic tools in the analysis and design process. The basic principles of structural reliability analysis (Thoft-Christensen and Baker 1982) (Ditlevsen and Madsen 1996) are applied here with special regard to the design of second generation lunar habitats. Based on an existing structural design of a lunar habitat, not only the fundamental requirements and benefits of a reliability-based design approach are described, but also the difficulties that are inevitably encountered are highlighted.

Structural Reliability

General

The assessment of the reliability of electrical systems used in battleships became a central issue during the 1940s (Faber 2006) and it can be seen as the starting point for reliability analysis methods. Thus, classical reliability analysis deals with technical components which are available in large numbers of the same type and which have usually only a single failure mode. The aim is to determine the failure rate in order to derive an estimation of the time until failure for each component. However, the characteristics of structural reliability analysis, which evolved from classical reliability analysis, is fundamentally different because normally a structural system is built up of unique components which have different failure modes. Here, failure occurs due to extreme events and the objective is to provide a measure of safety in order to prevent failure with a sufficiently small value of probability. Therefore, the estimation of failure probabilities of structural components requires a probabilistic modeling of both loads and resistances from which a reliability index β indicating the safety can be derived.

In the following, the reliability implemented in design concepts will be recalled shortly. On the basis of familiar code-based design concepts the theoretical background will be introduced which inherently leads to the reliability-based concepts.

Probability of Failure and Reliability Index

First of all, the uncertain resistance R and varying load S are expressed by probabilistic models. Then the probability of failure is the probability that the resistance R becomes smaller than the load S, which can be expressed in terms of a safety margin M (see Equation (37.1)).

$$M = R - S \le 0 \tag{37.1}$$

In order to determine the probability of failure, the probability density function of the safety margin M is evaluated for all possible realizations of the uncertain variables. Such a distribution of the safety margin M is shown in Figure 37.1. The number of standard deviations by which the mean value of the safety margin M exceeds zero defines the reliability index β and it gives also a good geometrical interpretation.

Thus, the higher the reliability index β is, the smaller is the probability of failure. The link between them is given by use of the standard normal distribution function in Equation (37.2):

$$P_F = \Phi\left(\frac{0 - \mu_M}{\sigma_M}\right) = \Phi(-\beta) \tag{37.2}$$

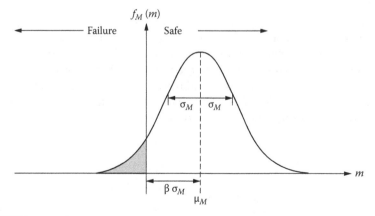

FIGURE 37.1
Reliability index β (Faber 2006).

However, in the general case resistances and loads may consist of several variables that altogether make up the limit state function. Depending on the problem, the limit state function can be linear in very simple cases or nonlinear in the general case.

Limit State Functions

The state of a structural component can be generally described in terms of a functional relationship $g(x)$ in which the resistances R are related to the loads S as introduced in Equation (37.1). This function $g(x)$ is denoted the limit state function because the determination of the result indicates the state of the structural component. For realizations of the basic random variables X, the component is in a safe state if $g(x) > 0$ and for $g(x) \leq 0$ the component is in a failure state, where x is a realization of X. The limit state function reads:

$$g(x) = R - S \leq 0 \tag{37.3}$$

With the definition of the failure state, the probability of failure can be determined by the integral

$$P_F = \int_{g(x) \leq 0} f_x(X) \, dx \tag{37.4}$$

in which $f_x(X)$ is the probability density function of the random variable X. From the evaluation of Equation (37.3) the reliability index can be derived which is directly connected to the probability of failure as shown above. In the following, first order reliability methods (FORM) will be used. These are widely used due to their efficiency in solving this equation. Generally it can be distinguished between linear and nonlinear limit state functions. It is assumed that the random variables are normally distributed in the considerations hereafter. In order to evaluate the reliability index these random variables have to be normalized into standardized normally distributed random variables according to Equation (37.4) so that the random variables have zero mean values and unit standard deviations,

$$U_i = \frac{X_i - \mu_{X_i}}{\sigma_{X_i}} \tag{37.5}$$

The limit state function then becomes:

$$g(x) \rightarrow g(u)$$

Linear Limit State Function

A linear limit state function is very simple and illustrative. It can be written as:

$$g(x) = a_0 + \sum_{i=1}^{n} a_i x_i \tag{37.7}$$

The mean value and standard deviation can be determined according to Equations (37.8) and (37.9):

$$\mu_M = a_0 + \sum_{i=1}^{n} a_i \mu_{X_i} \tag{37.8}$$

$$\sigma_M^2 = \sum_{i=1}^{n} a_i^2 \sigma_{X_i}^2 + \sum_{i=1}^{n} \sum_{j=1, j \neq i}^{n} \rho_{ij} a_i a_j \sigma_i \sigma_j \tag{37.9}$$

The reliability index β is calculated then according to Equation (37.10):

$$\beta = \frac{\mu_M}{\sigma_M} \tag{37.10}$$

In Figure 37.2 a geometrical interpretation of the normalized linear limit state function $g(u)$ and the reliability index β is given for the simple case with

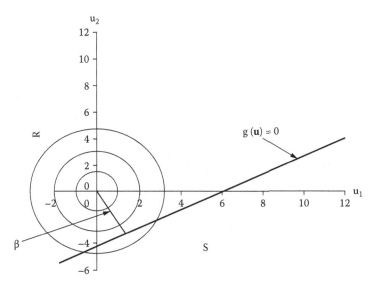

FIGURE 37.2
Characteristics of a linear limit state function (Faber 2006).

two basic random variables. The $u_1 u_2$ plane represents all realizations of the random variables. This plane is divided by the limit state function into a safe and a failure domain. As shown before, the reliability index is a measure of the distance between the mean value of the safe domain and the boundary of the failure domain. Of course, the smallest distance to the boundary has to be found which then defines the design point.

Nonlinear Limit State Function

In case the limit state function becomes nonlinear—which is the common case—the evaluation is not as easy as with a linear limit state function. In order to determine the reliability index, a linearization of the limit state function at the design point has been proposed by Hasofer and Lind (1974) for the normalized space. This is shown in Figure 37.3.

With this linearization the design point can be found. The result yields not only the distance β but also the direction is given by the normal vector α pointing into the failure domain. Due to the nonlinearity of the limit state

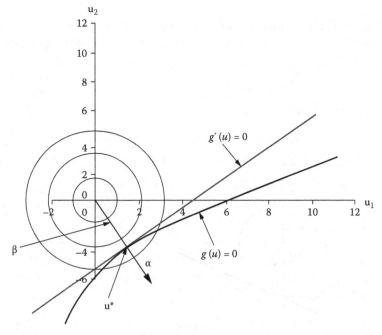

FIGURE 37.3
Characteristics of a linearized nonlinear limit state function (Faber 2006).

function the design point is not known in advance and needs to be determined by Equation (37.11):

$$\beta = \min_{u \in \{g(u)=0\}} \sqrt{\sum_{i=1}^{n} u_i^2} \tag{37.11}$$

In case the limit state function is differentiable the α- vector explained above can be determined by Equation (37.12):

$$\alpha_i = \frac{-\dfrac{\partial g}{\partial u_i}(\beta \cdot \alpha)}{\sqrt{\sum_{i=1}^{n}\left(\dfrac{\partial g}{\partial u_i}(\beta \cdot \alpha)\right)^2}}, i = 1, 2, \ldots n \tag{37.12}$$

so that the limit state function,

$$g(\beta \cdot \alpha_1, \beta \cdot \alpha_2, \ldots \beta \cdot \alpha_n) = 0 \tag{37.13}$$

can be solved as follows. First a design point has to be arbitrarily chosen and beginning with the evaluation of Equation (37.12) the solution can be iteratively found by calculating a new normal vector α and a new β value.

Reliability Implemented in Design Concepts

Modern design concepts use the partial safety factor concept in order to ensure the safety of a structure. The underlying idea is to define characteristic values of loads and resistances, which are combined with partial safety factors that take into account the different uncertainties of the loads and resistances. Due to this differentiation a more constant level of safety can be provided. This is the reason why the partial safety factor concept has outperformed the global safety factor concept, which have been used since the very beginning of the development of design standards.

Nowadays, in code-based design concepts such as the family of American standards or the Eurocodes, design equations are given for the verification of the resistance of structural members against a limit state, e.g., failure or excessive deformations. In general terms this design equation reads as follows:

$$\frac{R_k}{\gamma_M} - \gamma_F \cdot E_k \leq 0 \tag{37.14}$$

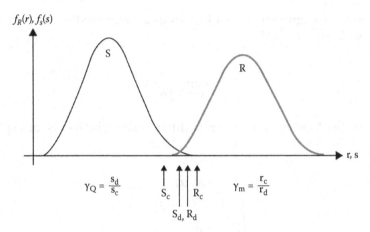

FIGURE 37.4
Code-based design concept with partial safety factors (Faber 2006).

where R_k = characteristic value of the resistance
         γ_M = partial safety factor for the resistance
         E_k = characteristic value of the load
         γ_F = partial safety factor for the load

In the codes, various values of partial safety factors exist that are attributed to different materials and load types. Together with the partial safety factors, the characteristic values of loads and resistances are introduced so that the partial safety factors ensure a specified minimum level of reliability for a structural design according to the code. Figure 37.4 illustrates the code-based design concept with partial safety factors.

The background of the safety factors used in the codes comes from an underlying reliability analysis from which the safety factors are derived. A calibration of the design equation can be made if a large number of buildings of similar type can be evaluated. However, this is only the case if the history of structures can be followed. In the case of a lunar habitat, it is a novel and unique structure with all new boundary conditions that have never been built before, so that applying design standards from Earth may seem to be one step too far. The first level in terms of a reliability analysis will be introduced hereafter.

Target Reliability

The building codes we are using on Earth are based on design criteria coming from reliability considerations. The safety levels of these standards have been well calibrated against target reliabilities, which have been derived from

TABLE 37.1

Society Perception of Hazards According to Keese and Barton (1982)

Probability of Failure PF	Acceptance in Society
10^{-3}	This level is unacceptable to everyone. When probability approaches this level, immediate action should be taken to reduce the hazard.
10^{-4}	People are willing to spend public money to control hazards at this level. Safety slogans popularized for accidents in this category show an element of fear.
10^{-5}	Though rare, people still recognize these hazards, warn children (e.g., drowning, poisoning). Some accept inconvenience to avoid such hazards (e.g., avoid air travel).
10^{-6}	Not of great concern to the average person. People are aware of these hazards, but feel it can never happen to them—a sense of resignation if they do (e.g., an act of God).

the history of operated structures over many decades. However, novel and unique structures for extreme environments, such as a lunar habitat, clearly exceed the scope of the existing standards and the most prominent question remains: how safe is safe enough? As introduced before, a reliability-based design concept seems best suited to fit the fundamental requirements of addressing this issue. Obviously it seems not reasonable to use the same target reliabilities as on Earth because the consequences of failure are much higher. This concerns not only loss of lives but also monetary reasons.

In Table 37.1 the perception of hazards by society is associated with a number for the probability of failure. Of course, more sophisticated approaches exist to define the target reliability but Table 37.1 can be easily understood by all readers. In the following, a target reliability of $\beta = 4.77$ corresponding to the probability of failure of 10^{-6} is chosen for component failure. This is well beyond the common target reliabilities used, e.g., in the Eurocodes $\beta = 3.70$.

Implications for Lunar Habitat Design

Without a large statistical base, design codes cannot be established and are therefore non-existent for lunar design. For the time being reliability analysis seems to be the best choice to design reliable and economical lunar structures. In the past global safety factor approaches have been used for preliminary lunar design. The main advantage of such approaches is a simpler structural analysis but, as was mentioned before, it is very difficult to judge how economical and safe a structure really is because the actual reliability remains unknown.

Structural Design Concepts of Lunar Habitats

Lunar habitat structures will have to satisfy very unique requirements. The environment on the Moon poses new challenges to the designer with respect to:

- Gravity: only 1/6 the gravity on Earth.
- Radiation/Shielding: Solar and cosmic radiation as well as micrometeoroid impacts endanger the inhabitants of a lunar base. Shielding is necessary.
- Vacuum/Internal Pressurization: The lunar structure must contain pressure and be failsafe against decompression. Material issues with exposure to vacuum.
- Dust: Highly abrasive and toxic.
- Temperature extremes: Issues for construction and insulation of the completed structure.

Therefore, structures have to be designed for:

- Structural adequacy: Optimized structures in terms of structural material use.
- Material properties: Stable mechanical properties like high strength, ductility, durability, stiffness, and tear and puncture resistance.

Maintenance: Minimum maintenance requirements

- Compatibility: With the life support systems.
- Transportation/Ease of erection: Minimum mass and volume, easy to assemble.
- Excavation/Foundations: Minimize requirements.
- Functionality: Low habitat volume to floor area ratios increase efficiency.
- Local materials: long-term goal to use as much indigenous material as possible.

A lunar base will go through evolutionary development, starting with limited capacity and expanding over time. Lunar development will stage in three main phases. In each of these phases a different generation of habitats will necessarily evolve (Benaroya 2002; Cohen 2002):

Generation I: Prefabricated and pre-outfitted hard shell modules

Generation II: Assembly of components fabricated on Earth with some assembly required

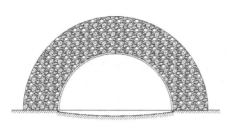 

FIGURE 37.5
Proposed structural design of a lunar habitat.

Generation III: Large-scale building structures comprised substantially of indigenous material

Second generation lunar habitats can be further divided into four main structural types:

1. Inflatable structures

2. Cable structures

3. Rigid structures

4. Underground construction

Many conceptual designs based on one or the other of these structural types exist today. It is evident that all the developed designs have different strengths and weaknesses. One way to determine if a design is generally suitable is the evaluation method described in Ruess (2004) and Ruess et al. (2006).

A number of existing designs passed this evaluation very well. Among them is a tied circular arch aluminum hard shell modular structure (Figure 37.5) (Ruess et al. 2006). This design is simple, robust and takes into account the lunar issues like temperature extremes, radiation, micrometeorites and construction on the lunar surface already mentioned above.

Structural analysis of this concept is straightforward in general. But safety issues are absolutely critical in lunar base design. First the different random variables (mainly loads and resistances) influencing structural design have to be known to the most exact level of detail possible. The more accurate the assumptions for the variables, the better the confidence in the safety level of the structure will be. There is definitely a need for further research in this area.

The eminent question is how safe is safe enough while being highly economical at the same time? As mentioned above, a global safety factor or a code-based design concept for lunar habitats cannot address safety neither in an appropriate nor quantitative manner for the time being. Thus, a structural reliability analysis is the best choice. But this analysis is again very sensitive to input values for the random variables. For the arch structure of the presented habitat design an example reliability analysis is shown in the following.

Example Calculation

In order to illustrate the principles described in the previous sections a step-by-step example calculation is described. The lunar habitat structure developed in Ruess (2004) is used in terms of structural system, geometry and material selection for the subsequent analysis and comparison.

Analysis Model

The tied circular arch structural system described in Ruess (2004) and Ruess et al. (2006) and shown in Figure 37.6 is the basis for all calculations. Only the arch section is analyzed here but the analysis and design of other structural members like the floor beam, end walls or connections can be performed accordingly.

Assumptions for Random Variables

Material Strength

A high strength aluminum alloy with a characteristic yield strength of $f_{y,k} = 50$ MPa is chosen. According to (JCSS 2001) the yield strength mean value can be calculated by Equation (37.15):

$$f_{y,mean} = 50 \cdot e^{(1.645 \cdot 0.07)} - 2 = 54.102 \; MPa \tag{37.15}$$

with the standard deviation according to Equation (16):

$$\sigma_M = COV \cdot f_{y,mean} = 0.07 \cdot 54.102 = 3.787 \; MPa \tag{37.16}$$

For this example it was decided to reduce the mean value by 25% to account for unknown material effects such as radiation exposure, welding in vacuum,

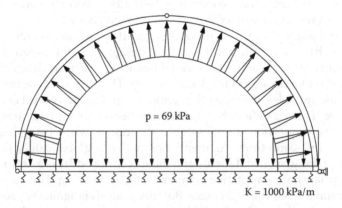

FIGURE 37.6
Analysis model.

etc. From the structural analysis performed in Ruess (2004) it is also known that the end wall loading of the structure introduces an additional transverse stress of 2.167 MPa into the arch section (for simplification assumed constant and directly superimposed). The mean value is then found to be:

$$f_{y,mean} = 0.75 \cdot 54.102 - 2.167 = 38.410 \; MPa \tag{37.17}$$

Loads

For the purpose of this example, only the two main loading conditions - internal pressure and regolith shielding loads - are considered.

Internal Pressure: an optimum for a mean value for lunar habitats is believed to be $p = 69$ kPa (Ruess 2004). Although technically the life support system will be able to guarantee this value with close tolerances, the maximum pressure value is assumed to reach 101 kPa (normal atmospheric pressure), resulting in a coefficient of variance (COV) of 0.3. The input for the limit state function as described in the above sections is needed in terms of axial force and bending moment of the structural member. The internal pressure only results in axial forces in the arch member and can be computed as follows (Ruess 2004):

$$N = 1.25 \cdot p \tag{37.18}$$

Regolith weight: An assumed regolith cover of 3 m to protect from micro-meteorites, radiation and provide thermal insulation results in a load of $q = 8.3$ kPa on the structure. The COV is assumed to be similar to the one for wood, a rather uncertain Earth material (in terms of weight). So the COV is taken to be 0.12. The regolith loading results in bending moment of the arch member only which is computed according to Ruess (2004):

$$M = 82.5 \cdot q \tag{37.19}$$

For the more complex example of a non-linear limit state function the cross section area and the section modulus are also introduced as random variables.

Cross section area: directly proportional to the flange thickness of the arch:

$$A = 6 + 5 \cdot t \quad [cm^2] \tag{37.20}$$

The COV is taken from (JCSS 2001, Faber 2006) to be 0.032.

Section modulus:

$$W = 7.5 \cdot A - 45 \quad [cm^3] \tag{37.21}$$

The COV is taken from (JCSS 2001, Faber 2006) to be 0.04.

Linear Limit State Function

The linear limit state function and its standard deviation with 3 random variables is straightforward:

$$g(x) = \mathbf{f}_y - \frac{\mathbf{N}}{A} - \frac{\mathbf{M}}{W} \leq 0 \tag{37.22}$$

$$dev = \sqrt{(\text{cov}\,\mathbf{f}_y)^2 - \text{cov}\,\frac{\mathbf{N}^2}{A} - \text{cov}\,\frac{\mathbf{M}^2}{W}} \tag{37.23}$$

With a flange thickness of only 1.62 mm the direct solution is:

Mean value: $g(x) = 38.41 - 6.11 - 11.26 = 21.04$
Standard deviation:

$$\sigma_M = \sqrt{(0.07 \cdot 38.41)^2 + (0.3 \cdot 6.11)^2 + (0.12 \cdot 11.26)^2} = 4.42$$

The resulting reliability index is

$$\beta = \frac{21.04}{4.42} = 4.76 \tag{37.24}$$

and the corresponding probability of failure

$$P_f = 9.69 \text{ E--7} \approx 1 \text{ E--6} \tag{37.25}$$

Nonlinear Limit State Function

If the cross sectional area and the section modulus are also assumed random, the limit state function becomes nonlinear. Its solution is then only possible by iterative methods as shown above. In order to get the same probability of failure as in the case above, the necessary flange thickness increases to 1.64 mm. Although the difference in this example is only small, one can see that with increasing degrees of uncertainty the dimensions of the structure inevitably increase, too.

Comparison of Different Design Approaches

The structural analysis performed in Ruess (2004) compared results with global safety factors of 4 and 5 respectively. A comparison of these design results with the results of the example reliability analysis above and an Earth-based design can be found in Figure 37.7 and Table 37.2.

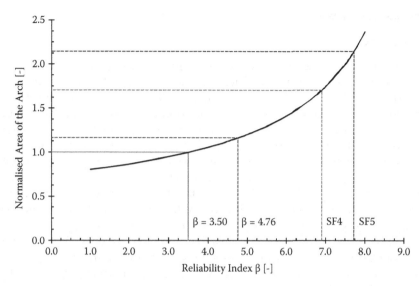

FIGURE 37.7
Arch area vs. reliability index.

Savings in Structural Mass

Obviously, a smaller flange thickness for the arch cross-section saves structural mass. Compared to the global safety factor design with a safety factor of 5, the reliability based approach saves 60% structural mass while providing a quantitative measure of the safety of the structure and still being more conservative than the Earth-based design approach.

With the cross-sectional areas of the global safety factor design it is possible to calculate the equivalent reliability index and corresponding probability of failure inversely. These values show that global safety factor design approaches result in a safety-level that is far beyond the desired (Keese and Barton 1982).

TABLE 37.2

Lunar Structure Design Results for Different Design Approaches and Safety Levels

	global SF = 5	global SF = 4	β (Earth) = 3.5	β (Moon) = 4.76
Flange thickness				
tf [mm]	4.03	2.96	1.24	1.64
A [cm²]	26.15	20.78	12.20	14.20
W [cm³]	151.10	110.82	46.52	61.49
$(A-A_{SF5})/A_{SF5}$ [%]	0.00	−45.70	−53.33	−20.54
$(W-W_{SF5})/W_{SF5}$ [%]	0.00	−26.65	−69.21	−59.31
$^2/\_$ [-]	7.72	6.90	3.50	4.76
γ_{glob} [-]	5	4	2.09	2.58
Structural mass [%]	100	73	30	40

Summary

So far, most of the known structural design proposals for lunar habitats have confined themselves to a conceptual design stage. With ongoing efforts to return to the Moon, these proposals have to face an assessment of their structural reliability with regard to lunar boundary conditions. This paper shows that common approaches from Earth using global or partial safety factors are not sufficient for lunar construction. Thus, the basic principles of structural reliability analysis were summarized firstly and applied here with special regard to the design of second generation lunar habitats. Based on an existing structural design of a lunar habitat, it was found that the reliability-based approach provides not only a quantitative measure of safety but also a highly efficient structure e.g., in terms of weight. This insight was only possible because reliability-based design criteria make it possible to compare different designs on a common basis. However, the accuracy of every reliability analysis increases with more statistical data about the characteristics of the basic variables.

For future lunar designs, it is of utmost importance to create a database of the mechanical properties of materials as well as the loading conditions on the lunar surface. As the design example shows, at the moment only engineering-type assumptions make the analysis possible at all resulting in an increased mass of the lunar habitat due to missing knowledge. But with enough construction experience and a large statistical database it will even be possible to develop a reliability-based "Lunar Building Design Code" similar to the Earth Design Codes, which standardizes and facilitates lunar structural analysis in the future.

Notation

M: Safety margin
R: Resistance
S, E: Load
β: Reliability index
P_F: Probability of failure
μ_M: Mean value
σ_M: Standard deviation
g(x): Limit state function
$f_x(X)$: Probability density function
α: vector
γ_M: Partial safety factor for resistances
γ_F: Partial safety factor for loads

COV: Coefficient of variance
f_y: Yield strength
p: Internal pressure
N: Axial force
M: Bending moment
A: Cross section area
q: Distributed (regolith cover) loads
t: Flange thickness
W: Section modulus
γ_{glob}: Global safety factor

References

Benaroya, H. (1994). "Reliability of structures for the Moon." Structural Safety, 15, 67–84.

Faber, M.H. (2006). Risk and Safety in Civil, Surveying and Environmental Engineering, Lecture Notes, Institute of Structural Engineering, Swiss Federal Institute of Technology, Zürich.

Ruess, F.; Schaenzlin, J.; Benaroya, H. "Structural Design of a Lunar Habitat." Journal of Aerospace Engineering 19 (2006), No. 3, pp. 133–157.

Ruess, F.: Structural Design of a Lunar Habitat. Diploma Thesis, University of Stuttgart & Rutgers University (2004).

Thoft-Christensen, P.; Baker, M.J.: Structural Reliability and its Applications. Springer Verlag, Berlin (1982).

Ditlevsen, O.; Madsen, H.O.: Structural Reliability Methods. John Wiley & Sons, Chichester (1996).

Benaroya, H.: An Overview of Lunar Base Structures: past and future, AIAA Space Architecture Symposium (2002).

Cohen, M.: Selected precepts in lunar architecture, 53rd International Astronautical Congress, The World Space Congress (2002).

Joint Committee on Structural Safety (JCSS): Probabilistic Model Code, 12th draft (2001).

Hasofer, A.M.; Lind, N.C.: An exact and invariant First Order Reliability Format. ASCE Proceedings, Journal of Engineering Mechanics, pp. 111–121 (1974).

Keese, D.L.; Barton, W.R.: Risk assessment and its application to flight safety analysis. Sandia National Laboratories (1982).

38

Cratering and Blowing Soil by Rocket Engines during Lunar Landings

Philip T. Metzger
NASA, KT-D3

John E. Lane
ASRC Aerospace

Christopher D. Immer
ASRC Aerospace

Sandra Clements
ASRC Aerospace
Kennedy Space Center, Florida

Philip T. Metzger is a research physicist and the founder of the Granular Materials and Regolith Operations Laboratory at NASA's Kennedy Space Center, where he has worked since 1985. He was a part of the Space Shuttle launch team and later with the International Space Station program. He is now performing research for lunar and Martian exploration and is leading the effort to understand how rocket exhaust interacts with soil. Dr. Metzger earned his doctoral degree in physics from the University of Central Florida and a bachelor's degree in electrical engineering from Auburn University. He grew up in Titusville, Florida and on Westport Island, Maine. He and his wife Gigi have four children and live in Orlando, Florida.

John E. Lane, PhD, received from the University of Central Florida, where his dissertation research supported NASA's Tropical Rainfall Measurement Mission (TRMM) ground validation. He received his BS and MS in physics from Florida Atlantic University. Currently an applications scientist with ASRC Aerospace, Kennedy Space Center, his function is to support applied technology activities, such as modeling and analysis of hydrometeors, development of photogrammetric algorithms to track birds using multiple cameras, modeling and analysis of high-speed jet-propelled lunar soil and dust particles, and development of software for simulation of electrostatic shielding of charged particle radiation.

ABSTRACT   This paper is a summary compilation of work accomplished over the past decade at NASA's Kennedy Space Center to understand the interactions between rocket exhaust gases and the soil of the Moon or Mars. This research is applied to a case study of the Apollo 12 landing, in which the blowing soil peppered the nearby Surveyor III spacecraft producing measurable surface damage, and to the Apollo 15 landing, in which the Lunar Module tilted backwards after landing in a crater that was obscured from sight by the blowing dust. The modeling coupled with empirical observations is generally adequate to predict the order of magnitude of effects in future lunar missions and to formulate a rough concept for mitigating the spray around a lunar base. However, there are many significant gaps in our understanding of the physics and more effort is needed to understand the problem of blowing soil so that specific technologies can be developed to support the lunar outpost.

Introduction

Without proper controls, the high temperature, supersonic jet of gas that is exhausted from a rocket engine is capable of damaging both the rocket itself and hardware in the surrounding environment. For about seven decades, NASA has invested significant effort into understanding and controlling these effects at the terrestrial launch pads [Schmalzer et al., 1998], and while the efforts have been largely successful, some damage to surrounding hardware still occurs on a routine basis. These challenges also exist when launching or landing rockets on other planetary bodies, such as the Moon, Mars, or asteroids. The exhaust gases of the landing or launching spacecraft could kick up rocks, gravel, soil, and dust. This can cause damage to the landing spacecraft or to other hardware that has already been landed in the vicinity. It can also spoof the sensors of the landing spacecraft and block visibility of natural terrain hazards, resulting in significant risk of an unsuccessful landing.

To date, humans have completed only 22 successful retro-rocket landings on other bodies. The United States landings included five robotic missions on the Moon in the Surveyor program, six human-piloted missions on the Moon in the Apollo program, two robotic missions on Mars in the Viking program, and one robotic mission on the asteroid Eros. At the time that this paper was written, the Phoenix mission was en-route for a retro-rocket landing on Mars. The successful Russian landings with retro-rockets have included seven robotic missions on the Moon in the Luna program and one robotic mission in the Mars program. Closely related to these, there were twelve terrestrial launches and landings of the DC-X rocket on the packed gypsum-powder surface of White Sands, New Mexico. The last of these missions resulted in the loss of the vehicle at landing, but not because of exhaust plume interactions. (The U.S. program has also landed three spacecraft on Mars using airbags rather than rockets at touchdown; the Russian program has landed ten spacecraft on Venus using parachutes and aerobraking; and the European program has landed one spacecraft on Saturn's moon Titan using a parachute.) There have been quite a few unsuccessful attempts to land with retro-rockets on other bodies, but so far none of these failures have been attributed to the exhaust plume's interaction with the surface

In the upcoming U.S. return to the Moon, there will be a greater concern with plume/soil interactions than in prior missions. That is because the landers will be larger with more thrust and because spacecraft will land and launch repeatedly in the vicinity of the lunar outpost, subjecting the hardware assets on the Moon to repeated high-velocity spraying of dust, soil, and possibly larger ejecta. Fortunately, there have been several cases of prior landings that provide significant insight into the possible effects of this spraying material. This paper analyzes the Apollo 12 and Apollo 15 landings as case studies in comparison with recent experiments and analysis.

The Problem

The Apollo 12 Lunar Module (LM) landed less than 200 meters away from the Surveyor III spacecraft as shown in Fig. 38.1. At the time, this distance was thought to be sufficient to minimize the effects that blowing soil might have upon the Surveyor spacecraft. The astronauts walked to the Surveyor, inspected it and removed portions for analysis on Earth in order to learn how the materials had been affected by the space environment (cosmic rays, micrometeoroids, vacuum, etc.). An interesting feature of the Surveyor hardware is that it had been sandblasted by a high-speed shower of sand and dust particles during the LM's landing [Jaffe, 1972]. The sandblasting cast permanent "shadows" onto the materials and these shadows were mathematically triangulated to a location on the lunar surface directly beneath the engine of the landed LM. Judging by the sharpness of the shadows and the lack of curvature allowable for the particles to fit the trajectory, investigators concluded that the particles must have been moving in excess of 100 m/s. Additional to this general scouring of the surface, there were discrete micro-crates or divots peppering its surface. Presumably the overall scouring was due to the large number of dust particles while the divots were due to the much smaller number of larger soil particles. Brownlee et al. [1972] studied the morphology of the resulting microscopic divots on the Surveyor's camera glass and estimated the particles were traveling between

FIGURE 38.1
Pete Conrad at Surveyor III with Apollo 12 LM.

300 and 2000 m/s. The authors have roughly estimated from the published reports and from the several boxes of engineering logbooks and documents at the lunar curation building at NASA's Johnson Space Center that there were on the order of 1.4 divots/cm2 on the side of the Surveyor camera cover facing the LM. Also, the Surveyor hardware had been injected by dust and sand particles that were blown into the tiny crevices and openings [Benson et al., 1972].

During the Apollo 15 landing, the crew reported that the blowing dust was visible from 46 m altitude and that from 18 m down the blanket of dust blowing across the field of view became so opaque that the landing had to be accomplished with zero visibility of the surface [Mitchell et al., 1972a]. On the other Apollo landings the visibility was not as bad [Mitchell et al., 1972b, Mitchell et al., 1973]. At footpad contact the LM rocked backward approximately 11 degrees from vertical before coming to rest [McDivitt et al., 1971], as shown in Fig. 38.2. One of the astronauts exclaimed "bam" over the radio coincident with the second contact event that terminated the backward rocking motion. It turns out that the LM had landed on the rim of a broad, shallow crater with two of its legs suspended in space over the crater and the other two legs resting on the soil outside the crater. It rocked backwards and to the left into the crater until three of the four legs were making contact with the soil, with the remaining leg of the LM

FIGURE 38.2
Apollo 15 LM tilted backwards 11 degrees into a shallow crater.

(the front leg, which was outside the crater) bearing no weight. The crater had not been visible to the astronauts during landing in part because it was shallow and hence inadequately shadowed in the center, and in part because the dense blanket of dust that was blowing over obscured it as illustrated by Fig. 38.3. As a result, the crew was not able to steer the LM past the crater to avoid the landing hazard. The resulting tilt angle of the LM was not so severe that it prevented successful completion of the mission, but it illustrates the potential problem of terrain features hidden by the dust.

The Apollo 12 experience illustrates that blowing material can damage nearby hardware. The Apollo 15 experience illustrates that it can pose a hazard to the lander, itself. In the context of the very successful Apollo program, these two situations were minor considerations to the respective missions and should not be exaggerated. In the context of returning to the Moon with

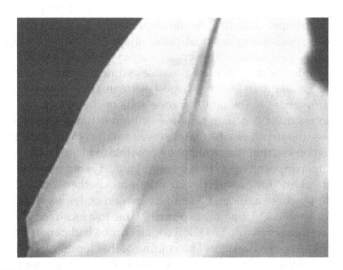

FIGURE 38.3
View from Apollo 15 LM descent imager camera, with distorted shadow of LM leg, footpad and soil contact probe draped across the blowing dust cloud (from the top center of the figure and pointing downward). Surface terrain features are not visible beneath the blowing dust.

multiple landings in the vicinity of a lunar outpost, they serve very usefully as case studies of the plume-soil interactions. From these measured effects, it is possible to calibrate a model of blowing soil and to gauge how much damage will be caused by future rockets as they launch and land in the vicinity of other hardware on the Moon.

To quantify the damage that may occur to surrounding hardware, it is crucial to quantify the erosion rate and total quantity of ejected soil. The estimates from the Apollo program did not agree with one another. One method to estimate the erosion rate was to first perform small scale experiments in vacuum chambers and measure the erosion rate [Clark and Land, 1963; Land and Clark, 1965; Land and Conner, 1967; Land and Scholl, 1969]. Then, Mason and Nordmeyer [1969] derived an empirical law for the erosion rate based upon these experiments, but calibrating the unknown effects of the lunar environment by the volume of the putative crater formed under the nozzle of vernier engine number three on the Surveyor V spacecraft, as seen in photographic images taken by that spacecraft. Mason [1970] used this erosion rate with the actual descent trajectory of the Apollo 11 spacecraft to calculate the expected soil erosion beneath the LM, and estimated that the crater depth would be in the range 1.3–2.0 cm (reported as 0.5–0.8 in.) and that the eroded volume would be in the range 36–57 liters (reported as 2200–3500 cu in.). A vastly higher erosion volume was estimated by R. F. Scott [1975] for Apollo 12 based on the number of particles required per square centimeter to scrub permanent shadows into the surface of the Surveyor III. Based on several assumptions, Scott estimated a removed soil depth of 18–25 cm (reported as 7–10 in.) over a radius of 2.3 m (reported as a diameter of 15 feet). The details of the calculation are not provided, but if he had assumed a conical crater shape, then this would represent a total eroded volume of 973–1390 liters, and if a cylindrical crater shape then this would represent 1460–2080 liters. A sphere-section crater shape would be intermediate to the cone and cylinder. Scott's smallest possible estimate of total erosion volume was 57 times greater than Mason's largest estimate. This cannot be attributed merely to differences in the landing zone soils or the trajectories of the two missions, so we must conclude that one or both estimates are not accurate. In both cases, the depth of soil removal was small compared to the natural terrain variations so that it would not be possible to identify a broadly tapering crater superimposed upon that terrain. Thus, it is not possible to directly measure its volume for any particular mission. Another comparison comes from Apollo 14 where a distinct, localized erosion crater was found near the nozzle of the landed LM, but it was probably due to a localized enhancement of the erosion rate where the LM's soil contact probe had penetrated and broken up the hard-packed surface. The volume of that localized crater was estimated to be 440 liters [Katzan and Edwards, 1991], and does not include any eroded soil over the broader region around the LM, so it does not provide an estimate of the natural erosion rate apart from the mechanical disturbance of the contact probe in this one case.

Another critical parameter to quantify is the ejection angle of the soil, because this will determine whether the soil will miss the surrounding hardware by flying over it or whether the soil can be blocked with a modest berm built by piling lunar soil around the landing zone. There was no clear consensus in the prior literature as to what determines the ejection angle. Roberts [1963a; 1963b; 1964; 1966] had assumed that aerodynamic forces do not significantly affect the ejection angle, so that the soil is ejected at the same angle as the local terrain slope, which acts as a ballistic ramp. Thus, the large and small particles will all be ejected into the same angle. However, we have observed in the Apollo videos that the dust blowing out from meter-scale impact craters on the lunar surface are ejected at an angle that modulates up or down coincident with the LM increasing and decreasing its thrust, and this indicates that the aerodynamics are a controlling factor and cannot be neglected. The scaled experiments discussed above did not measure ejection angles. A report on the conceptual design of a lunar base [Phillips et al., 1988; Phillips et al., 1992] used a plume flowfield calculated for free space [Alred, 1983], ignoring the presence of the lunar surface. This method ignores the all-important horizontal flow that develops across the lunar surface beneath the standoff shockwave and therefore cannot produce correct results.

This brief review indicates that neither the mass erosion rate nor the ejection angles have been adequately determined. The following sections of this paper describe additional methods to constrain these parameters.

Theoretical Background

Prior to each of the Surveyor, Apollo and Viking programs, NASA undertook a series of investigations to understand and quantify some of these physical phenomena to help ensure mission success. These studies discovered that the gas-soil erosion processes under a supersonic jet can be a complex set of solid/fluid interactions, depending upon the specific conditions of the jet and soil. To this day parts of the physics have not been accurately described or explained. Even a very basic, qualitative physical explanation has been lacking until recently for some aspects of a jet-induced cratering event. During the Apollo and Viking missions it was not necessary to fully understand these phenomena because the spacecraft engines were designed to prevent the most energetic of these processes from occurring. That is, the pressure developed upon the Lunar or Martian regoliths in the stagnation region of the impinging jets was kept sufficiently low to prevent the bearing capacity failure of the soil which otherwise may have occurred. This was possible in the lunar landings because the small mass of the LMs and the weak lunar gravity made it possible to use a lower thrust and because the unweathered lunar soil is very compacted with extremely high shear strength and very low gas permeability [Carrier et al., 1991]. However, the lunar regolith has a very loose layer of surface material (dust and sand-sized particles), just a few centimeters thick, and so the surface erosion of this loose material appears to be the primary effect in these landings.

Roberts [1963a; 1963b; 1964; 1966] developed a theory of this viscous erosion (VE) mechanism for lunar dust. His method was adopted by J. S. Dohnanyi [1966] to apply to the design of the LM engines. Roberts derived a set of equations which calculate the shear stresses on a flat, dust-covered surface, and calculated the quantity of material which would be entrained into the gas flow as a function of radial distance from the center of the plume. The region of maximum shear stress turned out to be a ring some distance out from the center of the exhaust because the gas velocity (v) increases radially while its density (ρ) decreases into the lunar vacuum and thus the dynamic pressure ($\rho v2/2$) is a maximum at some finite radius. Hutton [1968] compared the theory to the small-scale experiments in vacuum chambers by Clark et al., cited above, and found only limited correlation. We believe that this is partly because of the simplifications in Roberts' theory, but also partly because the experiments did not adequately simulate the lunar conditions. For example, Roberts' erosion rate equation omits the effect of particles eroded upstream in the flow upon the erosion rate of soil downstream in the flow. This is a good approximation only when the erosion processes occur over a small distance relative to the length scale of the flow field, but in the small scale experiments the flow field is very small and so this approximation is not appropriate. Furthermore, the volumetric erosion rate was so high in the experiments compared to the lunar case that the shape of the surface changed dramatically during the test, whereas Roberts assumed a flat surface.

Few studies have been done on the other exhaust cratering mechanisms besides VE. One such experimental study was performed by Alexander et al. [1966]. This study discussed VE but focused primarily upon bearing capacity failure (BCF) as a cratering mechanism, in that the stagnation pressure of gas directly beneath a jet may exceed the bearing capacity of the soil and mechanically push it downward, forming a crater under the jet. Cold gas jets and hot engine firings were used to create craters in sand and clay, and the resulting craters were measured for various dimensions. The data were compared to identify significant parameters and scaling relationships. The authors developed several methods to predict the approximate crater dimensions, including (1) an analogy to the classic cone penetrator test, (2) a refinement of the cone penetration model in which the diffusion of gas into the soil is assumed to have reached steady state to weaken the soil according to Terzaghi's effective stress hypothesis, (3) a purely elastic model of the sand to provide an order-of-magnitude estimate of the width of sand that would fail and be removed in the initial crater formation, and (4) a yield-strength analysis using the equations of soil mechanics to calculate the stresses as a function of distance beneath a point load to estimate crater depth. The experimental methods did not provide a direct view beneath the surface during or after the BCF event, so the major features of these models were untested.

Another study in this Apollo era by Scott and Ko [1968] identified the diffused gas eruption (DGE) mechanism. Whereas Alexander et al. were concerned only with how gas diffusion enhanced the BCF mechanism, Scott and Ko treated the gas diffusion as a distinct soil-moving mechanism in its own right. They fired rocket motors into soil and observed the results with a high-speed video camera. They discovered that radial diffusion of pressure could eventually blow out a toroidal region around the exhaust jet. This occurred because the high pressure gas diffusing into the soil beneath the engine would diffuse radially outwardly from the jet until the pressure of gas beneath the surface at some radial distance was sufficient to lift the overlying column of soil. They also found that when the rocket was shut off a spike of soil could blow up the center of the rocket nozzle as the gases quickly diffused back out from the soil. The investigators modeled these effects with a numerical, finite-difference algorithm. The model successfully predicted the DGE in the toroidal region during jet firing and also in the central region after jet cutoff. Hon-Yim Ko [1971] provided an improved analysis of how gas diffusion enhances BCF. That paper is presently inaccessible to the authors. Apparently, it describes a finite element program to analyze both gas diffusion and BCF, but the program did not produce sufficiently accurate results due to the limited computing capabilities available at the time.

During the lead-up to the Viking landings on Mars, a series of papers were authored with the interest in avoiding BCF altogether and keeping DGE to levels that could be safely ignored. In contrast to the lunar case, the thin Martian atmosphere will collimate rocket exhausts [Foreman, 1967] and focus the stagnation pressure onto a small portion of the regolith. Roberts' model therefore needed to be modified before it could be applied to Mars. Clark [1970] tested a scaled Viking lander in a 60-foot vacuum sphere, paying special attention to the cant angle of the nozzles on the multi-engine lander. Another Viking study [Romine et al. 1973] addressed exhaust cratering both theoretically and experimentally, showing that a conventional bell nozzle would affect the surface too much beneath the lander and making a number of significant contributions to our understanding of the physics. Finally, Hutton et al. [1980] described the observed disturbances that were actually caused by the Viking retro-rockets landing on Mars. These Mars studies provide some physical intuition of the physics for the lunar case, but cannot be directly applied to it due to the environmental differences.

To summarize, the investigations supporting the Apollo and Viking programs determined that there are several physical mechanisms of interaction between gas jets and soil. The identified mechanisms were viscous erosion (VE), diffused gas eruption (DGE), and bearing capacity failure (BCF). These will occur in varying proportions depending upon the particular conditions of the soil and the jet. Roberts' theory assumed implicitly that VE is the only mechanism capable of moving soil during the lunar landing. Experience

FIGURE 38.4
Area under Apollo 12 LM engine nozzle showing how surface has been "swept clean" of loose material. The narrow trench in the upper left part of picture was dug by the soil contact probe as it dragged beneath the descending LM.

shows that bearing capacity failure did not occur under the exhaust plumes in the Apollo program. Probably the bearing capacity of the lunar soil was sufficient to resist cratering because of its very high internal friction and its relatively low gas permeability. The area under the nozzle in each mission had a "swept clean" appearance as shown in Fig. 38.4, missing the loose layer of un-compacted soil and dust that was characteristic everywhere else on the Moon.

Because the soil around the nozzle was so undisturbed, it is unlikely that any DGE occurred after engine cutoff. On the landing videos, a thin, dusty mist is visible for a few seconds after engine cutoff, and this probably represents the entrainment of only very tiny dust particles as the regolith depressurizes. In light of these things, it would seem that the looser surface soil was swept away from beneath the nozzle but the deeper, more compacted layers remained in place. Nevertheless, it is problematic to explain this by Roberts' theory, because the shear stress of the gas is zero at the stagnation point under the center of the nozzle, and very low for a significant radius around that point until at higher distances the gas velocity becomes sufficiently high to move the soil. So what sweeps the soil away from the center-line of a jet? Similarly, in loose sand, why is a jet-induced crater deepest in the center where the gas velocity is zero? A simple test can show that a jet easily forms a crater even when its dynamic pressure is far below the pressure that the sand can support, and so BCF must not be the general explanation for the motion of sand directly under a jet. This is relevant to predicting the erosion rate of the Apollo missions since the combination of mechanisms that move the soil may predict a different rate than Roberts' theory, which assumes the mechanism to be VE, alone.

Experiments of Cratering Mechanisms

To gain more insight into the physics, tests were performed at ambient pressure with sand impinged normally by jets composed of different gases (nitrogen, carbon dioxide, argon and helium) to provide variations in gas density. The tests were performed with different exit velocities, different nozzle heights above the sand, and with different sized sand grains. The tests used two methods to identify soil behavior beneath the surface. In the first method, a sandbox was prepared with horizontal layers of different colored sand. A vertical jet was impinged upon the sandbox forming a vertical burst of sand that left a shallow, residual crater on the surface. This was filled in with black sand so that the crater would not slump while filling with epoxy and to provide color contrast as a record of the crater shape. Optically clear epoxy was diffused into the pore spaces of the sand and thermally cured so that it could be cut in half to reveal the deformation of the layers beneath the surface. The prediction from the model of Alexander, et al. was that the layers would be bent downward beneath the crater as they would be due to cone penetration. However, we found that the sand was pulled upward along the crater axis as shown in Fig. 38.5, quite the opposite of what we expected.

To explain this subsurface flow, the second test method performed the cratering on the edge of the sandbox with a clear window to see into the subsurface during the test. The top edge of the box was beveled outwardly to bisect the jet with minimal disturbance of the flow inside the sand, as shown in Figure 38.6. Two regimes of gas-sand interaction were observed as a function of jet velocity. For the higher velocity (but still subsonic) regime, the cratering was seen to consist of a very deep, very narrow, cylindrical hole that burrowed quickly to some (repeatable) depth and then abruptly stopped. While the jet remained, the hole maintained its steep sides and sand was being pulled up along the sides to deform the horizontal, colored layers of sand upward along both sides of the hole. When the jet was extinguished, the narrow hole collapsed leaving only a broad, shallow crater at the surface with slopes at the angle of repose. That dynamic is why the sand appeared to be pulled upwards toward the center of the shallow crater when examining the layers of sand after the test. The hole prior to collapse was much deeper

FIGURE 38.5
Cutaway view of sand layers (originally horizontal) as they were deformed beneath the surface of a crater.

and narrower than previously recognized. The crater dimensions measured by Alexander at al. [1966] were the conical, residual craters remaining at the surface and did not describe the hole prior to termination of the jet. While the jet was present, the motion of the sand exiting from the upper part of this hole may be properly characterized as turbulent aggregative fluidization

FIGURE 38.6
Test apparatus with window at front of sandbox. The curved shape above the sand is the beveled cutout in the window, intended to reduce the interference of the window with the gas jet while yet blocking sand from falling in front of the box and thus obscuring the view.

[Grace and Bi, 1997]. Near the bottom of the hole no sand was entrained and growth of the hold was entirely by motion of the bulk sand beneath the surface of the crater. From the video images, we tracked individual particles in the bulk to obtain the sand's velocity field beneath its surface. The analysis is described in detail by Metzger et al [2008a]. We found that sand flows in a thick band that is tangential to the surface of the crater, dragging it away from the tip of the hole so that the hole continues growing downward, and then dragging it up the sides of the hole creating the upward deformation of sand layers described above. This flow of sand is driven by the drag force of the gas diffusing through the sand, which creates a sufficient body-force distributed throughout the sand to setup a stress state that exceeds the soil's shear strength and initiates shearing. This mechanism of sand-gas interactions had not been previously described in the literature and we are calling it diffusion-driven shearing, or DDS. DDS differs from BCF because the sand moves tangentially to the free surface of the crater, not perpendicularly away from the surface as predicted by the BCF mechanism. DDS differs from VE because, although both mechanisms move the particles tangentially to the surface, DDS occurs in a thick band beneath the surface due to diffusive gas flow, whereas VE occurs only along the top layer of grains due to the free fluid flow in the boundary layer above the sand.

In the second regime of testing with slower jets of gas, the crater formed in a broad, conical shape as shown in Figure 38.6. With sufficient dynamic pressure of the gas the crater would also form a paraboloidal "inner" crater at the bottom of the conical crater as shown in Figure 38.7. The inner crater

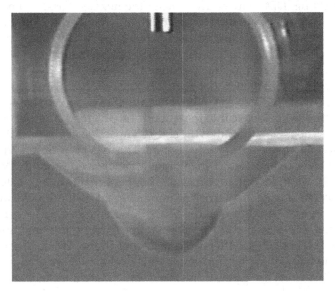

FIGURE 38.7
Crater formation with inner paraboloidal crater and outer conical crater.

was formed by the direct action of the jet whereas the outer conical crater was the result of slope failure, avalanching sand down into the inner crater and forming the outer slope at the angle of repose. The inner crater can be understood as a transitional form of the cylindrical hole that would occur in the faster-regime of cratering, described above. Diffusion-driven shearing was observed to occur just in the very tip of the inner crater, whereas viscous erosion was the predominant mechanism throughout the remainder of the inner crater, rolling grains uphill until they reached the inner crater's lip where they went airborne. A software algorithm was developed to automatically analyze the videos frame-by-frame throughout the duration of the tests to extract crater shape and related parameters and to perform volume integrals to calculate quantities of ejected sand. The analysis was complicated by the fact that sand recirculates in the crater multiple times: the crater widens and re-ingests sand deposits that had previously fallen around its perimeter; and some of the sand falls directly back into the crater from the air. The widening crater captures and recirculates an increasing fraction of the ejected sand, and this slows down the net growth rate. Compensating for this effect, we find that the ejection of sand is actually at a constant rate throughout the test [Metzger et al., 2008a]. Furthermore, it shows that erosion rate scales linearly with the dynamic pressure of the jet ($\rho v2/2$), which is consistent with the assumptions of Roberts' theory. In these tests, erosion occurred at the upper lip of the inner crater by VE. DDS only operated to deliver sand from the bottom of the crater up to the sides where the gas velocity was nonzero. The grains then rolled uphill under the increasing velocity of the jet to the point where VE was occurring right at the lip of the inner crater. Similarly, in a lunar landing, DDS may occur beneath the nozzle to assist in moving the loose top layer of soil outwardly, and then grains may roll along the surface to the regions of greater shear stress where lofting finally occurs. So VE may not be the only mechanism involved in the process, but ultimately it will still be VE that controls the rate of entrainment.

A fifth type of interaction between gas and soil has also been identified, occurring only when a rocket engine is ignited over soil so that the impinging gas sends a shockwave into the soil prior to the formation of the standoff shock. This shockwave modifies the soil's compaction as it passes through, as well as possibly breaking cohesive bonds. This has been observed in recent tests with solid rocket motors firing into a meter-deep sandbox [Metzger et al., 2007]. In these tests it appears that orders-of-magnitude greater surface erosion occurs during the transient impingement of the shockwave on the sand. A similar effect occurs at the Space Shuttle launch pad when concrete is excavated and blown out from the flame trench by the impinging shock [Lane, 2004]. These shock effects did not occur in the Apollo lunar landings because the stagnation pressure on the soil developed more gradually during descent. This effect must be considered in the future if we launch spacecraft directly from the lunar surface, unlike in Apollo where the descent stage was left behind, shielding the soil.

Dust Ejection Angles in Apollo Landing Videos

During the Apollo landings, the descent imager camera was a film camera mounted in the right-side window looking downward and forward from the LM. The videos show not only the cloud of blowing dust but also the shadow of the LM draped across that cloud as shown in Figure 38.3. From the distortion of the LM's shadow it is possible to measure the shape of the cloud and extract information about the ejection angle. To perform this analysis, we worked with a computer model of the LM developed by Sullivan [2004]. We also took physical measurements of an LM remaining from the Apollo program, located at the Kennedy Space Center. The three-dimensional measurements were accomplished using a photogrammetry system developed for the Space Shuttle *Columbia* investigation [Lane and Cox, 2007] in which a photogrammetry cube with reference markings is placed in the field of view and photographs are taken of the total scene from multiple perspectives. Software developed for this system is used to interpret the set of images three dimensionally and obtain measurements between pairs of points throughout the scene. Based on these LM dimensions, a geometric analysis of the shadows [Immer et al., 2008] indicates that the visible part of the dust cloud is usually leaving the vicinity of the LM at an ejection angle less than 3 degrees, as shown in Table 38.1. Several measurements were obtainable for most missions, depending on the number of usable images and the number of points where the LM altitude was audibly announced by the crew during the descent. For Apollo 12 the sun angle was too low to make measurements. For Apollo 15 the dust angles were remarkably higher. Examining the

TABLE 38.1

Dust Ejection Angles Measured from LM Shadows

Mission	Sun Angle	Ejection Angle
11	10.8	2.3
		2.3
12	5.1	—
14	10.3	2.5
		2.7
15	12.2	7.8
		7.2
		11.8
16	11.9	1.0
		1.4
		1.4
17	13.0	2.0
		1.6

landing terrain shows that the shallow crater beneath the cloud was probably responsible for this discrepancy, and in fact the high angle probably represents the real ejection angle of the dust leaving from the sloped forward bank of the crater. Since we lack a sufficient understanding of the erosion physics to model the ejection angle from first principles, we have used these empirical values for the subsequent modeling.

It should also be noted that in Apollo 15 the motion of the shadows in the final seconds of landing indicate that a "blowout" event occurred in which a high volume blast of soil was ejected at a much higher elevation angle. Unfortunately, it is impossible to measure this steeper ejection angle since the shadows are driven outside the field of view. Our best estimate, extrapolating the velocity of the shadows beyond the field of view, indicates the soil ejection angle was probably greater than 22 degrees for that brief moment. We believe that landing on a leveled and/or artificially stabilized surface may be required in the future if it is necessary to entirely eliminate these blowout events in the vicinity of the lunar outpost.

Modified Roberts' Model

To estimate the quantity and trajectories of soil and dust blown at the Surveyor III spacecraft, we have modified Roberts' model in the following ways [Metzger et al., 2008b]. First, we have integrated the equations over a realistic particle size distribution of lunar soil. To obtain an analytical form for this distribution we have measured a quantity of the lunar soil simulant JSC-1A using a Sci-Tec Fine Particle Analyzer to obtain very smooth statistics of the particle count distribution as shown in Fig. 38.8. An exponential decay fits the JSC-1A data sufficiently over the entire range above 10 microns. It is uncertain whether JSC-1A is representative of real lunar soil below 10 microns, so we believe that this functional form is, for the present, an adequate approximation over the entire range. The limits of integration over this size distribution were obtained using Roberts' equations. This predicts that the eroded particle sizes will be between 1 μm and 1 mm, with smaller particles being inseparable due to cohesion and larger particles being inerodible due to excessive mass relative to the aerodynamic forces. We have evidence in the Apollo landing videos that much larger particles than 1 mm are actually being eroded, and we suspect that Roberts' model is incorrect in this regard because it inadequately represents the aerodynamics forces in the boundary layer along the lunar surface. Nevertheless, this will not affect the estimate of the damage to the Surveyor because such large particles are relatively few and unless a remarkably large piece of gravel were to strike the Surveyor they would represent only a minor fraction of the total damage.

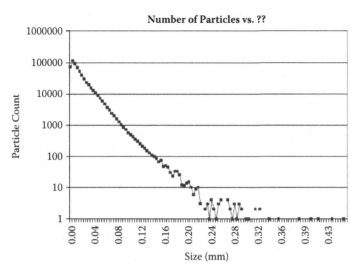

FIGURE 38.8
Number count distribution for JSC-1A particle sizes.

Second, we have replaced Roberts' estimate of ejection angle (based on the evolving crater shape of the soil) with a narrow distribution of angles clustered around the empirically-determined values. Third, we have integrated the resulting equations over the Apollo 12 LM descent trajectory as estimated from voice callouts of altitude by the astronauts during the landing, as shown in Fig. 38.9. An example of the soil flowfield predicted by the model is shown in Fig. 38.10. The model predicts the velocity of the eroded particles as a function of their size and of the LM altitude, as shown in Fig. 38.11. The highest velocities (for the smallest particles) are close to lunar escape velocity, 2.38 km/s. For example, a 10 μm particle may be blown at 1.9 km/s when the LM is near touchdown, and at that velocity and with a 3 degree ejection angle the trajectory will be as shown in Figure 38.12. This range of velocities agrees with the observation of Apollo 11 mission commander Neil Armstrong that the horizon became obscured by a tan haze [Armstrong et al., 1969]. This indicates that dust had sufficient velocity that the ballistics (with our empirically-measured 3 degree ejection angle) could take it beyond the horizon. At 24 m, the altitude when dust blowing began, the ballistics require a minimum velocity of 487 m/s. This is in the range of predicted velocities for the dust. The model also predicts that there would have been 3.1 divots/cm2 on the Surveyor due to the landing LM. This compares to the same order of magnitude as the estimated 1.4 divots/cm2 actually observed. In fact, considering the many large sources of error in the modeling at present, it must be admitted as coincidence that the comparison came out so well. Nevertheless, we take it as evidence that the model is good to the correct order of magnitude. The model also predicts that the total volume of soil

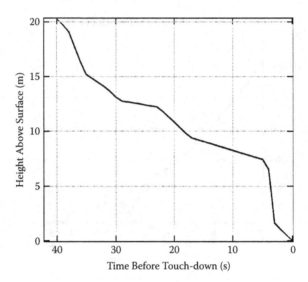

FIGURE 38.9
Apollo 12 LM descent trajectory.

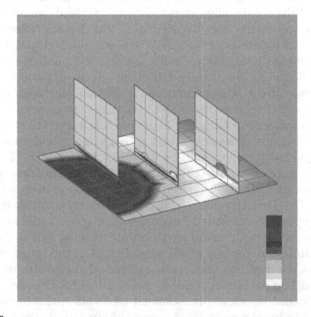

FIGURE 38.10
Example of type of output presented by modified Roberts' model, showing mass flux of blowing soil in a 3D map.

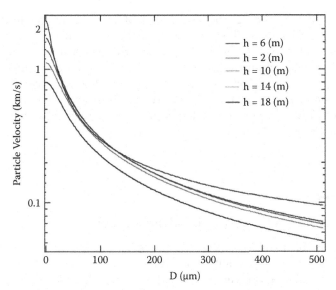

FIGURE 38.11
Predicted particle velocities as a function of diameter for several LM altitudes.

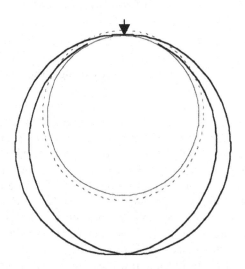

FIGURE 38.12
(Thin Solid Line) Circumference of Moon. (Dashed) Altitude of orbiting Command Module, for reference. (Arrow) landing site of LM. (Thick Solid Line) Trajectories of particles blown forward and backward from LM at 3 degree ejection angle and 1.9 km/s velocity.

removed in the Apollo 12 landing was 787 liters, intermediate to the values of Mason (36–57 liters) and Scott (1460–2080 liters, assuming Scott had used a cylindrical crater shape). The model predicts the maximum radius of erosion to be 7.57 m. Crudely estimating the erosion depth by assuming a conical crater shape, we predict only 1.3 cm at the center. Our model predicts a much wider erosion radius than the value used by Mason, and so despite the much larger erosion volume our predicted depth comes out comparable to the values of Mason (1.3–2.0 cm).

Optical Density in Apollo Landing Videos

It is also possible to measure the optical density of the dust cloud in the landing videos to extract information about the number of particles entrained in the gas. The calculation was performed by measuring the brightness of the image on a sunlit face of a rock and in its adjacent shadow, both when the dust cloud is present and when there is a momentary clearing of the cloud [Immer et al. 2008]. These four data points enable a calculation of the mass density from Mie scattering,

$$\rho = \frac{-m_g}{\pi a^2 Q_\lambda s} \ln \frac{I_{b,2,\lambda}(\textit{measured}) - I_{s,2,\lambda}(\textit{measured})}{I_{b,1,\lambda}(\textit{measured}) - I_{s,1,\lambda}(\textit{measured})} \tag{38.1}$$

where ρ is the mass density of the cloud, mg is the average mass of the dust grains, $\pi a2$ is the average cross sectional area of the particles, s is the pathlength of the light passing from the sun through the cloud to the ground and then back to the camera, $Q\lambda$ is the extinction coefficient of the mineral, assumed here to be unity for sufficiently large dust grains, and the four values of $I\lambda$ are the measured intensity of the image for the four cases, with the additional subscripts b or s representing "bright" and "shade" and 1 or 2 representing "without" or "with" the dust cloud, respectively. This calculation estimates that there are on order of 108 particles/m3 entrained in the cloud. This compares poorly with the modified Roberts' model describe above, which predicts only 106 particles/m3, an error of two orders of magnitude.

The underestimation of the Roberts' model is not hard to understand. The optical density is controlled primarily by the smallest erodible particle size, because the most surface area in the cloud is due to the smallest particles, which are more numerous and have the greatest area-to-mass ratio. The cohesion of lunar soil is still one of its least understood characteristics, and so Roberts' model made crude assumptions about the cohesive forces that would prevent the smallest particles from separating. A small error in that assumption produces a large error in optical density without greatly affecting the predicted number of divots (caused by the larger particles) or the total mass of eroded soil. In his final paper on the topic, Roberts [1964] wrote,

> ...there is negligible loss of visibility until the vehicle descends to this altitude [i.e., 20 feet, or 6m]; below 20 feet [6m], downward visibility may be reduced but lateral visibility will not be affected.

In contrast, the Apollo 12 mission report and crew debriefing say the following:

> On Apollo 12 the landing was essentially blind for approximately the last 40 feet. [McDivitt, 1970]
> ...the dust went as far as I [Pete Conrad] could see in any direction and completely obliterated craters and anything else. All I knew was there was ground underneath that dust. I had no problem with the dust, determining horizontal or lateral velocities, but I couldn't tell what was underneath me. I knew I was in a generally good area and I was just going to have to bite the bullet and land, because I couldn't tell whether there was a crater down there or not.... [After landing] it turned out there were more craters around there than we realized, either because we didn't look before the dust started or because the dust obscured them. [Conrad et al., 1969]

So it is not surprising that the measurement of optical density is a few orders of magnitude different than Roberts' predictions that were made with inadequate information on the cohesive forces.

Discussion

To-date, the best method for predicting the erosion rate of soil is still based on Roberts' method. Scott's calculation based on the total surface scouring of the Surveyor (rather than its divot count) assumed particles much larger than the dust fraction present in the soil, and thus overestimated the erosion rate. Mason's estimate was an order of magnitude smaller than ours and therefore does not agree well with the divot count on Surveyor III. We suggest that the use of small-scale testing in Mason's estimate may have contributed some error to the prediction. Much progress has been made in understanding granular media in the past several decades, and it is generally understood that granular phenomena are often unscalable. That is because, unlike ordinary fluids where the size of the molecules is irrelevant, the size of the sand grains is an important length-scale in the physics and so keeping all the important non-dimensional parameters constant requires the testing to be done only at full scale. For example, in the testing by Clark et al. discussed above, the Knudsen number was not kept constant, although it is important in determining the drag forces on the sand and thus the erosion rate. Also, the length scale of the diffused gas pressure field (e.g., the pressure divided by its own gradient) was not addressed in the small scale tests, although it is important in DDS to determine whether the soil will shear and form a deep crater (as seen in the small scale testing but not in the Apollo landings).

Roberts' method works from first-principles, assuming that the shear stress in the gas is exactly consumed by the change in momentum of the eroding soil, so that the erosion rate self-adjusts to the shear stress. To be more accurate, future modeling will need to account for the increasing shear strength of the soil with depth due to increasing soil compaction with depth [Mitchell et al., 1974]. It should account for the physical processes directly under the nozzle that push soil out to the annular region where VE occurs, since this soil will be uncompacted in contrast to the undisturbed soil in that region. It should also improve the model of aerodynamic forces on the particles. They are not well-understood in part because the structure of the boundary layer has not been characterized well for this supersonic, highly rarefied flow, and because the lift and drag coefficients around a tiny particle under the same conditions have not been studied in detail. Furthermore, the nature of turbulence and its effects in dispersing particles upward through this boundary layer are not well-known. Finally, the role of particle collisions in dispersing the dust cloud vertically and in transferring momentum between smaller and larger particles has not been determined. Preliminary modeling has been performed with Lagrangian calculation of the individual particle trajectories decoupled from an Eulerian calculation of the gas flow field [Lane et al. 2008, Lumpkin et al. 2007]. The results suggest that particle dispersion by turbulence and/or particle collisions is probably important because lift and drag alone are inadequate to explain the particle dynamics observed in the landing videos. For these reasons, we cannot yet predict the erosion rate with an expectation of accuracy, and neither can we predict the ejection angles as a function of particle size from first principles. It is quite likely that larger and smaller particles will be segregated into different ejection angles in this process (as suggested by preliminary modeling). Unfortunately, the measurement of ejection angles from the landing videos only tells us about the finest particles that have the greatest optical density. We do not know if the larger particles, say 100 microns and larger, go into a higher trajectory (as some preliminary simulations suggest). This is important because it was the larger particles that caused the divots in the Surveyor III, while the finer particles were responsible for scrubbing permanent shadows into its finish.

Despite these uncertainties, the work to-date suggests that a berm built out of lunar soil around the landing site may be highly effective at mitigating the damage to surrounding hardware. The berm could easily be built high enough to stop a 3 degree ejection angle of fine particles, and the large particles will be going sufficiently fast that even if they fly over the berm then they should pass right over the outpost, as well. The only concern would be the largest particles, such as gravel or small rocks, which might fly with sufficiently low velocities that they could be lofted over the berm and then arc downward to strike the outpost that is behind it. Further work is required to determine the maximum size particle that can be lofted, which is still uncertain as long as the aerodynamic forces are uncertain.

In order to support future lunar operations, a physics-based numerical model is being developed to incorporate all the known mechanisms of gas-soil interactions. If the unknown aspects of the physics are sufficiently characterized, and if the model is properly coded, then it will seamlessly predict all the mechanisms that may occur for the larger and multi-engine landers that may be used in the future, as a function of the propulsion system, trajectory, and soil characteristics.

Summary and Conclusions

Our present understanding of lunar plume effects is based on a synthesis of the astronaut observations, measured Surveyor III effects, analysis of Apollo videos and photographs, terrestrial experiments, and simulations of the physics. This synthesis demonstrates rough consistency between the various sources of knowledge. Some of the older methods developed to predict this problem (some of which were not reviewed here) are not adequate because they over- or under-predict the quantity of blown soil and predict incorrect ejection angles. More work is needed to be able to predict these things entirely from first principles. Left unchecked, the spray of soil will cause unacceptable effects upon the hardware and materials in the vicinity of the lunar outpost. The particles travel at such high velocity that it is not possible to get far enough away from the spray to prevent these effects. Because of the low ejection angle for most of this spray, it seems feasible to use a berm or other physical obstruction to block most of the material.

References

Alexander, J. D., W. M. Roberds, and R. F. Scott [1966], "Soil Erosion by Landing Rockets," Contract NAS9-4825, Hayes International Corp., Birmingham, AL.

Alred, J. W. [1983], "Flowfield Description for the Reaction Control System of the Space Shuttle Orbiter," AIAA-83-1548, Proc. of AIAA 18th Thermophysics Conference, Montreal, Quebec.

Armstrong, N.A., M. Collins, and E.E. Aldrin [1969], "Apollo 11 Technical Crew Debriefing," Vol. 1, NASA Johnson Space Center, Houston, TX, pp. 9.27 1 9.29.

Benson, R. E., et al. [1970]. "Preliminary Results from Surveyor 3 Analysis" in Apollo 12 Preliminary Science Report, NASA, Washington, DC, p. 219.

Brownlee, D., W. Bucher, and P. Hodge [1972], "Part A. Primary and secondary micrometeoroid impact rate on the lunar surface: a direct measurement," in Analysis of Surveyor 3 material and photographs returned by Apollo 12, NASA, Washington, DC, pp 143–150.

Carrier, W.D., III, G.R. Olhoeft and W. Mendell [1991], "Physical Properties of the Lunar Surface," in Lunar Sourcebook, A User's Guide to the Moon (G. H. Heiken, D.T. Vaniman and B.M. French, eds.), Cambridge University Press, Melbourne, Australia, pp. 475–594.

Clark, L.V. and N.S. Land [1963], "Dynamic Penetration and Erosion of Dust-Like Materials in a Vacuum Environment," A Compilation of Recent Research Related to the Apollo Mission, NASA Langley Research Center , Hampton, VA, pp. 145–154.

Clark, L.V. [1970], "Effect of Retrorocket Cant Angle on Ground Erosion–a Scaled Viking Study," TM X-2075, NASA Langley Research Center, Hampton, VA.

Conrad, C., R.F. Gordon, Jr., and A.L. Bean [1969], Apollo 12 Technical Crew Debriefing, NASA Johnson Space Center, Houston, TX, Vol. 1, pp. 9.11–9.12.

Dohnanyi, J. S. [1966], "Remark on the Rocket Plume–Lunar Surface Erosion Problem," TM-66-1011-2, Bellcomm, Washington, DC.

Foreman, K. M. [1967], "The Interaction of a Retro-Rocket Exhaust Plume with the Martian Environment," Grumman Res. Dept. Memorandum RM-354, Grumman Aircraft Engineering Corp., Bethpage, NY.

Grace, J.R. and H. Bi [1997], "Introduction to circulating fluidized beds," in Circulating Fluidized Beds (J.R. Grace, A.A. Avidan, and T.M. Knowlton, eds.), Blackie Academic and Professional, New York, NY.

Hutton, R.E. [1968], "Comparison of Soil Erosion Theory with Scaled LM Jet Erosion Tests," NASA-CR-66704, TRW Systems Group, Redondo Beach, CA.

Hutton, R.E., H.J. Moore, R.F. Scott, R.W. Shorthill, and C.R. Spitzer [1980], "Surface Erosion Caused on Mars from Viking Descent Engine Plume," The Moon and the Planets, Vol. 23, pp. 293–305.

Immer, C.D., P.T. Metzger, and J.E. Lane [2008], "Apollo Video Photogrammetry Estimation of Plume Impingement Effects," Earth and Space 2008, 11th Biennial ASCE Aerospace Division International Conference on Engineering, Construction and Operations in Challenging Environments, Long Beach, CA.

Jaffe, L. D. [1972], "Part I. Blowing of Lunar Soil by Apollo 12: Surveyor 3 Evidence," in Analysis of Surveyor 3 material and photographs returned by Apollo 12, SP-284, NASA, Washington DC, pp 94-96.

Katzan, C.M. and J.L. Edwards [1991], "Lunar Dust Transport and Potential Interactions With Power System Components," NASA Contractor Report 4404, Sverdrup Technology, Brook Park, OH, pp. 8–22.

Ko, H.-Y. [1971]: "Soil Properties Study," Viking Project Report VER-181, Martin Marietta Corp, Denver, CO.

Land, N.S. and L.V. Clark [1965], "Experimental Investigation of Jet Impingement on Surfaces of Fine Particles in a Vacuum Environment," TN-D-2633, NASA Langley Research Center, Hampton, VA.

Land, N.S. and W. Conner [1967]. "Laboratory Simulation of Lunar Surface Erosion by Rockets," Institute of Environmental Sciences 13th Annual Technical Meeting Proceedings, Vol. 1, Washington, DC.

Land, N.S., and H.F. Scholl [1966], "Scaled Lunar Module Jet Erosion Experiments," TN-D-5051, NASA Langley Research Center, Hampton, VA.

Lane, J.E. [2004], "Ground Camera Photogrammetry 3D Debris Trajectory Analysis," ASRC Aerospace, Kennedy Space Center, FL.

Lane, J.E. and R.B. Cox [2007], "Digital Image Inspection for Spacecraft Processing," ASRC Aerospace, Kennedy Space Center, FL.

Lane, J.E., P.T. Metzger, and C.D. Immer [2008], "Lagrangian trajectory modeling of lunar dust particles," Earth and Space 2008, 11th Biennial ASCE Aerospace Division International Conference on Engineering, Construction and Operations in Challenging Environments, Long Beach, CA.

Lumpkin, F., J. Marichalar, A. Piplica (2007), "Plume Impingement to the Lunar Surface: A Challenging Problem for DSMC," in Direct Simulation Monte Carlo Theory, Methods & Applications, Santa Fe, NM.

Mason, C.C. [1970], "Comparison of Actual versus Predicted Lunar Surface Erosion Caused by Apollo 11 Descent Engine," Geological Society of America Bulletin, Vol. 81, pp. 1807–1812.

Mason, C.C. and E.F. Nordmeyer [1969], "An empirically derived erosion law and its application to lunar module landing," Geological Society of America Bulletin, Vol. 80, pp. 1783–1788.

McDivitt, J.A., et al [1970], "Apollo 12 Mission Report", MSC-01855, NASA Johnson Space Center, Houston, TX, sect. 6.1.3.

McDivitt, J.A., et al [1971], "Apollo 15 Mission Report", MSC-05161, NASA Johnson Space Center, Houston, TX, pp. 62–63.

Metzger, P.T., B.T. Vu, D.E. Taylor, M.J. Kromann, M. Fuchs, B. Yurko, A. Dokos, C.D. Immer, J.E. Lane, M.B. Dunkel, C.M. Donahue, and R.C. Latta, III, [2007], "Cratering of Soil by Impinging Jets of Gas, with Application to Landing Rockets on Planetary Surfaces," Proceedings of the 18th Engineering Mechanics Division Conference, Blacksburg, VA.

Metzger, P.T., C.D. Immer, C.M. Donahue, B.T. Vu, R.C. Latta, III, M. Deyo-Svendsen, [2008a], "Jet-induced cratering of a granular surface with application to lunar spaceports," J. Aerospace Engineering (accepted for publication).

Metzger, P.T., J.E. Lane, and C.D. Immer [2008b], "Modification of Roberts' theory for rocket exhaust plumes eroding lunar soil," Earth and Space 2008, 11th Biennial ASCE Aerospace Division International Conference on Engineering, Construction and Operations in Challenging Environments, Long Beach, CA.

Mitchell, J.K., L.G. Bromwell, W.D. Carrier, III, N.C. Costes, W.N. Houston, R. F. Scott, H.J. Hovland [1972a], "Soil mechanics experiment", in Apollo 15 Preliminary Science Report, NASA Johnson Space Center, Houston, TX.

Mitchell, J.K., W.D. Carrier, III, W.N. Houston, R. F. Scott, L.G. Bromwell, H.T. Durgunoglu, H.J. Hovland, D.D. Treadwell, and N.C. Costes [1972b], "Soil mechanics experiment", in Apollo 16 Preliminary Science Report, NASA Johnson Space Center, Houston, TX.

Mitchell, J.K., W.D. Carrier, III, N.C. Costes, W.N. Houston, R. F. Scott, H.J. Hovland [1973], "Soil mechanics; characteristics of lunar soil from Apollo 17 flight lunar landing site", in Apollo 17 Preliminary Science Report, NASA Johnson Space Center, Houston, TX.

Mitchell, J.K., W.N. Houston, W.D. Carrier, III, N.C. Costes [1974], "Apollo Soil Mechanics Experiment S-200," NASA contract NAS9-11266, Space Sciences Laboratory Series 15, No. 7, University of California, Berkeley, CA.

Phillips, P.G., et al [1988], "Lunar Base Launch and Landing Facility Conceptual Design," NASA contract NAS9-17878, EEI Report 88-178, Eagle Engineering, Webster, TX.

Phillips, P.G., Charles H. Simonds, and William R. Stump [1992], "Lunar Base Launch and Landing Facilities Conceptual Design," in The Second Conference on Lunar Bases and Space Activities of the 21st Century (W.W. Mendell, ed.), NASA Johnson Space Center, Houston, TX, Vol. 1, pp. 139–151.

Roberts, L. [1963a], "The Action of a Hypersonic Jet on a Dust Layer," IAS Paper No. 63-50, Institute of Aerospace Sciences 31st Annual Meeting, New York, NY.

Roberts, L. [1963b], "Visibility and Dust Erosion During the Lunar Landing," in A Compilation of Recent Research Related to the Apollo Mission, NASA Langley Research Center, Hampton, VA, pp. 155–170.

Roberts, L. [1964], "Exhaust Jet—Dust Layer Interaction During a Lunar Landing," XIIIth International Astronautical Congress Varna 1962 Proceedings, Springer-Verlag, New York, NY, pp. 21–37.

Roberts, L. [1966], "The Interaction of a Rocket Exhaust with the Lunar Surface," in The Fluid Dynamic Aspects of Space Flight, Gordon and Breach Science Publishers, New York, NY, Vol. 2, pp. 269–290.

Romine, G.L., T.D. Reisert, and J. Gliozzi [1973]. "Site Alteration Effects from Rocket Exhaust Impingement During a Simulated Viking Mars Landing. Part I–Nozzle Development and Physical Site Alteration," NASA CR-2252, Martin Marietta Corporation, Denver CO.

Schmalzer, P.A., S.R. Boyle, P. Hall, D.M. Oddy, M.A. Hensley, E.D. Stolen, and B.W. Duncan [1998], "Monitoring Direct Effects of Delta, Atlas, and Titan Launches from Cape Canaveral Air Station," NASA TM-1998-207912.

Scott, R.F. [1975], "Apollo Program Soil Mechanics Experiment: Final Report", NASA CR-1444335, California Institute of Technology, Pasadena, CA.

Scott, R.F., and H.-Y. Ko [1968], "Transient Rocket-Engine Gas Flow in Soil," AIAA Journal, Vol. 6, No. 2, pp. 258–264.

Sullivan, S. P. [2004]. Virtual LM: a pictorial essay of the engineering and construction of the Apollo lunar module, the historic spacecraft that landed man on the moon. Apogee Books, Burlington, Ontario.

39

Lunar Habitats Protection Against Meteoroid Impact Damage

William P. Schonberg
Civil, Architectural, and Environmental Engineering Department
Missouri University of Science and Technology
Rolla, Missouri

R. Putzar
Fraunhofer Institute for High-Speed Dynamics
Ernt Mach Institute
Freiburg, Germany

F. Schäfer
Fraunhofer Institute for High-Speed Dynamics
Ernt Mach Institute
Freiburg, Germany

William P. Schonberg, PE, is professor and chair of the Civil, Architectural, and Environmental Engineering Department at the Missouri University of Science and Technology (formerly known as the University of Missouri-Rolla). Dr. Schonberg has over 20 years teaching and research experience in the areas of shock physics, spacecraft protection, hypervelocity impact, and penetration mechanics. He received his BSCE from Princeton University in 1981, and his MS and PhD degrees from Northwestern University in 1982 and 1986, respectively. The results of his research have been applied to a wide variety of engineering problems, including the development of orbital debris protection systems for spacecraft in low Earth orbit, kinetic energy weapons, the collapse of buildings under explosive loads, insensitive munitions, and aging aircraft. Since 1986, Dr. Schonberg has published over 60 papers in refereed journals on these topics, and has presented nearly 60 papers at a broad spectrum of international scientific and professional meetings, including several invited papers. In 1995 Dr. Schonberg received the AIAA's Lawrence Sperry Award for his work on the design of spacecraft protection systems. In 1998, Dr. Schonberg was promoted to the membership rank of Associate Fellow in the AIAA and in 2000 was selected to receive the Charles Beecher Prize for one of his recent papers on orbital debris protection systems from the Aerospace Sciences Division of the Institute of Mechanical Engineers in England. In 2004 and 2005 he was promoted to the member rank of Fellow of the American Society of Civil Engineers and the American Society of Mechanical Engineers, respectively. In 2009 he was asked to join another NASA Independent V&V Committee to review the MMOD risk assessment process for NASA's new Constellation program. In 2007 Dr. Schonberg received a Friedrich Wilhelm Bessel Research Award from the Humboldt Foundation in Germany. This award enabled him to spend 7 months at the Fraunhofer Ernst Mach Institute in Freiburg, Germany working on advanced MMOD protection systems for satellites and developing preliminary designs for safe lunar habitats using in-situ materials for protection against meteoroid impacts.

ABSTRACT The establishment of human habitats on the Moon and on Mars will require protecting them from the hazards of near-Earth and interplanetary space. In addition to solar radiation, another hazard to be faced by these habitats is the damage that can result from the high speed impact of a meteoroid on a critical structural component. Therefore, lunar habitats and their accompanying support facilities need to be designed with adequate levels of protection that will allow them to also withstand the damage that can result from a meteoroid impact. In this paper we discuss some approaches to shielding for lunar habitats, focusing on shielding that is intended primarily to provide protection against meteoroid impacts and on shielding approaches that use resources mined or extracted from the Moon. The Moon's mineralogy is discussed, and suggestions are presented for materials and material combinations that can be used to develop shielding for lunar habitats and which are comprised primarily or entirely of lunar materials. Several shielding mechanisms are also presented that could be effective against impacts by meteoroid particles having diameters on the order of that which are likely to strike a fairly large lunar habitat at least one or two times per year. The paper concludes with recommendations for continuing work in optimizing the design of meteoroid shielding for lunar habitats.

Introduction

The space-faring nations of the Earth are embarking on programs of space travel that we believe will ultimately lead to colonization of the Moon, Mars, and places beyond. The establishment of human habitats on such locations will require protecting them from the hazards of near-Earth and interplanetary space. In addition to the significant hazard of solar radiation, another hazard to be faced by these habitats is the damage that can be caused by the impact of a meteoroid on a critical habitat structure or structural component. Meteoroids can travel at speeds upwards of 20 km/s in the near-Earth region of space (see, e.g., [1, 2]), and at such high velocities can penetrate crew quarters, life support system facilities, for example. This can result in the shutdown of habitat operations, even loss of life. As such, lunar habitats and their accompanying support facilities need to be designed with adequate levels of protection that will allow them to also withstand the damage that can result from a meteoroid impact.

A common approach that solves both the radiation and the meteoroid impact problem is to bury the proposed lunar habitat beneath a substantial thickness of the lunar regolith (see, e.g., [3]). The use of the lunar regolith has the advantage of cost-effectiveness since this is a material that does not need to be transported to the Moon. In addition, because of its numerous layers, lunar regolith possesses a multi-shock capability. This property has been shown to enhance the destruction of impacting particles at very high impact velocities, thereby lowering their potential to inflict damage on the protected sub-structure.

However, initial estimates have shown that a regolith layer thickness of 1 m or more would be required to adequately protect inhabitants against the radiation emanating from a solar flare [4]. This is a very significant amount of mass, and brings to bear some significant quality of life issues. While initial occupants of such habitats may not object to living in such cave-like conditions, follow-on occupants, who might intend to be long-term residents or who would be part of a colonization effort, are likely to prefer less austere living conditions. As such, novel shielding concepts are required that will address radiation and meteoroid protection issues, and which will also provide a living environment that can be enjoyed by most people.

A recent effort aimed at assessing the meteoroid risk of lunar habitats demonstrated "the importance of a meteoroid threat analysis for any fixed surface base on the Moon" and concluded that there exists "the need to design the structure with this threat in mind." [5] In this paper we discuss some approaches to shielding for lunar habitats, focusing on shielding that is intended primarily to provide protection against meteoroid impacts and on shielding approaches that use resources mined or extracted from the Moon. The Moon's mineralogy is discussed, and suggestions are presented for materials and material combinations that can be used to develop shielding for lunar habitats and which are comprised primarily or entirely of lunar materials. Several shielding mechanisms are also presented that could be effective against impacts by meteoroid particles having diameters on the order of that which are likely to strike a fairly large lunar habitat at least one or two times per year. The paper concludes with recommendations for continuing work in optimizing the design of meteoroid shielding for lunar habitats.

Initial Considerations

Several possibilities exist for where to get the materials required for the construction of protective shielding for lunar habitats. Of course, one can build shields on the Earth using Earth-based materials and then ship them to the lunar destination. The primary advantage of this approach is that proven technologies can be used in shielding development and construction. Perforation resistant shielding is currently used on the International Space Station (see e.g., [6]), and can be said to be the culmination of over forty years of testing, design, analysis, and development work. This vast storehouse of information can certainly be brought to bear in the development, design, and construction of a shield that would be appropriate for the Moon and the meteoroid environment it encounters. However, the primary disadvantage of such an approach is the huge cost associated with transporting ready-made shields to their destination. The cost of transporting a pound or a kilogram of mass into low Earth orbit is already very expensive (approximately $10,000 per pound on the shuttle [7]), and so shipping a ready-made shield is prohibitively expensive.

The other alternative is to build the shields at the destination site using naturally-occurring resources. The advantage, of course, is that transportation costs will be significantly reduced. However, in this case extraction, refinement, and manufacturing issues come into play (see, e.g., [8, 9]). Significant energy sources would also be required to power these operations, although perhaps solar energy can be employed for these purposes. In addition, it may not be possible to refine/treat/mix the materials needed for shield construction to the same level of purity or temper as done on Earth. Quite simply, it just may not be possible to have all the processing equipment available at the destination site. However, it may not be necessary to do this. Perhaps the materials in their unprocessed or "more raw" condition are just as or nearly just as effective in providing protection against meteoroid impacts as their refined and processed counterparts. Most likely, a possible lesser grade can be compensated by a design tailored towards the specific material properties. Since such shields are comparatively low-cost, the easiest way to compensate lesser capabilities would be to use additional material to achieve the required level of protection. This issue will need to be addressed if this approach is to be considered as a viable alternative. Finally, simulant materials for Earth-based test programs to validate shields made of destination materials will need to be developed and made available in large quantities for test article manufacturing and testing.

Another alternative is, of course, some combination of the two above possibilities. In such a case, the best and worst aspects of the two options would be combined. An example is glass/epoxy, where the glass would come from heated lunar regolith while the epoxy would be brought to the destination from Earth (see, e.g., [10]). Final mixing, curing, and preparations would take place on the Moon in facilities that would need to be constructed for that

FIGURE 39.1a, b
Possible hybrid meteoroid shielding mechanisms.

purpose. One can also conceivably build an exo-skeleton for habitat using primarily Moon or Mars-based materials, then add an internal (perhaps inflatable) bladder brought from Earth (see Figure 39.1a). Or one can erect a two-layer structure made of Earth-based materials, and then insert regolith materials in between the two layers for shielding purposes (see Figure 39.1b).

The use of "sand bags," with the sand placed inside some flexible "cloth" container, is also a viable alternative. These sand bags can be either placed one on top of another, beginning on the lunar surface and continuing up to the top of a habitat or module, or draped down to the lunar surface from the top of a module or habitat in some inter-connected fashion (see Figures 39.2a,b).

Regardless of how and where the shield is built, it must have certain properties and characteristics to be useful against the meteoroid threat. These properties include impact resistance, durability, air tightness (a bladder of some sort will likely be needed), sustainability (shield construction and repair materials should be chosen on the basis of local lunar availability), reparability, commonality (it would be helpful if similar designs and materials could be used on the Moon as well as on Mars), and, if shipped from Earth, lightweight.

Sustainability is a key consideration in the design and establishment of human habitats on the Moon (and on Mars as well). The continued existence and operation of the habitat is dependent on the sustainability of its shielding. That is, the shielding must be easily maintained, repaired, and, if necessary, replaced whether it is using Earth-made materials and devices, or materials naturally occurring on the Moon. It can be significantly less expensive to repair, upgrade, and maintain existing facilities and shielding than to build or install new ones if existing structures or components are no longer usable. This benefit can be readily attained if the shielding is manufactured using materials found in-situ.

One disadvantage of using in-situ lunar resources for shield construction, however, is that construction and/or mining facilities need to be built for extraction and processing of in-situ resources, and these initial facilities will initially

FIGURE 39.2a, b
Possible use of sandbags for meteoroid protection.

not be protected. However, assuming that such mining, manufacturing, and construction facilities can be developed, the next question is what materials on the Moon or Mars can be used for meteoroid shielding purposes?

Use of Lunar Materials for Meteoroid Shielding

In order to determine what materials that are present on the Moon can be used to develop meteoroid shielding, it is first necessary to know what materials are in fact present on the Moon, and in what relative quantities. While lunar soil composition varies from location to location, composition studies performed using soil samples returned by various Apollo missions (see [11,12]) have revealed that the most common compound is SiO_2 (typically approximately 45% by weight). The compounds Al_2O_3, FeO, CaO, MgO, and TiO_2 are also fairly abundant: Al_2O_3, FeO, and CaO approx 15% by weight; MgO and TiO_2 approx. 8% by weight. To determine which of these compounds can be used for developing meteoroid shielding materials, either in existing form, or after being broken down and key elements have been extracted, we next look at the chemical composition of materials commonly used in meteoroid impact protection for spacecraft.

Although a wide array of materials choices exist for spacecraft shielding, the most common ones used are some sort of aluminum alloy or a composite material like Kevlar, Nextel, Glass, or Spectra (as woven fibers with or without epoxy). Tables 39.1 through 39.4 show the chemical composition of these various materials.

Comparing the information in Table 39.1 through 39.4 with the availability of compounds found in the lunar regolith, the following points become evident. First, regarding composite materials composition with respect to lunar mineralogy:

- C, H, and N are generally not found in lunar mineralogy.
- Nextel 550, 610, and 720 fibers are made of the two most common minerals found on the Moon.
- Fused silica and soda lime glasses are made of minerals commonly found on the Moon. However, soda lime glass requires a fair amount of Na, which is not that abundant on the Moon.

TABLE 39.1

Chemical Composition of Kevlar, Spectra, and Epoxy

Element	Kevlar	Spectra	Epoxy
C	✓	✓	✓
H	✓	✓	✓
N	✓		
O	✓		✓

TABLE 39.2

Chemical Composition of Various Nextel Fibers (% by wt) [13]

Compound	Nextel 312	Nextel 440	Nextel 550	Nextel 610	Nextel 650	Nextel 720
Al_2O_3	62	70	73	> 99	89	85
SiO_2	22	28	27	< 1	0	15
B_2O_3	14	2	0	0	0	0
ZrO_2	0	0	0	0	10	0
Y_2O_3	0	0	0	0	1	0

From these observations, it becomes clear that it may not be possible to "easily" manufacture Kevlar, Spectra, and Epoxy on the Moon. However, it should be noted that some of these elements do exist in the methane ice caps at the South Pole. It may also be possible to extract C, H from the refuse of preceding lunar missions (e.g., cellophane wrappers) [16]. It is also evident that Nextel fibers can be made entirely of materials already present in the lunar regolith. These fibers can be woven together to form blankets, which have already been shown to be highly effective orbital debris shields [17]. We can also conclude that fused silica will most likely be the glass manufactured from lunar regolith materials. Fibers made of fused silica, much like Nextel fibers, can also be woven together to form blankets.

However, the impact and damage resistant properties of glass fiber blankets have yet to be determined, and should be evaluated before this approach is taken. Of course, if a stiffer composite material is required, then epoxy will need to be imported from the Earth and mixed with the glass or Nextel fibers made from lunar material. However, it may be possible to manufacture an epoxy-like substance using materials found on the Moon so as to avoid importing Earth-made epoxy. On Earth we use the resin systems we have available and with which we are familiar. But on the Moon, perhaps a different solution might be possible and feasible.

TABLE 39.3

Chemical Composition of Various Glasses (% by wt) [14]

Compound	Fused Silica	Soda Lime	Borosilicate	Alumo-Silicate	Lead Borate
SiO_2	100	73	81	62	56
Al_2O_3	0	1	2	17	2
CaO	0	5	0	8	0
MgO	0	4	0	7	0
Na_2O	0	17	4	1	4
K_2O	0	0	0	0	9
B_2O_3	0	0	13	5	0
PbO	0	0	0	0	29

TABLE 39.4

Composition of Common Aluminum Alloys [15]. * Trace amount of these elements allowed.

Element	2017	2024	2219	3003	5056	6061	7075
Al	92–96%	91–95%	91–94%	97–99%	93–95%	96–99%	87–91%
Si	< 1%	*	*	*	*	< 1%	*
Fe	*	*	*	*	*	*	*
Mg	< 1%	1–2%	*		5–6%	1–2%	2–3%
Ti	*	*	*			*	*
Mn	< 1%	< 1%	< 0.5%	1–2%	< 0.2%	*	*
Cr	*	*			< 0.2%	< 0.4%	< 0.3%
Cu	3–5%	4–5%	6–7%	< 0.2%	*	< 0.4%	1–2%
Zn	*	*	*	*	*	*	5–6%
V			< 0.2%				
Zr			< 0.3%				

Second, regarding the composition of various aluminum alloys with respect to lunar mineralogy, the elements Cu, Zn, V and Zr are not naturally found on the Moon at all, while the elements Cr and Mn are found only in trace amounts. Because of these considerations, it appears that it will be very difficult to manufacture the aluminum alloys that are known to be effective as shield materials using just the materials found in the lunar regolith. However, of all the aluminum alloys considered, Al5056 appears to be the best choice because it requires the least number of additional materials that would need to be imported from Earth for its manufacture.

Using the Lunar Regolith as a Meteoroid Shield

An approach that would solve both the radiation and the meteoroid impact problem is to bury a proposed lunar habitat under a substantial thickness of the lunar regolith. Since this is a material that does not need to be transported to the Moon, the use of the lunar regolith in this fashion has the advantage of cost-effectiveness. In addition, because of its numerous layers, lunar regolith possesses a multi-shock capability. This property has been shown to enhance the destruction of impacting particles at very high impact velocities, thereby lowering their potential to inflict damage on the protected sub-structure [18].

However, initial estimates have shown that a regolith layer thickness of 1 m or more would be required to adequately protect human inhabitants against the radiation emanating from a solar flare [4]. This is a very significant amount of mass, and brings to bear some significant quality of life issues. While initial occupants of such habitats may not object to living in such cave-like conditions, follow-on occupants, who might intend to be long-term residents and are part of a colonization effort, are likely to prefer less austere living conditions. It would appear that some additional inquiry aimed at assessing the human-factors lessons learned from other long-duration habitations in remote and extreme environments would be appropriate to address this issue [19].

In addition to psychological reasons, windows are needed to provide direct sunlight to enhance photosynthesis [20]. Hence, a structure that provides viewing access for its inhabitants will also provide a means for sunlight to constructively affect one of the most basic processes that is required to sustain life on the Moon. Regolith also exhales Radon, which can be potentially life-threatening [21]. Perhaps the use of a simple barrier such as an air-tight bladder between whatever regolith is used and the livable space of the habitat where it is used could lower the Radon concentration to acceptable levels.

It is also important to note that not all facilities built on the Moon will be inhabited. As such, they need not necessarily have the same level of radiation protection as would habitable structures. However, if their contents were sufficiently valuable and needing protection against damage, such non-inhabitable facilities would likely require some sort of protection against meteoroid impacts. Hence, meteoroid protection could be a necessity in some cases where radiation protection through the use of lunar regolith may not be needed and not the appropriate solution.

Assessing the Effectiveness of Lunar Habitat Meteoroid Shielding

A multi-phase test program is required to assess the effectiveness of possible meteoroid shielding mechanisms that would involve the various design considerations mentioned in this paper. For example, an initial assessment of the effectiveness of the lunar regolith can be made using sand as a regolith stimulant target. Sand impact tests have been performed previously at NASA/Ames [22] and at the Ernst Mach Institute (EMI) in Freiburg, Germany [23]. The EMI tests have shown tunneling to occur at lower impact velocities. Hence, low velocity impacts might be just as much a concern as higher velocity impacts and should be investigated as well. After sand impact tests, target composition can be refined to be more representative of the lunar regolith.

Following some initial testing of sand to characterize its response against low and high velocity impacts, testing could begin on specimens representative of possible habitat wall configurations. Figure 39.3 below shows five possible configurations involving Nextel, aluminum, and sand. This would be an example of a hybrid protection system that involves materials brought from Earth (aluminum), materials made from in-situ lunar resources

(Nextel), and some regolith material as well (modeled by the sand). In all of the configurations, the "outside" is to the left, while "inside the habitat" is to the right.

Additional variations on these initial designs are varied and numerous. For example, the sand and/or sandstone can be placed loosely or can be compacted to study the effects of these variations in construction on impact response and protection. Low velocity tests should be performed to again assess how much channeling occurs for the various configurations, and whether or not such channeling could lead to a perforation of the rear or main wall of the target. High velocity tests can be used to assess whether there will be petalling or simply a through-hole if perforation occurs. Another issue to address is whether or not perforation occurs for mainly larger projectiles, or does the sand amplify some blast and shock loading effects that can also cause perforation to occur even for small projectiles.

In addition to perforation resistance, two other related aspects of the effectiveness of a candidate lunar habitat wall are (i) the ease of its reparability, especially through the use of in-situ resources, and (ii) its resiliency, that is, the level of protection it affords while in a damaged state. For example, considering the structural wall representations shown in Figure 39.3, for options (A), (B), and (E), if there is a hole "on the inside," it can be repaired and the wall resealed "on the outside" using techniques similar to those developed for the repair and resealing of a breached Space Station wall [24]. Resealing the "outside" wall in option (C) and (D) might be a little difficult since a patch will have to be sewn on somehow.

This, however, does lead to the question as to whether or not anything at all should be done about resealing a hole "on the outside," regardless of whether it is in an aluminum outer wall, or in a blanket made from Nextel. In some of the tests a projectile should be fired into the same location as a previous shot without resealing the outer wall or blanket or refilling the configuration

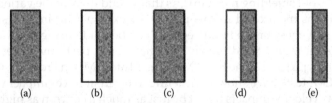

(a)   (b)   (c)   (d)   (e)

- Full (a), or half-full (b) "box" with all sides made from aluminum (black lines) and with sand or pieces of sandstone inside
- Full (c), or half-full (d) "box" with aluminum bottom and sides (black lines), and a Nextel blanket "top" with sand or sandstone inside
- A "box" with aluminum bottom and sides (black lines), and a combination of a Nextel blanket and aluminum "tops"

FIGURE 39.3
Possible test specimens simulating lunar habitat wall design.

with sand. In other tests, the sand and/or sandstone should be replaced, and again a projectile fired into the same location. The data obtained from such tests can be used to determine the extent to which external repairs must be performed on breached outer walls of lunar habitats, as well as the urgency with which such repairs must be made.

Another aspect of the test program that must be determined, in addition to impact velocities and target construction, is projectile dimensioning and material characterization. To determine an appropriate projectile size that should be used, it is first necessary to estimate how often a reasonably-sized particle can be expected to impact a reasonably-sized lunar habitat. Several previous studies that have proposed or considered some sort of lunar habitat have shown structures with exposed surface areas ranging from 500 m² to 750 m² (see [25-27]). Therefore, for the purposes of this study and to be somewhat conservative from an impact likelihood perspective, we assume an exposed surface area of 1000 m².

Using the current interplanetary meteoroid flux model [2], we calculate that a lunar habitat structure with an exposed surface area of 1000 m² can expect approximately 1.35 impacts per year by a 1 mm meteoroid particle, and approximately 0.0065 impacts per year (or 3.25 impacts every 500 years) by a 5 mm meteoroid particle. If we assume a 20 year life-span for the habitat, we can calculate the probability of an impact by at least one of each such particle size during those 20 years using the following equation:

$$Q = 1 - e^{-N}, \tag{39.1}$$

where

$$N = F \times A \times T, \tag{39.2}$$

with F being the number of impacts per unit area per year, A is the total exposed area, and T is the exposure time. For the 1 mm particle, we find that Q is very nearly 1, that is, an impact over 20 years by at least one 1 mm particle is a near certainty. However, for the 5 mm particle, $Q \sim 12.2\%$, that is, there is a greater than 10% probability that over 20 years there will be an impact by at least one 5 mm particle. Based on these considerations, projectiles not smaller than 1 mm and probably as large as 5 mm should be used in a test program to assess the protective capabilities of the proposed habitat wall configurations.

Regarding projectile shape and material, while meteoroids are far from round, it can be argued that they are "chunky." Hence, for the most part, to ensure consistency and repeatability, tests should be performed using spherical projectiles made from materials having a density approximately the same as the average meteoroid density in the size regime discussed previously. Although meteoroids are typically said to have an average mass density of

0.5 gm/cm³, the mass density of meteoroids is currently thought to be the following decreasing non-continuous function of mass [28]:

$$\rho = \rho(m) = \begin{cases} 2.0 \text{ gm/cm}^3 & for & m < 10^{-6} \text{ gms} \\ 1.0 \text{ gm/cm}^3 & for & 10^{-6} \text{ gms} < m < 10^{-2} \text{ gms} \\ 0.5 \text{ gm/cm}^3 & for & 10^{-2} \text{ gms} < m \end{cases} \quad (39.3)$$

Using Equation (39.3), we find that spherical particles having diameters between 1 and 3 mm would need to have a density of ~1.0 gm/cm³ (e.g., Lexan), while larger particles would require being made of a material having a density of ~0.5 gm/cm³ (e.g., aerated concrete [29]). Of course, current launcher technology limits the velocity at which a sizable projectile (on the order of 5 to 10 mm in diameter) can be fired at a target to approximately 7 to 8 km/s, while the average meteoroid velocity, as stated previously, is on the order of 20 km/s. To empirically assess the effects on target integrity of impacts at such high velocities, it is necessary to resort to scaling techniques, in particular, the scaling of impact energy.

One alternative is to keep the projectile diameter the same, and determine, based on an equal energy consideration, what material density would be needed to produce the same impact energy at a testable impact velocity (say 7 km/s) as that produced by a Lexan projectile at a "real" velocity of 20 km/s. Under these conditions we find that steel projected at 7 km/s could be used as a simulant material to model the impact of a Lexan projectile at 20 km/s.

Another alternative would be to assign a simulant material, and then calculate the diameter that would be required for a projectile of the specified simulant material to deliver an amount of energy at impact that would be the same as the amount of energy delivered by a Lexan projectile of a certain diameter at, say, 20 km/s. In this case, if we specify that the simulant material should be aluminum, we find that the assessment of the impact of a 2 mm Lexan projectile at 20 km/s can be performed using a 3 mm aluminum projectile fired at 7 km/s.

Final Thoughts

In addition to the more fundamental issues addressed in the preceding sections of this paper, the following considerations are offered for continuing work that would seek to refine the designs that may be developed after the fundamental issues have been addressed.

- Does the meteoroid flux have a local orientation dependence (i.e., near the Moon), similar perhaps to directionality of near-Earth orbital debris? If it does, perhaps the overall design of a lunar habitat can be tailored to take advantage of this orientation dependence.

- Does solar radiation also have orientation dependence as well? If so, it can be argued that radiation protection should be provided in the direction where it does the most good; the rest of the habitat might not need as much radiation protection, but should still have protection against meteoroid impacts.

- What role(s) can we expect to be played by next-generation materials, such as shape-memory alloys and self-healing materials? Although the self-healing property of cementitious materials has been known and studied for over twenty years [30], self-healing polymers are beginning to receive increasing attention in recent years (see [31, 32]).

- Is there a place for transparent shielding? How would a multi-layer glass window (e.g., three, four or five layers each ~1 cm thick and spaced over 10 to 20 cm) fare in the lunar environment? Additionally, transparent concrete is becoming a reality on Earth (see [33-35]), so it is certainly not too far a stretch of the imagination to envision transparent lunacrete, made with regolith materials, as an extension of the more 'traditional' formulation of lunar concrete (see [36-41]). This type of habitat wall or shielding would (1) protect against radiation, (2) protect against meteoroids, and (3) improve the quality of life since you can see through it.

Regardless of how they are built, habitats on the Moon will require protection from hazards of near-Earth and interplanetary space, including meteoroid impact on a critical habitat structure or structural component. As humans move to colonize and settle moons and planets, innovative techniques must be developed to take advantage of local resources to ensure that whatever shielding or wall construction is used is sustainable and repairable. This paper has presented some initial considerations in the design and development of such structures. Future test programs will provide data that can be used to assess the effectiveness of these designs, and to suggest design changes that will improve their effectiveness and even suggest new directions for exploration and discovery.

Acknowledgments

The authors are grateful for the support provided by the Humboldt Foundation through a Fraunhofer-Bessel Research Award that enabled this study, and for the helpful advice provided by their colleague, Dr. Eberhard Schneider of EMI.

This paper is an expanded version of a paper presented at the ASCE Earth and Space 2008 conference. A still more expanded version, with additional calculation details and other information, is being submitted for publication to the ASCE Journal of Aerospace Engineering.

References

1. Anderson, B., and Smith, R.E., *Natural Orbital Environment Guidelines for Use in Aerospace Vehicle Development*, NASA TM-4527, Marshall Space Flight Center, 1994.
2. Cour-Palais, B.G., Meteoroid environment model - 1969 [near Earth to Lunar surface], NASA-SP-8013, Johnson Space Center, 1969.
3. Bodiford, M.P., Fiske, M.R., McGregor, W., and Pope, R.D., "In-Situ Resource-Based Lunar and Martian Habitat Structures Development at NASA/MSFC", AIAA Paper No. 2005-2704, Proceedings of the 1st Space Exploration Conference, Vol. 2, Orlando, Florida, 2005.
4. Lindsey, N.J., "Lunar Station Protection: Lunar Regolith Shielding." Proceedings of the 2003 International Lunar Conference, S.M. Durst, et al, eds, American Astronautical Society, Science and Technology Series, Vol. 108, Springfield, VA, 2003, pp. 143–148.
5. Evans, S.W., Stallworth, R., Robinson, J., Stellingwerf, R., and Engler, E., "Meteoroid Risk Assessment of Lunar Habitat Concepts." Proceedings of the 2006 ASCE International Conference on Engineering, Construction, and Operations in Challenging Environments, Houston, Texas, 2006, pp. 543–550.
6. Destefanis, R., Schäfer, F., Lambert, M., and Faraud, M., "Selecting Enhanced Space Debris Shields for Manned Spacecraft." International Journal of Impact Engineering, Vol. 33, 2006, pp. 219–230
7. National Research Council, *Maintaining US Leadership in Aeronautics*, National Academy Press, Washington, DC, 1998.
8. Zing, X., Burnoski, L., Agui, J.H., and Wilkenson, A., "Calculation of Excavation Force for ISRU on Lunar Surface." AIAA Paper No. 2007-1474, *Proceedings of the 45th Aerospace Sciences Meeting*, Reno, Nevada, 2007.
9. Woodcock, G.R., "Economic Prospects for Lunar Industries." AIAA Paper No. 1995-4006, *Proceedings of the 1995 Space Programs and Technologies Conference*, Huntsville, Alabama, 1995.
10. Kaplicky, J., and Nixon, D., "A Surface Assembled Superstructure Envelope System to Support Regolith Mass Shielding for an Initial Operational Capability Lunar Base." in *Lunar Bases and Space Activities of the 21st Century*, W.W. Mendell, ed, Lunar and Planetary Institute, Houston, Texas, 1986.
11. Arnold, J.R., "Lunar Resource Surveys from Orbit, AIAA Paper No. 1977-0526, *Proceedings of the 3rd AIAA and Princeton University Conference on Space Manufacturing Facilities*, Princeton, New Jersey, 1977.
12. Taylor, S.R., *Lunar Science: A Post-Apollo View*, Pergamon Press, New York, 1975.
13. http://www.3m.com, accessed November 2007.
14. http://www.a-m.de/viskostaetglas-tafel1.htm, accessed November 2007.

15. http://www.matweb.com/, accessed November 2007.
16. Sanders, G., et al, "ISRU at a Lunar Outpost: Implementation and Opportunities for Partnerships and Commercial Development." *Proceedings of the 2007 International Lunar Conference*, to appear.
17. Cour-Palais, B.G., Theall, J.R., and Dahl, K.V., "Development of the Nextel Multi-Shock Shield, 1990-1993." AIAA Paper No. 1993-4083, *Proceedings of the 1993 Space Programs and Technologies Conference*, Huntsville, Alabama, 1993.
18. Cour-Palais, B.G., and Crews, J.L., "A Multi-Shock Concept for Spacecraft Shielding." International Journal of Impact Engineering, Vol. 10, 1990, pp. 135–146.
19. Sadler, P., et al, "The South Pole Greenhouse and Development / Construction of a Lunar Habitat Demonstrator." *Proceedings of the 2007 International Lunar Conference*, to appear.
20. de Vera, J.P.P., et al, "Photosynthesis of Eukaryotic Symbiotic Organisms in a Mars-Like Environment." *Proceedings of the 2007 International Lunar Conference*, to appear.
21. Levy, F., "Indoor Air Quality Implications of 22-Rn from Lunar Regolith." *Proceedings of the 2007 International Lunar Conference*, to appear.
22. Gault, D.E., Shoemaker, E.M., and Moore, H.J., Spray Ejected from the Lunar Surface by Meteoroid Impact, NASA TN D-1767, Ames Research Center, 1963.
23. Schneider, E., and Stilp, A., Verhalten von Sand als Auffangmaterial für Stahlsplitter, EMI Report No. E-13/77, July 1977.
24. Hall, S.B., "Marshall Researchers Developing Patch Kit to Mitigate ISS Impact Damage." Orbital Debris Quarterly, Vol. 4, No. 4, NASA Johnson Space Center, Houston, Texas, 1999.
25. Ring, C.B., Butterfield, A.J., Hynes, W.D., Nealy, J.B., and Simonsen, L.C., *Single Launch Lunar Habitat Derived from NSTS External Tank*, NASA TM-4212, Langley Research Center, 1990.
26. Connolly, J., "Overview of the Surface Architecture Elements Common to a Wide Range of Lunar and Mars Missions." AIAA Paper No. 1990-3847, *Proceedings of the 1990 Space Programs and Technologies Conference*, Huntsville, Alabama, 1990.
27. Rais-Rohani, M., *On Structural Design of Mobile Lunar Habitat with Multi-Layered Environmental Shielding*, NASA CR-2005-213845, Marshall Space Flight Center, 2005.
28. Drolshagen, G., and Borde, J., *ESABASE/DEBRIS: Meteoroid/Debris Impact Analysis – Technical Description*, Report No. ESABASE–GD–01/1, ESA/ESTEC, Noordwijk, The Netherlands, 1992.
29. http://www.matweb.com/search/PropertySearch.aspx, accessed November 2007.
30. http://www.rilem.net/tcDetails.php?tc=SHC, accessed November 2007.
31. Kessler, M.R., "Self-healing: A New Paradigm in Materials Design." Proceedings of the Institution of Mechanical Engineers, Part G: Journal of Aerospace Engineering, Vol. 221, No. 4, 2007, pp. 479–495.
32. Rule, J.D., Sottos, N.R., and White, S.R., "Effect of microcapsule size on the performance of self-healing polymers." Polymer, Vol. 48, No. 12, 4 June 2007, pp. 3520–3529.
33. Kroggel, O., and Grubl, P., "Ultrasonic Inspection of Concrete Structures - New Steps Towards 'Transparent' Concrete." *Proceedings of the International Conference on Repair and Renovation of Concrete Structures*, 2005, pp. 205–212.

34. Losonczi, A., and Bittis, A., "Transparent Concrete Causing a Sensation." Betonwerk und Fertigteil-Technik, Vol. 71, No. 3, 2005, pp. 66–69.
35. http://www.litracon.hu/
36. Kaden, R.A., "Methodology for Forming and Placing Lunar Concrete", in Lunar Concrete, R.A. Kaden, ed, SP-125, American Concrete Institute, Detroit, Michigan, 1991.
37. Toutanji, H., Fiske, M.R., and Bodiford, M.P., "Development and Application of Lunar 'Concrete' for Habitats." *Earth and Space 2006 - Proceedings of the 10th Biennial International Conference on Engineering, Construction, and Operations in Challenging Environments Earth and Space*, pp. 69ff.
38. Kanamori, H., "For the Realization of Lunar Concrete." *Proceedings of the 4th International Conference on Engineering, Construction and Operations in Space*, American Society of Civil Engineers, New York, New York, 1994, pp. 942–951.
39. Breyer, L.A., "Lunacrete: A Novel Approach to Extraterrestrial Construction." In *Space Manufacturing 5: Engineering with Lunar and Asteroidal Materials*, Proceedings of the 7th Princeton/AIAA/SSI Conference, Princeton, New Jersey, 1985, pp. 172–178.
40. Agosto, W.N. and Gadalla, A.M.M., "Lunar Cement Formulations for Space Systems Shielding and Construction." In *Space Manufacturing 5: Engineering with Lunar and Asteroidal Materials*, Proceedings of the 7th Princeton/AIAA/SSI Conference, Princeton, New Jersey, 1985, pp. 179–183.
41. Huston, H.L., Oishi, K., and Saito, T., "Radiation Shielding for Lunar Bases Using Lunar Concrete." Paper No. IAF-1992-0339, *Proceedings of the 43rd Congress of the International Astronautical Federation*, Washington, D.C., 1992.

40

Deployment of Greenhouse for Long-Term Lunar Base

Vadim Y. Rygalov
Human Factors & Environmental Design
The Department of Space Studies
John D. Odegard School of Aero-Space Sciences (JDOSAS)
University of North Dakota

Patrick Stoffel
Space Studies Program Distant Student
The Department of Space Studies
John D. Odegard School of Aero-Space Sciences (JDOSAS)
University of North Dakota

Vadim Y. Rygalov, PhD is a biophysicist and has worked in the area of Closed Ecological Systems (CES) for long-term Life Support (LS) since 1989. He received a PhD in multi-level systematic analysis of environment–sea organisms interactions from the Institute of Biophysics SB RAS (Krasnoyarsk-City, Siberia) in 1987. During his science-research career he was involved in a variety of projects related to multiple interactions between live organisms and their environments, including development of a Low Pressure Martian Greenhouse (MG) prototype at KSC NASA, 1999–2004. Currently he is an associate professor in the Department of Space Studies at University of North Dakota.

Patrick Stoffel is a resident of Madison, Wisconsin. He has a BS degree in biomedical science from St. Cloud State University, Minnesota, and an MS degree in space studies from the University of North Dakota, Grand Forks. This publication is resulting from his MS research project under advising of Dr. Vadim Y. Rygalov, his co-author. He got interested in space exploration at the age of 6 years, received a telescope for his 9th birthday, and after that spent nights on the family farm getting to know the stars. He enjoys working with animals and he is an animal welfare advocate. He spends time traveling, writing articles, and riding a motorcycle.

Introduction

A greenhouse, as an adjunct to the lunar base, will provide food, oxygen, water, and act as a sink for CO_2 and wastes (water and minerals). Having green edible plants at the lunar base will also have a huge positive impact on the psychology of the crews. (To see this, watch videos of the final crew of the Mir station consuming plants grown and cultivated on the station.) The long term goal for the greenhouse project is to provide the highest possible closure in the life support system loop (the goal is 100%) and to supply consumable resources for the lunar base. A greenhouse deployment timeline would have this goal at its endpoint. The start point has already begun–there are plants on board the International Space Station (ISS) and there were plants on the Mir station. These are mainly for research and psychological benefits; they are not part of a bioregenerative life support system. Growing plants on the Moon will probably begin in the same manner. Most plans and research into long distance space exploration (a lunar base, Mars exploration, etc.) conclude that in situ resource utilization (ISRU) and Closed Loop Life Support Systems (CELSS) are essential for success. Initial testing with CELSS based on plant-human interactions has already been done (see Appendix A). ISRU consumables and a high closure CELSS reduce the supply mass required for launch from Earth; the supply line is removed from Earth's gravity well. Producing food, oxygen, water, and other consumables "off the land" greatly lowers the cost of supporting long-distance space missions. The cost of getting through Earth's gravity well (in terms of energy, fuel, and funding) is the greatest impediment to the human exploration of space.

Scenarios of Greenhouse Deployment

Current NASA plans (Dale, 2006) call for lunar sorties to begin flights to the Moon by 2020 and the incremental establishment of a lunar outpost from 2020–2025. The outpost will serve as a test-bed for technologies to be used for missions to Mars (and beyond). It will most likely be positioned near the lunar south pole, on the rim of Shackleton crater; however that decision is not final. This area has the advantages of moderate temperature variation, access to solar energy (sunlight 70-80% of the time), and hopefully access to volatiles, such as water ice, in the crater regolith. Sometime between now and 2025 the decision will be made whether to eventually shut down the lunar outpost (between 2030 and 2035) and move the focus of the U.S. manned space program to the exploration of Mars, or to expand and make permanent the lunar "base" while simultaneously planning and initiating Martian sorties. Whether or not the Lunar Outpost will become permanent and the details and extent of Martian missions will depend on how successful ISRU technology becomes and the amount of supplies it can produce. A stable, working greenhouse can provide food; something that has yet to be achieved with physic-chemical (P/C) technology. The addition of a greenhouse to the lunar outpost can decrease the overall cost of resupply from Earth and will improve the possibility of keeping the lunar base operating after Martian sorties have begun.

There are three general visions for deploying a lunar greenhouse:

- Use a modular unit, similar in size and structure to a habitat module. Fabricated on Earth, it would require little or no effort to deploy on the lunar surface. No ISRU resources would be required for set up. Once landed, it could remain on the lander permanently, be moved closer to the habitat, or be connected directly to the other habitat modules. A Surface Mobility Carrier, looking like a spider with wheels, will be used to move modules from the landing pad to the habitat area. This greenhouse could grow in size by adding additional modules or hybriding it to an inflatable unit(s). These units may or may not use regolith for shielding. The resulting hybrid system would be a combination of P/C (from habitat) and Bioregenerative (from greenhouse) components.

- The use of inflatable units. Inflatable units will presumably be larger than modular habitat units and will be buried to utilize regolith for a more extensive radiation shield. These units will require work to deploy, inflate, and bury; internal construction may be required. This type of greenhouse may be moved to a pre-dug trench to allow easier covering of regolith. It may be directly connected to the habitat modules or remain a stand-alone unit, and it could easily be expanded by adding additional inflatable units. The combined habitat-greenhouse LSS would be hybrid but with expanded bioregenerative capabilities (either algae, higher plant life, or both). An Earth analog example would be something similar to BIOS-3.

- Building a large underground structure into an existing cavern— such as a lava tube or the bottom of a crater. This greenhouse would take advantage of the natural terrain to provide shelter from radiation and temperature variation. The unit would be built mostly with ISRU material such as regolith concrete (water based and waterless), glass and glass fibers, lunar "sandbags" (filled with regolith); melted and processed regolith (producing aluminum, titanium, magnesium, and iron) and whatever else can be created on site. The structure would be domed and pressure sealed. This structure would require the most work to construct and deploy but would provide the most volume, food production, buffer capacity, and stability. This structure would be largely bioregenerative with P/C for backup and to make corrections in the system. An Earth analog example would be something similar to Biosphere-2.

Life Support System Research

Some preliminary research has been done with maximizing closure levels in life support systems, using a combination of plants and human subjects. A complete listing and summarization of those accomplishments is provided for reference (Appendix A). The results of this comprehensive CES research, accomplished during the second half of twentieth century, confirmed the feasibility of life support systems based on the Closure Principle (CP):

- Humans/Photo-synthetic Organisms/Microbes are generally compatible in one combined ecological material turnover, including plant biomass production and human wastes treatment;
- It is quite enough to have ≥13–15 m^2 of planting area per person, in ~105 m^3 of closed environment, under ~150–180 W/m^2 of intensive illumination. This would provide a person with clean air (oxygen supply), water purification, and 30–40% of his food (vegetarian part of a diet) for long-term (~6 months) life support;
- Long-term existence (≥6 months) in an artificial CES generally does not have an adverse effect on human health, including "post-flight" adaptation to "normal" Earth ecology;
- Stability of this kind, LSS functioning, is completely determined by the level of active control by the crew;
- Active application of this LS approach for microgravity or planetary conditions requires development of technologies for self-sustainable autonomous maintenance.

On board the International Space Station (ISS), we have a small (~12 to 15 L volume) experimental greenhouse, *LASA*, built by collaboration between Utah State University and the State Research Center of the Russian Academy of Sciences' Institute of Bio-Medical Problems in Moscow, Russia. It is used for extensive research in higher plants' physiology, growth and development in microgravity, and the closed environment of spacecraft orbiting the Earth.

Deployment Rationale: Research, Logistics, and Economics

The NASA Exploration Systems Architecture Study (ESAS) final report was released in November 2005. This study reviewed all previous NASA plans for lunar and Mars missions going back to Apollo to formulate a workable architecture. This is the "blueprint" for the CEV, Ares I and V Launchers, Lunar Operations, and Mars Exploration plans (NASA, 2005). Lunar Missions Mode

Analysis and Lunar Surface Activities are described in detail (ch.4). This study helps provide an estimate for the best time to deploy a greenhouse.

The ESAS indicates a heavy emphasis on ISRU in both the initial lunar sorties and the lunar outpost to follow. In fact, the location and duration of the lunar outpost operations is dependent on early ISRU experiments deployed during the sorties. The location depends on ISRU experiments because it is hoped that significant quantities of water can be extracted from the dark crater floor of Shackleton (and other South Pole craters). The duration depends on ISRU success because the more supplies that can be produced on-site, the greater the probability that the lunar outpost will become permanent. These experiments will include O_2 production demos (using hydrogen reduction), H_2/H_2O extraction demos, and a demonstration of regolith excavation and transport (NASA, NASA's Exploration Systems Architecture Study, 2005). NASA will use the data accumulated in these demonstrations to plan and deploy lunar outpost ISRU pilot operations for regolith excavation and transport, O_2 production from regolith, and long term cryogenic storage and transfer of oxygen. Evolved capabilities from these operations will lead to larger-scale excavation and manipulation, propellant and consumable production, oxygen extraction using carbothermal ISRU, and surface construction (such as launch pads, roads, etc.). Physico-chemical systems have progressed and are reliable for water and atmospheric maintenance, but for a permanent lunar base and Martian missions, bioregeneration of consumables would provide a greater degree of stability in O_2 and H_2O supplies, and provide a dependable source of food. CES theory ensures that moving from a P/C system (Sabatiers, LiOH, scrubbers, electrolysis, etc.) to a bioregenerative system (algae, higher plants, greenhouse environments) would (Rygalov D. V., 2007):

- Increase the mass of the life support system,
- Increase the autonomy of the lunar base and its life support system,
- Increase the stability of the lunar life support system (by itself or as a backup to the physico-chemical system)—larger buffer, less accumulation of toxins and other unwanted materials,
- Increase reliability of the life support system employed–mechanical systems will age, breakdown, become degraded by lunar dust. Plants rejuvenate themselves and do not "breakdown or malfunction" easily, nor do they require spare parts,
- Have a beneficial psychological effect–humans love the green plants!

According to the Vision Architecture and its latest updates, NASA will build the transportation infrastructure to get to, land on, and return from the Moon. It will also provide the navigational and communication capabilities required for lunar transport and habitation. NASA envisions cooperation with other nations and corporations to build up the capabilities of the outpost and its operations (Coleman, Cooke, Yoder, & Hensley, 2008). Greenhouse deployment would fall into the cooperative category; facilitating long-term lunar occupation, and both utilizing and producing consumables via ISRU. Deploying a greenhouse on the Moon will provide time for research, construction, and operational experience to ensure successful implementation in missions to and bases on Mars.

Unilateral Capabilities of US/NASA	Cooperative Capabilities with US/NASA-ESA-RSA-JAXA-Corporations-Others
The Ares I and V Launch Systems	Equipment for Long-Term Outpost Operations
The Altair Lander and LOR Architecture	ISRU Production
Initial EVA and Surface Sortie Capabilities	Logistics and Resupply
Navigation and Communication Infrastructure	Pressurized and Unpressurized Rovers
Solar Power System	Augmented Power, Navigation, and Communication Systems

Financial and political constraints will make the ultimate determination whether the lunar outpost will be permanent. Shutting down the lunar base when it is producing a significant amount of consumables from lunar resources would be hard to justify. At $100,000/kg (based on shipping costs from Earth) the production of 10,000 kg of usable resources (life support needs and construction material–food, O_2, H_2O, shielding, glass, etc.) using ISRU would cost over $1 billion to ship from the Earth (Gaffey, 2007). Given that people consume ~0.62 kg food, ~28.75 kg of water (drinking, hygiene, cleaning, etc.), ~0.85 kg of oxygen per day, and assuming a permanent crew of four at the lunar outpost, the amount of consumables required is over 40,000 kg (40MT) per year. Based on this data (with 0% LSS closure and no ISRU), it would cost ~$4 billion (shipping costs) annually to supply consumables totally from Earth (Eckart, 1996).

Consumables Required (U.S. Space Station Standards Design Loads)

	Oxygen	Food	Water
1 person/1 day	0.85 kg	0.62 kg	28.75 kg
4 people/1 day	3.4 kg	2.48 kg	115 kg
4 people/1 year	1242 kg	906 kg	42,004 kg

Greenhouse Descriptions and Deployment

A comparison of the economics of the life support systems–the P/C units in the initial habitat module vs. adding a large bioregenerative component (greenhouse) will determine when and what type of greenhouse will be added (Rygalov D. V., 2007):

- Resource Costs: The cost of shipping a complete greenhouse module from Earth vs. building portions of it with ISRU; the cost of building such a unit (on Earth); the energy and manpower costs to set up and maintain vs. what is available; etc.

- System Autonomy: Spare parts, whether parts are made ISRU or shipped from Earth; the labor required to keep the system up and running; the amount of storage of O_2, H_2O, CO_2, and food; the crew/ staff ratios, etc.

- Reliability: Effect of aging on the components, the number of moving parts involved; the complexity of the system; the amount of exposure to the elements and its effect on the system; dust degradation comparisons; buffer capacity; etc.

- Intangibles: Astronaut preference; psychological benefits; contamination issues; politics; etc.

A smaller modular greenhouse, deployed on the surface, could test O_2, H_2O, CO_2, and food production in the lunar environment and would be a useful intermediate step before deploying a larger inflatable one. Similar in size and shielding to the outpost habitat module, it would allow plant growth and steady-state studies to commence. Steady-state studies will determine the length of time it will take for the module to achieve balance, determine how resources get distributed; determine toxin, metal, mineral, and gas build up; observe how much and where bacterial contamination occurs; and how to cope with the lunar dust. It will also provide ready, living plants that can be moved into the larger inflatable buried greenhouse once it comes online.

The size of the modular greenhouse will depend on the cargo capacity of the launcher and lunar lander. The U.S. launcher will presumably be the Ares V. The Ares V gross trans-lunar insertion (TLI) cargo capacity will be 63.5 MT (this compares with 48.6 MT for the Saturn V) (Sumrall, 2008). The lunar lander (Altair) design is not set in stone—a 2006 preliminary cost estimate was much greater than anticipated; alternate designs and approaches are being considered (Dorris, 2008). The three basic lander configurations will be: crew only, cargo only, and crew plus cargo; the configuration employed to land a greenhouse module would be the unmanned cargo version. The ESAS estimates a module of ~24 m³ pressurized volume, ~8 m³ of that for the airlock, could be mounted on the lander and brought to the lunar surface. By foregoing the airlock (the greenhouse would be directly connected to the habitat module or connected via causeway) and the ascent module (the initial deployment will be performed with an unmanned version of the lander) the entire ~24 m³ pressurized volume of the module could be utilized for plant growth. The amount of mass the LSAM (Lunar Surface Access Module) will deliver to the surface has been estimated between 1,800 kg to 4,300 kg for a crewed lander and over 17,000 kg for an unmanned version–based on a "minimum functional" design (Dorris, 2008). With modular dimensions of 6 m by 3 m, a cylindrical volume of 24m³, and subtracting space for humans to move around and for the plant life support systems, there is enough room to support food production to feed at least one and possibly two astronauts (NASA, NASA's Exploration Systems Architecture Study, 2005).

A greenhouse module could be set up and running autonomously before the first permanent crew arrives. Work is being done by NASA, the NSF, and the University of Arizona CEAC (Controlled Environment Agriculture Center) for just such a situation (NASA & UoA, 2007). Cable culture plant growth uses hydroponics and a NFT (Nutrient Filled Technique) system to automatically start and maintain plant growth. Pre-seeded nutrient membranes are Velcro sealed, but allow space at the top and sides for sprouts to emerge. A tube runs thru the membrane, bringing nutrients and water; the membranes are hung along a cable which stretches from one end of the greenhouse to the other. Many such cables would fill the module. This system is being tested at the South Pole Food Growth Chamber and has successfully grown strawberries, lettuce, tomatoes, potatoes, and other food. By deploying a greenhouse module filled with seeds embedded in this NFT system, and given the likely volume of such a module, food enough for at least one and maybe two astronauts (for eternity if maintained) could be waiting for them when the permanent occupation of the outpost begins. The food produced could reduce shipping costs from the Earth by at least $1 billion/yr (based on ~1000 kg of food, shipped at $100,000 per kg). In three to five years the greenhouse could pay for itself. The NFT system is being tested at the South Pole in inflatable habitats; NASA is seriously looking at inflatable units to be used for habitation and other space applications (such as greenhouses).

Much of the initial sortie demonstrations and pilot programs will be dedicated to excavating and transporting regolith. This includes building landing pads, blast protection berms, mining for O_2 production, mining regolith for H_2O production, electric cable trenches, and roads. These tasks were laid out at the SRR IX (Space Resources Roundtable) by the NASA Advanced Systems Division and the Colorado School of Mines in October 2007. Much of this work would be performed before the outpost is finished (2025). Dates given at the SRR IX indicate that these projects should be past the demo stages and into project phases by 2023. By 2023, "the Constellation Architecture should provide from lunar in-situ resources, no less than 1 MT (metric ton–1000 kg) of oxygen per year" (Mueller & King, 2007). By 2027 that increases to 10 MT annually. Once ISRU O_2 and H_2O production and the excavation and movement of regolith have been reliably demonstrated, the deployment of multi-unit and inflatable greenhouse modules, buried for radiation shielding, can commence. Given that the knowledge and experience of such deployment is important for Mars mission consideration, it should be implemented at the lunar base as soon as possible.

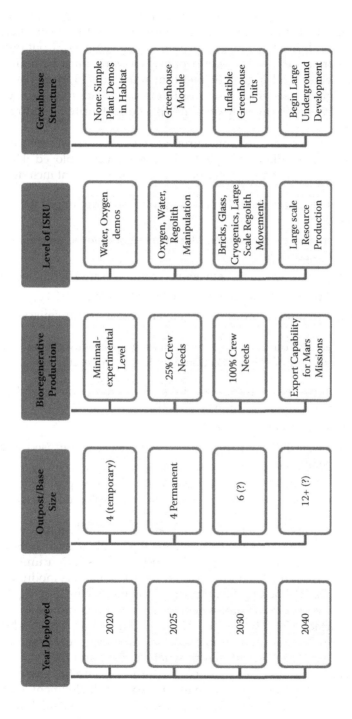

Once deployed, a greenhouse will require gas mixture controls and monitoring. The temperature must be maintained within a certain boundary, water and humidity must be provided and removed. Contamination safeguards must be in place. Atmospheric pressure must be maintained. Nutrients, water, and a growth medium must be supplied. Plants (and the human gardeners) must be protected from radiation (solar and GCR), large temperature changes, toxic buildup, and meteorite punctures (pressure loss).

The ESAS had established some ground rules for outpost construction in 2005–these rules would also apply to any greenhouse deployed. These considerations for outpost and greenhouse module deployment include:

- Landed elements should not be required to move unless absolutely necessary;
- Crew operations for outpost deployment should be as simple and limited as possible;
- Landed elements will be delivered on common cargo decent modules;
- The building of the lunar base is to be incremental;
- Using initial sortie missions to deliver cargo for future missions.

NASA has also laid out ISRU architectural assumptions to coincide with sortie missions and outpost construction (Sanders & Duke, 2005):

- Characterize lunar resources, the surface environment, and engineering unknowns as soon as possible;
- Develop one robust lunar outpost site;
- Demonstrate ISRU during Lunar Sortie phase to be used during Base operations;
- Develop lunar infrastructure and operations to support sustainable lunar operations and parallel Mars manned exploration missions.

Some of the equipment to be deployed on the sortie missions and outpost include: an ISRU Lunar Miner/Hauler, an O_2 pilot plant, and an ISRU Polar Resource Extractor (NASA, 2005). The ISRU demonstration units will be tele-operated from the Earth in-between sorties. A mobile, modular greenhouse could be attached to the outpost habitat module by moving it off the cargo lander with a crane and wheeling it over–a non-mobile unit would need the Surface Mobility Carrier. The greenhouse may be wheeled close enough to the habitat module to allow an inflatable cause-way structure to connect them together–similar to what subways and trains use to connect cars, and airports use to connect the terminal with the passenger jets. Benefits of this type of greenhouse include ease of deployment (compared to burying a large inflatable structure), similar radiation shielding (same as habitat module), and the option of directly integrating the greenhouse into the habitat life support system. This type of greenhouse may be easier to design and launch if it is essentially a hollowed-out copy of the habitat module structure.

An inflatable unit could be deployed in a similar manner or deployed as a separate stand-alone habitat. An example of an inflatable unit is the TransHab. The TransHab was developed as a habitation module for the ISS but was never deployed. The module consists of a central core surrounded by an inflatable shell. It is a huge unit, providing $342m^3$ of pressurized volume–this compares with $315m^3$ for BIOS-3, $150m^3$ for Mir, $227m^3$ for BioPLEX, and $75m^3$ for the ESA Columbus module deployed at the ISS. It measures 12.2m in length by 7.3m in diameter and is the shape of a cylinder, but rounded at the ends. To bury a TransHab-sized unit in regolith for radiation shielding, a trench at least 3.7m deep, 7.3m wide, and 12.2m long is required. This is $253m^3$ of regolith; at $1.5g/cm^3$ this equals a mass of 380MT. At least three meters of regolith covering is required for radiation shielding; this would require another 1267 MT. At the bare minimum, 1648 MT of regolith would need to be displaced. This compares with the expected 589MT to be displaced building a landing pad, the 283MT needed to construct a berm to surround the landing pad, and 358MT estimated to be excavated when the main habitat is deployed (Mueller & King, 2007). Using a smaller diameter inflatable module (rather than TransHab) would require much less regolith excavation and make radiation protection a less time consuming and energy intensive mission.

Potential Greenhouse Module Dimensions

Module of Unit Name	Diameter (meters)	Radius (meters)	Length (meters)	Volume meters³	Pressurized Habitable Volume meters³	Percent of Volume Habitable
Lunar Habitation Module	3.0	1.5	6.0	42.4	24.0	56.6%
ISS TransHab (not deployed)	7.3	3.6	12.2	507.8	342.0	67.3%
4m by 6m Habitation Module	4	2.0	6	75.4	50.7	67.3%
Genesis I (Bigelow Module)	2.54	1.3	4.4	22.3	11.5	51.6%
Large Spherical Habitat	16	8.0	16	2144.7	1501.3	70.0%

The Lunar Habitation Module dimensions are taken from the Exploration Systems Architectural Study (NASA, 2005)

The information on TransHab was taken from "Inflatable Habitats Technology Development" (Kennedy, 2000).

The Large Spherical Habitat concept (Roberts, 1988) was taken from "Structural Design of a Lunar Habitat" (Ruess, 2006).

Genesis I information was obtained from the Bigelow Aerospace website: www.bigelowaerospace.com

The 4m by 6m Habitation Module is a somewhat larger version of the Exploration Systems Architectural Study model (NASA, 2005).

Information in red are best estimates.

Potential Greenhouse Module Requirements for Excavation

| Module of Unit Name | Trench Excavation | | Radiation Shielding | | |
	Volume of Trench meters³	Mass of Regolith Excavated MT	Volume of Shielding Required meters³	Mass of the Shielding Regolith MT	Total Mass of Regolith Displaced MT
Lunar Habitation Module	N/A	N/A	N/A	N/A	N/A
ISS TransHab (not deployed)	253.9	380.9	844.9	1267.4	1648.2
4m by 6m Habitation Module	37.7	56.5	235.6	353.4	410.0
Genesis I (Bigelow Module)	11.1	16.7	126.0	189.0	205.7
Large Spherical Habitat	1072.3	1608.5	2787.6	4181.5	5790.0

The Lunar Habitation Module dimensions are taken from the Exploration Systems Architectural Study (NASA, 2005)

The information on TransHab was taken from "Inflatable Habitats Technology Development" (Kennedy, 2000).

The Large Spherical Habitat concept (Roberts, 1988) was taken from "Structural Design of a Lunar Habitat" (Ruess, 2006).

Genesis I information was obtained from the Bigelow Aerospace website: www.bigelowaerospace.com

The 4m by 6m Habitation Module is a somewhat larger version of the Exploration Systems Architectural Study model (NASA, 2005).

Settling into a Steady State

A number of considerations must be taken into account when integrating the greenhouse into the outpost, time must be allowed for the plants to grow to a productive size. P/C processes and Earth supply must provide consumables until the greenhouse reaches a stable state. Enough reserve supplies must be available to allow for failure or for over-estimating the resources that will be produced by the greenhouse. Previous experiments in CELSS have provided lessons that will be applied to any greenhouse deployed on the Moon. The BioPLEX facility (LSSIF) operated from 1995–1997 by NASA and the Johnson Space Center. Biosphere 2 operated from 1991–1993 and is located just northeast of Tucson, Arizona. The BIOS series–BIOS 1, 2, and 3 operated from 1965 thru 1996 out of the former Soviet Union and Russia at the Institute of Biomedical Problems. The BIOS-3 program was the most successful and achieved closure of 93–95%. All of these systems used either P/C by itself or a hybrid of P/C and Bioregenerative systems. A great deal of research data is available; major findings relevant to greenhouse deployment include:

- Off-gassing of equipment and materials in a closed module environment–ex. formaldehyde in the case of BioPLEX.
- Numerous power outages and equipment failures can occur despite the best efforts of planning.

- Provide enough lighting backup bulbs to replace burned out units.
- Beware of unimagined chemical reactions both within the enclosure and between the enclosure and the surrounding environment–ex. Biosphere 2 had a totally unexpected CO_2 drain into the foundation which caused the plants to decrease production of oxygen, thus oxygen levels dropped over time.
- Trace contaminants will be released from equipment and will accumulate in the LSS as it cycles thru the food chain–ex. Examination of BIOS-3 participants revealed increased levels of chromium, iron, nickel, zinc, copper being stored in their bodies introduced via life support systems.

See Appendix A for more information on past and current P/C and Bioregeneration analogs and simulators being researched for application on the Moon and Mars.

Deployment of Inflatable Units

Deploying an inflatable greenhouse module has advantages. It would provide a much greater volume per mass than a hard shell module. The light weight inflatable module would presumably cost less to ship from Earth than a hard-shelled module (because of the reduced mass). The deflated greenhouse module will have to fit on the Altair lander (either unmanned or within cargo area of a manned lander). The atmospheric pressure in a deployed inflatable module must be capable of withstanding the mass of the regolith shielding, however the one-sixth-g lunar gravity will reduce the structural requirements that a comparable unit deployed in Earth's gravity would need given the same need for radiation shielding. The TransHab module, in tests, successfully withstood up to four times Earth's atmospheric pressure. Inflatable greenhouse modules may be used at Earth standard pressures (sea-level 101 kPa) or at lower pressures (50–25 kPa) to simulate possible Martian greenhouse atmospheres (Rygalov V. , 2007).

NASA has researched and tested inflatable modules since the 1960s. Had lunar operations continued after Apollo 17, an inflatable "Lunar Stay Time Extension Module" would have been deployed. NASA considers inflatable technology, robotic construction, self-deploying structures, and self-healing structures enabling technologies, to be used to increase our ability to explore the solar system. Inflatable units are considered to be the leading candidates for lunar and Martian habitats and greenhouses. NASA extensively tested TransHab for the ISS; the design and functionality of the test modules were so successful that Bigelow Industries has purchased the rights to produce these modules for use as commercial space hotels. The use of inflatable technology for greenhouses has also been extensively researched–the South Pole Food Growth Chamber Project being the latest example.

Issues with Contamination

COSPAR policy for the Moon says that outgoing cargo does not require sterilization; forward contamination is not considered an issue because of the extreme environment of the Moon. However, experiments with the LDEF (Long Duration Exposure Facility) proved that spore forming bacteria can survive (up to 6 years on the LDEF) in the harsh vacuum of space (Glavin, 2007). The hostile environment of the Moon is similar in many respects to conditions found in the vacuum of space. Greenhouse deployment will involve plant life and the chances of these plants harboring bacteria or other organisms cannot be ruled out. Some studies have shown that bacteria grown in microgravity can have their gene expression altered. *Salmonella typhimurium* bacteria grown onboard STS-115 were compared to ground control bacteria and were found to be more virulent (Wilson, 2007). Over time the possibility of increased virulent bacteria infecting greenhouse life, crewmembers, or even people on Earth (upon return) increases. Studies with plant life on the ISS may quantify the risks involved. A greenhouse will not introduce a new risk–bacterial contamination is always possible–but it will increase the number of bacteria and undetected organisms gaining access to a microgravity environment.

Lunar research on bacterial propagation rates, the number and extent of contamination occurrences, contamination protocols, can be observed and applied to Martian mission studies. The sterile vacuum, radiation filled surface of the Moon could provide data on the degrees of contamination when acquiring surface samples; since these samples would be initially lifeless, any organic material found in the sample would be attributed to the sampling process itself. Missions to Mars, presumably a much more "bacteria friendly" environment, would benefit from lunar contamination research.

Summary: Greenhouse Deployment for the Lunar Base

The deployment of a modular or inflatable greenhouse is possible with the Constellation architecture. Deployment of a greenhouse with the initial Lunar habitat in or near 2025 would allow time to prepare for including them in missions to Mars—time to "get the bugs out" of the system. Given the current plans, this is very achievable. While a large underground structure built within a lava tube or crater bottom provides the greatest return of consumables, its deployment is unlikely before Mars missions begin.

NASA ESAS plans include all the technology, excavation equipment, experience, and resources that either inflatable or modular greenhouse deployment would require. It would add excavation work (in the case of an inflatable unit in a pre-dug trench covered with regolith) equivalent to at least the work required deploying the habitat module; wear and tear on the excavation equipment, and an increase in the amount of consumables used during deployment. Depending on the size, a greenhouse would save at least $1B/yr in food shipping costs; there would be oxygen production and water purification benefits that would also add to its economic value. The cost of developing, building, deploying, and maintaining the unit(s) must be factored into the economics. The larger the initial greenhouse volume, the greater the consumables produced, but the greater the initial cost of deployment.

Bibliography

Bigelow, R. (2008). *Bigelow Aerospace-Genesis I-General Specs*. Retrieved 2008, from Bigelow Aerospace: http://www.bigelowaerospace.com/genesis_I/?Genesis_I_General_Specs

Bodiford, M., Burks, K. H., Perry, M. R., Cooper, R. W., & Fiske, M. R. (2007). *Lunar In Situ Materials-Based Habitat Technology Development Efforts at NASA/MSFC*. Huntsville, AL: NASA/Marshall Space Flight Center.

Churchill, S. E. (1997). *Fundamentals of Space Life Science* (Vol. 2). Malabar, FL: Krieger Publishing Company.

Coleman, S., Cooke, D., Yoder, G., & Hensley, S. (2008). Lunar Architecture Update. *3rd Space Exploration Conference and Exhibit*. Denver, CO: NASA, JPL, 3rd Space Exploration Conference.

Cooke, D. (2007). Exploration Lunar Architecture. *Presentation to the Lunar Science Workshop*. Tempe, AZ: NASA, Exploration Systems Mission Directorate.

Dale, S. (2006). Exploration Strategy and Architecture. Implementing the Vision: 2nd Space Exploration Implementing Conference. NASA, Deputy Administrator.

Dorris, C. (2008). Altair Project. 3rd Space Exploration Conference. Denver, CO: NASA, Deputy Project Manager.

Eckart, P. (1996). Spaceflight Life Support and Biospherics. Torrance, CA, The Netherlands: Microcosm Press and Dordrecht, Kluwer Academic Publishers.

Gaffey, D. M. (2007, Fall). In Situ Resource Utilization for Life Support at Lunar and Martian Bases. Retrieved 2008, from Lecture/Seminar 45, Space Studies 410, UND.

Gitelson, I., Lisovsky, G. M., & MacElroy, R. D. (2003). Man-Made Closed Ecological Systems. London, New York: Taylor & Francis.

Glavin, D. E. (2007). In Situ Biological Contamination Studies on the Moon: Implications for Future Planetary Protection and Life Detection on Mars. Greenbelt, MD: NASA, Goddard Space Flight Center.

ISU. (2006). Luna Gaia: A Closed-Loop Habitat for the Moon-Final Report. SSP 2006 Team Project. International Space University.

Kennedy, K. J. (2000). Inflatable Habitats Technology Development. NASA, TransHab Project Office. Houston, TX: Johnson Space Center.

Larson, B., & Sanders, J. (2007). NASA In-Situ Resource Utilization (ISRU) Project And Its Linkage to Lunar Science. *Presentation to the Lunar Science Workshop.* Tempe, AZ: NASA, Lunar Science Workshop.

Mueller, R. P., & King, R. H. (2007). Criteria for Lunar Outpost Excavation. *Space Resources Roundtable-SRR IX* (pp. 1-39). Golden, CO: NASA/Colorado School of Mines.

NASA. (2005). *NASA's Exploration Systems Architecture Study.* National Aeronautics and Space Administration.

NASA, & Arizona, U. o. (2007). *Lunar Greenhouse (Video 2).* Retrieved 2008, from Brightcove: http://video.aol.com/video-detail/lunar-greenhouse-part-2/2348016504?

Roberts, M. (1988). Inflatable Habitation for the Lunar Base. *NASA Conference Publication No. 3166.* Washington, D.C.: NASA.

Ruess, F., Schaenzlin, J., & Benaroya, H. (2006). Structural Design of a Lunar Habitat. *Journal of Aerospace Engineering.*

Rygalov, D. V. (2007, Fall). *Hybrid LSS Approaches: System Integration.* Retrieved 2008, from Lecture/Seminar 38_1 and 38_2, Space Studies 410, UND.

Rygalov, V. (2007). *Low Pressure Plant Physiology: Martian Greenhouses (Low Pressure Greenhouses).* Grand Forks, ND: UND, Space Studies Dept. Vadim Rygalov.

Salisbury, F. B., Gitelson, J. I., & Lisovsky, G. M. (1997, October). *BIOS-3: Siberian Experiments in Bioregenerative Life Support.* Retrieved September 10, 2004, from BioScience: http://www.aibs.org/bioscience/bioscience-archive/vol47/oct97.salisbury.text.html

Sanders, G. B., & Duke, M. (2005). In-Situ Resource Utilization (ISRU) Capability Roadmap: Final Report (NASA). NASA, Johnson Space Center/Colorado School of Mines. Houston: NASA.

Schwartzkopf, S. H. (1992). Design of a Controlled Ecological Life Support System. BioScience , 42 (7).

Sumrall, P. (2008). Ares V Overview. 3rd Space Exploration Conference. Denver, CO: NASA, Ares Projects Advanced Planning Manager.

Wilson, J. E. (2007). Space flight alters bacterial gene expression and virulence and reveals a role for global regulator Hfq. Tempe, AZ: Center for Infectious Diseases and Vaccinology, The Biodesign Institute, ASU.

Terms

LSS: Life Support Systems
ISRU: In Situ Resource Utilization
CELSS: Controlled Ecological Life Support System
ISS: International Space Station
COSPAR: Committee on Space Research

LDEF: Long Duration Exposure Facility
ESAS: Exploration Systems Architecture Study
CEV: Crew Exploration Vehicle: NASA spacecraft in development
NASA: National Aeronautics and Space Administration
P/C: Physico-chemical (Life Support System)
IBMP: Institute for BioMedical Problems (Moscow, Russia)
TLI: Trans Lunar Insertion
MT: Metric ton = 1000 kg
NSF: National Science Foundation
NFT: Nutrient Filled Technique
CEAC: Controlled Environment Agriculture Center

Appendix A

TABLE 40.1

CES Research Historical Overview

Project Title, Country of Origin for Funding and Support	Years of Operation, Function	System Purpose; Bio-Engineering Characteristics and Outcomes	Key People Playing Integral Roles in the Project
Earth's Biosphere (Biosphere-1)	≥5.7 * 10^9 BC to present	Purpose unknown	Human history, leaders
		Volume ~5.1*10^{18} m^3; Planting area ~0.49 * 10 m^2	No human control on Biosphere-1 (Earth)
International, Global Life Support System		Complex life support of multiple affiliation	
Unique Natural Prototype for Artificial CES		Principle of Biosphere control: statistical regulation + evolution	
		Principle of control among human population: technological development	
Ground Based Experimental Complex (GBEC-1) (USSR)	1960–1961	Experiments linking micro-algae with rats and dogs for periods up to 7 days	Academician S. P. Korolev; Dr. Y. Shepelev et al.; of the Institute of Bio-Medical Problems (IBMP)
		It was shown that photosynthetic organisms can provide closed air loop together with animals	Institute of Plant Physiology (IPP), Moscow

Continued on next page

TABLE 40.1 (*Continued*)

Project Title, Country of Origin for Funding and Support	Years of Operation, Function	System Purpose; Bio-Engineering Characteristics and Outcomes	Key People Playing Integral Roles in the Project
Experiments at the United States Air Force School of Aviation Medicine USA)	1961	Gas exchange between monkeys and an algae tank occurred for up to 50 hours without any adverse effects	US Air Force
Micro-CES (m-CES) (USA, Russia, Europe)	~1965–Present	Micro-CES were created for fundamental research of closed environment functioning principles 100 mL – 5 L of micro-algae/bacteria/protozoa; Determined that materially closed m-CES can stay self-sustainable & functional for decades based on the simplest ecological structures	C. Folsome (University of Hawaii, Manoa) B. Kovrov et al. (Institute of Biophysics (IBP) RAS, Krasnoyarsk); Others
BIOS-1 (USSR)	1965–1968	First step to regenerate atmosphere for one human; Volume ~12 m^3; ~8m^2 of illuminated area for micro-algae *Chlorella* cultivation; Complete atmospheric closure; water and nutrition stored in advance; provided ~20% of material closure for essential substances; buildup of trace contaminants was noticed	Academician S. P. Korolev; Acting member of the Russian Academy of Sciences (RAS) J. I. Gitelson et al., IBP, Krasnoyarsk, Siberia
BIOS-2 (USSR)	1968–1972	Atmosphere & water regeneration Volume ~20.5 m^3 *Chlorella* + higher plants ~80 to 85% (Air + Water) of total material closure	Academician J. I. Gitelson, Prof. G. Lisovsky, Prof. B. Kovrov et al., IBP RAS, Krasnoyarsk, Siberia
GBEC-2 Ground Based Experimental Complex (USSR)	1969	Three people lived in small enclosed systems with physical-chemical + biological LS for one year Volume ~115 m^3 Food and water were stored ahead in time; several types of green plants were used to provide vitamins ~100% of oxygen and water regenerated	IBMP, Moscow

TABLE 40.1 (*Continued*)

Project Title, Country of Origin for Funding and Support	Years of Operation, Function	System Purpose; Bio-Engineering Characteristics and Outcomes	Key People Playing Integral Roles in the Project
BIOS-3 (USSR/Russia) As a result of this program, the Principle of Human Control for artificial CES self-sustainable functioning was formulated	1972–1991 (3 completely closed experiments; longest–½ year)	System built to achieve complete material isolation from outside and autonomy Volume ~315 m^3 Planting area ~120 m^2; 3 people + variety of higher plants ("conveyor cultures" for continuity in biomass supply) + micro-algae and physical-chemical waste regeneration units provided ~93 to 97 % (Air + Water + Vegetarian part of diet) of total material closure; trace contaminant buildup was noticed	Academician J. I. Gitelson, Prof. G. Lisovsky, B. Prof. Kovrov et al., IBP RAS, Krasnoyarsk, Siberia
BIOS-3M (Modernized) (Russia)	~1991–1996	Goal: further acceleration of material turnover to minimize physical dimensions BIOS-3 + Doubled Light Sources (40 xenon lamps with 12000 W/lamp) Fundamental instabilities noticed for artificial intensive CES, principle of Human Control as stabilizing factor for artificial LS was formulated	Prof. G. Lisovsky, and Dr. A. Tikhomirov et al., IBP RAS, Krasnoyarsk, Siberia; in collaboration with F. Salisbury et al. at Utah State University, Logan, USA
BIOS-3M Eco Ecological Research Facility (Russia)	~1991–1996	Investigation of CES eco-physiological stability limits; BIOS-3 + Doubled Light Sources (Intensive CES Stability); Volume ~157.75 m^3 Planting area ~20.5 m^2 Higher plants + test chemical toxicants (SO_2, NH_3, Ethylene) The concept of CES stability limits in combination with LSS physical dimensions and Human Control principle was formulated and confirmed in experiments	Prof. G. Lisovsky, and Dr. V. Rygalov et al., IBP, Krasnoyarsk, Siberia

Continued on next page

TABLE 40.1 *(Continued)*

Project Title, Country of Origin for Funding and Support	Years of Operation, Function	System Purpose; Bio-Engineering Characteristics and Outcomes	Key People Playing Integral Roles in the Project
Biosphere-2 (USA)	1991–Present	Unique simulation of Biosphere-1 and planetary LS base; completely closed in material exchange & open for natural sun-light and information flows Volume $\geq 210 * 10^3$ m^3; Area $\sim 1.28 * 10^4$ m^2	J. Allen, and M. Nelson et al., Institute Eco-techniques/Space Biospheres Ventures; USA/ Great Britain
Determined Eco & Physiological Processes, Statistical Regulation/Human Factors with the Intention of Self-Sustainability	(Closure experiment with 8 crew-members: 1991–1993)	Basic Earth biomes are included in the system + intensive agricultural (without chemicals) area and high-tech human habitat	
Breadboard Project, NASA Higher Plant-based Controlled Ecological Life Support System (CELSS) Program Biomass Production Chamber (BPC) (USA) (NASA, Kennedy Space Center)	1986–2004 (Currently reconfigured into Space Life Sciences Lab)	Facility total volume ~ 115 m^3 Plant area ~ 20 m^2 Goal is maximal facilitation of food production (mixed crop), air purification, water recycling and high material flux control based on advanced technologies	Dr. B. Nott, Dr. J. Sager, and Dr. R. Wheeler et al., Kennedy Space Center
Bio-Home (USA - NASA, Stennis Space Center)	1990–Present	Bio-Home is ~ 65 m^2 facility for testing the integration of specifically bio-regenerative technologies, including plants as indoor air purifiers and aquatic plant waste processing	Dr. B. Wolverton et al., NASA Stannis Space Center
Bio/Plex (USA) (NASA, Johnson Space Center)	1986–Present	Highly advanced technologies for Bio- and P/C-regeneration of the environment, integrated into one system for long-term human LS	Dr. D. Hanninger, and Dr. D. Barta et al., Johnson Space Center
CEEF (Japan)	1992–Present	Highly technologically advanced system for LS and Biosphere functions-Simulation includes basic Earth biomes and physical-chemical regeneration processes	Prof. K. Nitta et al., National Aerospace Laboratory

TABLE 40.1 (*Continued*)

Project Title, Country of Origin for Funding and Support	Years of Operation, Function	System Purpose; Bio-Engineering Characteristics and Outcomes	Key People Playing Integral Roles in the Project
Martian Greenhouse Prototype (USA) (NASA, Kennedy Space Center)	1998–Present	Highly automated system for plant growth under low atmospheric pressure ($\geq 1/10$ of Earth atm) oriented for use of natural space resources (In Situ Resources Utilization = ISRU concept) Volume ~0.417 m^3 Planting area ~0.8 m^2	Dr. R. Wheeler (KSC NASA), Dr. P. Fowler (Dynamac Inc.)
		Goal: automatic and autonomous higher plant growth and oxygen accumulation before crew arrival to the planet Mars	Dr. V. Rygalov (University of Florida)
GBEC-3 500 day experiment simulating small group interaction during Martian expedition	Plans are: 2006 to 2008	Physical/Chemical + Bio-Regenerative LS & Long-term social/ psychological effects of closed confined environment to specify workload, provide Equivalent System Mass (ESM) analysis, test equipment and specify LSS components	IBMP, Moscow
European Space Agency/Russian Space Agency/ Russian Academy of Sciences Potential participants: NASA, Canadian Space Agency, French Space Agency, National Space Agency of Japan, others			

Section VII

Lunar Soil Mechanics

41

A Constitutive Model for Lunar Soil

C.S. Chang
University of Massachusetts
Department of Civil and Environmental Engineering
Amherst, Massachusetts

Pierre-Yves Hicher
University of Nantes
Research Institute in Civil and Mechanical Engineering
Nantes, France

Pierre-Yves Hicher is currently professor of civil engineering at Ecole Centrale de Nantes, France and director of the Research Institute in Civil and Mechanical Engineering, UMR CNRS-Ecole Centrale Nantes-University of Nantes. His principal research activity and publications lie in the field of the mechanical behavior of soils and granular materials.

Introduction

The soil on the Moon formed by space weathering processes and dynamic impacts of micro-meteorites is composed of aggregates of minerals, rocklets and glasses, welded together into agglutinates. We are interested in the behavior of soils in a typical regolith of the Moon—a surface layer which consists of loose sand and rock fragments which overlies the solid rock. The fine particles of lunar soil are the products of the continual impact on the surface by meteoroids which smash and grind rocks into soil and weld soil into new rocks. The range of grain size distribution is summarized in a soil gradation curve given in Fig. 41.1 (Costes et al., 1970) based on the core samples obtained from the surface layer of the Moon (Apollo 11 and 12). According to the particle size distribution, the lunar soil can be classified as dry silty sand following the Unified Soil Classification System, ASTM standards.

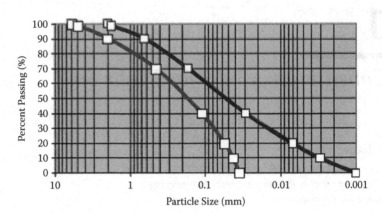

FIGURE 41.1
Approximated gradation range for Apollo 11 and 12 soil core samples.

However, observations and measurements conducted during Surveyor (1966–1968), Apollo (1968–1972), and Luna (1959–1976) missions indicate that lunar soil has an unusually high cohesion in comparison to the same type of soils tested under Earth conditions. According to Buzz Aldrin, astronaut on Apollo 11, the soil on the Moon "has the cohesive property that wet sand would have" (Costes et al., 1970). A stable vertical trench excavated during the Surveyor 7 mission is shown in Fig. 41.2a and a footprint left by Surveyor 3 demonstrates the fineness and cohesiveness of the lunar soil (Fig. 41.2b).

Our knowledge of engineering properties of lunar soil are derived mainly from operations performed during Surveyor and Apollo programs, including bulk and core samples returned to Earth, trench excavation, ground penetration, and some simple physical experiments. In laboratory studies, it was found that soils tested under a chamber with ultrahigh vacuum and high temperature had an increase of up to 13° in angle of internal friction and 1.1 kPa in cohesion (Bromwell 1966; Nelson 1967; Perko et al. 2001). This increase of soil strength is consistent with the observed stable vertical trend excavation in dry silty sand on the Moon.

For soil on Earth, cohesion is originated from electrostatic forces and surface-energy forces, which includes van der Waals forces (intermolecular potential energy). The surface-energy forces act at very short range. Compared to the gravitational forces, the surface-energy forces are negligible on Earth for soils with particle sizes larger than around 0.06 mm. Thus terrestrially, the forces are negligible for sand or silty sand. However, as particle size decreases to the size of clay, the surface-energy forces have a greater influence on its strength. The effect of surface energy forces on friction and adhesion properties of small particles can be found in the literature of powder technology, particulate technology and interface sciences (Derjaguin 1934, Derjaguin et al. 1975, Johnson 1971, Molerus 1978, Thorton and Ning 1998, Tomas 2001).

Due to various environmental differences on the Moon (such as the gravitational fields, temperature extremes, low to practically nonexistent

FIGURE 41.2a
Trench excavated by Surveyor 7 Spacecraft in Lunar subsurface soil.

FIGURE 41.2b
The Spacecraft (Surveyor 3–Apollo 12) had bounced upon landing, leaving a footprint.

atmospheric pressure, plus the space weathering which includes surface activating effects like UV radiation, the solar wind and galactic particle fluxes), it seems plausible to consider surface energy forces as a dominant cause for altering the mechanical behavior of lunar soil with the size of silty sand. An issue that may be addressed is to what order of magnitudes the surface-energy forces can influence the strength and deformation of lunar soils.

In this paper, a model is developed that accounts for the surface energy forces between particles and is capable of modeling its effect on the shear strength of the soil assembly. For this purpose, we adopt a microstructural modeling approach (Chang and Hicher, 2005; Hicher and Chang, 2005). Models using this approach can also be found in the work by Jenkins (1988), Walton (1987), Rothenburg and Selvadurai (1981), Chang (1988), Emeriault and Cambou (1996), Liao et. al (2000), Kruyt and Rothernburg (2002), among others. Elasto-plastic models using this approach can be found by Chang et al. (1992a, 1992b), Suiker and Chang (2003).

It is noted that the above mentioned microstructural continuum approach is not a "complete" micro-mechanical approach. Since the complete microstructural details for a particle assembly are not known, the complex deformation behavior of the particle assembly cannot be derived directly from the first-principles theory based solely on particle-level properties. Besides the parameters of particle-level properties, the present model adopts some mesoscale parameters because inter-particle behavior is not exclusively a local phenomenon; it is considerably influenced by the degree of interlocking and porosity of the surrounding particles. In the present model, these parameters are based on the critical void ratio concept, and are empirically determined from testing on soil specimens.

Therefore, the present model should be regarded as a model of semi-empirical nature. It is necessary to calibrate parameters from soil tests in order to predict complex behavior with good quantitative agreement. The main advantage of the present model over conventional continuum constitutive models is its microscale consideration, which allows one to conveniently extend the model for new phenomenon at the particle level, such as the surface energy forces.

However, adding the surface energy term from first principles physics into an otherwise semi-empirical model may elicit the following concern: How can one part of the physics be treated from first principles if the other parts of the physics are treated only empirically? The answer is that even though the surface energy forces are added from first principles physics, this does not alter the empirical nature of the model. The model is aimed to predict the "order of magnitude" of the surface energy effect.

In what follows, we first describe the formulation of this model, which takes account of the surface energy forces between particles. The model performance is then demonstrated through the results of a simulation of fine Hostun sand tested under terrestrial conditions (i.e., without surface energy forces). The fine sand has a mean particle diameter of 0.5 mm, loaded to failure in triaxial testing cells under different confining stresses. Finally, the effect of surface energy forces between particles is introduced to predict the stress-strain-strength behavior of lunar soil.

Stress-Strain Model

In this model, we envision a granular material as a collection of particles. The deformation of a representative volume of the material is generated by the mobilization of contact particles in all contacts. Thus, the stress-strain relationship can be derived as an average of the mobilization behavior of local contact planes in all contacts. For a contact plane in the αth orientation, the local forces f_i^α and the local movements δ_i^α can be denoted as follows: $f_i^\alpha = \{f_n^\alpha, f_s^\alpha, f_t^\alpha\}$ and $\delta_i^\alpha = \{\delta_n^\alpha, \delta_s^\alpha, \delta_t^\alpha\}$, where the subscripts n, s, and t represent the components in the three directions of the local coordinate system. The direction normal to the plane is denoted as n; the other two orthogonal directions, s and t, are tangential to the plane. Rotation of particles is not considered here.

The forces and movements at contact planes of all contacts are suitably superimposed to obtain the macroscopic stress strain tensors. The macroscopic stiffness tensor is obtained on the condition that the rate of energy dissipation expressed in terms of the macro stress and strain must be equivalent to that expressed in terms of micro forces and movements. In such formulation, it has usually been assumed that the microstructure is statically constrained, which means that the forces on each contact plane are assumed equal to the resolved components of the macroscopic stress tensor.

Inter-Particle Behavior

van der Waals Forces

The surface energy forces pull soil particles together, thus increase the shear strength of the soil. The physics of the surface energy forces between two particles is reviewed in this section. The adhesive forces between two solid grains result mainly from electrostatic forces and van der Waals forces. The electrostatic component in lunar soil is assumed to be negligable (Perko et al. 2001). Therefore, in this paper, the surface energy forces are calculated from van der Waals energy fields. For simplicity, lunar soil grains are represented by spheres of equal radii that are separated by thin layers of adsorbed molecules.

van der Waals forces between two bodies are derived from the dispersion interaction energy between two identical atoms or molecules (Israelachvili 1992):

$$W(r) = -\frac{C}{r^6} \tag{41.1}$$

where r is the distance between the two atoms and C is the London dispersion coefficient.

Then, with the assumption of additivity of these attractive forces, the interaction energy of a molecule located at a distance D of the planar surface of a solid made up of like molecules is the sum of its interaction with all the molecules in the body:

$$W(D) = -\frac{\pi C \rho}{6D^3} \tag{41.2}$$

where ρ = number of atoms per unit volume.

The corresponding van der Waals force is

$$F = \frac{\delta W(D)}{\delta D} = \frac{\pi C \rho}{2D^4} \tag{41.3}$$

It is then possible to calculate the interaction energy between two solids as, for example, two spheres:

$$W = -\frac{A R_1 R_2}{6D(R_1 + R_2)} \tag{41.4}$$

where D = distance between two spheres, R_1 and R_2 = radii of the two spheres, $A = \pi^2 C \rho_1 \rho_2$ is the Hamaker constant, ρ_1 and ρ_2 being the number of atoms per unit volume for the two bodies. For two flat surfaces, one obtains the interaction energy per unit area:

$$W = -\frac{A}{12\pi D^2} \tag{41.5}$$

The van der Waals force between two solids in contact can then be computed from the interaction energy. One obtains the following two cases:

Between two spheres:

$$f = 2\pi W(D) \frac{R_1 R_2}{R_1 + R_2} \tag{41.6}$$

Eq. (41.6) is called the Derjaguin approximation (Israelachvili, 1992). In the case of two identical spheres, Eq. (41.6) becomes:

$$f = \frac{A R_2}{12 D^2} \tag{41.7}$$

Between two flat surfaces, per unit area:

$$f = \frac{A}{6\pi D^2} \tag{41.8}$$

Let us now consider two soil grains as two identical elastic spheres. If subjected to an external force f, the two spheres will deform and create a flat circular contact area with a radius a:

$$a = \left(\frac{3(1-\upsilon_p^2)R}{4E_p} \right)^{1/3} f^{1/3} \tag{41.9}$$

E_p and υ_p are the Young's modulus and Poisson's ratio of the particles. The derivation of Eq. (41.8) can be found in Derjaguin et al. (1975), Johnson (1971), Dahneke (1972), Valverde et al. (2001). It is noted that, for simplicity, we consider only the case of elastic contact flattening (Eq. (41.9)). Under these simplified conditions, van der Waals forces acting between two particles can be considered as the sum of two terms, one due to the interaction between two flat surfaces of area $S = \pi a^2$ and one along the remaining surface of the two spheres. Using the Derjaguin approximation for the second term, we obtain the expression of the van der Waals force between two particles:

$$f = \frac{A}{6D^3} a^2 + \frac{AR}{12D^2} \tag{41.10}$$

where A = Hamaker coefficient; D = thickness of molecular layer between two particles; R = the radius of the particles. Hamaker constant A was estimated to be 4.3×10^{-20} J for lunar soil and 1.5×10^{-20} J for terrestrial quartz sand (Perko et al. 2001).

The thickness of the molecular layer between two particles D is highly dependent on the atmospheric pressure and composition. On the Moon, the atmospheric pressure is nearly zero, which can lead to a very thin layer of molecules between two particles compared to that under terrestrial environment. Therefore, according to Eq. (41.10), it is reasonable to expect that the surface energy forces between particles are much higher than those between particles under terrestrial environment.

It is noted that the radius a of the contact area in Eq. (41.10) increases with confining stress of a specimen. Thus the surface energy force also increases with confining stress, which indicates that the surface energy force will contribute to the shear strength not only on the cohesive component but also on the frictional component.

The orientation of a contact plane between two particles is defined by the vector perpendicular to this plane. On each contact plane, an auxiliary local coordinate can be established as shown in Fig. 41.3. The surface energy force f given in Eq. (41.10) represents only the force magnitude. Its direction is normal to the inter-particle contact plane given by n_i^α. Thus in a vector form, the surface energy force $f_i^{\alpha(SE)} = fn_i^\alpha$.

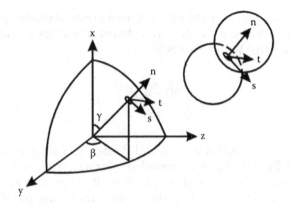

FIGURE 41.3
Local coordinate at inter-particle contact.

Between two particles, the inter-particle force f_i^α can be decomposed into two components: (1) due to applied load on the boundary of the soil assembly $f_i^{\alpha(A)}$, and (2) due to surface energy forces between particles $f_i^{\alpha(SE)}$. Thus

$$f_i^\alpha = f_i^{\alpha(A)} + f_i^{\alpha(SE)} \tag{41.11}$$

The surface energy force is a function of contact area as shown in Eq. (41.10). However, the contact area is in turn a function of inter-particle force (see Eq. (41.9)), thus the expression of Eq. (41.11) is implicit and nonlinear in nature.

Inter-Particle Force-Displacement Relationship

Elastic Part
The contact stiffness of a contact plane includes normal stiffness, k_n^α, and shear stiffness, k_r^α (assuming $k_r^\alpha = k_t^\alpha = k_s^\alpha$). The elastic stiffness tensor is defined by

$$f_i^\alpha = k_{ij}^{\alpha e} \delta_j^{\alpha e} \tag{41.12}$$

which can be related to the contact normal and shear stiffness

$$k_{ij}^{\alpha e} = k_n^\alpha n_i^\alpha n_j^\alpha + k_r^\alpha (s_i^\alpha s_j^\alpha + t_i^\alpha t_j^\alpha) \tag{41.13}$$

where n, s, t are three orthogonal unit vectors that form the local coordinate system (see Fig. 41.3). The vector n is outward normal to the contact plane. Vectors s and t are on the contact plane.

The value of the stiffness for two elastic spheres can be estimated from Hertz-Mindlin's formulation (Mindlin 1969). For sand grains, a revised form was adopted (Chang et. al, 1989), and is given by

$$k_n = k_{n0}\left(\frac{f_n}{G_g l^2}\right)^n \; ; \; k_r = k_{r0}\left(\frac{f_n}{G_g l^2}\right)^n \qquad (41.14)$$

where G_g is the elastic modulus for the grains, f_n is the contact force in the normal direction. l is the branch length between the two particles. k_{n0}, k_{r0} and n are material constants. For two spherical particles, the branch length is same as particle size $l = d$. If the Hertz-Mindlin's contact formulation is used, the value of k_{n0} in Eq. (41.14) can be expressed in the following form:

$$k_{n0} = G_g \frac{d}{2}\left(\frac{\sqrt{12}}{1-v_g}\right)^{2/3} \qquad (41.15)$$

Plastic Part

Plastic sliding often occurs with an upward or downward movement, thus shear dilation/contraction takes place. The stress-dilatancy is a well-recognized phenomenon in sand (see discussions in the work by Taylor 1948, Rowe 1962, Goddard 1990), and should be correctly modeled. The dilatancy effect has been described by Chang and Hicher (2005)

$$\frac{\dot{\delta}_n^p}{\dot{\Delta}^p} = \frac{f_r}{f_n} - \tan\phi_0 \qquad (41.16)$$

where ϕ_0 is a material constant which, in most cases, can be considered equal to the inter-particle friction angle ϕ_μ. Note that the shear force T and the rate of plastic sliding $\dot{\Delta}^p$ are defined as:

$$f_r = \sqrt{f_s^2 + f_t^2} \quad \text{and} \quad \dot{\Delta}^p = \sqrt{\left(\dot{\delta}_s^p\right)^2 + \left(\dot{\delta}_t^p\right)^2} \qquad (41.17)$$

The yield function is assumed to be of Mohr-Coulomb type:

$$F(f_i, \kappa) = f_r - f_n \kappa(\Delta^p) = 0 \qquad (41.18)$$

where $\kappa(\Delta^p)$ is an isotropic hardening/softening parameter. When $F > 0$, it indicates loading, otherwise unloading. The hardening function, defined as a hyperbolic relationship between κ and Δ^p, involves two material constants: ϕ_p and k_{p0}:

$$\kappa = \frac{k_{p0}\tan\phi_p \Delta^p}{|f_n|\tan\phi_p + k_{p0}\Delta^p} \qquad (41.19)$$

The value of κ asymptotically approaches $\tan \phi_p$. The initial slope of the hyperbolic curve is k_{p0}. On the contact plane, under a yield condition, the direction of plastic shear sliding $\dot{\Delta}^p$ follows the associated flow rule, thus is perpendicular to the yield surface. However, the plastic movement in the direction normal to the contact plane is governed by the stress-dilatancy equation shown in Eq. (41.16). Thus the overall flow rule is non-associated.

Interlocking Influence

One of the important elements to be adopted in granular modeling is the critical state concept. Under critical state, the granular material remains at a constant volume while it is subjected to a continuous distortion. The void ratio corresponding to this state is denoted as e_c.

The critical void ratio e_c is a function of the mean stress. The relationship has traditionally been written as follows:

$$ e_c = \Gamma - \lambda \log(p') \quad \text{or} \quad e_c = e_{ref} - \lambda \log\left(\frac{p'}{p_{ref}}\right) \tag{41.20} $$

Γ and λ are two material constants and p' is the mean stress of the packing, and (e_{ref}, p_{ref}) is a reference point on the critical state line.

The inter-particle friction angle ϕ_μ is a constant for the material. However, the peak friction angle, ϕ_p, on a contact plane between two particles is dependent on the degree of interlocking of neighboring particles, which can be related to the state of packing void ratio e by:

$$ \tan \phi_p = \left(\frac{e_c}{e}\right)^m \tan \phi_\mu \tag{41.21} $$

where m is a material constant (Biarez and Hicher, 1994).

For dense packing, the peak frictional angle ϕ_p is greater than ϕ_μ. When the packing structure dilates, the degree of interlocking and the peak frictional angle are reduced, which results in a strain-softening phenomenon.

Elasto-Plastic Force-Displacement Relationship

With the elements discussed above, the relations between the rate of force-displacement for two particles can be derived that includes both elastic and plastic behavior, given by

$$ \dot{f}_i^\alpha = k_{ij}^{\alpha p} \dot{\delta}_j^\alpha \tag{41.22} $$

The detailed expression of $k_{ij}^{\alpha p}$ can be found in Chang and Hicher (2005).

Stress-Strain Relationship

Macro Micro Relationship

The stress-strain relationship for an assembly can be determined by integrating the behavior of all inter-particle contacts. During the integration process, a relationship is required to link the macro and micro variables. Using the static hypotheses proposed by Liao et al. (1997), we obtain the relation between the macro strain rate and inter-particle displacement rate (here, we do not consider the finite strain condition)

$$\dot{u}_{j,i} = A_{ik}^{-1} \sum_{\alpha=1}^{N} \dot{\delta}_j^\alpha l_k^\alpha \qquad (41.23)$$

where $\dot{\delta}_j$ is the relative displacement rate between two contact particles, and the branch vector l_k is the vector joining the centers of two contact particles. It is noted that contact particles include both direct contact and indirect contact of neighboring particles associated with a Voronoi polyhedron as discussed by Cambou et al. (2000). For convenience, we let N be the total number of contacts. The variables $\dot{\delta}_j^\alpha$ and l_k^α are defined respectively as the values of $\dot{\delta}_j$ and l_k associated with the αth contact. The fabric tensor A_{ik} in Eq. (41.23) is defined as

$$A_{ik} = \sum_{\alpha=1}^{N} l_i^\alpha l_k^\alpha \qquad (41.24)$$

Using the principle of energy balance and using Eq. (41.23), the mean force rate on the contact is

$$\dot{f}_j^\alpha = \dot{\sigma}_{ij} A_{ik}^{-1} l_k^\alpha V \qquad (41.25)$$

In Eq. (41.25), the stress increment $\dot{\sigma}_{ij}$ can be obtained by the contact forces and branch vectors for all contacts (Christofferson et al., 1981, Rothenburg and Selvadurai, 1981)

$$\dot{\sigma}_{ij} = \frac{1}{V} \sum_{\alpha=1}^{N} \dot{f}_j^\alpha l_i^\alpha \qquad (41.26)$$

Applying the defined contact force in Eq. (41.25), Eq. (41.26) is unconditionally satisfied. Because of its approximation nature, Eq. (41.25) can be viewed as an averaged solution, in which the inter-particle force can be regarded as the mean value for forces on all contact planes of the same orientation. For convenience, we also regard the branch length as the mean value for all contact planes of the same orientation.

The inter-particle force in Eq. (41.26) consists of two parts: $f_i^{\alpha(A)}$ due to the applied load on the boundary of the soil assembly, and $f_i^{\alpha(SE)}$ due to the surface energy forces between particles (see Eq. (41.11)). Thus the stress can be separated into two parts:

$$\dot{\sigma}_{ij} = \frac{1}{V} \sum_{\alpha=1}^{N} \dot{f}_j^{\alpha(A)} l_i^{\alpha} + \frac{1}{V} \sum_{\alpha=1}^{N} \dot{f}_j^{\alpha(SE)} l_i^{\alpha} \qquad (41.27)$$

The second part of Eq. (41.27) represents the stress induced by surface energy forces, denoted as

$$\left(\dot{\sigma}_{ij}\right)^{SE} = \frac{1}{V} \sum_{\alpha=1}^{N} \dot{f}_j^{\alpha(SE)} l_i^{\alpha} \qquad (41.28)$$

It is noted that this term is not analogous to the usual concept of cohesion for bulk materials. In the present model, the surface energy stress depends on the packing structure and is a tensor rather than a scalar. Only for an isotropic distribution of the branch lengths l^{α}, the surface energy stress can be reduced to an isotropic tensor.

As mentioned earlier, the surface energy force is a function of contact area, which is in turn a function of inter-particle force. Thus all equations in Section 2 are nonlinear in nature. The set of nonlinear equations for stress-strain relationships is discussed in the next section.

Computation Scheme

Using Eqs. (41.22), (41.23), and (41.25), the following relationship between stress and strain can be obtained:

$$\dot{u}_{i,j} = C_{ijmp} \dot{\sigma}_{mp}; \quad \text{where} \quad C_{ijmp} = A_{ik}^{-1} A_{mn}^{-1} V \sum_{\alpha=1}^{N} \left(k_{jp}^{ep}\right)^{-1} l_k^{\alpha} l_n^{\alpha} \qquad (41.29)$$

The summation in Eq. (41.29) can be replaced by an integral over orientations. The integral can lead to a closed-form solution for the elastic modulus of randomly packed equal-size particles (Chang et al. 1995a). However, in elastic plastic behavior, due to the nonlinear nature of the local constitutive equations, a numerical calculation with an iterative process is necessary to carry out the summation in Eq. (41.29). In order to facilitate the numerical calculation, the orientations are selected to coincide with the locations of Gauss integration points in a spherical coordinate system. Summation over these orientations with the Gauss weighting factor for each orientation is equivalent to determining the integral over orientations. We found that the results were more accurate by using a set of fully symmetric integration points. From a study of the performance of using different numbers of orientations, we found $N \geq 74$ to be adequate (Chang and Hicher 2005).

Summary of Parameters

One can summarize the material parameters as:

- Normalized contact number per unit volume: Nl^3/V
- Mean particle size, d
- Inter-particle elastic constants: k_{n0}, k_{r0} and n;
- Inter-particle friction angle: ϕ_μ and m;
- Inter-particle hardening rule: k_{p0} and ϕ_0;
- Critical state for the packing: λ and Γ or e_{ref} and p_{ref}

Besides critical state parameters, all other parameters are for inter-particles. Standard values for k_{p0} and ϕ_0 are the following: $k_{p0} = k_{n0}$ and $\phi_0 = \phi_\mu$ and a typical ratio $k_{r0}/k_{n0} = 0.4$ can be generally assumed. Therefore, only six parameters have to be derived from experimental results and they can all be determined from the stress-strain curves obtained from drained compression triaxial tests.

Results of Numerical Simulation

Triaxial Tests on Terrestrial Soil without Surface Energy Forces

A series of drained triaxial tests on fine Hostun Sand were performed by Al Mahmoud (1999). The gradation curve tested sand is shown in Fig. 41.4. The mean size of the particles for fine Hostun sand is $d = 0.5$ mm. It is classified as uniform fine sand. The minimum and maximum void ratios for this sand are $e_{min} = 0.575$, and $e_{max} = 0.943$, respectively.

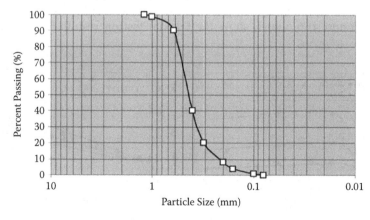

FIGURE 41.4
Gradation curve for fine Hostun sand.

TABLE 41.1

Model parameters for fine Hostun Sand

e_{ref}	p_{ref}(Mpa)	λ	ϕ_μ (°)	ϕ_0 (°)	m	k_{p0}/k_n
1.05	0.01	0.14	34	34	1.5	0.02

We assume here that surface energy forces can be neglected under terrestrial conditions. This assumption will be discussed in the next section. The model needs a small number of input parameters, such as mean particle size, particle stiffness, inter-particle friction, initial porosity tensor, and an initial degree of interlocking. The inter-particle elastic constant k_{n0} is assumed to be 61000 N/mm. Considering the grain size distribution curve, the contact numbers per volume Nl^3/V are 2 for dense sand and 0.9 for loose sand (Hicher and Chang 2005).

The value of k_{r0}/k_{n0} is commonly about 0.4, corresponding to a Poisson's ratio for Hostun sand $\nu = 0.2$ and the exponent $n = 0.5$ (Biarez and Hicher, 1994). From test results, we were able to derive the values of the two parameters corresponding to the position of the critical state in the e-p' plane: $\lambda = 0.14$ and $p_{ref} = 0.01$ MPa for $e_{ref} = 1.05$. The friction angle ϕ_μ was also determined from the stress state corresponding to the critical state: $\phi_\mu = 34°$. The equation governing the dilatancy rate requires the determination of the parameter ϕ_0. This parameter represents the concept of "phase transformation" as defined by Ishihara and Towhata (1983) or "characteristic state" as defined by Luong (1980). A value of $\phi_0 = \phi_\mu$ was retained.

The peak friction angle is not an intrinsic parameter, but varies with the void ratio according to Eq. (41.21). A value of $m = 1.5$ was determined from the test results. The value of k_{p0} was determined by curve fitting. The set of parameters for fine Hostun sand is presented in Table 41.1.

The tests were performed at different confining pressures on samples prepared at different initial void ratios. Typical results are presented in Figs. 41.5 and 41.6, which show the triaxial testing results for both dense

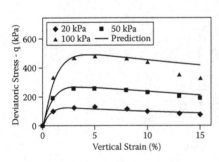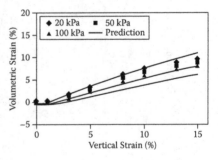

FIGURE 41.5

Stress-strain curves for dense sand.

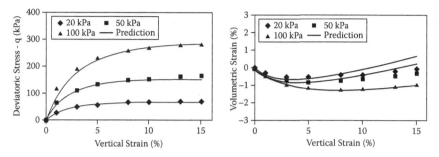

FIGURE 41.6
Stress-strain curves for loose sand.

and loose specimens made of Hostun sand. The stress strain curves are plotted for dense sand in Fig. 41.5, and loose sand in Fig. 41.6. The model performance can be demonstrated by comparing the predicted and measured macro behavior.

One can see the combined influence of the initial void ratio and of the mean effective stress on the stress-strain curves and the volumetric change. The stress-strain curve has a peak that increases with confining stress. For dilatant materials, the deviatoric stress reduces after peak and converges towards a constant state of stress corresponding to the critical state. In practice, it is difficult to reach this state because of strain localization, especially in dense materials. The critical state was estimated from the results on loose specimens.

Triaxial Tests with Consideration of Surface Energy Forces

In order to evaluate the possible effects of lunar environmental conditions on the stress-strain response of soils, the influence of the surface energy forces for different values of inter-particle distance D was examined. The stress-strain curves were predicted using the same soil parameters given in Table 41.1. Given the introduction of van der Waals forces, shear strength is expected to be higher. Fig. 41.7 shows the influence of the distance D between the particles on the shear strength for soil specimens under 20 kPa confining stress, where q_0 represents the shear strength for soil without surface energy forces and q represents the shear strength with the effects of these forces.

It is noted that when the distance D is greater than 2 nm, the effect of the van der Waals forces is practically negligible. This distance D is a function of the amount of molecules which can be adsorbed on the solid surface. The adsorbate thickness may be estimated by means of potential theory (Adamson 1990, Perko et al. 2001). Molecule adsorption on a solid surface is conditioned by temperature, gas pressure and atmospheric composition. On Earth, adsorption conditions are easily fulfilled due to the presence of a

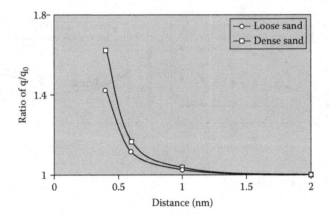

FIGURE 41.7
Effect of distance between particles on shear strength.

high atmospheric pressure. Therefore the distance D can easily exceed several nms such that the effect of the van der Waals force is negligible. Under lunar atmospheric condition, the thickness of the adsorbed molecule layer is likely to be very thin. For the case of 0.3 nm thickness ($D = 0.6$ nm), the comparisons of the stress-strain curves in Earth and in Moon atmosphere environments are given in Fig. 41.8 for dense soil and Fig. 41.9 for loose soil. Assuming the thickness D is 0.3–0.6 nm, the computed shear strength for lunar soils is 12 to 15% higher than that of Earth soil. The stiffness of the lunar soil is also higher. It is noted that the effect of van der Waals forces for this example would result in an increase of 2 to 3 kPa in shear strength, while an increase of 0.5 kPa in strength is sufficient to hold a 1 m trench cut in lunar gravity condition.

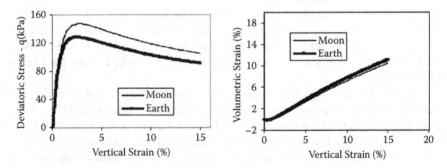

FIGURE 41.8
Predicted stress-strain curves for lunar and terrestrial dense soils.

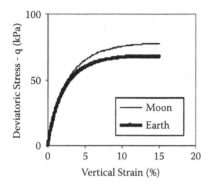 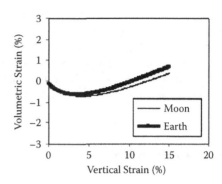

FIGURE 41.9
Predicted stress-strain curves for lunar and terrestrial loose soils.

Summary and Conclusions

In the present study we introduce surface energy forces between particles and estimate their effects on the shear strength of soil. The microstructural approach is adopted because it allows us to model interactions at the particle level.

In the model, a simple elastic-plastic behavior was assumed on each contact plane. The elastic part is based on the Hertz-Mindlin contact formulation, while the plastic part is based on a Mohr-Coulomb friction law with an isotropic hardening assumption and a non-associated flow rule. The interaction among inter-particle planes is assumed to be dependent on the degree of interlocking and the porosity of the assembly of soil. The effect of porosity is modeled by a phenomenological approach using the concepts of critical state such that strain softening behavior can be modeled for dense materials. On the whole, the model requires a limited number of parameters, which can easily be determined from conventional triaxial tests.

The ability of the model to reproduce the main features of sand behavior has been demonstrated. Model simulations were compared with drained triaxial test results at different initial void ratios and different confining stresses leading to contractant or dilatant behavior of the sand specimens. The comparison demonstrated that the model is capable of reproducing the general trend for both loose and dense sands.

The model was applied to predict the shear strength of lunar soil under extremely low atmospheric pressure. Even if the predicted shear strength for lunar soil obtained by including surface energy forces is only a hypothesis, which must await further testing on the Moon for validation and calibration, the numerical simulations seem to give reasonable results.

Model predictions indicate that soil under extremely low atmospheric pressure has an increase of shear strength by several kPa higher than the one which would be present under Earth's atmospheric pressure. The influence is more pronounced in dense specimens than in loose specimens, due to the fact that the particles are arranged more closely to each other in dense packing. This result is in accordance with the general trend observed in-situ. The magnitude is on the same order as the measured increase of shear strength for lunar soil simulants tested under the usual atmospheric pressure and under a chamber with ultrahigh vacuum.

References

Adamson, A. W. (1990) Physical Chemistry of Surfaces, 5th Ed., John Wiley & Sons, New York.

Al Mahmoud, M. (1999) "Etude en laboratoire du comportement des sables sous faibles contraintes", Thèse de doctorat de l'université des sciences et technologies de Lille, p. 141.

Biarez J. and Hicher, P.Y. (1994) Elementary Mechanics of Soil Behaviour, Balkema, p. 208.

Cambou, B., Dedecker, F., and Chaze, M. (2000), "Relevant local variables for the change of scale in granular materials," Constitutive Modelling of Granular Materials (Dimitrios Kolymbas, Ed.), Springer, Berlin, 275–290.

Chang, C. S. (1988), "Micromechanical modeling of constructive relations for granular material," Micromechanics of granular materials (Satake, M. and Jenkins, J. T., eds.), 271–279.

Chang, C. S., Chao, S. C., and Chang Y. (1995a), "Estimates of Mechanical Properties of Granulates with Anisotropic Random Packing Structure," International Journal of Solids and Structures, Vol. 32, No. 14, pp. 1989–2008.

Chang, C. S., and Gao, J. (1995b), "Second-gradient constitutive theory for granular material with random packing structure," International Journal of Solids and Structures, 32 (16), 2279–2293.

Chang, C. S., and Gao, J. (1996), "Kinematic and Static Hypotheses for Constitutive Modelling of Granulates Considering Particle Rotation," Acta Mechanica, Springer-Verlag, Vol. 115, No. 1–4, 1996, pp. 213–229.

Chang, C. S., and Hicher, P.-Y., (2005) "An elasto-plastic model for granular materials with microstructural consideration." International Journal of Solids and Structures, Vol. 42, No. 14, pp. 4258–4277.

Chang, C. S. and Kuhn, M. R. (2005), "On Virtual Work and Stress in Granular Media," International Journal of Solids and Structures, Vol. 42, pp. 3773–3793.

Chang, C. S., and Liao, C. (1990), "Constitutive Relations for Particulate Medium with the Effect of Particle Rotation," International Journal of Solids and Structures, Vol. 26, No. 4, pp. 437–445.

Chang, C. S., Kabir, M., and Chang, Y. (1992a), "Micromechanics modelling for the stress strain behavior of granular soil II: evaluation," Journal of Geotechnical Engineering, ASCE, 118 (12), 1975–1994.

Chang, C. S., and Ma, L. (1992), "Elastic Material Constants for Isotropic Granular Solids with Particle Rotation," International Journal of Solids and Structures, Pergamon Press, Vol. 29, No. 8, pp. 1001–1018.

Chang, C. S., Misra, A., and Acheampon, K. (1992b), "Elastoplastic deformation of granulates with frictional contacts," Journal of Engineering Mechanics, ASCE, 118 (8), 1692–1708.

Chang, C. S. and Misra, A. (1990), "Application of uniform strain theory to heterogeneous granular solids," Journal of Engineering Mechanics, ASCE. 116 (10), 2310–2328.

Chang, C. S., Sundaram, S.S., and Misra, A. (1989) "Initial Moduli of Particulate Mass with Frictional Contacts," International Journal for Numerical and Analytical Methods in Geomechanics, John Wiley & Sons, Vol. 13 (6), pp. 626–641.

Christoffersen, J., Mehrabadi, H. M. and Nemat-Nasser, S. (1981) A micromechanical description of granular material behavior. Journal of Applied Mechanics, ASME 48(2), 339–344.

Costes, N. C., Carrier, W.D., Mitchell, J.K. and Scott, R.F. (1970) "Apollo 11: soil mechanics results." Journal of Soil Mech. and Found. Div., ASCE, vol. 96, n° 6, pp. 2045–2080.

Dahneke B. (1972) "Influence of Flattening on Adhesion of Particles," Journal of Colloid and Interface Science 40 (1): 1.

Derjaguin B (1934) "Analysis of friction and adhesion IV The theory of the adhesion of small particles," Kolloid-Zeitschrift 69 (2): 155–164.

Derjaguin B. V., Muller V. M., Toporov Y. P. (1975) "Effect of Contact Deformations on Adhesion of Particles," Journal of Colloid and Interface Science 53 (2): 314–326.

Emeriault, F., and Cambou, B. (1996), "Micromechanical modelling of anisotropic non-linear elasticity of granular medium," International Journal of Solids and Structures, 33 (18), 2591–2607.

Goddard, J. D., and Bashir, Y. M. (1990), "On Reynolds dilatancy," Recent Development in Structured Continua (D. De Kee and P. N. Kaloni, Eds.), Vol. II, Longman's, London, 23–35.

Hicher, P-Y. (1998), "Experimental behavior of granular materials," in Behavior of Granular Materials, Ed. By B. Cambou, Springer Wien, New York, 1–97.

Hicher, P.-Y., and Chang, C. S., (2005) "Evaluation of two homogenization techniques for modelling the elastic behaviour of granular materials." J. Eng. Mechanics, ASCE, Vol. 131, n° 11, pp. 1184–1194.

Ishihara, K., and Towhata, I. (1983), "Cyclic behavior of sand during rotation of principal axes, " Mechanics of Granular Materials, Ed. Elsevier, 55–73.

Israelachvili, J. (1992), "Intermolecular and surface forces," 2nd Ed., Academic Press, San Diego, p. 450.

Jenkins, J. T. (1988) "Volume change in small strain axisymmetric deformations of a granular material," Micromechanics of granular materials (Satake, M. and Jenkins, J. T., Eds.), 143–152.

Johnson K. L. (1971), "Surface Energy and Contact of Elastic Solids," Proceedings of the Royal Society of London Series A- Mathematical and Physical Sciences 324 : 301.

Kruyt, N. P., and Rothenburg, L. (2002), "Micromechanical bounds for the effective elastic moduli of granular materials," International Journal of Solids and Structures, 39 (2): 311–324.

Kruyt, N. P. (2003), "Statics and kinematics of discrete Cosserat-type granular materials," International Journal of Solids and Structures, 40: 511–534.

Liao, C. L., Chan, T.C., Suiker, A.S.J., and Chang, C.S. (2000), "Pressure-dependent elastic moduli of granular assemblies," International Journal for Analytical and Numerical Methods in Geomechanics, 24, 265–279.

Luding, S., Latzel, M., Volk, W., Diebels, S., and Herrmann, H. J. (2001), "Computer Methods in Applied Mechanics and Engineering," v 191, n 1-2, Micromechanics of Brittle Materials and Stochastic Analysis of Mechanical Systems, p. 21-28.

Luong, M. P. (1980), "Stress-strain aspects of cohesionless soils under cyclic and transient loading," Int. Symp. on Soils under Cyclic and Transient Loading, Swansea, 353-376.

Mindlin, R. D. (1969), "Microstructure in linear elasticity," Arch. Rational Mech. Anal., 16, 51-78.

Mitchell, J.K. (1972), "Soil mechanical properties at the Apollo 14 site," J. Geophys. Res., vol. 77 (29), pp. 5641-5664.

Molerus O. (1978), "Effect of Interparticle Cohesive Forces on Flow Behavior of Powders," Powder Technology, 20: 161.

Nemat-Nasser S. and Zhang, J. (2002), "Constitutive relations for cohesionless frictional granular material," International Journal of Plasticity, 18, 531-547.

Perko, H., Nelson, J. and Sadeh, W. (2001), "Surface cleanliness effect on lunar soil shear strength," Journal Geotechnical Engineering, ASCE, Vol. 127, No. 4, 371-383.

Rothenburg, L., Selvadurai, A. P. S. (1981), "Micromechanical definitions of the Cauchy stress tensor for particular media," Mechanics of Structured Media, (Selvadurai, A. P. S., Eds.), 469-486. Amsterdam: Elsevier.

Rowe, P. W. (1962), "The stress-dilatancy relations for static equilibrium of an assembly of particles in contact," Proc. Roy. Soc. London A 269, 500-527.

Scott, R.F., (1973), "Lunar soil mechanics." Proc. 8th Int. Congr. On Soil Mech. and Found. Engrg., Moscow, pp. 177-190.

Suiker, A.S.J., and Chang C.S. (2003), "Modelling failure and deformation of an assembly of spheres with frictional contacts," Journal of Engineering Mechanics, ASCE, (Accepted).

Taylor, D. W., (1948), Fundamentals of soil mechanics. J. Wiley and Sons, New York, N.Y.

Tejchman, J. and Bauer, E. (2005), "Modeling of a cyclic plane strain compression-extension test in granular bodies within a polar hypoplasticity," Granular Matter, v 7, n 4, p. 227-242.

Thornton C, Ning Z. M. (1998), "A theoretical model for the stick/bounce behaviour of adhesive, elastic-plastic spheres," Powder Technology 99 (2): 154-162.

Tomas J. (2001), "Assessment of mechanical properties of cohesive particulate solids. Part 1: Particle contact constitutive model," Particulate Science and Technology, 19 (2): 95-110.

Valverde J.M., Castellanos A., Watson P.K. (2001), "The effect of particle size on interparticle adhesive forces for small loads," Powder Technology 18 (3): 236-241.

Wallton, K. (1987), "The effective elastic moduli of a random packing of spheres," Journal of Mechanical and Physical Solids, 35, 213-226.

42

Geotechnical Engineering Properties of Lunar Soil Simulants

Haydar Arslan

Laboratory of Atmosphere and Space Physics
University of Colorado, Boulder

Susan Batiste

Laboratory of Atmosphere and Space Physics
University of Colorado, Boulder

Stein Sture

Vice Chancellor and Dean of Graduate School
University of Colorado, Boulder

ABSTRACT   The goals of lunar soil mechanics studies are both to improve scientific knowledge about the properties of lunar soil and to provide the engineering knowledge needed to plan and perform lunar surface activities. To enhance the scientific understanding of the behavior, and the mechanisms responsible for shear failure of lunar soil, geotechnical properties of different simulants have been evaluated. Mechanical and engineering properties of a simulant UNB-AN-1 were determined experimentally. Conventional tri-axial compression experiments were conducted on UNB-AN-1 at a density of 1.7 g/cc which is close to the average density of the upper few centimeters of the lunar surface. The shear strength dependence on grain size distribution of simulants has been analyzed experimentally. The small amount of difference in grain size distribution results in different shear strength properties for the simulants. The effect of particle shape on the behavior of lunar soil was simulated with an illustrative microstructure deformation mechanism for angular materials as lunar soil. The effect of particle size and particle shape is more dominant at low confinement. Thus, more consideration is needed for the microstructural properties of lunar soil.

Introduction

Investigation of mechanical properties of lunar soil is needed for the landing of spacecraft and to understand engineering properties for future operations including mobility, construction, mining and foundation design. Results of the investigation can be used to assess hazards such as landslides and compressible soils for future landing site selection and to provide engineering input for the design of Moon landing vehicles, lunar rovers, sampling devices, and other equipment. The rovers planned for future missions have increased in capability to travel further from the landing site and into more diverse terrain. The evaluation of soil mechanical properties of regolith simulant is of increased importance for these long distance or duration excursions.

Lunar soil mechanics studies started in 1964 by Surveyor and Apollo missions. Scott (1965) presented a report on lunar soil properties. Prior to 1969, when 13 kg of regolith fines were returned on Apollo 11, no lunar regolith was available to model a simulant after. Apollo missions brought back a total of approximately 115 kg of regolith by the end of 1972, and Luna missions brought back 321 g between 1970 and 1976. While these samples remain extremely significant, this is not sufficient for performing complete soil mechanics studies. Instead, the returned samples were used to evaluate physical, chemical, and limited geotechnical properties. That information was then used to select terrestrial soils, which would sufficiently mimic the lunar regolith.

Lunar soil properties such as density, friction angle cohesion are published by Christensen et al. (1967), Jaffe (1969) and Scott (1969). Costes et al. (1970) and Costes and Mitchell (1970) studied the resistance of lunar soil to penetration for different densities. Carrier et al. (1991) studied density measurements and settlement of lunar soil. Klosky (1997) used Minnesota Lunar Simulant (MLS-1) to study a comprehensive geotechnical engineering investigation about lunar soil. Willman et al. (1995) studied grain size distribution of JSC-1. Klosky et al. (2000), Arslan et al. (2008) investigated some geotechnical engineering properties of JSC-1.

Engineering Properties of Different Simulants

JSC-1 was the one of the first lunar soil simulants produced in 1980s. It was produced to a better understanding on future human activities such as material handling, construction, excavation, and transportation on the Moon. The simulant is available in large quantities to any qualified investigator. JSC-1 was mined from a volcanic ash deposit located in the San Francisco volcano field near Flagstaff, AZ. One basalt flow from a nearby vent has a K-Ar age of 0.15 ± 0.03 million years. The source quarry is within an area mapped as "slightly porphyritic basalt" (Moore and Wolfe, 1987). Lunar simulant MLS-1 is produced by the University of Minnesota (Weiblen et al., 1990). MLS-1 is derived from a high-titanium basalt hornfels which approximates the chemical composition of Apollo 11 soil. The material source for MLS-1 was crystalline. Table 42.1 summarizes engineering properties of both simulants and lunar soil studied by different researchers.

TABLE 42.1

Summary of Geotechnical Engineering Properties of JSC-1, MLS-1 Simulants and Lunar Soil

	MLS-1	JSC-1	Lunar Regolith
Passing #200	43% [Perkins and Madson, 1996]	36% [Perkins and Madson, 1996]	≈52% [from Carrier, 2003]
Cu	16 [Perkins and Madson, 1996]	7.5 [Perkins and Madson, 1996]	16 [Carrier, 2003]
Cc	1.1 [Perkins and Madson, 1996]	1.12 [Perkins and Madson, 1996]	1.2 [Carrier, 2003]
D_{50}	≈0.095 mm [from Perkins and Madson, 1996]	≈0.11 mm [from Perkins and Madson, 1996]	0.072 mm [Carrier, 2003]
Specific Gravity	3.2 [McKay et al., 1994]	2.91 [Willman et al., 1995]	2.9–3.2, 3.1 recommended [Carrier et al., 1991]
Void Ratio max	$e_{max} = 1.05$ [Perkins and Madson, 1996] $\rho_{min} = 1.56$ g/cc	$e_{max} = 1.18$ [Perkins and Madson, 1996] [Klosky, 2000] $\rho_{min} = 1.33$ g/cc $\rho_{min} = 1.43$ g/cc	Apollo 11 [Cremers et al., 1970] 1.39 1.26 Apollo 12 [Jaffe, 1971] 1.15 Apollo 14 [Carrier et al., 1973] 2.26–2.37 0.87–0.89 Apollo 15 [Carrier et al., 1973] 1.94 1.10
Void Ratio min	$e_{min} = 0.45$ [Perkins and Madson, 1996] $\rho_{max} = 2.20$ g/cc	$e_{min} = 0.61$ [Perkins and Madson, 1996] $\rho_{max} = 1.80$ g/cc [Klosky, 2000] 1.83 g/cc	Apollo 11 [Costes et al., 1970] 0.67 1.80 Apollo 12 [Jaffe, 1971] 1.93 Apollo 14 [Carrier et al., 1973] 0.87–0.94 1.55–1.51 Apollo 15 [Carrier et al., 1973] 0.71
Elongation		1.69 [Willman et al., 1995]	1.31–1.39 [Mahmood, 1974]
Aspect Ratio		0.68 [Willman et al., 1995]	0.4–0.7 [Gorz, 1972]
Glasses	No	Yes	Yes
Agglutinates	No	No	Yes

(Continued)

TABLE 42.1 *(Continued)*

	MLS-1	JSC-1	Lunar Regolith
Shear Strength	φ = 58° for c=0 at p=10kPa [Perkins and Madson, 1996]	φ = 64° for c=0 at p=10kPa [Perkins and Madson, 1996]	φ = 30-50° c=0.1-1.0 kPa [Mitchell, 1974]
Residual Strength	44° [Perkins and Madson, 1996]	42° [Perkins and Madson, 1996]	
E, MPa Young's modulus	[Perkins, 1991] Dr=37% E= 4.60 MPa Dr=66% E= 7.99 MPa Dr=97% E= 7.92 MPa Increases with mean stress Slightly higher than JSC-1 values	[Klosky, 2000] Dr= 40%, E=18-60 MPa Dr=60%, E=65-110 MPa Increases with mean stress Slightly lower than MLS-1 values	
K, MPa, Bulk Modulus	[Perkins, 1991] Dr=37% K=9.63 MPa Dr=66% K=7.69 MPa Dr=97% K= 12.1 MPa Increases with mean stress Slightly lower than JSC-1 values	[Klosky, 2000] Dr= 40%, K=35-60 Dr=60%, K=75-110 Increases with mean stress Slightly higher than MLS-1 values	

Lunar Regolith — Shear Strength:

	ρ, g/cc	φ, deg	c, kPa	[Carrier et al., 1991]: Depth, cm	φ, deg	c, kPa	
McKay, 1994	1.50, 1.60, 1.65	45.0°	≤ 1.0	0-15	42°	0.52	
Perkins, 1991	1.9	49°	0.2	0-30	46°	0.90	
Carrier, 1991		52-55°	2.4-3.8	30-60	54°	3.0	
Klosky, 1996	1.62	44.4°	3.9	0-60	49°	1.6	
Klosky 1996	1.72	52.7°	13.4				

Lunar Regolith — Dilatancy:

	ρ, g/cc	Dil. Angle, deg	Conf. Stress, kPa
Klosky 1996	1.62	44.0°	1 kPa
		40.5°	10 kPa
Klosky 1996	1.72	65.0°	10 kPa

Geotechnical Engineering Properties of a Simulant UNB-AN-1

The major component of soil strength and deformability is derived from interparticle friction at particle contacts. These properties are highly dependent on the size, texture, and orientation of contacts. The particle size distribution in an unconsolidated material, such as lunar soil, is a variable that controls to various degrees the strength and compressibility of the material, as well as its optical, thermal, and seismic properties. Grain size analysis is an integral part of soil classification in that it enables determination of the exact composition and distribution of particles sizes. Grain size distribution consists of separating size classes by sieving for coarse-grained particles and by using the hydrometer for fine-grained particles. Designation D422 (ASTM 1995) covers the sieve and hydrometer procedures required for particle analysis, and dry sample preparation should be used. Particle size distribution consists of quantifying the size classes by percentages based on weight from the graphical results of the grain size analysis. The weight percentages are calculated for those particles passing (or finer than) designated sieve sizes and hydrometer results and displayed as gravel percentage; coarse, medium, and fine sand percentages; silt percentage; and clay percentage.

Grain size distribution of a simulant UNB-AN-1 was studied to compare this simulant with others and lunar soil. Figure 42.1 shows results and compares with grain size distribution of lunar soil regolith as reported by

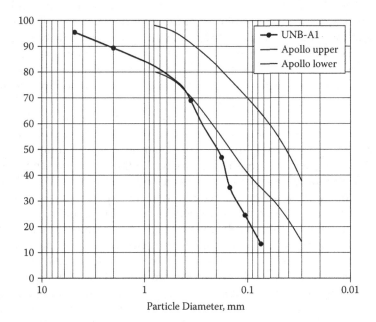

FIGURE 42.1
Grain size distribution of UNB-AN-1.

Carrier et al. (1973), illustrated with upper and lower curves representing the range of grain size distributions found among the tested return samples. As can be seen, UNB-AN-1 is significantly more coarse-grained than and not as well-graded as a typical lunar soil.

Strength and volume change characteristics are key to understanding and predicting the mechanical behavior of the regolith simulant. The conventional triaxial compression test was used to evaluate the shear behavior of UNB-AN-1. There is a breadth of techniques (and variations thereof) for sample preparation in the laboratory, and also no standardization in the terminology (i.e. similar techniques appear with different names). Common methods are using dry material, like the dry deposition method, which has the benefit of forming uniform samples by adjusting the height of pluviation, while some variations of the method include additional compaction (e.g. Yoshimine et al. 1998). If compaction of the sample is performed by use of a thin rod, then the gravitational (transverse isotropic) fabric of the soil is presumably destroyed leading to an isotropic fabric (Oda, 1981). On the other hand, tapping on the mold is thought to destroy less of the gravitational fabric leading to intermediate levels of anisotropy (Miura and Toki 1982).

The triaxial test system, developed by Alshibli (1995), consisted of a standard Brainard-Kilman triaxial test cell, a standard loading frame, a fluid system comprised of pipes, connections and pressure reservoirs, controlled by air pressure regulators, and a control and data collection system. A Brainard Kilman 5600 Load Frame was used to compress the specimens, and a 5600 Electronics Module provided a user interface to control the displacement distance and rate. A linear voltage differential transducer (LVDT) hardmounted on the load frame recorded the axial displacement of the top test cell endcap. The confining pressure was controlled by an air regulator connected to a house air pressure supply, using a dial gauge for first order approximation setting. The regulated air was then routed to a sealed cylindrical Plexiglas pressure reservoir partially filled with water, which was connected by plastic pipes to a differential pressure transducer, with one side open to the atmosphere. This was used for fine pressure reading and was connected to the test cell which provided confining pressure to the specimen. The reservoir was divided into two sections, each connected to the test cell line, but with valves to open or close the connection. During experimentation, only one section of the reservoir remained open to the test cell, and allowed water to be provided to or collected from the cell as the volume decreased. It also allowed the change in pressure between the two sides to be measured, and thus the change in height, which was then converted to volume, of the reservoir side open to the test cell. As the system worked by allowing the height of water in the reservoir to change in order to measure volume changes, the system did allow the confining pressure in the test cell to increase or decrease slightly (up to 1.75 kPa, or 0.25 psi) as the volume of the specimen increases or decreases, respectively.

Specimens tested were cylindrical in shape, 74.6 mm in diameter by 150 mm. In order to prepare a sand specimen for testing, a latex membrane was first folded around a six-piece mold with internal dimensions the size of the sample as stated above, with a small vacuum placed on the mold to assure the latex membrane conformed with the mold. This was then secured onto the bottom platen of the test cell (Figure 42.2a). The rest of the test cell is assembled (cylindrical water jacket, tie rods, upper plate) and the space between the specimen and the water jacket is filled with water and pressurized. The vacuum internal to the specimen is then removed (pore space is left open to the atmosphere, at gauge pressure), with the surrounding water left to support the specimen. The test cell is then placed on the load frame, connected with measurement apparatuses and the experiment is started (Figure 42.2b).

A latex membrane is used to contain the specimen, providing a barrier between the sand and the confining water. Initially, the membrane is unstretched, but as the specimen is compressed and expanded radially the membrane is deformed and therefore it applied additional confinement on the specimen. In order to account for this effect, the stiffness of the membrane must be investigated. Results of this stiffness are used to calculate the induced radial confining pressure on the specimen.

 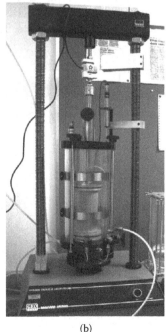

(a)   (b)

FIGURE 42.2
Triaxial test sample preparation.

Triaxial Test Results

Stress-strain and volumetric behavior of the UNB-AN-1 samples under three different confinements are given in Figure 42.3 and Figure 42.4. The stress ratio data will be used to find shear strength parameters cohesion and friction as will be given later. From these data, it is determined that the average internal friction angle of the UNB-AN-1 and the average cohesion is 1.4 kPa and the friction angle is 41.2° for the density of 1.7g/cm^3.

As Figure 42.4 illustrates, UNB-AN-1 is highly dilatant. The simulant showed strong dilatancy after a small amount of initial elastic compression. It must be noted that strong sample compression accompanies unloading of the sample, indicating reverse dilatancy. This is a result of the higher angularity of the simulant.

The cohesion value, 1.4 kPa is a reasonable value for a lunar soil simulant (Perkins, 1991; McKay, 1994; Klosky, 1996) and indicates the particulates are probably sufficiently angular, thus the basic method used to create the material is probably acceptable. The friction angle is low however; a value between 45–55° would be preferred (Perkins, 1991; McKay, 1994; Klosky, 1996). This may be due to the coarser grain size distribution of the UNB-AN-1 simulant, as reported above.

It must be noted that high dilatancy is one of the dominant properties of the simulants and lunar soil too. This dilatancy is highly related with angularity of lunar soil and the simulants. The effect of angularity or asperities can be

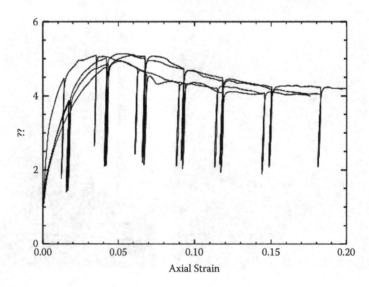

FIGURE 42.3
Behavior of stress ratio as a function of axial strain for density of 1.7 g/cc at 30 kPa confining stress.

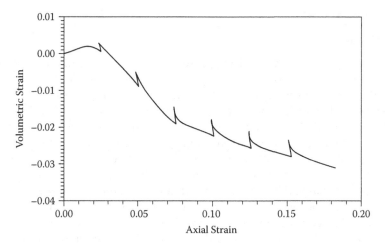

FIGURE 42.4
Volumetric behavior of UNB-AN-1 for density of 1.7 g/cc at 30 kPa confining stress.

explained with a block representation of granular materials. As Figure 42.5 illustrates, higher angularities (·) results in more vertical deformation (dh) due to horizontal or shear deformation (dx). This dilatancy will be a more dominant behavior under low confinement in a lunar environment due to low gravity.

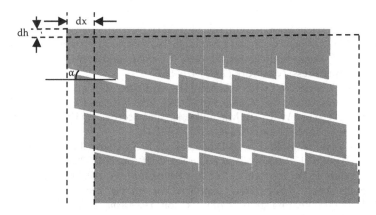

FIGURE 42.5
Representation of angularity-dilatancy relation.

Conclusion

Engineering and mechanical properties of different simulants have been presented and discussed. A sample preparation technique for triaxial testing and effect of grain size distribution on engineering properties of UNB-AN-1 simulant has been explained. The grain size distribution of the simulant was slightly out of the upper and lower bound range proposed for lunar soil. This leads to different shear strength properties. Based on experiments conducted for this study and previous studies, the size and shape of lunar soil are perhaps the most important factors affecting the strength of granular soils. The mechanical behavior of lunar soil is dependent on the properties of the grains from which they are constituted. As was shown experimentally, the dilative behavior of the lunar soil simulants is significant under shear loading. High angularity of the particles results in a dilative response of lunar soil which is more dominant under low confinement.

References

Alshibli, K.A. (1995). Localized Deformations in Granular Materials. Ph.D. dissertation, University of Colorado at Boulder.

Annual Book of ASTM Standards, Soil and Rock, American Society for Testing and Materials, ASTM, Philadelphia, PA (1995).

Arslan, H., Sture, S., Batiste, S. (2008) Experimental simulation of tensile behavior of lunar soil simulant JSC-1, *Materials Science and Engineering* A volume 478, pp. 201–207.

Carrier W. D. III, Mitchell J. K., and Mahmood A. (1973a) The nature of lunar soil. *J. Soil Mech. Found. Div, Am. Soc. Civ. Eng.*, 99, 813–832.

Carrier W. D. III, Mitchell J. K., and Mahmood A. (1973b) The relative density of lunar soil. *Proc. Lunar Sci. Conf. 4th*, pp. 2403–2411.

Carrier, W. D., Olhoeft, G. R., and Mendell, W., "Physical Properties of the Lunar Surface," Lunar Sourcebook, Heiken, Vaniman, and French, eds., Cambridge University Press, New York (1991) 475–567.

Costes N. C., Carrier W. D. III, Mitchell J. K., and Scott R. F. (1970a) Apollo 11: Soil mechanics results. *J. Soil Mech. Found. Div., Am. Soc. Civ. Eng.*, 96, 2045–2080.

Costes N. C., Carrier W. D. III, Mitchell J. K., and Scott R. F. (1970b) Apollo 11 soil mechanics investigation. *Science*, 167, 739–741.

Cremers C. J., Birkebak R. C., and Dawson J. P. (1970) Thermal conductivity of fines from Apollo 11. *Proc. Apollo 11 Lunar Sci. Conf.*, pp. 2045–2050.

Görz H., White E. W., Johnson G. G., and Pearson M. W. (1972) CESEMI studies of Apollo 14 and 15 fines. *Proc. Lunar Sci. Conf. 3rd*, pp. 3195–3200.

Jaffe L. D. (1971a) Bearing strength of lunar soil. *The Moon*, 3, 337–345.

Jaffe L.D., Recent Observations of the Moon by Spacecraft, Space Science Reviews, V. 10 N.4 (1969) pp. 491–616.

Klosky, J.L., S. Sture, H.-Y. Ko, F. Barnes (1996) "Mechanical Properties of JSC-1 Lunar Regolith Simulant", Proceedings of Space V: Engineering, Construction and Operations in Space '96, Vol 1, pp. 680–688 ASCE.

Klosky, J.L, "Behavior of Composite Granular Materials and Vibratory Helical Anchors", Ph.D. Dissertation, University of Colorado at Boulder (1997).

Klosky, J.L., S. Sture, H.-Y. Ko, F. Barnes (2000) "Geotechnical Behavior of JSC-1 Lunar Regolith Simulant", *Journal of Aerospace Engineering*, ASCE 13:4. pp. 680–688.

Mahmood A., Mitchell J. K., and Carrier W. D. III (1974) Particle shapes of three lunar soil samples. Unpublished report, available from W. D. Carrier III.

McKay, D.S., J.L. Carter, W.W. Boles, C.C. Allen, J.H. Allton (1994). "JSC-1: A New Lunar Soil Simulant", Proceedings of Space IV: Engineering, Construction and Operations in Space, V. 2, pp. 857–866.

Miura, S. and Toki, S. (1982): A simple preparation method and its effect on static and cyclic deformation-strength properties of sand, *Soils and Foundations*, 22(1), 61–77.

Mitchell J. K., Houston W. N., Carrier W. D. III, and Costes N. C. (1974) *Apollo Soil Mechanics Experiment S-200*. Final report, NASA Contract NAS 9–11266, Space Sciences Laboratory Series 15, Issue 7, Univ. of California, Berkeley.

Miura, S. and Toki, S. (1982): A simple preparation method and its effect on static and cyclic deformation-strength properties of sand, *Soils and Foundations*, 22(1), 61–77.

Oda, M. (1981): Anisotropic strength of cohesionless sands, *Journal of the Geotechnical Engineering Division*, ASCE, 107(9), 1219–1231.

Perkins, S.W. (1991) Modeling of Regolith Structure Interaction in Extraterrestrial Constructed Facilities, Ph.D. thesis, University of Colorado, Boulder.

Perkins, S.W., C.R. Madson (1996) "Mechanical and Load-Settlement Characteristics of Two Lunar Soil Simulants", *Journal of Aerospace Engineering*, ASCE 9:1, 1–9.

Scott R.F. Lunar Problems in Soil Engineering, ASCE J. Soil Mech. And Fnd. Div. Vol. 91 (1965) pp. 1–14.

Weiblen P.W., Murawa M.J., and Reid K.J. (1990) Preparation of simulants for lunar surface materials. *Engineering, Construction and Operations in Space II*, American Society of Civil Engineers, New York, pp. 428–435.

Willman, B.M., W.W. Boles, D.S. McKay, C.C. Allen (1995). "Properties of Lunar Soil Simulant JSC-1" ASCE *J. of Aerospace Engineering*, 8:2, 77–87.

Yoshimine, M., Ishihara, K. and Vargas, W. (1998): Effects of principal stress direction and intermediate principal stress on undrained shear behavior of sand, *Soils and Foundations*, 38(3), 179–188.

43

Tension of Terrestrial Excavation Mechanics to Lunar Soil

Jason R. Florek

Baker Engineering and Risk Consultants, Inc.
Washington, DC

Jason R. Florek received his doctorate in mechanical engineering in 2007 from Rutgers University, where his research focused on the large deformation of thin-walled structures to both uniform and non-uniform blast loading. He is a member of ASME and is currently working as a project consultant for Baker Engineering and Risk Consultants, Inc. (BakerRisk) in their Washington, D.C. office.

ABSTRACT   This paper focuses on extending models for cutting terrestrial soil for use with lunar soil. This area is of particular importance since nearly all lunar base designs call for some form of regolith shielding. It has been shown that Earth-based analyses do not directly translate for use with lunar soil. Required forces do not just simply scale by one-sixth due to the reduced lunar gravity. Papers that make similar extensions typically assume a single value for the important lunar soil properties. Most studies do not take into account the variability of these parameters with depth or the uncertainty associated with the property values. In contrast, those studies that account for this variability tend to ignore how the parameters relate with one another.

As such, here, appropriate ranges of parameter values are considered for input to various two-dimensional excavation models. Parameter dependencies are also accounted for, resulting in a more realistic calculation of the forces required to cut and move lunar soil. Comparisons are made between these forces and those required to move lunar simulants in 1/6 g and 1 g environments. Additionally, recommendations are made for which contributing factors can be ignored in a simplified analysis.

Introduction

Modeling the cutting and moving of lunar soil is of particular design importance for future habitation of the Moon. Nearly all lunar base designs call for some form of regolith shielding for protection against radiation and micrometeorite impact [1]. Additionally, digging regolith is critical to activities such as paving roads and collecting natural resources. Before any of these projects can begin, the forces associated with digging and excavating regolith must be well understood so as to minimize the associated power and weight costs.

It has been shown that Earth-based analyses do not directly translate for use with lunar soil. Ettouney and Benaroya [2] outlined numerous key differences associated with reduced gravity, including reduced stiffness, wave propagation speed, damping and relative bearing capacity for lunar soils as compared to their analogs on Earth. As experimentally shown by Boles et al. [3], required forces do not just simply scale by one-sixth due to the reduced lunar gravity. Such a scaling would only give a lower bound on these forces, with 1 g results on Earth providing an upper bound. Meanwhile, Willman and Boles [4] extended four terrestrial excavation models for use on the lunar simulant MLS-1. They found all the models to be inaccurate in predicting experimentally-obtained results. However, only single values for important soil properties (e.g., cohesion and internal friction angle) were assumed.

Like other studies (e.g., Ref. [5]), Willman and Boles [4] did not take into account the variability of these parameters with depth or the uncertainty associated with the property values. In contrast, Wilkinson and DeGennaro [6] accounted for this variability in their parameter study using the two- and three-dimensional excavation models previously examined in Ref. [7]. Blouin et al. [7] considered four different two-dimensional cutting models, namely that of Gill and Vanden Berg [8], McKyes [9], Swick and Perumpral [10], and Osman [11]. More recently, Wilkinson and DeGennaro [6] examined these four models along with an empirical model used to design the excavation arm of the Mars Viking lander [12]. While Refs. [6, 7] both provide equations for horizontal drawbar for each model in terms of soil and tool parameters, Wilkinson and DeGennaro [6] distinguish themselves by performing a parameter study, plotting drawbar force as a function of one variable parameter while keeping all others constant, to determine which terms have the largest effect on drawbar. These plots allow for a more direct comparison between individual models than by merely presenting the relevant equations. However, these plots ignore how certain parameters physically relate with one another. For example, soil density is a function of depth [13], while failure surface angle is a function of the rake and friction angles [14]. Wilkinson and DeGennaro [6] made these quantities essentially independent of one another.

The current paper extends the comparisons provided by Wilkinson and DeGennaro [6]. First, their treatment of linear failure models is summarized, considering only tool depth or rake angle variable. Then, Osman's model [11], which is given only a superficial treatment in Ref. [6] due to its relative complexity, is thoroughly examined. The forces predicted by each of these models are compared in terms of moving lunar soil in a 1/6 g environment or simulant MLS-1 on Earth. Finally, the models of Gill and Vanden Berg [8] and McKyes [9] are modified so as to account for a density varying with depth. Cohesion and internal friction angle are also considered to be depth-dependent per collected lunar soil data [15]. This interdependence allows for a more realistic parameter study. Recommendations are made for which contributing factors (e.g., adhesion, surcharge) can be ignored in a simplified analysis.

It should be noted that Refs. [10, 12], both considered by Wilkinson and DeGennaro [6], are not developed here. In Ref. [6], it is shown that Swick and Perumpral's model [10] yields matching force values to that of McKyes [9] despite having different formulations. Moreover, Swick and Perumpral's general equation also appears in McKyes' later text [14]. To avoid redundancy, only McKyes' formulation [9] is examined here. In addition, the Lockheed-Martin/Viking model [12] is ignored here due to the non-physical nature of its empirically-derived terms. It is uncertain how modifying certain terms of Ref. [12] will affect the validity of the model.

Penetration and Excavation Models

Blouin et al. [7] list the three basic earthmoving processes as penetrating, cutting and excavating the soil. Figure 43.1 provides a clear representation of all the forces that come into play for the middle process assuming a linear failure model. Here, l and w are the length and width of the blade, d is the depth of the cut, v is the tool speed, T is the magnitude of the required tool force, to adhesion along the blade and cohesion along the failure plane. As different authors tend to use different notations for these dimensions, forces and angles, the current paper generally adopts those used in Refs. [6, 7] as per Fig. 43.1. Discrepancies with notations in the other papers can be resolved by consulting Table 43.1.

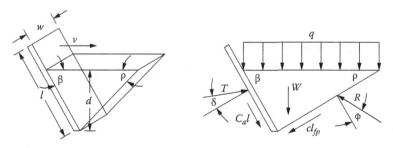

FIGURE 43.1
Parameter designations and applied forces for a linear failure model.

TABLE 43.1

Parameter Values and Notations from Various Sources

Parameter	Units	MLS-1	Min	Avg	Max				
Reference Number		[4]	[6]	[6]	[6]	[4]	[6]	[11]	[16]
Adhesion	N/m²	0.9	200	1930	5000	C_a	C_a	c_a	c_a
Cohesion	N/m²	900	68	170	4500	c	c	c	c
Density	kg/m³	1920	1200	1680	3500	γ/g	γ	—	γ/g
Failure Plane Angle	degrees	—	20	30	55	β	ρ	$45-\varphi/2$	β
Friction Angle (Ext.)	degrees	24	0	10	50	δ	δ	δ	δ
Friction Angle (Int.)	degrees	37	20	35	50	φ	φ	φ	φ
Rake Angle	degrees	60	5	45	90	α	β	α	ρ
Surcharge	kg/m²	0	0.5	1	100	q/g	q	q/g	q/g
Tool Depth	m	var	0.05	0.5	1	z	d	z	d
Tool Length	m	var	0.1	0.7	1	—	l	a	L_t
Tool Width	m	0.13	0.3	1	3	b	w	—	—

Lineal Failure Model

References [8, 9] assume a linear failure plane as shown in Fig. 43.1. Although this triangular shape diverges from the actual failure geometry, the procedure for determining required drawbar is much more efficient than for an assumed logarithmic spiral, which is discussed in the next section. Balancing the forces shown in Fig. 43.1 yields the general Fundamental Earthmoving Equation

$$F = (\gamma g d N_\gamma + c N_c + q N_q + C_a N_a) w d \qquad (43.1)$$

where the N factors are unique for each author as some include the effects of surcharge and adhesion, while others do not. The other parameters in Eq. 43.1 are as defined in Table 43.1.

Gill and Vanden Berg Model

Per Ref. [6], the horizontal drawbar of Gill and Vanden Berg [8] is of the general form

$$H^* = H - kw = T_{ts} + T_d + T_c + T_k \qquad (43.2)$$

where Tts accounts for tool-soil interactions, Td for changes in depth, Tc for soil cohesion, and Tk for kinetic effects. These terms are defined, respectively, by

$$T_{ts} = F_M l_{tg} \frac{sin(\beta + \rho)}{sin \rho}$$

$$T_d = \frac{1}{2} F_M d_{tg} \frac{sin(\beta + \rho)}{sin^2 \rho} [cos(\beta + \rho) + sin(\beta + \rho)tan \beta]$$

$$T_c = \frac{F_M c}{sin \rho(\rho + \phi \cos \rho)}$$

and

$$T_k = \frac{F_M \gamma v^2 sin \beta}{sin(\beta + \rho)sin(\rho + \phi \cos \rho)}$$

where

$$F_M = \frac{wd(sin \beta + \delta \cos \beta)sin(sin \rho + \phi \cos \rho)}{sin(\beta + \rho)(1 - \phi\delta) + cos(\beta + \rho)(\phi - \rho)}$$

and v is the tool speed. Wilkinson and DeGennaro [6] vary this speed between 0.01 and 0.3 m/s, with an average of 0.1 m/s. Ranges and averages for the other parameters in these expressions can be found in Table 43.1. Surcharge is not included in this model. It should be noted that, in contrast to other papers, Wilkinson and DeGennaro [6] do not use weight density in their equations, rather they substitute mass density multiplied by gravitational acceleration. Similarly, they use a mass surcharge in place of a weight surcharge when applicable. This may seem like a trivial substitution, but it makes the effect of changing gravity much clearer. This change is very important in applying earthmoving models for use on the Moon.

Meanwhile, the cutting resistance term *kw* in Eq. 2 is removed from the total horizontal force since, according to Blouin et al. [7], it is only important when an obstacle or significant wear are present. Actually, Wilkinson and DeGennaro [6] retain this term. The effects of doing so are shown in Fig. 43.2. At shallow depths, the constant cutting force is the most significant of all the terms in Eq. 43.2. As depth increases, its contribution lessens in favor of the depth (weight) and tool-soil terms. Similar trends are seen when a parameter other than depth is varied. The fact that this cutting resistance is constant for all depths, as well as rake and friction angles, makes it seem rather arbitrary. It appears that the term, not present in any of the other models in Refs. [6, 7], is artificially increasing the drawbar. As such, it should be ignored in a simplified analysis.

Furthermore, the cohesion term in Eq. 43.2 only becomes important for *c* values much greater than the average 170 N/m² given in Table 43.1. Also, the kinetic term seems to have very little effect in all cases. As a result, it appears both T_c and T_k can be ignored in a simplified analysis, making drawbar

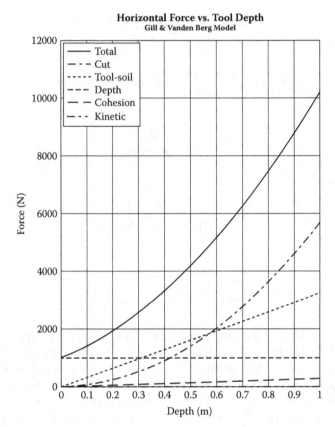

FIGURE 43.2
Horizontal drawbar force as a function of tool depth per Ref. [8].

completely dependent on soil weight and tool-soil interaction. This assertion though will be revisited when property interdependencies are considered.

Moreover, the Gill and Vanden Berg model [8] can be applied to the experimental set-up of Willman and Boles [4]. In their study, drawbar forces of 192 N, 522 N and 825 N were measured for three different cut depths into a container of MLS-1 simulant. These measurements did not compare well with four three-dimensional cutting models, one of which overpredicted the force by at least a factor of 2. Meanwhile, the other models all yielded similar results, but this time they underestimated the required force by at least a factor of 4. Substituting the average values for Ref. [4] from Table 43.1 into Eq. 43.2, the three forces are matched by the Gill and Vanden Berg model [8] when the shear plane failure angle (unspecified in Ref. [4]) takes on values of 10.4°, 10.6° and 11.5°, respectively. However, these angular values are well below the minimum 20° given in Ref. [6]. For more typical failure plane angles, the model of Ref. [8] would tend to overpredict the required forces.

This is not entirely surprising as the dominant depth and tool-soil terms in Eq. 43.2 are both functions of gravity. Using 1 g to replicate the experimental results of Ref. [4] then makes a simplified Gill and Vanden Berg model [8] an upper bound to the true required drawbar per Ref. [3].

McKyes Model

In contrast to Gill and Vanden Berg [8], McKyes [9] incorporates surcharge into his model so that

$$H = T_q + T_{ts} + T_d + T_c + T_k,$$ (43.3)

where

$$T_q = F_M q g (\cot \beta + \cot \rho)$$

$$T_{ts} = F_M C_\alpha [1 - \cot \beta \cot(\rho + \phi)]$$

$$T_d = \frac{1}{2} F_M d \gamma g (\cot \beta + \cot \rho)$$

$$T_c = c[1 + \cot \rho \cot(\rho + \phi)]$$

$$T_k = F_M \gamma v^2 \frac{\tan \rho + \cot(\rho + \phi)}{1 + \cot \beta \tan \rho}$$

and

$$F_M = \frac{wd}{\cos(\beta + \delta) + \sin(\beta + \delta)\cot(\rho + \phi)}$$

From Fig. 43.3a, it appears again that depth and tool-soil terms are most significant, with cohesion becoming more relevant as tool depth increases. Furthermore, it is apparent that the surcharge of 1 kg/m² from Table 43.1 is insufficient to produce a visible change in drawbar.

Comparing Figs. 43.2 and 43.3a, it is clear that Gill and Vanden Berg's model [8] predicts drawbar forces more than double that of McKyes' model [9]. Indeed, comparing all charted data in Ref. [6] for these two models, regardless of variable parameter, the former model predicts a horizontal force between two to four times larger than the latter model of McKyes [9]. The difference is most striking with regard to varying rake angle. While Gill and Vanden Berg [8] predict a drawbar monotonically increasing with β, due

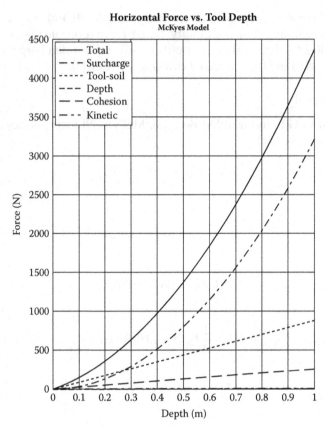

FIGURE 43.3a
Horizontal drawbar force as a function of a) tool depth and b) rake angle per Ref. [9].

to his formulation, McKyes' model [9] shows the behavior in Fig. 43.3b. For small rake angles, the tool-soil term actually subtracts from the other terms. Furthermore, the depth (weight) term shows a local minimum near $\beta = 30°$. As rake angle is typically greater than this value, these trends do not cause a serious problem.

Despite their differences in magnitude, the predictions of Refs. [8, 9] show similar trends. In a simplified analysis, it appears that cohesion, kinetic effects and surcharge can generally be ignored. However, unlike the model of Gill and Vanden Berg [8], McKyes' model [9] cannot be made to fit the experimental values from Ref. [4] by simply varying a quantity such as the failure plane angle as the predicted forces are too low. These lower forces arise due to the tool-soil term in Eq. 43.3 being related to adhesion, as opposed to weight as in Eq. 43.2. Note that the experimental adhesion of 0.9 N/m^2 is particularly low in comparison with the values of Ref. [6] in Table 43.1.

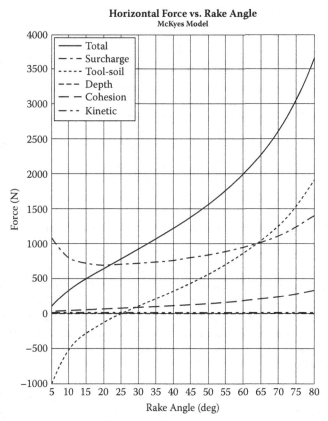

FIGURE 43.3b
Horizontal drawbar force as a function of a) tool depth and b) rake angle per Ref. [9].

Logarithmic Spiral Failure Model

Meanwhile, Osman's model [11] incorporates a logarithmic failure plane as shown in Fig. 43.4. This failure profile is formed by arbitrarily locating the logarithmic spiral center (on, with $n = 1, 2, 3$ in Fig. 43.4) on line ad, which makes an angle $\rho = 45 - \phi/2$ with the horizontal, where ϕ is the internal friction angle of the soil. The spiral sweeps an angle w′ from point b on the tip of the cutting tool to line ad per the equation

$$r_1 = r_0 e^{w' \tan \phi}$$

where r_0 is the distance from the spiral center to point b. By geometric construction, line *de* is tangent to spiral arc *bd*, also running at an angle ρ below the horizontal. Therefore, triangles *adf* and *edf* are mirror images of each other.

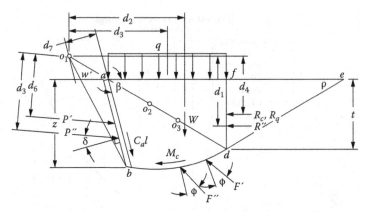

FIGURE 43.4
Parameter designations and applied forces for a logarithmic spiral failure model.

The necessary drawbar force for equilibrium is found per a moment balance of the various forces acting on wedge *abdf*, making use of moment arms d_n with respect to the chosen spiral center (o_1 in Fig. 43.4). The actual drawbar force for a given set of soil and tool parameters is the minimum value for all possible centers o_n, as it will be this force that initially fails the soil. The failure surface in Fig. 43.4 is very similar in shape to what would be seen experimentally.

However, the above procedure is relatively complicated and no explicit formulations exist. The seven distances *dn* must be calculated for each spiral center location until a minimum drawbar force is attained. As a result, Wilkinson and DeGennaro [6] only superficially examined this model. All *dn* are arbitrarily set to a unitless value of 1 in Ref. [6]. From Fig. 43.4, these distances clearly have a dimensional length and are unequal.

Moreover, Refs. [6, 7] both incorrectly substitute the failure plane angle (angle *def* in Fig. 43.4) for internal friction angle ϕ in their presented equations. Actually, such a substitution is not surprising since many papers use different notations for their parameters as shown in Table 43.1. The chance of a substitution error increases when parameter designations are interchanged such as rake angle and failure plane angle in Refs. [6, 16]. The former reference uses β and ρ, respectively, for these parameters, while Luengo et al. [16], who extended Eq. 43.1 for automatic excavations, reverse these designations.

The total drawbar acting on the cutting tool can be calculated by summing forces P' and P'' from Fig. 43.4. These forces are the cohesive and frictional components of the total force, respectively. Per a moment balance about the spiral center,

$$P' = \frac{w}{d_g}\left[\frac{c}{2\tan\phi}(r_1^2 - r_0^2) + 2ct\tan\left(45 + \frac{\phi}{2}\right)d_4 + qgt\tan\left(45 + \frac{\phi}{2}\right)d_5 + C_\alpha l d_\tau\right] \quad (43.4)$$

where t is the Rankine passive zone depth as depicted in Fig. 43.4 (see Ref. [17]). The first term is due to cohesion acting on arc bd (Mc in Fig. 43.4), the second due to cohesion in Rankine zone def (Rc in Fig. 43.4), the third due to surcharge on line af, and the fourth due to adhesion along the blade length. It should be noted that Refs. [6, 7] not only incorrectly substitute failure plane angle ρ for ϕ in Eq. 43.4 (ρ is actually equal to $45 - \phi/2$), but they also give the wrong expression for length af over which surcharge q acts ($t/\sin(45-\rho/2)$ as opposed to $t/\tan(45 - \phi/2)$). Art was here, but was deleted.

Meanwhile, per another moment balance,

$$P'' = \frac{w}{d_3}\left[\frac{1}{2}\gamma g t^2 \tan^2\left(45+\frac{\phi}{2}\right)d_1 + \gamma g A d_2 + \gamma g t \tan^2\left(45+\frac{\phi}{2}\right)d_4\right] \qquad (43.5)$$

where A is the area of wedge abdf. Both $\tan 2(45 + \phi/2)$ terms (R'' and Rq in Fig. 43.4) arise from a principal stress analysis via Mohr's circle [17]. Refs. [6, 7] incorrectly substitute this area for the area of the spiral sector obd, namely

$$A_{sector} = \frac{r_0^2}{4\tan\phi}[e^{2w'\tan\phi} - 1]$$

The total area A should consist of the sum of A_{sector} and the area of triangles adf and oab. For the case when the spiral center is below the surface ae (like o_2 in Fig. 43.4), these triangular areas add together. In contrast, when the center is above the surface (like o_1 in Fig. 43.4), areas adf and oab subtract to obtain the correct area abdf.

In order to calculate P' and P'', and therefore the total drawbar force, all seven distances d_n must be known. Wilkinson and DeGennaro [6] claim that these distances are indeterminate. In actuality, distances d_1, d_4 and d_5 can be readily evaluated from geometry, distances d_3, d_6 and d_7 from the elementary moment equation

$$|\vec{M}| = |\vec{r} \times \vec{F}| = |\vec{F}|d_\perp$$

The remaining distance d_2, which spans from the spiral center to the centroid of wedge abdf, is the most difficult distance to calculate. This wedge is a composite of three distinct areas as noted above. As such, the centroid of each of these areas must be found in order to determine the position of the wedge centroid. Conversely, a numerical scheme can be employed once the shape of the failure plane is known.

Again, each location of the spiral center yields unique values of the seven d_n. The actual drawbar is the one that minimizes the sum of Eqs. 43.4 and 43.5. The horizontal component of this drawbar is found by multiplying the result by $\sin(\beta + \delta)$.

However, there is a major problem with this procedure as described. Since the terms in Eqs. 43.4 and 43.5 arise from a moment balance about the spiral center, they do not inherently add together. For spiral center o1 in Fig. 43.4, the sign notations in these equations are indeed correct as only P′ and P″ produce counterclockwise moments. Yet, for any center located below the surface ae, the adhesion force along the tool length also produces a counterclockwise moment. For a center such as o3, the d4 and d5 terms also subtract from P′ and P″. Moreover, when the rake and external friction angles are small ($\beta + \delta < 90°$), the orientation of P′ and P″ is more akin to force T shown in Fig. 43.1, which may further alter the signs in Eqs. 43.4 and 43.5. This behavior is reminiscent to that shown in Fig. 43.3b, where the adhesion (tool-soil) term subtracts from the net drawbar at small rake angles. However, in contrast to McKyes' model [9], where the minimum total drawbar is zero, the end result for Osman's model [11] could be the prediction of a negative drawbar force, depending on the individual parameter values. To alleviate this problem, the above procedure is simplified as per Ref. [17]. Hence, the adhesion and surcharge terms in Eqs. 43.4 and 43.5 are ignored in the moment balance. After the minimum failing force is determined, the horizontal component of that force is added to the horizontal component of the adhesive force in order to obtain the required drawbar.

This simplified procedure was used to generate the curves in Fig. 43.5 for rake angles of 30° and 60°. These curves show the total predicted failing force, along with the contribution of the adhesive force to the total drawbar. It is clear that adhesion is dominant for the case of the smaller rake angle, which provides for a longer tool length for the adhesive force to act over at any given depth. However, when $\beta = 60°$, the total horizontal force is larger than for the shallower rake angle for cuts at least 35 cm deep. In that case, the weight term is larger and the moments in Eqs. 43.4 and 43.5 tend to be additive as per the discussion above. Fig. 43.5 also shows how these presently-generated curves relate to Wilkinson and DeGennaro's formulation [6] of the Osman model [11] when $\beta = 45°$. The Ref. [6] curve has a starting value of over 4000 N due to all distances dn being assumed equal to 1. Intuitively, a non-zero starting value does not make sense since all physically-acting moments approach zero as depth decreases. Still, the total force at a 60° rake angle approaches the overestimating curve at a depth of 1 m.

Using the Ref. [4] input from Table 43.1 (recalling that failure plane angle is predetermined as a function of internal friction angle), the simplified Osman model [11] can be used to predict the required forces to fail the stimulant MLS-1. Doing so, values of 60 N, 167 N and 298 N are respectively generated at depths of 4 cm, 7 cm and 9.5 cm. These values are roughly 1/3 of those obtained experimentally by Willman and Boles [4]. Like for McKyes' model [9], the low adhesion of 0.9 N/m2 makes Osman's otherwise important tool-soil term insignificant. From Fig. 43.5, a more typical adhesion value contributes to over half of the total drawbar for depths under 10 cm when $\beta = 60°$. Still, the vastly underpredicting results cited

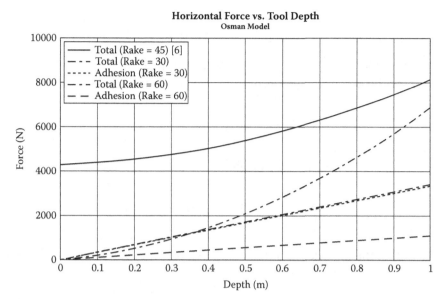

FIGURE 43.5
Horizontal drawbar force as a function of tool depth per Ref. [11].

above give a closer estimate than any of the three-dimensional predicting models considered in Ref. [4]. But due to the lack of experimental cor-
roboration, it appears that the use of neither Osman's model [11] nor even
more complicated three-dimensional models can be justified in the current
application.

Property Interdependencies

According to the models of Gill and Vanden Berg [8] and McKyes [9], the
forces due to the depth (weight) of the soil and the interaction between the
cutting tool and soil are the most critical in evaluating required drawbar
force. In their current state, both models assume a constant mass density.
However, per figures and charts in Refs. [13, 15], this assumption is far from
accurate for lunar soil. It has been shown that the density of lunar soil can be
well-represented by the following curve-fits:

$$\gamma(z) = 1.89\left(\frac{z + 1.69}{z + 2.9}\right)$$

from Ref. [13], and

$$\gamma(z) = 1.92\left(\frac{z + 12.2}{z + 18}\right)$$

from Ref. [15], where z is depth in cm and $\gamma(z)$ is in g/cm^3.

Taking the depth (weight) term from McKyes' model [9], namely

$$T_d = \frac{1}{2} F_M d\gamma g(\cot\beta + \cot\rho)$$

considering a variable density yields

$$T_d = F_M g \int_0^H \gamma(z)(H - z)dz(\cot\beta + \cot\rho)$$

where H is the total depth of the cut (previously d). Evaluating the integral with $\gamma(z)$ of the form $\gamma_{max}(z + c_1)/(z + c_2)$,

$$\int_0^H \gamma(z)(H - z)dz = \gamma_{max}\left[-\frac{1}{2}z^2 + (H + c_2 - c_1)z \right.$$
$$\left. + \frac{3}{2}c_2^2 + (H + c_2)(c_1 - c_2)\ln(z + c_2)\right]\Bigg|_{z=0}^{z=H}$$

With a constant density, Eq. 43.8 merely equates to $\gamma H^2/2$. As such, an effective average density can be calculated by comparing the result of Eq. 43.6 with $\gamma H^2/2$.

Figure 43.6 shows that using the average density of 1680 kg/m^3 from Table 43.1 overestimates the maximum density of Eq. 43.6 for about the first 8 cm, and increasingly underestimates the maximum for deeper cuts. This threshold depth is roughly 35 cm for the density of Eq. 43.7. In contrast, considering an effective average density, Eq. 43.6 does not achieve 1680 kg/m^3 until cuts past 40 cm. Equation 43.7 does not reach this value until about 1.3 m. From Fig. 43.6, it appears that the average density should be closer to 1400 kg/m^3 for cuts less than 10 cm deep, while closer to 1500 kg/m^3 for cut depths near 20 cm. Therefore, depending on how deep the desired cut depth, the given predictive models may need to be modified to account for variable density. Otherwise, large deviations can arise.

Similarly, cohesion and internal friction angle have been shown to depend on the lunar soil's relative density [13, 15]. This relative density is itself a function of bulk density and, therefore, also of depth. Figure 43.6 shows values of cohesion obtained from plotted data in Refs. [13, 15] that correspond to the effective average densities per Eqs. 43.6 and 43.7. The average cohesion of 900 N/m^2 from Table 43.1 overestimates these curves up until depths of 22 cm and 63 cm,

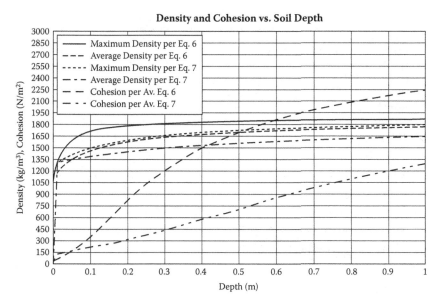

FIGURE 43.6
Density and cohesion as a function of lunar soil depth.

respectively. An underestimation occurs thereafter, particularly for the Eq. 43.6 case. Cohesion appears to be highly sensitive to density as large deviations exist between the two cohesion curves in Fig. 43.6 (roughly a factor of 2-3 for cuts deeper than 10 cm) despite the corresponding densities differing by only about 10% over the same depth range. Furthermore, the failure plane angle is not arbitrary. It is predetermined in Osman's model [11], while it is a function of rake and friction angles in the other considered models [14]. The parameter study of Wilkinson and DeGennaro [6] does not take into account any of these interdependencies.

Figure 43.7 compares the baseline depth-varying curves from Ref. [6] for the models of Gill and Vanden Berg [8] and McKyes [9] with alternate curves that consider density, cohesion and internal friction angle all functions of depth. The relation between average density, cohesion and depth for the alternate curves is as shown in Fig. 43.6. Internal friction angle, meanwhile, is varied per plotted data in Refs. [13, 15] with values ranging between 35° at zero depth to 52° at a depth of 1 m for the effective average density distribution calculated from Eq. 43.6.

All six curves in Fig. 43.7 are relatively close to one another for tool depths up to 20 cm. This is with the cutting force term removed from the Gill and Vanden Berg curves as per Eq. 43.2. For deeper cuts, the differences in required drawbar become evident as density and cohesion drastically increase. Indeed, at a depth of 1 m, the highest and lowest curves in Fig. 43.7 deviate by over a factor of 3. Moreover, using the density of Eq. 43.6 with the noted property

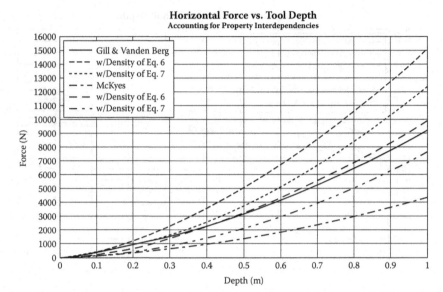

FIGURE 43.7
Horizontal drawbar force as a function of tool depth accounting for property interdependencies.

interdependencies, the McKyes' model [9] actually predicts a higher drawbar than the standard Gill and Vanden Berg model [8] for tool depths greater than 45 cm. This behavior can be explained as follows. For shallow cuts, the tool-soil term dominates for all models. At a depth of 20 cm, the depth (weight) term becomes significant for all models, eventually becoming the most dominate as depth increases. By 20 cm, the cohesion term is also significant for all models, but much more so for the McKyes' formulations [9], where it eventually overtakes the tool-soil term as the second-largest contributor to drawbar. Actually, with cohesion considered variable with depth, this contribution ranges between 20–40% of the total drawbar of the McKyes' model [9] regardless of depth or density distribution used. It appears then that for cuts less than 20 cm, a simplified model that includes only weight and tool-soil terms may be adequate using a constant density value. For deeper cuts, a separate cohesion term should also be included in the analysis, along with density and cohesion values that are functions of depth.

Conclusions

For establishment of a lunar base to be successful, the properties of regolith and forces associated with moving the lunar soil must be well understood. Different published models show a wide range of calculated drawbar forces and relative complexities. While some three-dimensional formulations correlate poorly with limited experimental results, certain two-dimensional models compare surprisingly well. It appears that from the better-correlating models, soil weight and tool-soil interactions are most important for the design of first-generation cutting machines, where cutting depths may not be particularly deep. Gill and Vanden Berg's model [8] may be used to produce a conservative estimate of required drawbar in this situation. For later designs, where cuts will need to be deeper than 20 cm, soil cohesion should also be considered a major factor. Additionally, density and cohesion should both be considered variable with depth. Experimental testing of lunar soil samples obtained from initial landing missions should be used to validate and refine these assertions, and to possibly generate other important interdependencies.

References

1. P. Eckart. *The Lunar Base Handbook*. The McGraw-Hill, New York, NY, second edition, 2006.
2. M. E. Ettouney and H. Benaroya. Regolith mechanics, dynamics, and foundations. *J. Aerospace Engng.*, 5(2):214–229, 1992.
3. W. W. Boles, W. D. Scott, and J. F. Connolly. Excavation forces in reduced gravity environment. *J. Aerospace Engng.*, 10(2):99–103, 1997.
4. B. M. Willman and W. W. Boles. Soil-tool interaction theories as they apply to lunar soil stimulant. *J. Aerospace Engng.*, 8(2):88–99, 1995.
5. L. E. Bernold. Experimental studies on mechanics of lunar excavation. *J. Aerospace Engng.*, 4(1):9–22, 1991.
6. A. Wilkinson and A. DeGennaro. Digging and pushing lunar regolith: Classical soil mechanics and the forces needed for excavation and traction. *J. Terramech.*, 44:133–152, 2007.
7. S. Blouin, A. Hemami, and M. Lipsett. Review of resistive force models for earthmoving processes. *J. Aerospace Engng.*, 14(3):102–111, 2001.
8. W. R. Gill and G. E. Vanden Berg. Agricultural Handbook 316: Soil Dynamics in Tillage and Traction. Agricultural Research Service, U.S. Department of Agriculture, Washington, DC, 1968.
9. E. McKyes. *Soil Cutting and Tillage*. Elsevier Science Publishers B.V., Amsterdam, The Netherlands, 1985.
10. W. C. Swick and J. V. Perumpral. A model for predicting soil-tool interaction. *J. Terramech.*, 25(1):43–56, 1988.
11. M. S. Osman. The mechanics of soil cutting blades. *J. Agricultural Engng. Res.*, 9(4):313–328, 1964.
12. T. Muff, R. H. King, and M. B. Duke. Analysis of a small robot for Martian regolith excavation. In AIAA Space 2001 Conference and Exposition. AIAA, 2001.
13. S. W. Johnson and K. M. Chua. Properties and mechanics of the lunar regolith. *Appl. Mech. Rev.*, 46(6):285–300, 1993.

14. E. McKyes. *Agricultural Engineering Soil Mechanics.* Elsevier Science Publishers, B.V., Amsterdam, The Netherlands, 1989.

15. W. D. Carrier III, G. R. Olhoeft, and W. Mendell. Physical properties of the lunar surface. In G. Heiken, D. Vaniman, and B. M. French, editors, *Lunar Sourcebook: A User's Guide to the Moon,* pages 475–594. Cambridge University Press, New York, NY, 1991.

16. O. Luengo, S. Singh, and H. Cannon. Modeling and identification of soil-tool interaction in automated excavation. In Proceedings of the 1998 IEEE/RSJ International Conference on Intelligent Robots and Systems, pages 1900–1906, Victoria, BC, Canada, 1998.

17. K. Terzaghi and R. B. Peck. *Soil Mechanics in Engineering Practice.* John Wiley & Sons, New York, NY, 1948.

44

Lunar Concrete[1]

Martina Pinni

ESA-EAC
Astronaut Training Division
Linder Höhe, Köln, Germany

Martina Pinni is a registered architect with an interest in high-tech buildings, extreme environments, and spinoff applications of space technology in building construction. Her thesis in architecture focused on an interior design for a space habitation module (2000). Her professional experience includes 6 years of technical and commercial positions in façade engineering, architectural design, and research (2001–2007). Teaching experience includes three semesters as faculty assistant of building technology at the University IUAV of Venice (2006–2007), and current position in astronaut training—structures, thermal control, life support system, emergencies—at ESA-EAC Cologne for the International Space Station project (2008–present). She earned a *Laurea* (M. Arch.) from the University IUAV of Venice (2000), an MS from the University of Houston, Space Architecture (2004), and attended the ISU SSP'07 at BeiHang University in Beijing (2007).

Introduction

Concrete is an ancient building material. So primitive and heavy that Buckminster Fuller, overeager of achieving structural efficiency per unit of weight, preferred by a long way the new aeronautical alloys to build his super lightweight "ephemeralized" high-tech domes. Fifty percent of concrete's weight is (structurally) redundant, wrote Fuller in an article in 1961 (Tensegrity). He repeatedly stated throughout his life that unnecessary weight and gravity structures are popular only because we still use "stone age logic" (Lightful Houses, 1928; Designing a New Industry; 1946, Synergetics, 1975).

The pristine reasons for its use are rediscovered today by space scientists, who imagine that engineered concrete could be well suitable for extraterrestrial constructions of a later generation, those for which planetary material

---

[1] An early version of this article featured in the Italian construction magazine *Costruire* N° 299, April 2008, pp. 80–84. Published with permission.

FIGURE 44.1
Jim Irwin, the Lunar Roving Vehicle and the Landing Module with Mount Hadley Delta in the background (courtesy of NASA).

will be used. All in all, future lunar houses could be built almost like those on Earth. Therefore, in the international context of a new race to space, the space agencies of the various countries, especially in the United States, are wiping the dust-covered researches made in the Seventies on extraterrestrial mineral and energetic resources, mostly coming from the Apollo missions (1969–1972), almost 35 years ago.

Some of these studies and knowledge gained from missions to the Moon have been discussed during the Rutgers Symposium on Lunar Settlements, organized in June 2007 by Prof. Haym Benaroya, Director of the Center for Structures in Extreme Environments (CSXE) at Rutgers University, New Jersey. The topics that have become of major interest lately, in the context of the planning for future human-inhabited outposts on the Moon, concern the In-Situ Resource Utilization (ISRU) of the rocks and minerals forming the regolith of the Moon, the nearest of our celestial bodies.

The themes that concern construction on the Moon are complex and diverse, and they range from energy to structures, to thermal isolation, to seismic protection, covering all of the fields of traditional building, with the addition of particular environmental matters, such as raised by abrasive dust, galactic and solar radiation, vacuum conditions (10^{-15} bar), and reduced gravity (1/6 g). The building of future habitats became a stimulating discussion among the scientists that took part in the meetings. And, among the most "trendy" issues and novelties, they talked about lunar concrete, waterless concrete, and lunar soil sintering/melting.

Until recently, studies of extraterrestrial construction involving the possibility of making lunar concrete had focused on the Portland cement variety, whose extraterrestrial applications had been studied extensively in the eighties. Lunar concrete was first prepared by Professor Tung Dju (T.D.) Lin, a Taiwanese born scientist and U.S. citizen, with samples brought to Earth by the astronauts of the Apollo 16 mission (the next to last mission in early 1972).

Dr. Lin, who is presently a Research Professor at National Cheng Kung University in Taiwan and a Research Consultant for the Portland Cement Association (PCA), USA, presented the results of his fundamental research and experiments in various publications starting from 1985, in a study endorsed by NASA. Using 40 grams of real lunar soil (regolith) from the lunar highlands at the Apollo 16 site, Dr. Lin prepared a lunar concrete sample.

However, he initially experimented with a lunar regolith simulant, JSC-1, prepared by Johnson Space Center, Houston, using a volcanic rock from Arizona, in order to have similar engineering properties to lunar regolith and have a composition similar to certain terrains of the Moon. With the simulant, Dr. Lin, as well as other engineers, experimented with several techniques for

FIGURE 44.2A
Harrison Schmitt, Apollo 17, collects soil samples near a crater at Station 5, Camelot (courtesy of NASA).

FIGURE 44.2B
Picking up rocks at Station 2, Apollo 15 (NASA).

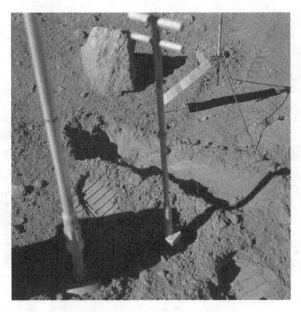

FIGURE 44. 2C
Apollo 16, station 4, soil sample 399 (NASA).

FIGURE 44. 3
Lunar concrete production tests: 1/2″ cube and 1″ cube samples (Courtesy of Dr. T.D. Lin, Research Professor at National Cheng Kung University in Taiwan and a Research Consultant for the Portland Cement Association, USA).

the casting of lunar concrete; among them, the one entitled "dry mix/vapor injection" proved very effective. The dry-mix/steam-injection method was developed to overcome casting concrete in a vacuum environment on the lunar surface. Soil that had been properly sized was placed in sealed forms, dry mixture, and pressurized steam was added. After shaping little cubes through dry mixing the components, pressurized steam flows through for 18 hours, giving concrete samples with excellent characteristics, both those made with the simulant and real lunar regolith.

In future lunar constructions, the binder could be transported from Earth or even obtained from the soil. Dr. Lin's theory was that the lunar ground already contains everything needed to make a good concrete and that it would only take transporting some hydrogen in form of methane or ammonia in a liquid state to get water, the only component completely missing in the lunar environment.

Lunar regolith, as we now know, can produce a better concrete than terrestrial soil, because of the conditions of the high vacuum, which makes small particles aggregate into larger ones, resulting in a very resistant aggregating material. The resistance to compression of the samples proved quite good. The tiny sample prepared with the 40 grams of highlands regolith has shown a resistance to compression of 75 MPa, and exposed to vacuum for long periods, it has maintained its strength at 80%. Much more testing would be needed to see how this material really compares to high resistance terrestrial concrete. But, as Dr. Lin puts it, the 1/6 g lunar gravity will certainly benefit the design of flexible structural components, such as beams/ slabs and the supporting operations in construction.

According to Dr. Lin, the shattered lunar rocks, the regolith, could produce, therefore, a thin, light, and extremely resistant cement. By additional processing, other materials could be extracted from the soil, among which would be metallic iron to produce bars. Iron would be easily workable under

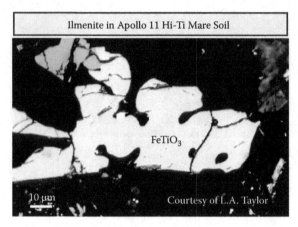

FIGURE 44.4
Ilmenite in Apollo 11 Hi-Ti Mare Soil (Courtesy of Larry Taylor).

the low gravity conditions. So, there are plenty of theoretical possibilities. The very real problem on the Moon is how to produce water. As has been demonstrated several years ago in studies by Professor Larry Taylor from Tennessee, at temperatures of 800–900°C, hydrogen reacts readily with the iron-titanium-oxide mineral, *ilmenite,* to break the iron-oxygen bonds thereby releasing the oxygen to combine with the hydrogen to produce water.

Hydrogen should be transported from Earth in percentages of 3 in 1,000 of the final mass of the concrete to be produced. As Taylor explains: "Solar Wind consists of about 95% protons (H^+), and the remainder being alpha (He^+), carbon, nitrogen, etc. particles. Due to the lack of an atmosphere, these particles impinge on the lunar soil with high velocities. In this manner, the particles become implanted within the outer few 10s of nanometers of each soil grain. As such, the lunar soil contains 50–100 parts per million (0.01 wt %) of these particles. At first thought, this seems trivial. However, taking the hydrogen from the lunar soil grains in the outer few meters of the regolith over an area of some five soccer fields would give us about twenty tonnes of liquid hydrogen, capable of producing 180 tonnes of water. Obviously the amount of soil to be handled to get all this is large, but not unrealistic."

There are other challenges. In fact, in such a high vacuum as that on the Moon, clear water would evaporate as soon as it could be formed because of its low boiling point, which is so greatly lowered by the extremely reduced pressure on the Moon. Therefore, it is necessary to look for alternative materials to cement the soil grains together to form concrete. In other words, an alternative binder needs to be used, something that can undergo phase changes at very low temperatures. In the rocks and soils of the Moon is the mineral troilite (FeS), which might be used to recover native sulfur as alternative binder for lunar cement.

Sulfur-based concrete, using sulfur as the cement, is commonly used in harsh terrestrial environments. It has properties different from Portland

FIGURE 44.5
Comparative deterioration after immersion in 20% sulphuric acid of samples of sulfate-resistant Portland cement concrete (after three weeks), and STARcrete™ (after 3 years) (Courtesy STARcrete™ Technologies, Inc.).

cement, in that it doesn't have capillary porosity, and this makes it impermeable to fluids. The concrete would be composed of mineral aggregates cemented together with highly polymerized native sulfur, thereby making a durable concrete. The typical percentages of ingredients for the terrestrial sulfur concrete are around 80% of aggregates, 12% of sulfur, and 8% of fly ashes. It is already used in some civil construction because of its properties, especially in chemically aggressive environments and in the presence of salts. It is capable of being operational within 24 hours from the time of casting and also has the possibility of being cast at temperatures well below 0°C. Although for most applications it cannot compete commercially with Portland cement concrete, its unique characteristics make it a viable construction material for extreme environments such as with extraterrestrial applications.

Compared to the complex operations needed to work with cement from the lunar soil and water, the process required to produce sulfur concrete appears to have a good level of feasibility. The sulfur would be produced by effective heating of the troilite to about 1200°C, which could be obtainable with a simple solar-mirror concentrator. Professor Houssam Toutanji, from the University of Alabama in Huntsville, proposed to make it even stronger, to reinforce the concrete with glass fibers that could be obtained by melting the lunar regolith. Professor Toutanji used sulfur powder and JSC-1 lunar regolith simulant to cast plates (4x10x0.5 in). The sulfur concrete mixes consisted of 35% sulfur and 65% JSC-1 regolith simulant by mass. Sulfur binder and JSC-1 aggregate mixtures reinforced with short and long glass fibers, both derived from regolith simulant.

Lunar regolith simulant was melted at melting temperatures of 1450 to 1600°C for periods of 30 min to one hour. Glass fibers were pulled from the melt. For glass fiber pulling, the crucible was placed onto a refractory brick and an alumina ceramic rod was used to pull the fibers, as shown in Fig. 44.6.

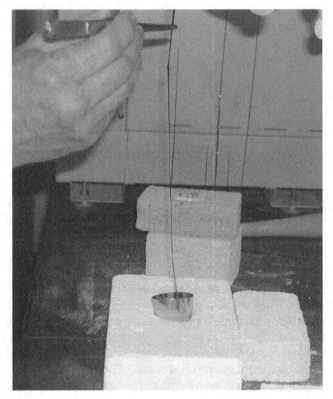

FIGURE 44.6
Hand-draw of glass fibers (Toutanji et al. 2006).

These fibers were long fibers with an average length of 8 inches and a diameter of about 1 mm, as shown in Fig. 44.7.

The results showed that with the addition of fibers the ultimate flexural strength was increased significantly. Sulfur concrete strengthened with short and long glass fibers (glass fibers made from regolith simulant) showed an increase of about 40 to 50%. Professor Toutanji indicated that this was a preliminary data and he is currently conducting a study to investigate the effect of the glass fibers on the ductility and strain energy capacity of the sulfur concrete.

No Portland concrete in space, then? Maybe yes, but for other applications and with a new recipe. Recent studies conducted at the University of Alabama in Huntsville have shown how the Portland cement can be mixed with glass microbeads (microscopic hollow spheres), latex, acrylic fortifier, and water, all in the appropriate proportions. The resulting material, dried for twelve hours, is so light that it floats on water, so flexible it bends without breaking, and so inert it withstands radiation. In comparison with other types of concrete, especially in cost savings compared to the "traditional"

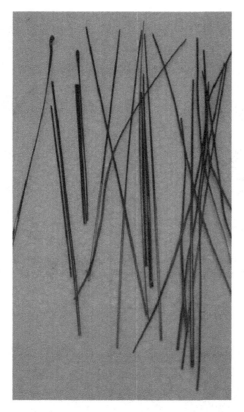

FIGURE 44.7
Glass fiber drawn from lunar regolith simulant (Toutanji et al. 2006).

graphite epoxy, this "new" type of concrete should have several good space applications, for example in tanks and rocket fuselages. For the time being, Professor John Gilbert from UAH says, with this flexible concrete, reinforced with glass fibers, they have built a canoe for the traditional concrete canoe competition. This type of canoe is difficult to ride in, because its elasticity makes it move in resonance with the waves (Figs. 44.8–44.9).

An entirely startling discovery about lunar soil recently has been made by Professor Larry Taylor, Director of the Planetary Geosciences Institute at the University of Tennessee. As he explains, "If a pile of lunar soil is placed in your kitchen microwave oven, it will melt at more than 1200°C, before your tea-water will boil (100°C)." Quite unbelievable until Taylor explains why. The major weathering and erosional process on the Moon is meteorite and micro-meteorite (less than 1mm) impacts, at velocities of 30,000 to 150,000 km/hr. This is the major process responsible for the formation of the soil on the Moon. With such high velocities, the micro-meteorites not only smash rocks and soil particles into minute sizes, but most importantly, they melt large portions of the soil, which produces glass. A good half of mature lunar soils are this

FIGURE 44.8
Several batches of concrete were prepared and students used drywall knives to place the mix over three layers of graphite fiber mesh (courtesy Team University of Alabama Huntsville).

FIGURE 44.9
Students built the canoe on the mold (courtesy University of Arizona).

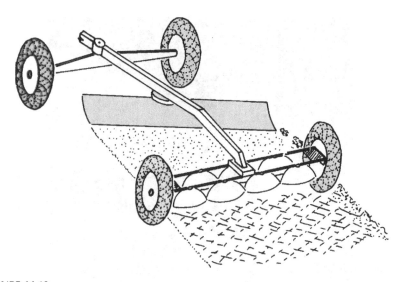

FIGURE 44.10
Microwave wagon designed by Larry Taylor. It is equipped with a magnetron with the primary cone of microwaves, and with a parabolic metal reflector, to reflect waves on the soil for homogeneous sintering (courtesy of Taylor and Meek, 2005).

impact glass. Part of this impact melt actually vaporizes and is then deposited as myriads of tiny metallic iron particles in the impact glass.

It is the nano-sized metallic iron that couples so highly with the microwave energy. Using but one magnetron out of a microwave oven, which normally has four, Prof. Taylor's research team has been able to fully focus the single 2.45 GHz microwave beam such that it has extreme concentrations of energy. This single magnetron can melt real lunar soil at about 1750°C in some 10 seconds.

Taylor has designed a wagon (Fig. 44.10) whereby a series of magnetrons, at a certain frequency and power, will sinter the lunar soil to a depth of 30–40 cm. Another row of magnetrons, at another frequency and power, can melt the upper 3–5 cm of the soil, which will rapidly quench to a glass.

With such a piece of microwave equipment, it should be possible to create effective roads and more importantly, rocket landing pads, which will keep lunar dust from being thrown all over, possibly settling on solar cells. The lunar dust, in fact, is highly abrasive, can make moving parts useless, and damage delicate mechanisms of the rovers, space suits and spacecraft, and it can be very dangerous if inhaled by the astronauts.

All these solutions have advantages and drawbacks. Perhaps they will all be used, but for different applications. For paving there are other proposals, like the simple removal of the dust layer, which was created above the bedrock by meteoroid impact. Special machines and equipment will need to be designed for that. The ultimate decision on which technology to use for which application will be based on improved knowledge, and in-situ assessment and verification.

FIGURE 44.11
Astronaut Gene Cernan test-driving Lunar Rover, raising abrasive dust, December 11, 1972.

The first exploration settlements needed for the exploitation of all these resources, though, may require preliminary non-concrete structures. As Professor Haym Benaroya from Rutgers University says: "[I believe that] the early settlements would not be concrete-based because we will still need some infrastructure to create the concrete building blocks and erect the structure. I believe that the sequence of structural types will be: prefabricated -inflated/rigidized -solid, perhaps sintered regolith or some concrete-based building blocks." He continues: "Probably, many different technologies will merge as our infrastructure develops on the Moon. The linking of sintering/ microwave technologies may allow for the initial formation of building materials. Furthermore, the development of robotic fabricators that can prepare, mix, and build the structural components autonomously may allow for the preparation of structures long before astronauts land."

Acknowledgments

Many thanks to Haym Benaroya, Tom Gilbert, T.D. Lin, Larry Taylor, and Houssam Toutanji for kindly providing information and images.

45

Silicon-Utilizing Organisms May Be Used in Future Terraforming of the Moon

Satadal Das

Department of Microbiology,
Peerless Hospital & B. K. Roy Research Centre,
Kolkata, India
President, National Space Society-Kolkata Chapter

Satadal Das is a consultant microbiologist at Peerless Hospital and B.K. Roy Research Centre at Kolkata, India. His main areas of research at present are on silicon utilizing microorganisms, isolation of newer antibiotics from microorganisms in a barren island of the Sundarbans—the world's largest mangrove forest, and the effect of titanium on bacterial growth and nutrition. He has published 39 papers in various journals and presented more than 100 papers in various international and national conferences. He is a member of 11 scientific societies and is the secretary of Indian Association of Medical Microbiologists, West Bengal Chapter. At present he is the president of National Space Society-Kolkata Chapter.

ABSTRACT Silicon-utilizing organisms are well known to tolerate different physical, chemical and biological stresses. Thus there is a scope to use them in selected batches on lunar surfaces for initial terraforming—at least for caeliforming. Although they produce aberrant morphological forms on silica based medium containing extremely trace amounts of carbon and nitrogen—somewhat resembling lunar surface constituents, but even in these altered forms they are quite able to initiate a micro environment leading to formation of complex micro ecosystem which may be remodeled later for the benefit of mankind. Once this initial protocol is standardized, we may use it on other extraterrestrial surfaces to create similar micro environments in future. Silicon-utilizing organisms will help in terraforming the Moon.

Introduction

Silicon-utilizing organisms have been defined as organisms with high silicon content (≥1% dry weight) and which can metabolize silicon with or without demonstrable silicon transporter genes (SIT) in them [1]. They can survive in extremes of temperature, pressure, radiation, pH, salinity and nutrient conditions. Thus it is important to see whether there is any possibility to use silicon-utilizing organisms for initial terraforming process on the Moon. A classification of silicon utilizing organisms has already been described [1].

Although there are some similarities of silicon with carbon, carbon is the key element in the development of living creatures on earth and in no way silicon is comparable to carbon regarding this. However, in the early desolate days in the life history of earth when conditions were not suitable for survival of the presently omnipotent carbon-based life, silicon may play a role in the initiation of life on earth and a "nutrient profile" of silicon similar to nitrate and phosphate in ocean may be traced even today.

Among different minerals only N, K, Si, Ca and Na are found more than 1% in plants, but it is unfortunate to know that while ecological aspects of N, K, Ca, and Na are well known, the role of silicon in terrestrial ecosystems is not sufficiently understood. However, silicon-utilizing organisms can tolerate radiation stress, water stress, oxygen lack, micro gravity etc. which are prevalent on lunar surface. Except the two most important elements essential for life on earth—carbon and nitrogen which are present in trace amounts, all other chemicals are present in sufficient amounts on lunar surface. The Moon contains oxygen (42.6%), magnesium (20.8%), silicon (20.5%), iron (9.9%), calcium (2.31%), aluminum (2.04%), nickel (0.472%), chromium (0.314%), manganese (0.131%), titanium (0.122%), many chemicals in trace amounts and possible presence of water in polar craters. The dark colored lunar basalts, which are igneous rocks, are mainly silicates and they also contain iron, titanium, zirconium, uranium and lanthanum. Lunar sedimentary rocks are equivalent to regolith, which also contains large amounts of silicon. Titanium present on the Moon may also help growth of silicon utilizing organisms on solid surfaces (unpublished data).

Silicon biomineralization occurs as a biologically controlled process or by a biologically induced process [2]. In biologically controlled silicon mineralization, there is formation of intracellular silica deposition vesicle (SDV) and silicate is sequestered and transferred to the mineralization site by energy driven pump mechanism (Fig. 45.1) in presence of specific transporter protein [3]. In biologically induced silicon mineralization at first there is nucleation of silicon followed by gradual transformation of the initial amorphous phase into a crystalline phase. There are also other methods of silicon deposition mainly by conversion of monomeric silica into oligomers and ultimately are converted into silanol and siloxane. Other methods like hydrogen bonding, cation bridging and direct electrostatic interactions have also been described [4,5,6]. There are at least five SIT genes described in diatoms and following this, SIT genes of some other silicon-utilizing organisms have also been described [7,8]. Diatom silicon transporters (SITs) are membrane-associated

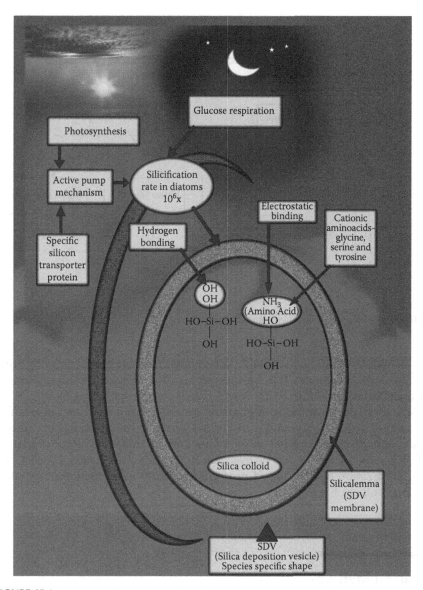

FIGURE 45.1
General mechanism of silicification of diatoms.

proteins that directly transport silicic acid [3]. Different types of silaffins and long-chain polyamines (LCPA) are found in embedded proteins of silica matrix of diatoms and many silaffins and LCPA (Fig. 45.2) can promote rapid precipitation of silica [9,10,11,12]. Why some groups of organisms prefer to use silicon is very difficult to explain; probably it was developed in more silica rich hydrosphere during the Cambrian [13] mainly to construct cell

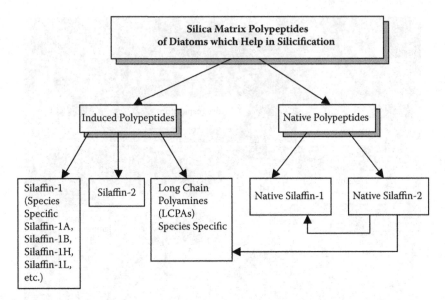

FIGURE 45.2
Polypeptides which help in silicification.

walls with less energy consumption and helping photosynthesis by forming $CO2$ from bicarbonates [14], besides acting as armor against predation by zooplankton [15].

Whether silicon-utilizing organisms can survive in such extraterrestrial situations where carbon and nitrogen present in extremely trace amounts is the main aim of the present study. Other aims of this study are formulation of a basal media which may be used as a replica of extraterrestrial "soils" and standardization of a proposed protocol for future terraforming of extraterrestrial bodies.

General Methods

Formulation of a Basal Medium to Study Whether a Microorganism Can Grow on Extraterrestrial "Soils"

The "extraterrestrial" basal medium (ETBM) comprising 3 mL of a solution containing biologically active chemicals of a known extraterrestrial body composition other than silicate, sodium and phosphates (concentrated 2.5x) plus 3 mL sodium metasilicate (analytical, Fluka at present Sigma-Aldrich, linear formula Na_2SiO_3, formula weight 122.06; 11.8 g/dL) was used as a solidifying base, with the addition of 1.5 mL phosphoric acid (analytical, Fluka at present Sigma-Aldrich, linear formula H_3PO_4 formula weight 98.00; 16%v/v). For a general study where particular composition of extraterrestrial body is unknown an "extraterrestrial" medium (ETM) comprising of Na_2HPO_4, 0.375 g; KH_2PO_4, 0.5 g; $MgSO_4$.0.075 g and distilled water 50 mL; sterilized by passing through G-5 sintered glass filter, was solidified with sodium metasilicate and phosphoric acid as described for ETBM above. Sodium metasilicate and phosphoric acid solutions were sterilized by autoclaving and kept for at least 48 hours at 6–8°C before preparation of the medium. Media prepared in this way showed less release of water. Carbon and nitrogen were not added as in extraterrestrial situations they are usually present in trace amounts and these trace amounts were available through "carry over" in cultures or through natural contaminations.

Microorganisms

The strains of microorganisms used were locally isolated Gram positive soil microorganisms - *Mycobacterium smegmatis, M. terrae, M. triviale, M. gordonae, Bacillus subtilis, Aspergillus fumigatus, Rhizopus stolonifer.*

Cultivation Procedures

The stock cultures of mycobacteria, *Bacillus* and fungi were available on Lowenstein-Jensen (L-J) medium, nutrient agar (NA) medium and Sabouraud dextrose (SD) medium. For different microorganisms inoculates were prepared by transferring 5–6 colonies of bacterial strains or thoroughly fragmented 1–2 colonies of fungal strains to glass homogenizers, followed by grinding and suspension in 5 ml of phosphate buffer (pH 7.4) saline and homogenization by shaking with glass beads, next followed by a low speed centrifugation (10 min) to give uniform homogeneous supernatant. These were matched and standardized with reference to McFarland standard 1.0 and their viable counts determined, so as to contain 50–100 isolated colonies on L-J medium, NA medium and 10–50 isolated colonies on SD medium in 0.01 mL of each standardized suspension. The inoculation of different ETM and conventional media were done accordingly each in five lots and they were observed for 60 days. The time taken for growth on a particular medium at 37°C, its increase with the duration of incubation, colonial morphology, pigmentation, effects of serial transfer of growth, microscopic morphology with different types of strains were studied in this experiment, and electron microscopy of *M. smegmatis* growths were studied separately. The growth characteristics were also examined with respect to the capacity for reversion to the original morphological characters, when returned to the conventional media from ETM medium.

Electron Microscopical Study

Electron microscopy observations of *M. smegmatis* grown on ETM and conventional media were done with the help of Jeol JEM-200 CX electron microscope (Jeol Ltd, Tokyo, Japan) after staining with 1% solution of uranyl acetate on a carbon coated copper grid (300 mesh).

Enrichment of Silicon-Utilizing Microorganisms from the Environment in Sodium Metasilicate Solutions

Many silicon utilizing microorganisms can thrive in sodium metasilicate (SM) solution as high as up to 4% concentration [1]. Details of this technique have been described elsewhere [1]. To confine common silicon utilizing organisms from the environment SM solutions of four different concentrations—0.5%, 1%, 2% and 4% were prepared with tap water. SM solutions were kept in wide mouth plastic containers in different lots (10 each), they were exposed to air for 24 hours then they were closed with airtight caps and kept in dim lighted areas for as long as 5 years in ambient temperature for "enrichment" of a selective group of stress adaptive silicon utilizing organisms. This five-year "enrichment" time was calculated to allow proper adaptation in a long period of stress following my experiences in previous works. After five years of stress the organisms present in different solutions were exposed to daylight (~500 µmol m-2s-1 PPF 12 hours each day) for one month following diurnal rhythm. After this macroscopic and microscopic studies of the growths were done directly and after standard staining and cultural procedures for different microorganisms as described elsewhere [16,17,18,19].

Results

Growths of Different Microorganisms on ETM

Mycobacterium Smegmatis

On L-J medium tiny white colonies appeared within 24 hours, their size and number increased gradually till they became confluent on 5th day after inoculation. After 15 days the colonies were grayish white in color, they were dry and mammillated. On ETM tiny colonies appeared on second day after inoculation which became prominent within 5th day and after that it remained unchanged till 60th day when size of the colonies were significantly increased and the original dry colonies became moist. In Ziehl-Neelsen (Z-N) staining the bacilli grown on ETM showed many clumps with penetration in the deeper layers of the medium, branchings were frequently seen along with pseudomycelial forms. In Gram's staining beading appearance of the bacilli were more on ETM. All these changes were not seen in the bacilli grown on L-J medium. After repeated subcultures on ETM growth rate of the bacilli was increased, colonies were white, smooth, circular and low convex in type and when examined microscopically they showed marked branching pattern with almost mycelia like structures. After re-cultivation of this altered growths on L-J medium original appearances were restored.

Mycobacterium Terrae

On L-J medium initial appearance of growths occurred on 5th day after inoculation as tiny white moist colonies, which began to increase in size and number from 10th to 12th days and became confluent by 30th day. On ETM almost similar colonies appeared on 5th day after inoculation and slowly increased in size up to 60 days. In Z-N staining and Gram's staining of the bacilli grown on ETM, bacilli were very much elongated with branchings showing club shaped ends and sometimes they were arranged in chains. After repeated subcultures on ETM branchings were more with occasional mycelial patterns. Original morphological patterns were restored after subculture of the altered growths on L-J medium.

Mycobacterium Triviale

Dry tiny white colored colonies appeared on L-J medium on 5th day after inoculation which began to increase in size in between 12th to 21st day and became confluent by 25th day. On ETM although similar but moist colonies appeared on 5th day after inoculation they remain unchanged even up to 60 days of follow up period. In Z-N staining and in Gram's staining the bacilli grown on ETM showed only cocci forms without any evidence of any branchings or chains which were frequently seen in the bacilli grown on L-J medium. After repeated subcultures on ETM, very tiny, colonies appeared within 24 hours showing only coccoid forms in microscopical study.

Mycobacterium Gordonae

Tiny dry yellowish colonies appeared on L-J medium after 3rd day of inoculation which gradually increased in size, became orange-yellow in color and were confluent by 10th day. Although similar colonies were observed on ETM after 3rd day of inoculation there was mild increase of their size up to 10th day and after that there was no change. In Z-N staining and Gram's staining the bacilli grown on ETM were longer with frequent branchings and also in clumps. After repeated subcultures on ETM growths appeared with 24 hours, microscopically branchings were more prominent. Original morphological characters were reverted back when these altered growths were subcultured on L-J medium.

Bacillus Subtilis

Prominent colonies were seen on NA medium within 24 hours which became confluent within 48 hours. On ETM tiny colonies appeared within 24 hours and they became moderately increased in size within 5th day and after which there was no change. In Gram's staining of the growths on ETM spores were less in comparison with the growths on NA medium.

Aspergillus Fumigatus

On SD medium tiny colonies appeared within 24 hours which became greenish powdery like along with patches of white cottony growths and white reverse side within 3rd day, the growths were deep green in color within 25 days. On ETM although tiny growths appeared within 24 hours they were mildly increased in size and showed licheniform appearance within 25 days after inoculation. In Lactophenol cotton blue (LCB) staining of the growths on ETM, it was found that vesicles of conidiophores were extremely scanty in number when compared with the growths which were observed on SD medium. When re-cultivated on SD medium original morphological forms were seen.

Rhizopus Stolonifer

Tiny colonies appeared within 24 hours of inoculation on SD medium which became white cottony growth with yellowish back within 72 hours, gradually the color of the reverse side of the growths changed to deep orange yellow color. On ETM, white cottony growths appeared on 10th day after inoculation and slightly increased in size up to 15th day without any change of reverse side color of the growths. There was no further change. In LCB staining plenty large columellae with very thick hyphae were found in growths on ETM. Repeated cultures on ETM showed mild increase of growths in size and in number. Original morphological forms were reverted back after subculture on SD medium.

Electron Microscopy of the Growths of
M. Smegmatis on ETM and on L-J Medium

When growths of *M. smegmatis* on ETM and L-J medium were studied under electron microscope many differences were noted as given in Table 45.1.

Growths in Different SM Solutions

Different varieties of organisms grew in different concentrations of SM solutions- light green color growth in 0.5% SM solution, yellow color growth in 1% SM solution, orange color growth in 2% SM solution and a scanty whitish color growth in 4% SM solution. In comparison to the control medium

TABLE 45.1

Growths	On ETM	On Lowenstein-Jensen Medium
Length	Normal	Decreased
Thickness	Normal	Decreased
Plasma membrane	Normal	Thickened
Surface glycolipid	Prominently present	Not so prominent
Fibrous rope like structures	Scanty in number	Present in large numbers
Mesosome	Present	Present
Dense granules	Moderately present	Markedly increased
Lipoidal bodies	More	Less
Metachromatic granules	Less	More

(tap water only) diatoms were markedly increased (~4 times) in both 0.5% and 2.0% SM solutions. Phytoplanktons other than diatoms were more in the control medium than that in silicate solutions (Graph-1). Acid-fast bacteria belonging to Runyon Group II variety grew abundantly in 1.0% silicate solutions. *Aspergillus* spp. predominates in 0.5%, 1.0% and 2.0% silicate solutions while scanty fungi of diverse varieties were found in 4% SM solution.

Discussion

Limited Growth of Some Gram Positive Microorganisms on Carbon and Nitrogen "Free" Medium

It is very important to note that there should be no model culture medium mimicking extraterrestrial condition which may be used to study growths of earthly carbon-based life forms. It is not only due to the fact that there are extreme variations in composition of extraterrestrial bodies but it is also due to the low level of our knowledge of the Universe. However, for initiation of our terraforming attempts on the near extraterrestrial objects of the Solar System this ETM/ETBM medium may be helpful.

To study the growth patterns of silicon-utilizing organisms on ETM initially some soil Gram positive microorganisms were selected other than diatoms, experiments on which is now going on in our laboratories. Gram positive microorganisms were selected because most of them satisfy the criteria of silicon-utilizing organisms.

It is not that we shall spread this organism on extraterrestrial surfaces and the organisms will grow and produce the micro-ecosystem instantaneously. In fact, it will never be so simple. At first we should select most probable places for their possible growth, e.g. on polar craters and within lava tubes on the Moon. After spread, the microorganisms may grow on some patchy

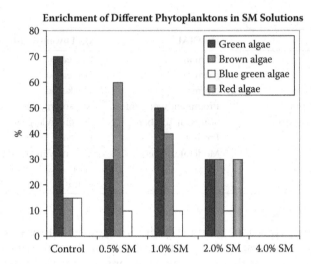

FIGURE 45.3
Graph 1: Growth of phytoplanktons in different silicate solutions. SM-Sodium metasilicate.

small areas only; but once they grow on those difficult sites, further enhancement of the growths could be done easily.

Initial experiments with these Gram positive microorganisms isolated from soil showed that there were limited growths of these microorganisms on ETM probably with the help of some carry over carbon and nitrogen during cultivation procedures. However, increase in growth rate after repeated subcultures could not be explained at present. In a previous study by us when silicon level was studied in such grown up cells on carbon "free" silicate medium by electron prove microanalyser following thorough washing procedures, it was found that silicon in cells grown on carbon "free" silicate medium was much higher (24.9%) than those grown on conventional carbon based medium (0.84%) [20]. Experiments on diatoms—which are excellent stress tolerated silicon-utilizing organisms—are going on and are found to show better growths on ETM than these tested organisms (unpublished data). However, these initial findings are encouraging for our future application of these groups of organisms on extraterrestrial surfaces for artificial micro-ecosystem formation.

Enrichment of Silicon-Utilizing Microorganisms in Silicate Solutions from Environment for Future Use in Initial Terraforming Procedures

In this study silicon-utilizing organisms from the environment were allowed to grow selectively in large numbers in different SM solutions. This simple experiment showed us a way to produce silicon utilizing microorganisms in large quantities from the environment for their possible use on lunar surface or on other extraterrestrial surfaces. Silicon-utilizing organisms were selected for initial terraforming process as their important ecotypes particularly diatoms have already developed as stable mutants against different kinds of stress throughout their long stay on earth for more than 200 millions of years.

Constantly changing environment in recent times and a critical change in sea level as was found before occurrence of all mass extinctions on earth in the Phanerozoic [21], indicates that our world may be a hostile place to us in coming years. Again there are possibilities of an impact of a meteorite on earth in near future. Considering of all these aspects it is obvious that mankind should find out other areas away from earth for habitations.

A Protocol for Use of Silicon-Utilizing Organisms in Terraforming/Caeliforming on Lunar Surfaces

A simple protocol may be followed to use silicon-utilizing organisms on lunar surfaces. After providing minimum essential requirements for life on lunar extraterrestrial situation, these organisms may be utilized. Otherwise this protocol may be followed directly on a lunar crater or in lava tubes to allow the organisms to find out a suitable zone for their growth.

Microterraforming Phase

In the initial venture antibiosis between various species to be used on extraterrestrial surfaces should be avoided. Thus phytoplankton should be used before zooplanktons. Diatoms of Eu-eurytherm variety of *Nitzschia* and *Chaetoceros* group may be used initially and they should be kept for 3–12 months. Then red algae grown in 2% SM solution should be scattered and kept for 3–12 months and finally yellow green algae grown in 0.5% silicate solutions may be scattered to boost up the algal inhabitants and it also should be kept for 3–12 months on the selected site. All selected microorganisms should be tested for any pathogenecity and toxigenecity by standard procedures before their application. After successful completion of this phase microbial mats may be developed on lunar surface. Sub-cultivations may be done if necessary in between steps because active and passive dispersal mechanism will be less on lunar surface.

Selection of diatoms for the initiation of microterraforming process was mainly based on their survival in the last mass extinction on earth—the K-T event, in which only 20–25% extinction occurred in planktonic diatoms while 80% and 90% extinctions occurred in radiolarians and in foraminifera respectively [22].

Macroterraforming Phase

In this phase different lichens may be used initially and they should be kept for 3–12 months. Then important silicon utilizing plants and following that other organisms (only extremophile variety) like rotifers, tardigrades, nematodes along with protozoa, fungi and bacteria may be added which will live in close association of silicon utilizing plants and this process may continue for 1–5 years, following this food related silicon accumulator plants may be used for another 1–5 years.

Follow-Up Phase

After the initial terraforming process one should monitor the biomass on the lunar surface. A follow up analysis of the micro-climate is necessary to understand changes in the complex variables including temperature, radiation, humidity, "wind," etc. Although there are various methods of measurement of biomass e.g. drying, ashing, determination of energy content, carbon analysis, etc., remote sensing data combined with ground based measurements is probably the best method to be used. Carbon partitioning will be highly effective on lunar surface. Micro-meteorological techniques like aerodynamic and eddy covariance methods may also be tried. One can also measure Adenylate Energy Charge (AEC) ratio. This measurement is simple to perform and is extremely sensitive [23]. As there is no evidence of life in any of the materials, which were brought from the moon, there is no chance of obliteration or interference of any *in situ* "biosphere" on it.

Conclusion

Some Gram positive microorganisms showed limited growth on a silicon based solid medium without any carbon- and nitrogen-based ingredient. This indicates that they may be used on extraterrestrial surfaces for initiation of terraforming process. A simple method for production of stress tolerated silicon-utilizing microorganisms for use in terraforming process has been described along with a protocol consisting of orderly arrangement of such organisms for use on extraterrestrial surfaces. Follow-up studies are necessary to understand the kinetics of such artificial micro-environment on the extraterrestrial surfaces.

Acknowledgments

I would like to acknowledge Professor Haym Benaroya, Department of Mechanical and Aerospace Engineering and Director, Center for Structures in eXtreme Environments, Rutgers University, NJ, USA for allowing me to deliver my lecture on a related topic at the "Rutgers Symposium on Lunar Settlements" which was held in between 3 June to 8 June, 2007 at Rutgers

University. I thank Werner Grandl, Architect and Civil Engineer, Tulln, Austria and Lee Morin, Astronaut Office, Johnson Space Center, for taking interest in my work and discussions.

References

1. S. Das, A simple differential production method of silicon utilizing organisms for future use in lunar settlements, Rutgers symposium on lunar settlements, 3–8 June, 2007, Rutgers University. www.lunarbase.rutgers.edu/papers/Das_paper.pdf
2. H. A. Lowenstam, Minerals formed by organisms, *Science* 211(1981)1126–1131.
3. M. Hildebrand, B. E. Volcani, W. Gassmann, J. I. Schroeder, A gene family of silicon transporters, *Nature* 385 (1997) 688–689.
4. V. R. Phoenix, R. E. Martinez, K. O. Konhauser, F. G. Ferris, Characterization and implications of the cell surface reactivity of *Calothrix* sp. Strain Kc97, *Applied and Environmental Microbiology* 68 (2002) 4827–4834.
5. V. R. Phoenix, K. O. Konhauser, F. G. Ferris, Experimental study of iron and silica immobilization by bacteria in mixed Fe-Si systems: implications for microbial silicification in hot-springs, *Canadian Journal of Earth Sciences* 40 (2003) 1669–1678.
6. S. V. Lalonde, K. O. Konhauser, A. L. Reysenbach, F. G. Ferris, Thermophilic silicification: the role of Aquificales in hot spring sinter formation, *Geobiology* 3 (2005) 41–52.
7. K. Thamatrakoln, A. J. Alverson, M. Hildebrand, Comparative sequence analysis of diatom silicon transporters: towards a mechanistic model for silicon transport, *Journal of Phycology* 42 (2006) 822–834.
8. J. F. Ma, K. Tamai, N. Yamaji, N. Mitani, S. Konishi, M. Katsuhara, M. Ishiguro, Y. Murata, M. Yano, A silicon transporter in rice, *Nature* 440 (2006) 688–691.
9. N. Kroger, R. Deutzmann, M. Sumper, Polycationic peptides from diatom biosilica that direct silica nanosphere formation, *Science* 286 (1999) 1129–1132.
10. N. Kroger, R. Deutzmann, M. Sumper, Silica-precipitating peptides from diatoms. The chemical structure of sulfating-A from *Cylindrotheca fusiformis*, *Journal of Biological Chemistry* 276 (2001) 26066–26070.
11. N. Kroger, R. Deutzmann, C. Bergdorf, M. Sumper, Species specific polyamines from diatoms control silica morphology, *Proceeding of National Academy of Sciences, USA* 97 (2000) 14133–14138.
12. M. Sumper, A phase separation model for the Nanopatterning of diatom biosilica, *Science* 295 (2002) 2430–2433.
13. K. Kurt, *Introduction to geomicrobiology*, first ed., Blackwell Publishing, 2007, 180.
14. A. J. Milligan, F. M. M. Morel, A proton buffering role for silica in diatoms, *Science* 297 (2002) 1848–1850.
15. C. E. Hamm, R. Merkel, O. Springer, P. Kurkoje, C. Maler, K. Prechtel, V. Smetacek, Architecture and material properties of diatom shells provide effective mechanical protection, *Nature* 421(2003) 841–843.

16. A. Vonshak, Microalgae: laboratory growth techniques and the biotechnology of biomass production. Editors: D. O. Hall, J. M. O. Scurlock, H. R. Bolhar-Nordenkampf, R. C. Leegood, S. P. Long, In *Photosynthesis and production in a changing environment: A field and laboratory manual*, Chapman & Hall, London 75 (1994) 347–349.

17. B. A. Forbes, D. F. Sahm, A. S. Weissfeld, *Bailey & Scott's diagnostic microbiology*, eleventh ed., Mosby, St. Louis, 2002.

18. ASTM (American Standard for Testing and Materials), Standard specification for Substitute Ocean Water (Designation D) 15(1975)1141–1175.

19. APHA-AWWA-WPCF (American Public Health Association; American Water Works Association; Water Pollution Control Federation), In: L. S. Clesceri, A. E. Greenberg, R. R. Trussel (Eds.), *Standard Methods for the Examination of Water and Wastewater*, 17th ed., 1989.

20. A. N. Chakrabarty, S. Das, K. Mukherjee, S. G. Dastidar, D. K. Sen, Silicon (Si) utilization by chemoautotrophic nocardioform bacteria isolated from human and animal tissue infected with leprosy bacillus, *Indian Journal of Experimental Biology*, 26 (1988)839–842.

21. A. Hallam, *Phanerozoic Sea-level Changes*, first ed., Cambridge University Press, New York, 1992.

22. J. A. Kitchell, D. L. Clark, A. M. Gombos, Biological activity of extinction; a link between background and mass extinction, *Palaios* 1 (1986) 504–11.

23. A. M. Ivanovici, R. J. Wiebe, Towards a working 'definition' of 'stress': a review and critique. In *Stress Effects on Natural Ecosystems*, ed. G. W. Barrett, R. Rosenberg, John Wiley & Sons, New York, 1981, 13–27.

RUTGERS

THE STATE UNIVERSITY
OF NEW JERSEY

Haym Benaroya

Professor
Department of Mechanical & Aerospace Engineering
Director
Center for Structures in eXtreme Environments
Rutgers, the State University of New Jersey

American Institute of
Aeronautics and Astronautics

Rutgers Symposium on Lunar Settlements
3–8 June 2007 Rutgers University

Acknowledgments

There are many people whose efforts have made this Symposium an excellent meeting. Certainly first on this list must be Mrs. Patricia Mazzucco, whose tireless efforts in working with our vendors and local organizational matters, as well as with many of the attendees truly made this meeting come together in such a nice way. We are grateful to Vice President for Academic Affairs Philip Furmanski for his enthusiasm as well as financial support for the Symposium. Similarly we are grateful to Rutgers University, the institution, for being the exciting and dynamic place that it is. Elan Borenstein is thanked for his significant efforts on creating our website and making sure we were properly set up for the Symposium. Tushar Saraf is appreciated for his work in preparing this Abstract Book. Paul Bonness is appreciated for his efforts at pulling together Symposium materials. Shefali Patel and Helene Press are thanked for their assistance and support with the Symposium preparations. Aiesha Jenkins is sincerely thanked for her supportive efforts during the Symposium. Professor Yogesh Jaluria, Chair of Mechanical and Aerospace Engineering, is acknowledged for his assistance and support for this endeavor. Kendra Cameron is thanked for her assistance in helping us gather promotional items for the attendees. Of course we are truly grateful to all the presenters who took time and expense to come to Rutgers and offer us some of their expertise. Finally, a personal thanks to Ana Benaroya, my daughter, for her illustration of a future lunar settlement that has become our Symposium logo.

Thank you, Haym Benaroya

© Ana Benaroya 2007

	SUN 3 JUNE 07	MON 4 JUNE 07	TUES 5 JUNE 07	WEDS 6 JUNE 07	THUR 7 JUNE 07	FRI 8 JUNE 07
8 – 9		8:00 Registration & Opening Ceremonies	8:00 Registration 8:30-9:15 James Logan	8:30 Registration	9:00-9:45 Edwards	
9 – 10		Harrison Schmitt	9:15-10:00 Launius	9:00-9:45 Gerzer Bishop	9:45-10:30 Eckert	9:00 SYMPOSIUM SUMMARY PANEL
10 – 11		Benaroya Grandl 10:40 COFFEE BREAK	Howerton Rodiek 10:40 COFFEE BREAK	Smirnov 10:40 COFFEE BREAK	Han Palaia 10:40 COFFEE BREAK	CLOSING THOUGHTS BY REMAINING ATTENDEES
11 – 12		Freeman Lundquist LUNCH	Pearson Smithers LUNCH	Das Giacomelli Banerji LUNCH	Heiss Levy Lewandowski LUNCH	Adjournment

Continued

	SUN 3 JUNE 07	MON 4 JUNE 07	TUES 5 JUNE 07	WEDS 6 JUNE 07	THUR 7 JUNE 07	FRI 8 JUNE 07
1–2		1:10-1:50 Bell 1:50-2:30 Taylor	1:00-1:45 Sherwood 1:45-2:30 Hart	1:30-2:15 Nield	1:15-2:00 Janes	
2–3		2:50 Konesky (1)	2:40 Rhoades	2:20-3:00 Rowe	2:00 Maccone	
3–4		Brandhorst 3:40 COFFEE BREAK	Fisher Ruess 3:40 COFFEE BREAK	3:20 Taylor 3:40 COFFEE BREAK	Konesky (2) 3:40 COFFEE BREAK	
4–5	4PM – 7PM COFFEE & Registration	Zacny Lowman	Dunne Florek Gulak	Durst Spell Pimenta	Toklu Shevchenko	
5–6		Tirat-Gefen				
6++		6 – 8 OPENING RECEPTION	6 – 11 SPECIAL EVENT EVE	6 – 9 ZIMMERLI BANQUET	ON YOUR OWN	

DAY 1: Monday – June 4, 2007

Morning Session

Plenary Speaker

Harrison H. Schmitt, PhD

Apollo Astronaut, U.S. Senator, Entrepreneur
"Return to the Moon—Expanding the Earth's Economic Sphere"

Day 1 Morning Schedule:

- 8:00–9:00 am   Registration and Opening Ceremonies
- 9:00–10:00 am   Harrison Schmitt
- 10:00–10:20 am   Haym Benaroya
- 10:00–10:40 am   Werner Grandl
- 10:40–11:00 am   COFFEE BREAK
- 11:00–11:20 am   R.D. Richards
- 11:20–11:40 am   Marsha Freeman
- 11:40–12:00 pm   Charles Lundquist
- 12:00–1:00 pm   LUNCH

Return to the Moon: Expanding the Earth's Economic Sphere

Harrison Schmitt

It has been suggested by the President's Science Advisor and the Administrator of NASA that it is time to include the Moon in the "economic sphere of the Earth." Both history and comparative analysis indicate that a privately financed and managed initiative would be the most efficient and productive approach to returning to the Moon in the foreseeable future and to accomplishing this long-term economic goal. Any large scale private initiative focused on a Return to the Moon will have as its ultimate aim a return on investment from production and sale of lunar resources, in particular helium-3 for fusion electric power plants on Earth. In addition to helium-3 for fusion power, sales of by-products from its production, such as hydrogen, water and oxygen to customers in space, will add to bottom line income as well as to investor return. The same can be said of ancillary services based on the existence of an investor-financed lunar settlement and the new space transportation systems required to establish and service that settlement.

Lunar Structures

Haym Benaroya

While interest continues to increase with the nation's return to the Moon by astronauts for a permanent settlement there, the question of why we need to do this continues to be asked, even though answers abound. The brief list of whys follows:

- Safeguarding the species, by locating humans in off-Earth locations and for planetary defense for asteroid early warning and targeting;
- Attracting young Americans to the study of engineering, science and mathematics;
- Manifest destiny, competition against China, Europe and other space faring nations, staking a claim;
- Economic rewards such as resources, extraterrestrial site-specific benefits for industry, tourism, dual-use technologies;
- Science: astronomy, space and earth sciences, biological sciences;
- The Moon as a base for the development of "space legs" so humans can go to Mars and beyond, learn more about human and plant physiology, human psychology, in the space (radiation) and low-g environment, to learn how to "live off the land" in space;
- Creating an epic and positive vision for humanity as a balance to ward off societal pessimism, in particular for the coming generations, much as the New World and the West did during the Age of Exploration and the Westward Expansion in the United States. I believe this is the most forceful reason!

Concepts for lunar base structures have been proposed since long before the dawn of the space age. We will abstract suggestions generated during the past quarter century, as these are likely to form the pool from which eventual lunar base designs will evolve. Significant studies have been made since the days of the Apollo program, when it appeared likely that the Moon would become a second home to humans. Such studies continue today. While many ideas are futuristic and exciting, we must recognize the serious engineering and physiological issues that must be surmounted before a permanent and manned settlement exists on the Moon, and therefore that the first settlement is likely to be a very simple set of structures, but still very exciting.

Lunar Base 2015: A Preliminary Design Study

Werner Grandl

The Lunar Base 2015—design study is a concept for the return of humans to the moon from 2015 to the end of the century. The proposed lunar station (Stage 1) is built of 6 cylindrical modules, each one 17m long and 6m in diameter. 8 astronauts or scientists can live and work in the station. Each module is made of aluminium sheets and trapezoidal aluminium sheeting and has a weight of approximately 10.2 tonnes, including the interior equipment and furnishing. The outer wall of the cylinders is built as a double-shell system, stiffened by radial bulkheads to provide structural redundancy and for shielding. To protect the astronauts from micrometeorites, radiation and extreme temperatures, the caves between the two shells of the outer wall are filled with a 0.6m thick layer of regolith in situ by a small teleoperated digger vehicle. Using lunar material for shielding the payload for launching can be minimized. For launching the ARIANE 5 ESC-B rocket or a similar US or Russian launcher can be adapted. (12 tonnes payload required) For the flight from earth orbit to lunar orbit the modules can either be propulsed by a small rocket engine attached to each module or be moved by a "space tug" (one more flight is necessary). To land the modules on the lunar surface, a teleoperated "rocket crane" is used. This vehicle will be assembled in lunar orbit and is built as a structural framework carrying rocket engines, fuel tanks and teleoperated crawlers to land and move on the lunar surface. To establish the basic Stage 1 on the moon 11 flights are necessary: 1 flight- lunar orbiter, a small manned spaceship; 1 flight- manned lander and docking module for the orbiter; 1 flight- teleoperated rocket crane; 6 flights- lunar base modules; 1 flight- machinery: teleoperated digger vehicle, regolith ripper-excavator; 1 flight- scientific equipment, lunar rover, etc.

Extended version (future): Due to its modular design the Lunar Base can be enlarged in stages, finally becoming an "urban structure" for dozens of astronauts and tourists, always using the same launchers and machinery with current technology.

The New Race to the Moon: Old World Ideas versus New World Opportunities

R.D. Richards
Excalibur Moon LLC
Ontario, Canada

Between 1969 and 1972, twelve humans walked on another world. After the most awesome technological and psychological achievement of the human species, they left, never to return. Why? Was it a useless waste of human ingenuity on an Olympian folly? What caused humanity to abandon its first toehold on another world? More importantly, what is the basis to go back? What new forces and motivators are at play today that make the story a different one than the Apollo dead end?

Today there is a rebirth of interest in going back to the Moon among many nations. As co-chair of the International Lunar Conference in 2005, Bob Richards assembled the world's foremost scientists and policy makers to discuss humanity's return to the Moon. However while nations plan and strategize how to navigate the political minefields and conflicting national priorities that justify the value of the Moon to the everyday tax payer, there are some new kids on the block not so constrained. They are the privateers; visionaries too, however their driving metric for going to the Moon is sustainable business and commerce.

In this talk Bob Richards outlines how a carefully planned private Moon mission could set in motion the technological, political, legal and regulatory precedents that will allow humanity to rationally and peacefully embrace and develop the Moon as the world's eighth continent.

Krafft Ehricke's Moon: The Extraterrestrial Imperative

Marsha Freeman
Associate Editor
21st Century Science and Technology
Washington, DC

German-American space visionary Krafft Ehricke is well known for his statement: "It has been said, 'If God wanted man to fly, He would have given man wings.' Today we can say, 'If God wanted man to become a spacefaring species, He would have given man a moon.' "

Over the course of more than 30 years, Krafft Ehricke laid the philosophical basis for man's exploration of the Solar System, and created exquisitely detailed plans for the human settlement and industrial development of the Moon. In 1957, in his "Anthropology of Astronautics," Ehricke proposed that only man places limits on himself; that not only the Earth, but the entire Solar System are man's rightful field of activity; and that by "expanding through the universe, man fulfills his destiny as an element of life, endowed with the power of reason and the wisdom of moral law within himself."

Krafft Ehricke saw the exploration of space as based on an Extraterrestrial Imperative; that for mankind to grow and develop, a new "open world" was needed, not limited to the confines of Earth. Such an effort was not voluntary, but mandatory, he posited, because a growing world population and increasing standards of living would require new worlds and resources to explore and exploit. The only alternative would be increasing shortages of strategic materials and energy, increasing geopolitical conflicts, and, eventually, war.

A multi-decade approach to lunar development was proposed by Ehricke, who described the Moon as Earth's Seventh Continent. The most advanced technologies, based upon the use of nuclear fission power, automated and robotic systems, laser, and other directed-energy applications, would make industrial activity on the Moon, as, or more, productive than that on the Earth. With thriving lunar industries, mankind would no longer be limited to the resources of his home planet, and would have created the basis for the next steps in exploration, to Mars.

Krafft Ehricke's vision, his attention to engineering detail, and his incorporation of the most advanced technologies becoming available, should inform the methods being considered today for the approaching intensive robotic and then manned return to the Moon.

One can only imagine Ehricke's excitement were he alive today, at seeing the wide array of nations now planning the exploration and development of the Moon.

Apollo Knowledge Transfer

Charles Lundquist
Skycorp
Huntsville, Alabama
wingod@earthlink.net

Dennis Wingo
Research Institute
University of Alabama, Huntsville

A perplexing social issue is how to convey knowledge and experience from the Apollo Program in a way that is efficiently helpful to present teams planning return missions to the moon. Such a transfer of know-how is hard even when there is a continuity of work-force, but the transfer is exceedingly difficult when an interval of some forty years must be bridged. Surely historical documents from the Apollo Program exist in many places and forms that are accessible with enough effort. Realistically, the present government and contractor team members are so pressed with current issues that they feel they can devote but little time to primary literature searches. They can be aided by their social community.

Recognizing its long-standing relationship with the NASA Marshall Space Flight Center and the Huntsville contractor community, the University of Alabama in Huntsville, UAH, has accepted a role as a repository for space information deserving preservation in a publicly accessible archive. One mode of operation has been acceptance of personal materials donated by retirees from the Marshall Center and other individuals. Many video oral-history interviews have been conducted. The collections at UAH related to lunar exploration include a Saturn V collection, a Lunar Roving Vehicle Collection, an Apollo mission collection, and documents from the activities of the Apollo Group for Lunar Exploration Planning, GLEP. The GLEP considered many factors that influenced landing site selection and investigation plans for each Apollo mission.

Fortunately, present lunar team members, and students who will become team members, have grown up in the era of computer data bases and they are skilled at accessing such information. This suggests that one obvious aid to them is to provide Apollo knowledge and experience in a computer searchable format. The Archives and Special Collections Department in the library at UAH is among many entities that recognized and implemented this option. While it is prohibitive to immediately scan in full all the available documents, a selection of the most pertinent ones can be provided online. The others can have an online finding guide and abstracts to allow needed documents to be identified and retrieved.

An issue faced by all archives, but particularly space archives, is the circumstance that some old records exist in electronic formats that have become obsolete. The archiving entity then has the task of transforming these records into a currently used format, preferably one that facilitates online access.

The authors fully recognize that providing online and in-library access to information is only one option in preserving and conveying the Apollo experience. Scholarly surveys of Apollo material is another. Given the scope of the past, present and future lunar exploration programs, and recognizing the large number of organizations involved, the information preservation and transfer task is indeed a challenging social problem.

DAY 1: *Monday – June 4, 2007*

Afternoon Session
Plenary Speakers

Prof. Larry Bell

Director, SICSA, University of Houston
"Surface Infrastructure Planning and Design Considerations for
Future Human Lunar and Mars Habitation"

Prof. Lawrence A. Taylor

Director, Planetary Geosciences Institute, Department of Earth and Planetary Sciences
University of Tennessee, Knoxville
"Surface Infrastructure Planning and Design Considerations for
Future Human Lunar and Mars Habitation"

Day 1 Afternoon Schedule:

- 1:10–1:50 pm   Larry Bell
- 1:50–2:30 pm   Lawrence A. Taylor
- 2:30–2:50 pm   Päivi Jukola
- 2:50–3:10 pm   Gregory Konesky
- 3:10–3:30 pm   Henry W. Brandhorst
- 3:40–4:00 pm   COFFEE BREAK
- 4:00–4:20 pm   K. Zacny
- 4:20–4:40 pm   Paul D. Lowman Jr.
- 4:40–5:00 pm   Bernard H. Foing
- 5:00–5:20 pm   Yosef G. Tirat-Gefen
- 6:00–8:00 pm   OPENING RECEPTION

Surface Infrastructure Planning and Design Considerations for Future Lunar and Mars Habitation

Larry Bell

Professor/Director of SICSA
Sasakawa International Center for Space Architecture (SICSA)
University of Houston, Texas

This presentation discusses and illustrates ways that planning and design for habitable structures and human operations on the Moon and Mars differ in fundamental aspects from terrestrial circumstances on Earth. Key among these differences are severe launch and landing constraints upon equipment and element mass and volume; minimal or nonexistent availability of surface construction and site preparation systems, tools and labor resources; limited and potentially periodic power tied to solar source conditions; and temperature extremes and dust impacting equipment reliability and maintainability.

The presentation also offers a variety of facility types and configurations that respond to different design strategies and mission applications. Included are both conventional and expandable (such as inflatable and telescopic) pressurized structures, each type correlated with special advantages and limitations. These respective benefits and constraints influence how they can be landed on the surface, maneuvered and deployed; alternative ways they can be grouped together to meet evolutionary site development requirements, and radiation protection countermeasure options. Examples draw upon numerous research and design studies undertaken by the Sasakawa International Center for Space Architecture (SICSA) over a period spanning more than two decades.

In-Situ Resource Utilization on the Moon: A Marriage of Science and Engineering

Lawrence A. Taylor

Planetary Geosciences Institute
Earth & Planetary Sciences
University of Tennessee, Knoxville

The establishment of a base on the Moon, with human occupancy, will necessitate the use of the indigenous resources on the Moon. Virtually everything we know about the Moon and its rocks and soils comes from the science that evolved from the Apollo Program, continuing today with sample studies and refinements of earlier remote-sensing data. It is this knowledge of the physical and chemical properties of lunar regolith that forms the basis for the engineering endeavors to make it possible to effectively "live off the land" in our exploration endeavors.

"Science enables Exploration, and Exploration enables Science." It is this synergy that must be fully appreciated, realized, and utilized for this massive lunar endeavor to be successful. The properties of the lunar rocks and soils that make up the regolith vary widely over the lunar surface. The spatial distributions of the regolith, mainly as to its chemistry, have been forthcoming from the successful remote-sensing efforts of the Lunar Prospector and Clementine Missions, using the Apollo samples as ground truths.

Lunar architecture plans are to produce supplies of liquid oxygen (LOX) and liquid hydrogen on the Moon. The sources of these volatiles on the Moon are well-known; different processes for recovery of oxygen necessitate special feedstocks, and our science has already given us much of this knowledge. Regions of ilmenite-rich mare are well known, with such a feedstock being one of the best for hydrogen-reduction to release oxygen from this $FeTiO_3$ mineral. The presence of solar-wind particles on the surfaces of all lunar regolith grains has also been known since Apollo. The step-wise heating of lunar soil will easily release the hydrogen, helium, carbon, nitrogen, and other elements from the soil. And because the solar-wind implantation is a surface phenomenon, the finer fractions of the lunar soil possess the largest amounts of solar wind/mass.

These examples illustrate the synergies and symbiotic relationships between lunar exploration, with its largely engineering clientele, and lunar science, with its effective geologic and geochemical community. The major endeavor before us is tremendous and will only be possible by the effective union of these two divisions of knowledge, funding, and expertise. Apollo came about so successfully and quickly because there was no large division in personnel and administration between the engineers and scientists. Post-Apollo evolution of NASA has promoted the separation of these expertises. It behooves us to start thinking more like "materials scientists," who epitomize the effective marriage of engineering and science. Working together, as per Apollo days, is called for to accomplish our goals of a lunar base within our lifetimes.

Urban Innovations for the Moon

Päivi Jukola

Helsinki University of Technology, Finland

People as decision-makers in complex systems are the most critical element for system safety, reliability and performance. Their creativity, adaptability and problem solving capabilities are keys to effective performance. Design guidelines that decrease psychological and social stress factors can be critical to mission success even during short-term lunar missions. The importance of enjoyable home and work atmospheres, places for sport and cultural activities, increase during long-term assignments in isolated hostile environments. Sustainable urban development is a necessity for a larger population working and living on site. This article explores alternative strategies to create the first lunar technology park benchmarking similar investment projects on Earth.

According to the study in order to solve the financial, technical and cultural space exploration challenges it would be beneficial to direct more attention to human factors, behavioral finance theories and user-oriented design-principles. The most important customer is the taxpayer. Investors are more eager to invest when risks, profits and the outcome of the project are simple enough to understand and persuasively communicated. There is a need for detailed urban design drawings by architects to engage the public in interactive participation. Real estate development is a timely process that includes stakeholders from government officials to private investors and end-users, from engineers to architects and brand identity marketing experts. Feasibility studies and master plans are valuable tools when evaluating future scenarios for the experimental station concept. However, without timely discussion on international treaties and financial policies the public at large is less likely to give support for lunar projects.

The most severe environmental conditions cause death to the poorest on Earth every day. Lack of clean water and air, dust, extreme temperatures, waste management and energy efficient transportation are similar infrastructure challenges on the Moon. Sincere determination to solve ecological problems on Earth paves the way to environmental innovations in the lunar gravity. Recent virtual reality and technological innovations that allow end-users to manipulate elements of interior atmosphere according to one's own personal taste are valuable for astronauts in remote locations, or hospital patients and elderly living in institutions. Research on 3D visual simulation of real world environments and motion flight simulators is useful for training and virtual tourism to Moon and Mars. Thus, technological transfer opportunities are significant either way.

Settlement Site Selection and Exploration Through Hierarchical Roving

Gregory Konesky

SGK Nanostructures, Inc.
Hampton Bays, New York

While orbital reconnaissance is useful for initial lunar settlement site selection, it is no substitute for in-situ ground truth, which could be easily accomplished with teleoperated roving. The Mars Exploration Rovers (MER) experience with Sojourner in 1997, and Spirit and Opportunity in 2004 show how the selection of rover size influences the capabilities and nature of their respective missions. Smaller rovers can more closely explore a complex environment and tend to be more nimble, but at a cost of reduced payload capability. Larger rovers enjoy enhanced payload capabilities, but at a cost of being somewhat ponderous and difficult to maneuver in complex environments. The best of these extremes can be optimized by a hierarchical roving approach wherein a large rover carries a hierarchy of smaller specialized rovers. The large rover, in addition to serving as a transport for the collection of smaller rovers to a remote deployment site, also acts as a communications relay link and power recharge source for the smaller rovers. In a typical operational scenario, the smaller specialized rovers are deployed at a given site and execute their collective mission. They are then recovered by the large carrier rover and transported to the next site. Some of the benefits of hierarchical roving include greater situational awareness, redundancy, spatially distributed capability, and the opportunity for self-rescue. The large rover can also serve as an anchor point for tethered roving, permitting smaller rovers to negotiate steep slopes or down-hole exploration that would otherwise be inaccessible to a wheeled vehicle.

Design and operating experience with a test model carrier vehicle containing three smaller specialized rovers is discussed, as are the design tradeoffs. Test results from tethered rover operations are also presented.

A Solar Electric Propulsion Mission with Lunar Power Beaming

Henry W. Brandhorst, Jr.
Space Research Institute
Auburn, Alabama

Julie A. Rodiek
Space Research Institute
Auburn, Alabama

Michael S. Crumpler
Space Research Institute
Auburn, Alabama

Mark J. O'Neill
ENTECH, Inc.
Keller, Texas

As the NASA Vision for Space Exploration evolves, a key issue that affects lunar exploration is the ability to provide electric power at various surface locations. This power should be available through daylight times as well as at night. It is the purpose of this paper to describe an electric propulsion mission to the moon that will use laser power beaming to provide power to multiple locations on the lunar surface.

The major benefit of solar electric propulsion (SEP) is that more payload can be delivered to the moon for less cost than by chemical means. In addition, SEP allows orbital adjustment to permit a range of characteristics to fit the mission requirements at small fuel expenditures. However, one disadvantage of SEP is that it takes longer to reach the moon, but this is not a limiting factor for this case. This paper will describe a solar electric propulsion mission to the moon, insertion into an elliptical orbit and beaming laser power to the surface.

Many options exist for orbits around the moon that could be used for power beaming. Beaming power from the L1 point leads to a beaming distance of about 56,000 km. The constraints on laser power beaming over this distance lead to substantial losses. If a Molniya-type, highly elliptical orbit were chosen for the power beaming location, the apogee may be only about 12,000 km

712

which substantially reduces beaming distance, hence losses. However the length of time the lunar surface site is in view becomes important in order to keep the mass of the energy storage system on the surface small. In the same way, circular orbits of varying heights will encounter the same view time issue. So maximum elevation of the beaming spacecraft, the precession of orbits around the moon and the perturbations of lunar gravity all combine to complicate the analysis, and the results of these options will be presented.

For both the laser beaming spacecraft and the lunar surface receiving photovoltaic array, the Stretched Lens Array (SLA) on the SquareRigger platform design will be used. For the orbiting spacecraft, triple junction cells will be used in the array. A single cell test module with a triple junction cell and overall efficiencies of 29% has been demonstrated for this case. For the surface array, GaAs cells will be used to receive the beamed laser power. Testing of GaAs solar cells with a ~800 nm laser under the SLA has yielded efficiencies over 45% at room temperature.

This equates to over 800 W/kg and 800 W/m^2 at a 70–75°C operating temperature that is typical of a solar array in GEO. Temperature of an array orbiting the moon will depend upon its altitude and view angle of the lunar surface. Of course, the lunar surface temperature will markedly impact the surface array and those results are included in this study. As cell efficiency increases, the amount of waste heat decreases thus leading to an overall temperature reduction for a lunar surface array. This paper will present the results using one and two beaming spacecraft that will beam power only when the target site(s) are in darkness and the satellite is in sunlight. The impact of the Van Allen trapped radiation belts on the solar array power output will also be presented. The amount of power delivered to the surface is dependent upon the power level of the SEP spacecraft and will be presented for a nominal 100 kW BOL array.

Drilling and Excavating Technologies for the Moon

K. Zacny

J. Craft

S. Frader-Thompson

K. Davis

B. Glass
NASA Ames Research Center
Moffett Field, California

C. Stoker
NASA Ames Research Center
Moffett Field, California

Reaching the cold traps at the lunar poles and directly sensing the subsurface regolith is a primary goal of lunar exploration, especially as a means of prospecting for future In Situ Resource Utilization efforts. As today's missions to space are highly cost driven, flight systems must deal with modest limits to mass and power. This means that robotic systems must become more "intelligent" and capable of performing difficult tasks autonomously.

For the past 20 years Honeybee Robotics has been developing various drills and other excavating systems for extraterrestrial applications (Mars, Moon, asteroids, etc.). These systems differ based on the required depth of penetration, size, mass, required power and the level of autonomy. There is no doubt that any drill system can be scaled down in size and mass, however, the most difficult part is drill autonomy.

Deep drilling with limited power is certainly a difficult task to automate, however, not impossible. This has been proven by two robotic drill systems the MARTE and the DAME drills. Together, these drills have proven the technologies necessary for low-powered, fully autonomous deep drilling on any planet or moon.

The presentation will focus on describing various drill technologies with a wide range of autonomy, as well as reporting drilling tests in lunar soil simulants. In addition an innovative method of mining lunar top soil currently being investigated by Honeybee will also be presented.

Malapert Mountain: A Recommended Site for a South Polar Outpost

Paul D. Lowman, Jr.
NASA Goddard Space Flight Center
Greenbelt, Maryland

The Moon's poles have emerged as scientifically and strategically valuable areas for a new lunar program as proposed by President George W. Bush in 2004. The Clementine missions showed that there are large areas of permanently-shaded terrain at both poles, but the southern one is more concentrated and well-defined. Consequently most interest has been focused on the South Polar Region and specifically on the crater Shackleton because it is thought to be in continual sunlight, and is close to the hydrogen-bearing shaded areas. However, Malapert Mountain appears to have several advantages over Shackleton.

It was shown by D. Schrunk and B. Sharpe that Malapert Mountain is sunlit 90% or more of the lunar year, whereas Lunar Orbiter 4 pictures show that the rim of Shackleton is only partly illuminated at any one time. Furthermore, the illuminated area varies over the lunar year, and no one site is continually sunlit.

Malapert Mountain is old, pre-Nectarian terrain, probably with a thick regolith saturated with implanted hydrogen and helium. It offers a broad and smooth landing area, demonstrably in continuous microwave visibility of Earth (for tracking and communications). It is close to permanently-shaded areas to the south, which should be easily reached by a vehicle driven down the south flank of Malapert Mountain.

Malapert Mountain, in summary deserves careful study in light of mission safety, scientific importance, and evaluation of lunar resource.

SMART-1 Highlights and Lunar Settlements

Bernard H. Foing & SMART-1 Team

SMART-1 is the first of Small Missions for Advanced Research and Technology as part of ESA science programme "Cosmic Vision." Its objective is to demonstrate Solar Electric Primary Propulsion (SEP) for future Cornerstones (such as Bepi-Colombo) and to test new technologies for spacecraft and instruments. The spacecraft had been launched on 27 Sept. 2003, as Ariane-5 auxiliary passenger, and spiraled out towards lunar capture on 15 November 2004. It then spiraled down towards lunar science orbit (300-3000 km) until March 2005. The SMART-1 mission orbited the Moon for a nominal period of six months, with 1 year extension until end of mission impact on 3 September 2006.

The spacecraft has carried out a complete program of technology and science measurements. There is an experiment (KaTE) aimed at demonstrating deep-space telemetry and telecommand communications in the X and Ka-bands, a radio-science experiment (RSIS), a deep space optical link (Laser-Link Experiment), and the validation of a system of autonomous navigation (OBAN). For lunar science, the payload includes a miniaturized high-resolution camera (AMIE) for lunar surface imaging, a near-infrared point-spectrometer (SIR) for lunar mineralogy investigation, and a very compact X-ray spectrometer (D-CIXS) with a new type of detector and micro-collimator which will provide fluorescence spectroscopy and imagery of the Moon's surface elemental composition.

We shall also present the highlights of lunar science results from SMART-1 payload, featuring many innovative instruments and advanced technologies with a total mass of some 19 kg. SMART-1 lunar science investigations include studies of the chemical composition of the Moon, of geophysical processes (volcanism, tectonics, cratering, erosion, deposition of ices and volatiles) for comparative planetology, and high resolution studies in preparation for future steps of lunar exploration. The mission addresses several topics such as the accretional processes that led to the formation of rocky planets, and the origin and evolution of the Earth-Moon system.

We shall discuss ongoing SMART-1 collaborations with upcoming missions, lessons for the future exploration, and results relevant to preparing a human lunar settlement.

Validation of Mission Critical Power and Control Systems for Lunar Settlement

Yosef G. Tirat-Gefen

Aerospace Division, Castel Research Inc.
and George Mason University
Blacksburg, Virginia

Julio C. G. Pimentel

Department of Electrical and Computer Engineering
Laval University
Quebec, QC, Canada

A future settlement on the Moon would use a massive number of semi-automated or fully-automated complex mission-critical systems dealing with issues in areas as diverse as environmental control, air quality management, radiation protection, and biomedical monitoring. These systems will be very expensive and difficult to test. In addition to that, no validation methodology would completely preclude hardware-in-the loop testing of many of these subsystems. Power and control subsystems will be the backbone of many of these safety-critical systems. Therefore, the system degradation in the presence of transient or permanent faults should be fully accessed before final deployment. Also, the time and testing budget spent to set up and to fine-tune these systems will affect the overall economic feasibility of a mission. This work discusses the use of real time simulation for testing of power and control systems to be deployed in a lunar settlement. This simulation technology is capable of hardware-in-the loop validation. In other words, the system under test is able to operate in parallel and real-time with the simulator, where the latter drives and receives signals from the former through its analog and digital interfaces. Preliminary applications include subsystems in power electronics and biomedical monitoring. The simulator is based on a low cost reconfigurable computing infrastructure (e.g. embedded processors and field programmable gate arrays—FPGAs) and it is capable of having simulation steps on the order of 0.5 microseconds, which is enough to model several key electromechanical and power system modules. This work presents the simulator architecture and its implementation. We conclude with a discussion of how this technology could be leveraged with other approaches in safety-critical validation such as fault modeling and formal verification methods.

DAY 2: Tuesday – June 5, 2007

Morning Session
Plenary Speakers
James S. Logan, MD, MS

Space Medicine Associates, Belcamp, MD
"The Elephant in the Room: Biomedical Show-Stoppers
for Long Duration Human Lunar Habitation?"

Roger D. Launius, PhD

Chair, Division of Space History
National Air and Space Museum, Smithsonian Institution
"Why Go to the Moon? The Many Faces of Lunar Policy"

Day 2 Morning Schedule:

- 8:00–8:30 am   Registration
- 8:35–9:20 am   James S. Logan
- 9:20–10:05 am   Roger D. Launius
- 10:05–10:25 am   Alex Howerton
- 10:25–10:45 am   Julie Rodiek
- 10:45–11:20 am   COFFEE BREAK
- 11:00–11:40 am   Jerome Pearson
- 11:40–12:00 pm   Gweneth A. Smithers
- 12:00–1:00 pm   LUNCH

The Elephant in the Room: Biomedical Showstoppers for Long Duration Lunar Habitation?

James S. Logan
Space Medicine Associates
Belcamp, Maryland

The Vision for Space Exploration (VSE) envisions "permanent human presence" on the moon, first by establishing an "outpost" capable of supporting seven-day missions in 2020, then incrementally extending mission duration to as long as six months. NASA's Global Exploration Strategy team distilled the reasons for returning to the moon into six major themes (www.nasa.gov). The first theme is "Human Civilization: Extend Human Presence to the Moon to Enable Eventual Settlement," a truly daring goal considering the relative paucity of human lunar experience.

Six Apollo missions (1969–1972) cumulatively logged slightly less than 300 hours on the lunar surface including 81 total hours of lunar EVA. Since each Lander had a two-person crew, human beings now have almost 600 man-hours of lunar surface experience, approximately 162 of which are lunar EVA. Although the final and longest mission (Apollo 17) spent a little more than three days on the moon, the average lunar time per astronaut was only 2.08 days and the average lunar EVA time per astronaut was only 13.5 hours. While significant, these exposure times are not compelling from a biomedical perspective.

In contrast the current lunar concept of operations (CONOPS) consists of a crew of four placed on the lunar surface for variable durations living in a habitation "element" performing frequent lunar EVA—two teams of two crewmembers doing 6–8 hour EVAs every other day for six days (on the seventh day the crew presumably rests). It is astonishing to realize the VSE CONOPS will surpass cumulative Apollo time on the moon early on day 7 of the first mission and surpass cumulative Apollo lunar EVA time by the middle of week two.

Even if NASA can return to the moon for significantly less than the cost of Apollo ($105–160 billion in inflation-adjusted 2007 dollars), the magnitude of the investment merits early, frequent and exhaustive analyses of VSE strengths, weaknesses, opportunities and threats.

Three questions are germane to assessing potential threats.

First, if the goal of VSE is long-term lunar habitation or even settlement, are there biomedical showstoppers that could potentially threaten the VSE or the current CONOPS? Second, in the past 35 years (since the return of Apollo 17), what have science and our operational space experience taught us that could better qualify or quantify potential threats? Third, what are the implications of the answers to the above on the viability and eventual success of the VSE or of permanent human lunar settlement itself?

Why Go to the Moon?: The Many Faces of Lunar Policy

Roger D. Launius

Division of Space History
National Air and Space Museum, Smithsonian Institution
Washington, DC

What is it about the Moon that captures the fancy of humankind? A silvery disk hanging in the night sky, it conjures up images of romance and magic. It has been counted upon to foreshadow important events, both of good and ill, and its phases for eons served humanity as its most accurate measure of time. This paper discusses the Moon as a target for Human exploration and eventual settlement. This paper will explore the more than 50-year efforts to reach the Moon, succeeding with space probes and humans in Project Apollo in the 1960s and early 1970s. It will then discuss the rationales for spaceflight, suggesting that human space exploration is one of the least compelling of all that might be offered. The paper will then discuss efforts to make the Moon a second home, including post-Apollo planning, the Space Exploration Initiative, and problems and opportunities in the 2004 Vision for Space Exploration.

The Human Factor

Alex Howerton

Business Development Manager
NASTAR Center

Returning to the Moon is a noble and timely goal. There are many institutional and engineering challenges that have to be addressed, but no less important, and ultimately most important, are the human factors of the equation. If a successful Return to the Moon program is to be instituted, the human equation should be blended in at the outset.

The NASTAR Center is the National Aerospace Training and Research Center, located north of Philadelphia. Amongst other activities, we evaluate, train, and adapt people for spaceflight. This includes pilots, crews, and passengers, both government-sponsored and private. In addition, we support research into human factors and aeromedical issues of space launch physiology.

The NASTAR Center can support a Return to the Moon program in at least three distinct ways:

- We can model the specific flight profile, including G exposure, of any flying vehicle, including the proposed CEV for President Bush's VSE. Pilots can train for launch with the precise sensory inputs, including G exposure, they will actually experience upon launch and reentry.

- We offer training to private individuals planning for sub-orbital flights. The more people that experience spaceflight, the more will experience what Frank White has called "The Overview Effect." This can lead to deeper general public support for all space activity, including the Return to the Moon proposal.

- We can support research and data collection into the human factors involved in the Return to the Moon program, including optimal flight characteristics, ergonomics, biomedical monitoring, aeromedical issues, and related projects.

Both from a technical point of view as well as a cultural/political point of view, The Human Factor has to be addressed early, as an integral part of any plan to Return to the Moon. This can help assure the long-term success of such a venture.

Performance Projections for Solar Array Power Options on the Lunar Surface

Henry W. Brandhorst, Jr.
Space Research Institute
Auburn, Alabama

Julie A. Rodiek
Space Research Institute
Auburn, Alabama

Mark J. O'Neill
ENTECH, Inc.
Keller, Texas

Michael Eskenazi
ATK-Space, Inc.
Goleta, California

In NASA's ambitious vision for space exploration, return visits to the moon are the initial focus. The Lunar Architecture Team has concluded that the first objective is a polar outpost site, the Shackleton Crater rim being a notable option. This is due to the high percentage of sunlight throughout the year, greater than 70%, allowing solar power to become the major power source.

Lunar solar arrays should have the following characteristics: high efficiency, light weight, high packaging density and be able to withstand the broad temperature swings on the moon. In addition, for those robotic missions that will explore the permanently dark polar craters, it is possible that beamed laser power may be an option to radioisotope powered rovers. Of course beamed laser power may also be applicable to providing power over the nighttime.

This study will demonstrate the capability of the Stretched Lens Array on the SquareRigger platform (SLASR) as the basis. The Stretched Lens Array (SLA) developed by ENTECH is a space solar array that uses refractive concentrator technology to collect and convert solar energy into useful electricity. At the present time this design has the following characteristics: specific power – 300 W/kg, areal power density – 300 W/m2, stowed power – 80 kW/m3 and capable of high voltage (>600 V) operation.

One critical aspect of this study is the operating temperature on the moon and how it affects performance projections. Although a polar region seems certain for the first outpost, this study will take into account the wide temperature swings on the equator, and demonstrate that solar power is applicable for all lunar outpost sites. For equatorial sites the orientation of the solar array and the possible need to reduce the surface background will be included. Several surface treatments have been described in the past and will be used in this study.

The projected performance of a 25–30 kW lightweight, high efficiency SLASR array using multijunction solar cells expected to be available in 2010 time frame will be determined for a lunar polar region with high daylight during the year, an equatorial location during the day and an array in a permanently shadowed crater relying on laser illumination. The latter array will have GaAs solar cells matched to a nominal 800 nm wavelength laser and be sized for about 500 W.

Energy storage issues will also be discussed along with how much power can be delivered to the lunar surface. A detailed plan of how to build up a lunar base incrementally will be demonstrated. Power projections will show that the SLA is a lightweight, reliable, and cost effective power option for all locations on the lunar surface.

The Vision for Lunar Exploration: Results from the 2005 International Lunar Conference

R.D. Richards

Co-Chair, ILC 2005
Optech Inc.
Ontario, Canada

The 2005 International Lunar Conference on the Exploration and Utilization of the Moon took place in Toronto on September 18–23, 2005. Leading scientists and space industry professionals from around the world gathered in Toronto to discuss and plan the world's return to the Moon. ILC 2005 featured presentations of senior representatives from NASA, ESA, and CSA as well as the commercial sector to work toward a collaborative international lunar exploration framework.

ILC 2005 was the 7th conference of the International Lunar Exploration Working Group (ILEWG), a public forum sponsored by the world's space agencies to support "international cooperation towards a world strategy for the exploration and utilization of the Moon—our natural satellite" (International Lunar Workshop, Beatenberg (CH), June 1994).

The event was a catalyst and forum for senior stakeholders responsible for laying the framework for international space exploration. ILC 2005 cultivated ideas and inputs of scientists, engineers, policy makers, program managers and entrepreneurs from many nations and put these ideas into motion with concrete plans and roadmaps for a human and robotic future embracing the Moon, Mars, and beyond.

This paper summarizes the proceedings of ILC 2005 and the resulting declaration of the world's collaborative Vision for Lunar Exploration.

Lunar Frontier Transportation Options

Jerome Pearson

Star, Inc.
Mount Pleasant, South Carolina

John Oldson

Star Inc.
Mount Pleasant, South Carolina

Harry Wykes

Star, Inc.
Mount Pleasant, South Carolina

An integrated transportation system is proposed from the lunar poles to the equator, to L1, to Earth orbit, using a lunar space elevator, a system of lunar tramways and highways, and electric vehicles. The system could be used to supply lunar equatorial, polar, and mining bases with non-time-sensitive cargo, and could transport large amounts of lunar resources to Earth orbit for construction, radiation shielding, and propellant depots. We present options for long-range lunar surface transport, including "cars, trucks, and trains," along with their infrastructures of "roads, highways, and tramways" that can provide the interface between the local transportation around a lunar base and the space transportation of rockets and space elevators. The Apollo Lunar Rovers proved that wheeled vehicles could move fairly efficiently over the unprepared lunar surface, and they demonstrated a battery-powered range of nearly 50 kilometers. However, they also demonstrated the problems of lunar dust. For building dustless highways, it appears particularly attractive to create paved roads by microwaving the lunar dust into a hard surface. For tramways, it would be possible to erect towers to support high-strength ribbons that would carry cable cars over the lunar craters; the ribbon might even be fabricated from lunar materials. Such a transportation system could connect the polar mines with the equatorial lunar space elevator base, send cargo capsules up the lunar space elevator to beyond the L1 Lagrangian point, and from there rockets could take the cargo capsules with lunar resources into Earth orbit for building space habitats and space hotels for cislunar space development. We examine the power and energy storage requirements for lunar surface vehicles, the design and effectiveness of lunar tramways, and the materials requirements for the support ribbons of lunar tramways and lunar space elevators. This transportation system concept is adapted from an invited presentation at the Moonbase Workshop in January in Washington, DC, sponsored by the High Frontier and Jamestown on the Moon organizations, and is based on the results of a NIAC-funded study on the lunar space elevator. A 4-minute video is also available.

A One-Piece Lunar Regolith-Bag Garage Prototype

Gweneth A. Smithers

NASA
Marshall Space Flight Center, Huntsville, Alabama

Mark K. Nehls

NASA
Marshall Space Flight Center, Huntsville, Alabama

Mary A. Hovater

NASA
Marshall Space Flight Center, Huntsville, Alabama

Steven W. Evans

NASA
Marshall Space Flight Center, Huntsville, Alabama

J. Scott Miller

Qualis Corp.
Marshall Space Flight Center, Huntsville, Alabama

Roy M. Broughton, Jr.

Auburn University, Auburn, Alabama

David Beale

Auburn University, Auburn, Alabama

Fatma Killinc Balci

Auburn University, Auburn, Alabama

Shelter structures on the moon, even in early phases of exploration, should incorporate lunar materials as much as possible. We designed and constructed a prototype for a one-piece regolith-bag unpressurized garage concept, and, in parallel, we conducted a materials testing program to investigate six candidate fabrics to learn how they might perform in the lunar environment. In our concept, a lightweight fabric form is launched from Earth to be landed on the lunar surface and robotically filled with raw lunar regolith.

In the materials testing program, regolith-bag fabric candidates included: VectranTM, NextelTM, Gore PTFE FabricTM, ZylonTM, TwaronTM, and NomexTM. Tensile (including post radiation exposure), fold, abrasion, and hypervelocity impact testing were performed under ambient conditions, and, within our current means, we also performed these tests under cold and elevated temperatures. In some cases, lunar simulant (JSC-1) was used in conjunction with testing. Our ambition is to continuously refine our testing to reach lunar environmental conditions to the extent possible.

A series of preliminary structures were constructed during design of the final prototype. Design is based on the principles of the classic masonry arch. The prototype was constructed of KevlarTM and filled with vermiculite (fairly close to the weight of lunar regolith on the moon). The structure is free-standing, but has not yet been load tested. Our plan for the future would be to construct higher fidelity mockups with each iteration, and to conduct appropriate tests of the structure.

DAY 2: Tuesday – June 5, 2007

Afternoon Session
Plenary Speakers
Brent Sherwood
Manager, Opportunities Development
NASA Jet Propulsion Laboratory
"What Will We Actually Do on the Moon?"

Prof. Terry Hart
Professor, Lehigh University and past NASA Astronaut
"Working in Space"

Day 2 Afternoon Schedule:

- 1:00–1:45 pm   Brent Sherwood
- 1:45–2:30 pm   Terry Hart
- 2:40–3:00 pm   Carlton L. Rhoades
- 3:00–3:20 pm   Gary C. Fisher
- 3:20–3:40 pm   Florian Ruess
- 3:40–4:00 pm   COFFEE BREAK
- 4:00–4:20 pm   M. Dünne
- 4:20–4:40 pm   Jason R. Florek
- 4:40–5:00 pm   Yuriy Gulak
- 6:00–11:00 pm   SPECIAL EVENING EVENT (NEW YORK HARBOR TOUR)

What Will We Actually Do on the Moon?

Brent Sherwood

NASA Jet Propulsion Laboratory
Pasadena, California

Descriptions are provided for eleven specific, representative lunar activity scenarios selected from among hundreds that arose in 2006 from the NASA-sponsored development of a "global lunar strategy." The scenarios are: pave for dust control; establish a colony of continuously active robots; kitchen science; designer biology; tend the machinery; search for pieces of ancient Earth; build simple observatories that open new wavelength regimes; establish a virtual real-time network to enable public engagement; institute a public-private lunar development corporation; rehearse planetary protection protocols for Mars; and expand life and intelligence beyond Earth through settlement of the Moon. Evocative scenarios such as these are proposed as a communications tool to help win public understanding and support of the Vision for Space Exploration.

Working in Space

Terry Hart

Professor, Lehigh University and Former NASA Astronaut

Since the first cosmonauts and astronauts walked in space, we have been climbing a continuing learning curve of how people can work productively in space. The effects of weightlessness and the physical limitations of pressure suits and spacecraft designs continue to challenge crews as ever-more sophisticated tasks are being accomplished.

And while we have come a long way in our ability to work in space, much needs to be done if we are to return to the moon with a permanent presence and venture on to Mars. Such long-duration missions will put new challenges on engineers and crews to adjust to the physical and psychological demands of these missions. With international cooperation, these challenges will be met and crews will learn to work effectively as we establish a permanent presence in space.

The Moon: First Line of Asteroid Defense

Carlton L. Rhoades

Millions of asteroids orbit the sun and, during each orbit, may intersect Earth orbit. Hundreds of thousands of these are of the "small" 50 to 100 meter size that could provide another Meteor Crater example. The same event at that site today would put at risk Interstate 40, the Burlington Northern and Santa Fe railroad, major electrical power transmission lines, a major natural gas transmission line, and over 10,000 people. Other effects would include overloading of alternate routes between Greater Los Angeles and the Midwest, and interruption of the electrical and gas service to the large areas served. The cost to recovery could be in the tens of billions. The event at or near a metropolitan center anywhere in the United States would be a catastrophe with orders of magnitude greater casualties and costs. Less than one percent of these "small" asteroids have been identified. Those identified as Earth impactors must have an orbit alteration to mitigate the threat.

The Moon is the ideal platform to expedite identification of Earth orbit intersecting asteroids. The Lunar sky obstructed only by the Earth, the Sun, the zodiacal light, and the local horizon can be searched 24/7.

This paper will expand on development of a Moon based, remotely operated, asteroid search system. The expansion will consider use or upgrades of existing subsystem elements before considering new subsystem elements. The expansion will consider shielded and pressurized accommodations for use by occasional on site personnel. The elements are:

Determine optimal search qualities and select site(s) having those qualities.

Determine requirements for the system including Logistics, Search, and Support Equipment.

Design and construct the Logistics, Search, and Support Equipment.

Transport Logistics Equipment to selected site(s) and prepare the site(s) for use.

Transport the Search and Support Equipment to the selected site(s) for installation and checkout.

Horizontal or Vertical Cylindrical Habitat?

Gary C. Fisher
The Mars Foundation
Bryn Athyn, Pennsylvania

For the purposes of creating in situ habitable spaces for a Lunar settlement, cylinder-shaped structures covered with regolith present the best near-term option over alternatives, such as: spherical or domed shaped structures, excavated structures, lined lava tubes, and regolith covered masonry vaulted arches.

Cylindrical structures have various options besides length and diameter. For example: hemispherical or flat end caps; rigid or inflatable? Another option is whether you create a true cylinder, or go with a more ellipsoid or flattened cylinder. The primary consideration, however, is whether to stand the cylinder up vertically on an end, or lay it down horizontally on its side. This paper is primarily concerned with deciding between these two options.

While this paper is part of the ongoing research of the Mars Foundation to design the first permanent settlement on Mars, the subject matter is equally relevant to a Lunar base or settlement.

Structural Reliability Considerations for Lunar Base Design

Florian Ruess

Benjamin Braun

The National Aeronautics and Space Administration (NASA) recently announced to build a permanent lunar base by the year 2020 [1]. The erection of such an outpost on the Moon is not only a question of how to choose the most suitable structure for a lunar habitat as summarized and presented in [2]. But also the risk and safety of such structures need to be assessed in an appropriate manner. For the design on Earth, standards and codes (e.g. the family of American Standards or Eurocodes) exist that help the designer to analyze and design a safe structure. However, special uncertainties exist in lunar base design which can be assessed as a first step only on the basis of a decision analysis, namely a structural reliability analysis. It offers a way to allocate the available resources both most efficiently and in a sufficiently safe manner. Fundamentally, uncertainties can arise from an inherent randomness in loads and resistances, inadequate knowledge (e.g. influence of radiation on material properties) and statistical uncertainty due to sparse information. Such uncertainties imply the need to use statistical and probabilistic tools in the analysis and design process. The basic principles of structural reliability analysis [3], [4] are introduced here with special regard to the design of second and third generation lunar habitats. A further key aspect in designing robust structures for the lunar surface is redundancy i.e. component failure should not govern system failure. In this respect the reliability analysis is finally discussed on the level of the overall structural system.

1. Dale, S.: Exploration Strategy and Architecture. Opening Keynote, 2nd Space Exploration Conference, Houston, Texas, December 4–6, 2006.
2. Ruess, F.; Schaenzlin, J.; Benaroya, H.: Structural Design of a Lunar Habitat. Journal of Aerospace Engineering 19 (2006), No. 3, pp. 133–157.
3. Thoft-Christensen, P.; Baker, M.J.: Structural Reliability and its Applications. Springer Verlag, Berlin (1982).
4. Ditlevsen, O.; Madsen, H.O.: Structural Reliabilty Methods. John Wiley & Sons, Chichester (1996).

AstroHab: A Multidisciplinary Payload for a Lunar Precursor-Mission

M. Dünne

OHB-System AG
Bremen, Germany

K. Slenzka

OHB-System AG
Bremen, Germany

For humans to live in extreme environments such as Moon and Mars requires sustainable closed ecological life support systems (CELSS) as well as knowing more about the impact of different environmental conditions (mainly gravity, magnetism, and radiation) on biological processes in larger time scales.

Essential steps before longer explorative missions are investigations in extreme environments on Earth, in LEO (ISS) and also a series of missions to the Moon. Developed support technology can be verified without having exceeded the point of no return of astronauts. A first precursor-mission to the Moon will demonstrate the efficiency of developed life support technologies and increasing the knowledge about existing environmental conditions on the Moon (which are more stable and relevant than at the ISS) as well of their impact on biological and other processes.

OHB-System, Bremen, Germany, is currently analyzing in the Mona Lisa Initiative of German Space Agency DLR such a first precursor-mission to push the lunar settlement significantly. Analysis is performed together with a scientific team, whose members come from several disciplines. R&D activities were started to develop main elements of CELSS. The general concept for AstroHab—the autonomous biological payload for this first precursor-mission—was developed. AstroHab will consist of a biosolar energy supply unit as first CELSS-element as well as an experiment bioreactor and a sensor unit, which will monitor the environmental conditions at the landing site as well as further experiment conditions in the bioreactors.

Additional to contributing to humans' ambition of exploring existing frontiers, activities serve as a technology and scientific driver in general.

An overview of AstroHab and the status of current R&D activities will be given.

Extension of Terrestrial Excavation Mechanics to Lunar Soil

Jason R. Florek

Department of Mechanical Engineering
Rutgers University
Piscataway, New Jersey

This presentation focuses on extending models for cutting and moving terrestrial soil for use with lunar soil. This area is of particular importance since nearly all lunar base designs call for some form of regolith shielding for protection against radiation and micrometeorite impact. Before such shielding can be constructed, the forces associated with digging and excavating regolith must be well understood.

It has been shown that Earth-based analyses do not directly translate for use with lunar soil. Required forces do not just simply scale by one-sixth due to the reduced lunar gravity. Furthermore, papers that make similar extensions typically assume a single value for the important lunar soil properties (e.g., cohesion or internal friction angle). Most studies do not take into account the variability of these parameters with depth or the uncertainty associated with the property values. In contrast, those studies that account for this variability tend to ignore how the parameters relate with one another. For example, the aforementioned properties are both related to soil density, while failure surface angle is a function of the rake and friction angles. Published parameter studies tend to make these quantities independent of one another.

As such, here, appropriate ranges of parameter values are considered for input to various two- and three-dimensional excavation models. Parameter dependencies and uncertainties are also accounted for, resulting in a more realistic calculation of the forces required to cut and move lunar soil. Comparisons are made between these forces and those required to move lunar simulants in both a 1 g and 1/6 g environment. Additionally, recommendations are made for which contributing factors can be ignored in a simplified analysis.

Heat Pipes: How to Increase the Capillary Heat Transfer Limit?

Y. Gulak

CSXE, Rutgers University

Heat pipes are popular heat transport devices in the aerospace and ground-based applications that provide high efficiency in transferring the thermal energy. They operate on a closed two-phase cycle, in which the heat of evaporation of the working fluid is carried out between the heat source and the heat sink. Typically, heat pipes are light-weight and do not require external power.

The performance of low and moderate temperature heat pipes might be limited due to several well-known factors, the most important of which are the capillary and boiling limitations. In the talk, we discuss the possibility of maximizing the capillary transfer limit by designing the wick whose porosity is allowed to vary along the heat pipe length. An optimal porosity distribution is then calculated as a solution of a non-smooth optimization problem for zero, Moon, and Earth gravity heat pipe's working conditions.

DAY 3: *Wednesday – June 6, 2007*

Morning Session

Plenary Speaker

Prof. Rupert Gerzer
Director of the Institute of Aerospace Medicine
German Aerospace Center (DLR), Cologne-Porz
"Travel Medicine: Medical Suggestions for Trips to Moon and Mars"

Day 3 Morning Schedule:

- 8:30–9:00 am   Registration
- 9:00–9:45 am   Rupert Gerzer
- 9:50–10:10 am   Sheryl L. Bishop
- 10:10–10:30 am   Igor Smirnov
- 10:40–11:00 am   COFFEE BREAK
- 11:00–11:20 am   Satadal Das
- 11:20–11:40 am   Gene Giacomelli
- 11:40–12:00 pm   Prasanta Banerji
- 12:00–1:00 pm   LUNCH

Travel Medicine: Medical Suggestions for Trips to Moon and Mars

Rupert Gerzer

Director, Institute of Aerospace Medicine
German Aerospace Center
Cologne, Germany

Going to Moon and staying on a Moon habitat for an extended time period or even going to Mars poses several challenges for medicine: keeping the crews healthy despite weightlessness or reduced gravity, keeping them motivated, protecting them from hazards like radiation or fine dust, supporting them during emergencies and providing an effective and affordable habitat including closed loop regeneration systems.

Presently, astronauts on a long term mission in weightlessness are supposed to do physical exercise for about 150 min per day in order to counteract negative weightlessness effects. Still, they have problems post flight like orthostatic intolerance and bone and muscle loss. Thus, novel countermeasures are needed. One such method might be artificial gravity: a short arm centrifuge with a radius shorter than 3 m could be very useful. The astronaut – with head in the centre – would spin, and at the same time, he/she could do additional exercise like leg vibration and/or ergometer or treadmill training. Due to the gravity vector, especially the lower body would be accelerated, thus strongly stimulating the vascular system in the lower body and training vascular resistance to pressure. At the same time working and being vibrated would give a combined stimulus to the musculoskeletal and cardiovascular systems. Thus, such a method might be able to reduce daily training to less than an hour and be more effective than present methods.

Special emphasis should be given to radiation protection. Both, external shielding by new materials and technologies and internal shielding by supporting molecular radiation protection mechanisms of the human body are necessary. This will also include selection of those astronauts who are most likely to have high physiological resistance to cancer.

The biggest challenge in medicine for the future of human spaceflight is the development of a "digital astronaut" system that involves an intelligent storage of personal medical information of the respective astronaut on the one side, and up-to-date medical knowledge on the other side, and that is able to give individualized support in case of an emergency. Such systems are needed in terrestrial medicine as well, will enable individual support of people wherever needed and improve homecare of aged people and long term patients dramatically.

Habitats for astronauts will make the development of affordable bioregeneration systems important. Results can also be applied on earth for many applications.

In summary, medical care for astronauts on long term missions in space or on stations on Moon or Mars requires the development of many new technologies and applications that are also urgently needed on earth. Due to the many unresolved tasks, we should initially focus on the development of systems that help to keep astronauts on such missions healthy and thus contribute to the task to improve the possibilities for human presence in space.

Here to Stay: Designing for Psychological Well-Being for Long Duration Stays on Moon and Mars

Sheryl L. Bishop

University of Texas
Galveston, Texas

Current psychological and sociocultural considerations for Moon and Mars bases are embedded within a complex matrix of long duration issues that have been demonstrated to significantly impact on human behavior and performance. Planning for psychological health for long duration stays rather than short duration task accomplishment profiles is further complicated by the need to address both flight issues as well as station issues. Long duration Moon missions will allow us to develop and test effective structural and procedural countermeasures for group and individual well-being that address both in-flight and planetside factors related to: 1) a reliance on technology for life support and performance; 2) physical and social isolation and confinement; 3) high risk and associated cost of failure; 4) high physical/physiological, psychological, psychosocial, and cognitive demands; 5) human-human, human-technology, and human-environment interfaces; and 6) requirements for team coordination, cooperation, and communication.

The fundamental challenge to developing effective countermeasures for long duration missions is that space missions are not truly psychologically comparable to any other undertaking humans have ever attempted differing most notably in the enormous distance to travel, the unique separation from the rest of humanity and the extraordinary environmental demands. Most current support strategies used to foster crew morale and psychological well-being were developed on Mir and ISS and rely on intensive ground-based support which may be ineffective for missions whose profiles will be as characterized by the length of stay as it is the list of tasks to accomplish.

A broad spectrum of social psychological and behavioral research has contributed to the emerging realization that many of the negative psychological factors of long duration Moon/Mars missions (e.g., prolonged isolation, confinement, exposure to unpredictable and unknown extreme environment, reliance on closed loop environmental system) could be mitigated by designing habitats that thoughtfully countered some of these impacts. Instead of a focus on mere survivability, efforts to promote and sustain elements that contribute to human thriving appears to be far more productive. The intersection of psychology and psychosocial factors with habitat design allows us to implement psychological support in a non-intrusive, holistic environmental modality that is preventive in orientation rather than palliative. The present paper discusses those psychological factors amenable to a thoughtful and proactive integration into support systems and countermeasure environmental elements, including enabling technologies on the horizon that would significantly contribute to the successful psychological adaptation of long duration space inhabitants.

MRET Activated Water and Its Successful Application for Prevention Treatment and Enhanced Tumor Resistance in Animal Oncology Models

Igor Smirnov, PhD

Global Quantech, Inc.
San Marcos, California

Objectives:
The goal of this investigation was to study the effect of MRET water for the prevention and treatment of two kinds of oncology diseases on mice (laboratory models of Ehrlich's ascites tumor and Sarcoma ascites form). MRET Water is produced with the help of patented (US Patent No. 6,022,479), non-chemical Molecular Resonance Effect Technology (MRET). The anomalous electrodynamic characteristics and viscosity of MRET water provide some evidence regarding the possible effect of MRET water on electrical activity and proper function of the cells.

Methods:
The ability of animals for tumor resistance was studied in the experiments conducted on 500 mice (22 groups with 20 mice in each and 10 groups with 5 mice in each group) with the help of the following methodology: a) study of possible anti-tumor effectiveness of "preventive" administration of different fractions of MRET water; mice received MRET water during 2 weeks before tumor cell transplantation and after transplantation; b) study of possible anti-tumor effectiveness of "therapeutic" administration of different fractions of MRET water; mice received MRET water after tumor cell transplantation; c) investigation of functional citotoxic activity of lymphocytes containing natural killer cells (NK-cells) isolated from spleens of mice (without tumors) which received MRET water; lymphocytes were incubated with tumor target cells.

Results:
The experimental results confirm that consumption of all types of MRET water leads to the significant inhibition of tumor growth and suppression of mutated tumor cells. The best results were observed in the groups of mice on MRET water activated for 30 minutes (optimal regime). The resulting decrease of the Total Number of Viable Tumor Cells was 76% in "preventive

treatment" group and 55% in "therapeutic treatment" group. The observed average survival time of mice which received optimal activated water in "preventive treatment" regime increased by 61.7% compare to the control group. The increase of **cytotoxic index** in both regimes (**21 days and 14 days** of application of activated water for mice without tumors) **by 26% and 10%** respectively was observed only in the groups of mice under MRET water activated for 30 minutes.

Conclusions:

The significant positive effect of MRET activated water on tumor resistance of animals was observed in the process of this investigation in vivo in all groups of mice on different fractions of activated water. The significant anti-tumor effect of MRET Activated distilled water on mice was close to the action of the chemotherapy agents and allowed to avoid the side effects that typically follow chemotherapy treatment of oncology. The application of activated water can be a quite promising approach for non-drug stimulation of NK-cells immunization vaccines.

A Simple Differential Production Method of Silicon Utilizing Organisms for Future Use in Lunar Settlements

Satadal Das

Roy Research Centre
Peerless Hospital & B.K.
Kolkata, India

Silicon utilizing organisms are probably the fittest living creatures having a capacity of survival in extraterrestrial situations where they can tolerate more environmental stress and strain than their equals on Earth. One can also classify them according to their silicon utilizing capacity very easily.

Silicon utilizing organisms can thrive in sodium metasilicate (SM) solution as high as up to 4% concentration. To confine common silicon utilizing organisms from the environment for future use in lunar settlements one has to prepare SM solutions of four different concentrations: 0.5%, 1%, 2% and 4%. After preparation of such solutions in plastic containers one has to keep them in a greenhouse for as long as 5 years. Different varieties of organisms will grow in different concentrations- from a light green color growth in 0.5% SM solution, yellow color growth in 1% SM solution, orange color growth in 2% SM solution and a scanty whitish color growth in 4% SM solution. Besides many unknown microorganisms, algae are present in every solution but are of different kinds. Diatoms of diverse varieties are found in profound numbers in 0.5% and 2% SM solutions; plenty nocardioforms are also found in 1% SM solution and scanty fungi are usually present in 4% SM solution.

During growth of silicon utilizing organisms in SM solutions there are many biochemical changes in the medium. While hardness of the water cannot be measured in silicate solutions, pH is almost neutral in 2% solution, while it is always higher even after 5 years in other silicate solutions. Chlorides are very high in 4% solution. There is about 50% increase of sulfate and 25-50% increase of nitrate in all the SM solutions, marked increase and decrease (both about 4 times) amount of iron in 2% and 4% solutions respectively. These changes are probably due to disparity growth of organisms in different concentrations of silicate.

A simple protocol may be followed to use these silicate-utilizing organisms in lunar settlements. After providing minimum essential requirements for life in lunar extraterrestrial situation, these organisms may be utilized. In the initial venture antibiosis between various species should be prevented. Thus phytoplankton should be used before zooplanktons. Diatoms of Eu-eurytherm variety of Nitzschia and Chaetoceros group may be selected initially. Then golden algae grown in 2% and then in 0.5% SM solutions may be scattered to boost up the algal inhabitants. Then important silicon utilizing plants (specific silicon utilizing strains) like horsetails, grasses, lilies, silver vase, spider plant and following that organisms (only extremophile variety) like rotifers, tardigrades, nematodes, protozoa, fungi and bacteria may be added which will live in close association of small silicon utilizing plants and this process may continue. As it is not practicable to carry all essential nutrients for lunar settlements, creation of such biosphere is essential for future survival of inhabitants in lunar settlements.

Development of a Lunar Habitat Demonstrator

Phil Sadler

CEO
Sadler Machine Company
Tempe, Arizona

Gene Giacomelli

Director
University of Arizona
Controlled Environment Agriculture Center
Tucson, Arizona

Lane Patterson

Graduate Student
University of Arizona
Agricultural & Biosystems Engineering
Tucson, Arizona

Roberto Fufaro

Asst. Research Professor
University of Arizona
Aerospace & Mechanical Engineering
Tucson, Arizona

The NSF Amundsen-Scott South Pole Station is recognized as one of the best analogs for a future Lunar or Martian surface effort. The station and crew of over 50 individuals being physically isolated for 8 months at the coldest and most remote environment on Earth is a valuable asset with a high degree of mission fidelity for future Lunar and Martian surface efforts. We at the University of Arizona's Controlled Environment Agriculture Center (UA-CEAC), in cooperation with Sadler Machine Co. (SMC) have constructed, delivered (in 2004), and continue to remotely support operation of the South Pole Food Growth Chamber (SPFGC) for the NSF's Office of Polar Programs civilian operations contractor, Raytheon Polar Services Company. The SPFGC provides the isolated winter crew with a continuous supply of a multitude of fresh hydroponic grown produce and salad crops from an artificially lit (24 square meter) growth chamber located inside the station building. From

748

lessons learned from this effort we developed the Cable Culture growing system for use in future inflatable membrane structure Lunar and Mars greenhouse modules, while trying to achieve the lowest ESM number. In 2005 UA-CEAC/SMC fabricated a conceptual Mars Greenhouse to demonstrate the deployment and operation of this Cable Culture growing system. With the change in focus from a Mars Mission to a Lunar Base, we are constructing 4 demonstration Lunar Greenhouse modules to further develop the Cable Culture growing system and demonstrate water recycling and air revitalization using ALS/CELSS technology. This life support/greenhouse component is highly integrated with the total habitat and required us to develop an entire conceptual Lunar habitat, which is in progress. Our conceptual Lunar habitat consists of six module trains radiating from a central hub and is designed to be deployed autonomously with crops growing by the time the human crew arrives.

Possible Use of Ultra-Diluted Medicines for Health Problems During Lunar Missions

Prasanta Banerji
PBH Research Foundation
Kolkata, India

Satadal Das
PBH Research Foundation
Kolkata, India

Gobinda Chandra Das
PBH Research Foundation
Kolkata, India

Pratip Banerji
PBH Research Foundation
Kolkata, India

Conventional medicines are probably not suitable for addressing health problems in lunar missions due to the fact that they are not utilized by the system properly in their optimum dose which is necessary in space.

Thus one should investigate whether ultra-diluted medicines, with proven efficacy, may be used without these problems in space. Ultra-diluted medicines are non-toxic, with extended shelf-life, non-addictive, with negligible weight and volume, low cost, and easily administrable. We have classified ultra-diluted medicines for use in lunar missions into different groups according to their roles in different health problems. These classifications have been done as per their proven actions, on various health problems matched with our long experience on different patients with similar health problems, in the earthly environment.

Following our classification we prepared one combination medicine (PBHRF-1), which may be administered to the astronauts from one week before the start of the mission, containing Lycopodium clavatum 30c, Symphytum officinalis 200c, Berberis vulgaris 200c, Nicotiana tabacum 200c, Fluoricum Acidum 200c, Coffea arabica 200c, Ruta graveolens 6c, Calcarea Phosphorica 3X, Kali Muriaticum 3X, and Ferrum Phosphoricum 3X. These medicines are incorporated in lactose globules (5 grain ~ 0.324 g) and are to be taken 3 to 4 doses per a day.

This combination medicine will prevent or alleviate different health problems in space such as: "fluid shift" (stuffy nose, headache, puffy face, facial edema), bone loss, renal stone formation, destruction of anti-gravity muscles, early motion sickness in space, protection from radiation, destruction of RBC, immunosuppression (due to reduced action of lymphocytes), mental stress, insomnia etc. Similarly, another medicine, PBHRF-2 (a combination of Aconite napellus 200c and Crataegus oxyacantha 3X in same dosage) may be used to prevent health problems such as cardiac deconditioning during return to earth. These ultra-diluted medicines can be used easily in space as they are least affected by gravity, radiation, and thermal changes during space missions.

DAY 3: Tuesday – June 5, 2007

Afternoon Session

Plenary Speaker

George C. Nield

Deputy Associate Administrator for Commercial Space Transportation
FAA/AST-2
"Commercial Spaceports—An Overview and Status Update"

Day 3 Afternoon Schedule:

- 1:30–2:15 pm   George C. Nield
- 2:20–3:00 pm   William J. Rowe
- 3:00–3:40 pm   Tom Taylor
- 3:00–3:20 pm   COFFEE BREAK
- 4:00–4:20 pm   Steve M. Durst
- 4:20–4:40 pm   Chester S. Spell
- 4:00–4:20 pm   M. Dünne
- 4:40–5:00 pm   Manny Pimenta
- 6:00–9:00 pm   BANQUET at Zimmerli Museum

Commercial Spaceports:
An Overview and Status Update

George C. Nield
Deputy Associate Administrator for Commercial Space Transportation
Federal Aviation Administration/AST-2
Washington, D.C.

Although near-term missions to the moon are likely to be conducted by NASA or comparable government agencies from other nations, commercial entities will one day provide the majority of the necessary transportation services. Even today, non-federal spaceports play an important role in supporting our country's space programs. The purpose of this presentation is to review the statutory authority for the regulation of spaceports, to summarize the applicable federal regulations, and to provide a status update on the activities of current and proposed commercial launch and reentry sites.

Moon Dust May Simulate Vascular Hazards of Urban Pollution

William J. Rowe

Former Assistant Clinical Professor of Medicine
Medical University of Ohio at Toledo

A long duration mission to the moon presents several potential cardio-vascular complications. To the risks of microgravity and hypokinesia, and the fact that pharmaceuticals cannot be always depended upon in the space flight conditions, there is a possible additional risk due to inhalation in the lunar module of ultra fine dust (<100nm). This may trigger endothelial dysfunction by mechanisms similar to those shown to precipitate endothelial insults complicating ultra fine urban dust exposure. Vascular constriction and a significant increase in diastolic blood pressures have been found in subjects inhaling urban dust within just two hours, possibly triggered by oxidative stress, inflammatory effects, and calcium overload with a potential magnesium ion deficit playing an important contributing role. Both Irwin and Scott on Apollo 15 experienced arrhythmias, and in Irwin's case associated with syncope and severe dyspnea with angina during reentry. After the mission both had impairment in cardiac function, and delay in cardiovascular recovery, with Irwin in addition having stress test-induced extremely high blood pressures, with no available stress test results in Scott's case for comparison. It is conceivable that the chemical nature or particle size of the lunar dust is sufficiently variable to account for these complications, which were not described on the other Apollo missions. This could be determined by non-invasive endothelial-dependent flow-mediated dilatation studies in the lunar environment at various sites, thereby determining the site with the least endothelial vulnerability to dysfunction. These studies could be used also to demonstrate possible intensification of endothelial dysfunction from inhalation of ultra fine moon dust in the lunar module.

Lunar Commercial Logistics Transportation

Walter P. Kistler

President
Lunar Transportation Systems, Inc.
Bellevue, Washington

Bob Citron

CEO
Lunar Transportation Systems, Inc.
Bellevue, Washington

Tom Taylor

Vice President
Lunar Transportation Systems, Inc.
Las Cruces, New Mexico

This paper offers a commercial perspective to new lunar transportation and proposes a logistics architecture that is designed to have sustainable growth over 50 years, financed by private sector partners and capable of cargo transportation in both directions in support of lunar resource recovery. The paper's perspective is from an author's perspective of remote sites on Earth and some of the problems experienced in logistics that didn't always work. The planning and control of the flow of goods and materials to and from the Moon's surface may be the most complicated logistics challenge yet to be attempted. The price paid if a single system does not work well is significant. On the Alaskan North Slope, we had four different logistics transportation systems and none worked successfully all the time. The lessons learned will be discussed and solutions proposed. The industrial sector has, in the past, invested large sums of risk money, $20 billion for example, in resource recovery ventures like the North Slope of Alaska, when the incentive to do so was sufficient to provide a return on the risk investment. Stimulating an even larger private investment is needed for the Moon's resource development. The development of the Moon can build on mankind's successes in remote logistics bases on Earth and learn from the $20 billion in private sector funds used to recover oil assets above the Arctic Circle.

The Moon is estimated to be 50 times more remote than Prudhoe Bay, Alaska, the early transportation to the Moon is 100 to 1,000 times more expensive than to the Arctic and the lunar environment is more severe than the Arctic, but some of the logistics lessons learned in the Arctic can potentially work again on the Moon. The proposed commercial lunar transportation architecture uses new innovations for modularity and flexibility leading to reduced development and logistics costs, faster development schedule, and better evolvability. This new trade lunar route for mankind utilizes existing Expendable Launch Vehicles (ELVs) available and a commercially financed small fleet of new trans-lunar and lunar lander vehicles. This architecture is based on refueling a fleet of fully reusable spacecraft at several locations in cislunar space, which creates a two-way highway between the Earth and the Moon. This architecture offers NASA and other exploring nations more than one way to meet their near term strategic objectives with commercial space transportation, including sending small payloads to the lunar surface in a few short years, sending larger payloads to the lunar surface in succeeding years, and sending crews to the Moon and back to the Earth by the middle of the next decade. Commercially, this new lunar logistics route permits capability and technology growth as the market grows, offers affordable transportation for the commercial sector and the later recovery of lunar resources. After NASA moves on to other destinations in our solar system, commercial markets and this "in place" commercial logistics system can service, stimulate and sustain a lunar commercial market environment.

An Analysis of the Interface between Lunar Habitat Conditions and an Acclimatized Human Physiology as Defined by the Digital Astronaut Program

Richard L. Summers

University of Mississippi Medical Center
Jackson, Mississippi

Thomas G. Coleman

University of Mississippi Medical Center
Jackson, Mississippi

Background:
The physiologic acclimatization of humans to the lunar environment is complex and requires an integrative perspective to fully understand the requirements for settlement habitat conditions. A large computer model of human systems physiology (Guyton/Coleman/Summers Model) provides the framework for the development of the Digital Astronaut used by NASA in the analysis of biologic adaptive mechanisms. The model provides a means for the examination of the interface between a lunar adapted human physiology and potential habitat environments.

Methods:
The current model Digital Astronaut Program contains over 4000 equations/variables of biologic interactions and encompasses a variety of physiologic processes of interest to humans during spaceflight. The model is constructed on a foundation of basic physical principles in a mathematical scheme of interactions with a hierarchy of control that forms the overall model structure. Physiologic relationships derived from the evidence-based literature are represented as function curves within this structure. Different physiologic systems and body organs are connected through feedback and feedforward loops in the form of algebraic and differential equations to create a global homeostatic system. The model also contains a biologic-environment interface with external conditions such as temperature, barometric pressure, atmospheric gas content and gravity. During computer simulation studies,

the predicted physiologic responses to a habitat environmental change are tracked over time.

Results:
Computer simulations using the model have been found to accurately predict the physiologic transients seen during entry into, prolonged exposure to, and return from the microgravity and bed rest environments. Computer simulation studies suggest that humans with a lunar adapted physiology would be more vulnerable and less tolerant to extreme changes in habitat temperature, humidity and atmospheric oxygen content as compared to an equivalent earth-based setting.

Conclusions:
An analysis of the interface between proposed lunar habitat conditions and an acclimatized human physiology as defined by the Digital Astronaut Program may be important to reduce potential health risks. This system can be used as a tool in the technical planning and design of lunar settlements.

The Mental Health Implications of Working in a Lunar Settlement

Chester S. Spell

School of Business—Camden
Rutgers University, Camden, New Jersey

One of the feasibility issues of a lunar settlement concerns the effect of working in such an environment on people. The focus here is on the long-term implications for the mental health of the base workforce and how working in isolation for extended periods might influence their overall depression and anxiety levels. This is important because a wealth of research in the psychological, management and occupational health literature has found clear correlations between mental health of workers and productivity (National Mental Health Association, 2005) and that poor mental health over the long term is associated with cardiovascular disease and other physical problems (Suls & Bunde, 2005).

Most of the prior research on mental health and working conditions has not examined situations similar to the isolated and otherwise extreme working conditions of a lunar settlement. The research that exists concerns environments like remote mining towns in Australia, where for decades it has been known that women suffer high rates of neurotic problems (Sharma & Rees, 2007). A study of workers at McMurdo Station in Antarctica and the Amundson-Scott South Pole Station revealed higher levels of depression after one year of working in the confined and isolated conditions (Palinkas, Johnson, & Boster, 2004).

Recent research (Spell & Arnold, in press) found that anxiety and depression of individuals working in teams was related to what co-workers thought about their working conditions, above and beyond their own feelings. In other words, attitudes can spread among group members like a "social contagion" and potentially lead to reduced mental health among other team members. While this research was not conducted in an isolated environment, under such conditions it is likely that social interaction among team members is even more critical since team members are the only source of support. While relatively scant attention has been paid to this issue, the studies to date suggest that the link between isolation and worker mental health may be a critical one for a lunar base.

Malapert Base

Manny Pimenta

President, Lunar Explorer

The design and construction of the first large scale Lunar base is explored.

Current designs for Lunar outposts tend to focus on the near-term Return To The Moon missions and are therefore limited in capacity, scope and vision.

The ultimate stated goal of all space activity should be the evolution of Humanity into a Space Faring civilization. The clear demarcation point that we have achieved Space Faring Civilization status is the establishment of the first large scale, permanent off planet colony. The logical location for this first space colony is the Moon.

Malapert Base is intended to show what is technologically possible to achieve within the next two decades in terms of our first true extra terrestrial colony.

It is further intended as a bold, compelling, and inspiring vision of our future in Space; one which will connect with individuals in a deep personal way, giving them a stake in its achievement and opportunities to contribute and participate.

Malapert will be based on existing or near, term technologies, and it is intended as an economically self-sufficient venture—a critical requisite for permanency.

The only critical path assumption made is that the cost of launching material to LEO will be drastically reduced to some arbitrary enabling level within the next 10 to 15 years.

DAY 4: Thursday – June 7, 2007

Morning Session

Plenary Speakers

Bradley Edwards
Carbon Designs, Incorporated
"Space Elevator for the Moon"

Paul Eckert
International and Commercial Strategist, Boeing IDS—Space Exploration
"Attracting Private Investment for Lunar Commerce: Toward
Economically Sustainable Development"

Day 4 Morning Schedule:

- 9:00–9:45 am   Bradley Edwards
- 9:45–10:00 am   Paul Eckert
- 10:00–10:20 am   Seon Han
- 10:20–10:40 am   Joseph E. Palaia
- 10:40–11:00 am   COFFEE BREAK
- 11:00–11:20 am   Klaus P. Heiss
- 11:20–11:40 am   Francois Lévy
- 11:40–12:00 pm   B.E. Lewandowski
- 12:00–1:00 pm   LUNCH

Space Elevator for the Moon

Bradley C. Edwards

Black Line Ascension
Seattle, Washington

The Apollo program was one of man's most impressive achievements. However, in spite of this the program failed to become self-sustaining. Since Apollo our technology has advanced in many areas but we have been unable to reduce the cost of getting to space. Considering these realities the Moon/Mars initiative has an ambitious set of goals. The United States has tried extremely similar programs before (1970s, 1989) and failed. This paper summarizes a NASA proposal for an exploration program based on an innovative Earth-to-space transportation system, the space elevator. This program would enable inexpensive (over 90% savings on launch costs), high-capacity (3000 tons/year) delivery of cargo and crews to the moon, Mars and other solar system destinations. The proposed program has the potential to create a sustainable and affordable program for exploring the solar system. The spin-off from this program will be commercially valuable infrastructure with existing, ready customers, international interest and endless possibilities for scientific studies.

In the proposed program costs are broken down into system and delivery costs. From Koelle's lunar model, a lunar laboratory with an average crew of 69 would cost $57B over 40 years. Delivery costs run $20B in capital expense and $1B/year in operating costs for two space elevators capable of delivering 3000 tons per year to the moon. This capacity is more than sufficient for the lunar laboratory scenario. The large mass capacity of the elevator will allow crewed components to be over-designed, and carry ample propellant surpluses, allowing for greater mission flexibility, robustness, safety and ultimately, sustainability. In our proposed effort, redundant fuel, supplies, habitats, rovers, parts and CEVs can be placed in geosynchronous and lunar orbits and on the lunar surface to provide multiple back-ups for the exploration endeavors. In addition, we have investigated the possibility of a commercial revenue generation from the excess launch capacity. Program expenditures are estimated to be $68B primarily spent between 2005 and 2023 but with revenue generation these can be reduced substantially.

Attracting Private Investment for Lunar Commerce: Toward Economically Sustainable Development

Paul Eckert

Space Exploration, IDS, The Boeing Co.
Arlington, Virginia

Established and startup companies are collaborating in an international consortium designed to promote the growth of space commerce, extending from Earth to the Moon. Firms engaged over the past two years have included The Boeing Company, Lockheed Martin Corporation, Northrop Grumman Corporation, Honeywell International, Raytheon Company, Mitsubishi Corporation, and Alcatel Alenia Space, as well as startups including Transformation Space Corporation and Lunar Transportation Systems. These companies have recognized that entrepreneurial innovation by startup companies–profitably applying existing technology to meet customer needs in a marketplace–is at least as important as efforts to develop new technologies in the laboratory. The global industry consortium is also working to encourage development of truly commercial, self-sustaining space activity involving non-government customers, rather than simply extending government contracting. Because government and established company resources are usually insufficient to ensure the initial viability of new firms, early funding must generally come from angel investors—wealthy individuals for whom creating a legacy of innovation is often more important than making a profit. Personal relationships are a far more potent means of gaining the interest of such individuals than are more impersonal methods. In addition, description of new space-related opportunities should be in terms investors can understand. It is particularly important to attract investment for relatively low-cost commercial Earth-to-orbit transportation systems serving the International Space Station (ISS), because such systems may constitute the first step toward development of commercial Earth-to-moon (i.e., cislunar) transportation capabilities. Affordable and reliable cislunar transportation services are among the most important factors enabling sustainable lunar development. International interoperability of systems is an equally vital factor in assuring commercial sustainability.

Design of a Space Elevator

Seon Han

Department of Mechanical Engineering
Texas Tech University
Lubbock, Texas

The purpose of this work is to present the design issues that must be considered in designing a space elevator from dynamic standpoint. A space elevator is modeled as a long cable that is anchored on the Earth. The dynamic forces that the elevator is subject to are the tidal forces due to the Sun and the Moon, the gravitational attraction from the Earth, and the environmental load due to wind modeled as a point load near the bottom of the elevator. In this study, the two counterweights are added. The first one is placed at the outer end, and the second at geosynchronous orbit. The first one is added so that a shorter cable can be used, and the second to make the system stable. The basic design parameters are the total length of the cable which subsequently determines the counterweight, the counterweight at the geosynchronous orbit, the cross-sectional area, and the constant stress level if the cross-sectional area is tapered.

In this work, both the tapered and un-tapered cables are considered. It was suggested in the past that the cross-sectional area be tapered (largest cross-sectional area at geosynchronous orbit) so that the stress level along the cable is constant. This is done so that any material can support its own weight under gravity. However, not any material with tapered area can be used because the required taper ratio (the area at geosynchronous orbit equals the area at Earth's surface) can be unreasonably large for materials with low specific strength and low constant stress level. Carbon nanotubes have a high specific strength so that taper ratio for the carbon nanotubes can be as low as 1.5, and they are strong enough so that un-tapered cable can support itself under gravity.

So far, the dynamics of a space elevator have rarely been considered in its design. Considering from statics and quasi-statics points of view, the cross-sectional area and stress level may be determined from the desired carrying capacity of the elevator. The static analysis provides no criterion for setting the counterweight at the geosynchronous orbit and the cable length except that it must extend beyond geosynchronous orbit. One obvious limiting factor is that it is difficult to send a large counterweight into space using traditional rockets. Other design constraints on the basic design parameters come from dynamics. In this work, three criteria will be considered:

1. The fundamental frequencies must be such that they are well away from the forcing frequencies.
2. The dynamic load due to moving elevator must not induce resonance.
3. The fundamental frequency when an elevator car is parked along the cable must be away from the forcing frequencies.

Stepping Stones to Mars Settlement

Joseph E. Palaia IV
VP Operations/R&D and Co-Founder
4 Frontiers Corp.
Hudson, Florida

The 4 Frontiers Corporation is an emerging space commerce company focused on Mars settlement as well as other inner solar system development. Early activities include the development of profitable space technologies (including those applicable to the lunar domain), consulting for key manufacturers and government agencies, and engaging in public entertainment and education. These expanding activities will lead to the creation of a Mars settlement mockup/tourist attraction which will introduce the public to the company's vision as well as its space technology innovations. Early emphasis will be placed on developing those technologies having common application to both the Lunar and Mars systems.

Private Property Rights and the Economic Exploration of the Moon: Hammurabi, Tyrolean Homesteads and the Outer Space Treaty

Klaus P. Heiss
Executive Director
High Frontier, Inc.

There is a common objective by all participants in the discussion of the Outer Space Treaty ("OST") and related issues: assured benefits to all mankind. The question is: how does one best assure this. The answer advanced herein is the affirmation and assurance of unencumbered private property—rather than central, national or global committee planning constructs of national (sometimes secret) and international Space bureaucracies, with economic opportunities postponed ad infinitum. Precedents are given throughout history, chief amongst them the Code of Hammurabi and the Tyrolean Höferecht tradition of homesteading.

Mankind has a moral obligation to bring about economic and beneficial uses of Space so that indeed mankind can benefit, as without such uses nobody will benefit. The historically proven way is through private property rights.

We have the further moral obligation to bring about the earliest possible settlement of and in Outer Space, if only for the "Precautionary Principle" increasingly advocated by some in other contexts, so as to assure the very survival of mankind, given the great unknown cataclysmic disasters that periodically hit Earth over eons past. Herein we should follow the example set by Jamestown and the spectacular economic development of the North American continent—in contrast to the barren latifundi of aristocratic, royal or nationalized precedents elsewhere.

The purpose of "laws" is to enable useful human activities, not to prohibit or prevent them. Indeed, law and order are the very foundations of freedom and free markets, including the enforcement of laws, if necessary by force: the enforcement of freely entered contracts is the enabling instrument of the working of markets and the freedoms to which all of us aspire and which some of us already cherish.

Indoor Air Quality Implications of ^{222}Rn from Lunar Radon

F. Lévy

MS Student
Civil, Architectural & Environmental Engineering Department
University of Texas at Austin
Austin, Texas

J. Fardal

Austin, Texas

Recently, interest has grown in resuming lunar exploration with the possible establishment of long-term bases. Due to payload costs and the scale of permanent bases, there is a compelling need to employ in situ resources, leading to extensive use of lunar soil, or regolith, in such bases. Regolith is prone to 222Rn exhalation. We modeled two scenarios for radon exhalation in regolith-based lunar construction. We examined the potential for human health risks due to 222Rn decay-product exposure in such closed, hermetically sealed environments where lunar inhabitants would potentially have direct, long-term contact with regolith. We found the potential for significant health concerns, but more detailed data on the physical properties of regolith-aggregate concrete and its 222Rn exhalation rates is required to accurately determine radon emanation and diffusion coefficients.

Human Power Generation to Augment Lunar Settlement Power Sources

B.E. Lewandowski

Bioscience & Technology Branch
NASA Glenn Research Center
Cleveland, Ohio

K.J. Gustafson

Department of Biomedical Engineering
Case Western Reserve University
Cleveland, Ohio

D.J. Weber

Dept of Physical Medicine & Rehabilitation
University of Pittsburgh
Pittsburgh, Pennsylvania

Electrical power is a critical issue as the lunar settlement is designed and developed. A plentiful power source is necessary in order to carry out the mission, but at the same time, uploading consumable power sources must be minimized due to the significant cost. This conflict in requirements will be particularly evident at the initial stages of settlement development. Dependence on vehicle power will be the greatest prior to when a permanent power grid can be established at the settlement. Even as permanent power is established, there will continue to be a need for mobile power sources. Examples of items needing mobile power include electronic sensors within EVA suits, mobile communication devices and power tools. Recharging methods will be needed for the power sources of these mobile devices.

A significant amount of energy is dissipated by the human body as it interacts with the environment. For example, heat is radiated from the skin. Mechanical energy is dissipated into the ground during walking and into structures and equipment that is touched. Arm and leg joints in motion house kinetic energy. During each breath the weight of the chest wall is displaced over a distance and air is expired at a higher temperature and pressure than the surrounding atmospheric air. It is possible to harness this dissipated energy and convert it into electricity through low mass energy conversion methods, such as thermoelectric, electromagnetic, piezoelectric and electrostatic methods. Humans are also capable of doing work on objects

769

and can generate electricity through hand cranking and shaking motions, during bicycling and by manipulating objects in other ways.

Some examples of Earth based human power generation concepts include heel strike generators [1; 2] and inductive backpack generators [3] that harvest energy during walking or running. Research exists on the conversion of human kinetic motion to electricity [4; 5] and bicycle power generators are commercially available products [6]. An activity performed in space by the astronauts that is well suited for energy harvesting with these methods is exercising. Astronauts exercise in space to counteract the deleterious effects of the reduced gravity environment, such as loss of bone and muscle mass and reduction of aerobic capacity [7]. Energy harvesting methods such as heel strike generators and generators embedded into the exercise equipment could be used to generate significant amounts of power. This power could augment the consumable power sources that must be expensively uploaded by extending their life time or by increasing their reliability, or it could be used to recharge the power sources of mobile electronic devices. This additional power would be particularly useful during the initial stages of lunar settlement development.

1. Kymissis, J., Kendall, C., Paradiso, J., and Gershenfeld, N., "Parasitic power harvesting in shoes," Wearable Computers, Second International Symposium, 1998, pp. 132–139.
2. Shenck, N. S. and Paradiso, J. A., "Energy scavenging with shoe-mounted piezoelectrics," *Micro, IEEE,* Vol. 21, No. 3, 2001, pp. 30–42.
3. Rome, L. C., Flynn, L., Goldman, E. M., and Yoo, T. D., "Generating electricity while walking with loads," *Science,* Vol. 309, No. 5741, 2005, pp. 1725–1728.
4. Niu, P., Chapman, P., Riemer, R., and Zhang, X., "Evaluation of motions and actuation methods for biomechanical energy harvesting," Power Electronics Specialists Conference, IEEE 35th Annual, Vol. 3, 2004, pp. 2100–2106.
5. von Buren, T., Lukowicz, P., and Troster, G., "Kinetic energy powered computing—an experimental feasibility study," Wearable Computers, Seventh International Symposium on, 2003, pp. 22–24.
6. Anonymous, 2007, http://www.windstreampower.com/
7. Buckey, J. C., *Space Physiology,* Oxford University Press, New York, 2006.

DAY 4: Thursday—June 7, 2007

Plenary Speaker

Harry W. Janes
Research Professor, Cook College, Rutgers University
"Bioregenerative Life Support: Closing The Life-Support Loop: What Is Stopping Us?"

Day 4 Afternoon Schedule:

- 1:15–2:00 pm   Harry W. Janes
- 2:00–2:30 pm   Claudio Maccone
- 3:00–3:20 pm
- 3:20–3:40 pm   Gregory Konesky
- 3:40–4:00 pm   COFFEE BREAK
- 4:00–4:20 pm   Y. Cengiz-Toklu
- 4:20–4:40 pm   V. V. Shevchenko
- 6:00–   ON YOUR OWN

Bioregenerative Life Support: Closing the Life-Support Loop: What Is Stopping Us?

Harry W. Janes

Research Professor
Department of Plant Biology and Pathology
Rutgers University
New Brunswick, New Jersey

Life support designs for short-term manned space flight missions have relied primarily on storage of materials before launch, physicochemical technologies, and on re-supply. Longer missions leading to manned exploration of the solar system with bases established on the moon and missions to Mars require a life support system that cannot be based solely on these physicochemical technologies. When re-supply is impossible and long-term storage is impractical, these technologies simply cannot regenerate food from waste. Therefore, any life support system for extended missions must include green plants capable of generating food, oxygen and potable water. Additionally, systems to degrade waste and re-supply minerals must be developed around and integrated with the plant growing system. These systems will require use of microorganisms and will probably be aided by physicochemical processes. The goal of bioregenerative life support is to emulate in space the life-sustaining processes of earth. The challenge in developing this system is to not only understand the subsystems involved but to blend them in such a way that we can create a model of earth's system that is both reliable and small.

For long-term space exploration, where stand-alone habitat systems are absolutely required, the U.S. has recently halted the process of developing the enabling technologies. For almost 30 years, NASA has funded bioregenerative life-support research for space applications, but because of recurring policy and funding changes, the agency has not been able to bring to fruition breakthroughs that are needed for sustained human survivability in isolated or extreme environments. Management decisions have resulted in a disjointed program with frequent stops, restarts, and direction changes. Faculty from seven leading universities, with over 150 years of advanced life support experience are currently developing the Habitation Institute as a multi-institutional scientific partnership with the objective of advancing the basic research necessary to develop the technologies required for long-duration space travel.

Protecting the Farside of the Moon for the Benefit of all Humankind

Claudio Maccone

International Academy of Astronautics
Torino (Turin), Italy

The need to keep the Farside of the Moon free from man-made RFI (Radio Frequency Interference) has long been discussed by the international scientific community. In particular, in 2005 this author reported to the IAA (International Academy of Astronautics) the results of an IAA "Cosmic Study" where he reached the conclusion that the center of the Farside, specifically crater Daedalus, is ideal to set up a future radio telescope (or phased array) to detect radio waves of all kinds that it is impossible to detect on Earth because of the ever-growing RFI. In this paper we propose the creation of PAC, the Protected Antipode Circle. This is a large circular piece of land about 1820 km in diameter, centered around the Antipode on the Farside and spanning an angle of 30 deg in longitude, in latitude and in all radial directions from the Antipode, i.e. a total angle of 60 deg at the cone vertex right at the center of the Moon. There are three sound scientific reasons for defining PAC this way:

1. PAC is the only area of the Farside that will never be reached by the radiation emitted by future human space bases located at the L4 and L5 Lagrangian points of the Earth-Moon system (the geometric proof of this fact is trivial);
2. PAC is the most shielded area of the Farside, with an expected attenuation of man-made RFI ranging from 15 to 100 dB or higher;
3. PAC does not overlap with other areas of interest to human activity except for a minor common area with the Aitken Basin, the southern depression supposed to have been created 3.8 billion years ago during the "big wham" between the Earth and the Moon.

The International Lunar Observatory Association (ILOA) 2007 and Lunar Settlement

Steven M. Durst

Space Age Publishing Company
Kamuela, Hawaii

Originating in Hawaii near the center of the Pacific hemisphere, the ILOA in 2007 has been endorsed by and seeks membership from institutes, individuals and enterprises to realize, place and operate a multifunction astrophysical observatory near the Moon's south pole as early as 2010. The ILOA also seeks to help support a follow-on human service mission to that facility and to parellel robotic village facilities that constitute the emerging lunar base settlement. The ILOA is an Earth–Moon interglobal enterprise with projected membership from major spacefaring powers Canada, China, India, Japan, Europe, Russia, Brazil, Crescent Moon Countries, United States and others representing the great majority of the planet's people.

Primarily an observatory for radio, submillimeter, infrared and visible wavelength astrophysics, for other non-astronomical observations, and for some geophysical science, the ILO also will function as a solar power station (with silicon photovoltaic research), communications center (with varied commercial broadcast possibilities), site characterizer (solar wind, radiation, temperature, duration; micrometeorites, ground truth), property claim agent, virtual dynamic nexus, toehold for lunar base build-out and settlement, and Hawaii astronomy booster.

Facilitating the continuing rise of excellence for Mauna Kea observatories through interglobal interaction, the ILOA is incorporating in Hawaii as a 501 (c) (3) non-profit to serve as the enabling, executive, governing, directing vehicle for the ILO. ILOA initial assets consist of, at least, four professional technical feasibility research studies, space/lunar flight-tested instruments, industrial partner service advances, two international astronomy center MOUs, ILOA News, Hawaii/ Mauna Kea office maintenance, and directors' employment through financial reserves. ILOA 2007 progress and developments will be updated through at least six major ILOA presentations in China, America, India, and Europe, including the ILOA Founders Meeting on Hawai`i Island 4–8 November and the ILOA Founders Meeting Preliminary Session at the International Astronautical Congress in Hyderabad 26 September. Supporting humanity's ascent to multiworld species and to interstellar, galaxy exploration, the ILOA has a promising outlook and future well worth advancing and pursuing.

ILEWG Roadmap from Precursors to Lunar Settlements

Bernard H. Foing

ILEWG Executive Director
Noordwijk, Netherlands

We discuss the rationale and roadmap from robotic precursors to lunar settlements.

We report on recent activities, and on previous ILEWG conferences, recommendations and declarations, including from the last ILEWG 8th conference at Beijing in July 2006. We shall cover issues debated by task groups within the ILEWG forum:

Science opportunities: Clues on the formation and evolution of rocky planets, accretion and bombardment in the inner solar system, comparative planetology processes (tectonic, volcanic, impact cratering, volatile delivery); records of astrobiology, survival of organics; astronomy and space science; past, present and future life; early Earth samples

New instrumentation: Remote sensing miniaturised instruments; surface geophysical and geochemistry package; instrument deployment and robotic arm, nano-rover, sampling, drilling; sample finder and collector.

Technologies for robotic and human exploration: Mecha-electronics-sensors; tele control, telepresence, virtual reality; regional mobility rover; autonomy and navigation; artificially intelligent robots, complex systems, man–machine interface and performances.

Living off the Land: Establishment of permanent robotic infrastructures; environmental protection aspects; solutions to global Earth sustained development; life sciences laboratories; support to human exploration; permanent lunar settlements.

Lunar Optical Data Links

Gregory Konesky
SGK Nanostructures, Inc.
Hampton Bays, New York

Ongoing future lunar activities and settlement efforts will require substantial bandwidth both for the transmission and reception of large volumes of data, imagery, communications, remote teleoperation, etc. High bandwidth lunar optical data links can be broadly grouped into two categories: inter-lunar site, and Moon-to-Earth.

Inter-lunar site optical communications benefit from the absence of atmospheric effects that plague terrestrial counterparts with attendant absorption, scattering, beam wander, and pulse spreading. However, lunar dust presents its own set of concerns due to its highly abrasive nature, wide particle size distribution, and strong electrostatic attraction to surfaces. Long distance communication is limited, after local topology, by the surface curvature of the Moon, whereas atmospheric effects predominate in terrestrial optical links. Various design examples of inter-lunar optical data links, for both stationary and mobile sites, are presented as are dust mitigation approaches and the general rigors of the lunar environment.

A Moon-to-Earth optical data link can be established using only a 1 Watt laser on the Moon, transmitted through a 1 meter aperture (at 830 nm) and received on the Earth also by a 1 meter aperture, and yet produce a net positive link margin. In practice, however, a smaller aperture would be used on the Moon, and supplanted by a larger aperture on the Earth for obvious economic reasons. Terrestrial atmospheric effects are fortunately limited to only the last few kilometers of beam propagation and can be largely negated, within limits, by adaptive optics. Alternatively, optical data link reception can be affected in Earth orbit. The design tradeoffs between these two approaches are considered in terms of the overall system complexity and link availability. Design examples of Moon-to-Earth optical links with various transmitter powers, apertures, and operating wavelengths are considered in terms of link budget. The origins of atmospheric effects on beam propagation will be considered, as will the extent to which these can be dealt with by adaptive optics. Demonstrated design examples of adaptive optic systems, and their applicability to a Moon-to-Earth data link are discussed. Difficulties of an Earth-to-Moon optical data link are also considered.

Comparison between Terrestrial and Lunar Mining

Y. Cengiz Toklu

Mechatronics Engineering Department,
Bahcesehir University
Istanbul, Turkey

One big issue in lunar settlement is the supply of all types of necessities on the Moon. The difficulties in supply from the Earth make it obligatory to build and operate plants on the Moon to produce most of the necessities using lunar resources as raw material. This makes mining on the Moon an operation of primordial importance.

Mining on the Moon will have its own characteristics as compared to mining on the Earth. The first difference will come from the fact that substances are likely to be more uniformly distributed on the surface of the Moon as compared to the Earth. Thus it is probable that there will not be different mines for different raw materials and almost all substances will be extracted from the same source with serial or parallel processes. The other differences between lunar mining and terrestrial mining will originate mainly from environmental differences like those in gravitation, and existence and non-existence of atmosphere.

These differences will result in new advances in mining technologies, mining equipments and extraction processes which, at the end, will affect positively the way of life on the Earth and elsewhere.

Lunar Resources for Rescue of Mankind in XXI Century

V. V. Shevchenko

Sternberg Astronomical Institute
Moscow University
Moscow, Russia

At the outside of the atmosphere, the Earth receives about 1/2,200,000,000 of the solar radiation, about 1.8×10^{17} J/s (joule per second) or 180×10^{15} W. About 50% of the incoming solar energy is received by the Earth's environment. It equals about 90×10^{15} W or 90000 TW. The essential increasing of the additional energy will destroy system of our environment completely. In results of many ecological investigations it has been found that the permissible level of the energy production inside Earth's environment is about 0.1% of solar energy received by Earth's surface. The value is about 90 TW (90×10^{12} Watt). Any estimation of future electricity requirements depends on the assumptions of the population growth and energy usage. An analysis of the known data indicates that simple representation of the population growth tendency is an exponential regression. In near future expected world's population will become level at about 10 billion people. From different sources the world energy use per capita at present corresponds to about 1.7 kW of power per person and general world energy consumption is about 3.3×10^{20} J/y or ~10 TW of capacity. The general prognosis shows that total energy use (and production) in the world will increase by a factor of two (to about 20 TW) by 2040. If this tendency is preserved, the total energy production in the world will approach about 55 TW by the end of the century. A more complicated calculation has been carried out in which allowance is made for the effect of energy usage growth (per capita) with time. Simple estimates are that in the 2050 timeframe the world will require nearly 40 TW of total capacity, and more than 110 TW will be required in 2100. It means the permissible level of the energy production inside Earth's environment will be exceeded in the near future. But it is obvious that the processes destroying Earth's environment in global scale will begin before it—after middle of century. Hence, the first result of the practical actions for rescue of the Earth's environment must be obtained not later than in 2020–2030. The only way to resolve this problem consists in the use of extraterrestrial resources and industrialization of the space. The nearest available body—source of space resources and the space industrial base is the Moon. The most known now space energy resource is lunar helium-3. Very likely, the lunar environment contains new resource possibilities unknown now. So, the lunar research space programs must have priority not only in fundamental planetary science, but in practical purposes too.

Possible Applications of Photoautotrophic Biotechnologies at Lunar Settlements

I.I. Brown
Jacobs Sverdrup ESCG

D. Garrison
Jacobs Sverdrup ESCG

S. Sarkisova
Jacobs Sverdrup ESCG

J.A. Jones
NASA Johnson Space Center

C.C. Allen
NASA Johnson Space Center

D.S. McKay
NASA Johnson Space Center

D. Bayless
University of Ohio

A major goal for the Vision of Space Exploration is to extend human presence across the solar system. With current technology, however, all required consumables for these missions (propellant, air, food, water) as well as habitable volume and shielding to support human explorers will need to be brought from Earth. In situ production of consumables (In Situ Resource Utilization-ISRU), such as propellants, life support gas management, as well as habitat and support system construction materials, will significantly facilitate human hopes for exploration and colonization of the solar system, especially in reducing the logistical overhead such as recurring launch mass.

The life support, fuel production and material processing systems currently proposed for spaceflight are not completely integrated. The only bioregenerative life support system that has been evaluated on a habitat scale by

NASA employed only traditional crop production. This has been proposed as a segment for bioregenerative life support systems, even though the efficiency of higher plants for atmospheric revitalization is generally low. Thus, with the release of the NASA Lunar Architecture Team lunar mission strategy, the investigation of air bioregeneration techniques based on the activity of photosynthetic organisms with higher rates of CO_2 scrubbing and O_2 release appears to be very timely and relevant. Future systems for organic waste utilization in space may also benefit from the use of specific microorganisms. This janitorial job is efficiently carried out by microbes on Earth, which drive and connect different elemental cycles. It is possible that bioregenerative environmental control and life support systems will be capable of converting both organic and inorganic components of the waste at lunar settlements into edible biomass.

The most challenging technologies for future lunar settlements are the extraction of elements (e.g. Fe, O, Si, etc) from local rocks for life support, industrial feedstock and the production of propellants. While such extraction can be accomplished by purely inorganic processes, the high energy requirements of such processes drive the search for alternative technologies with lower energy requirements and sustainable efficiency. Well-developed terrestrial industrial biotechnologies for metals extraction and conversion could therefore be the prototypes for extraterrestrial biometallurgy.

Despite the hostility of the lunar environment to unprotected life, it seems possible to cultivate photosynthetic bacteria using closed bioreactors illuminated and heated by solar energy. Such reactors might be employed in critical processes, e.g. air revitalization, element extraction, propellant (oxygen and methane) and food production. The European Micro-Ecological Life Support System Alternative (MELiSSA) is an advanced idea for organizing a bioregenerative system for long term space flights and extraterrestrial settlements.

We propose additional development and refinement of the MELiSSA system by the employment of Spirulina strains with increased productivity of essential amino acids, immunomodulators as well as by the addition of biometallurgy and fuel production to the life support cycle. Such a synthesis of technological capability, as embodied in a lunar surface ISRU bioreactor, could decrease the demand for energy, transfer mass and cost of future lunar settlement.

Scaling and Sizing Aspects for Resource Utilization Devices to Support Development of the Moon and Mars Exploration

Bee Thakore

International Space University

With the new wave of exploration targeted at extending lunar missions to support humans and preparing towards a permanently occupied human outpost, the role of in-situ resources utilization (ISRU) is becoming apparent. Several studies have indicated that use of ISRU devices may provide significant cost effectiveness over a given time of operation. With the lack of demonstration mission, one link still remains to be investigated further: what is the best means for such devices to be scaled up to cater with an increased demand of products.

This paper mainly focuses on use of space resources to harness oxygen, hydrogen and water on the Moon, as the processes, technologies and economics for these near-term 'derived products' are better understood.

Strategies for extending a new-sufficient supply of space resource derived products to those that can be commercially mined is not the main focus, but are included for comparison of increased energy, mobility and operating robustness required. Finally, a role of such a strategy in preparing a demonstration mission for Moon and that for an ISRU device for large scale water production on Mars is discussed.

International Cooperation Models for the Development of Settlements on the Moon

Ryan Zelnio

George Mason University

It is widely recognized that the development of a lunar settlement cannot be done by the United States alone due mainly to the costs associated with such a large endeavor. However, getting the international community to join such an effort will be a challenging endeavor. This paper examines four frameworks for international cooperation that have been used in past international space endeavors. Frameworks examined will include cooperative models like CEOS, GEOS, & ILEWG, augmented models like Cassini, interdependence models like the International Space Station and integrated models like Intelsat and the European Space Agency. Examples of each framework used in prior cooperative international space ventures will be examined with a discussion on their pros and cons. This paper shall then discuss the application of the frameworks to lunar development.

There have been many changes in the political landscape since the last major space cooperative project, the ISS, has been put together in the early 1990s. Russia's economic and industrial base is now on much better footing, the technology maturity of the Europe Union and Japan is much greater than it was before, and China has emerged as space power. What this means and how it impacts future cooperative models will be discussed.

International Cooperation
Models for the Development or
Settlements on the Moon

Printed in the United States
by Baker & Taylor Publisher Services